D1715361

Woody Plants of the Southeastern United States

Woody Plants of the Southeastern United States

A WINTER GUIDE

Ron Lance

The University of Georgia Press

Athens and London

© 2004 by the University of Georgia Press
Athens, Georgia 30602
All rights reserved
Designed by Louise OFarrell
Set in 9/10 Times by Pracharak Technologies (P) Ltd, Madras, India
Printed and bound by Thomson-Shore, Inc.
The paper in this book meets the guidelines for permanence and durability
of the Committee on Production Guidelines for Book Longevity
of the Council on Library Resources.
Printed in the United States of America
08 07 06 05 04 C 5 4 3 2 1

Library of Congress Cataloging-in-Publication Data
Lance, Ron.
Woody plants of the southeastern United States : a winter guide / Ron Lance.
p. cm.
Includes bibliographical references (p.) and index.
ISBN 0-8203-2524-4 (hardcover : alk. paper)
1. Woody plants—Southern States—Identification. 2. Woody plants—Southern States—Pictorial works. I. Title.
QK124 .L25 2003
582.16'0975–dc21
2002156525

British Library Cataloging-in-Publication Data available

For Margeau Jennifer Lance

Contents

Preface — ix

Introduction — 1

Master Keys to Genera — 15

Descriptive Text and Keys to Species — 43

Glossary of Terms Used in This Text — 391

References — 405

Index — 411

Preface

This book was prepared as a guide for winter identification of woody plants in the southeastern United States, excluding peninsular Florida. Toward this purpose, the species accounts are brief and offer primarily condensed diagnostic information useful in identification, as well as a summary of the range and habitat. The illustrations were similarly prepared to show diagnostic features, as a supplement to the keys. For further reading about the species included in this text, see the list of references near the end of this book.

For many years, botanical bait has lured me across the southeastern United States. This pull has been on many occasions a consuming quest to find and inspect, to collect and study, to photograph, illustrate, and record, as nearly as personally possible the totality of our native woody flora. Doing this amid and between jobs, during vacations, and at nearly every other available opportunity, I ultimately came to the realization that this assignment would have no end. Therefore, to see finally at least part of a lifelong interest rendered into this publication is a joy and a relief. This is but a meager token of knowledge applied toward a vast world of creation, but the study of plants offers rewards on all scales. For that minority of people that marvel and cheer internally each time they see a species in its place in the world for their very first time, this book is presented not only as a guide of what can be seen but also as an enticement to see differently. Each reader will no doubt find unfamiliar plants among these pages, as well as familiar plants described in unfamiliar ways. Bare twigs and dormant buds are merely another dimension in the appearance of our woody plants, a dimension often overlooked yet perfectly distinctive if viewed with an eye for detail. I hope that this book will inspire, rather than intimidate.

Across the many years of intentional and unintentional progress toward this publication, I have been honored by making many helpful acquaintances, too numerous to ascribe here from memory. To all such people who read or use this book, I express thanks for their part in my botanical journeys. This book is inconspicuously composed by their inspiration, education, direction, support, and companionship.

There are a few individuals that deserve particular gratitude from me for major contributions toward my completion of this book. The biological sciences staff of Haywood Community College, in North Carolina, propelled my early career in biology and plant interests and provided a means to test twig keys and identification guides over many subsequent years. I am therefore indebted to Doug Staiger, Richard Lindsay, and John Palmer for encouragement and support and to the students of theirs and mine, who helped shape the twig keys of this book in its primordial stages. For their help during the many years of plant-finding expeditions and specimen collecting, I am also grateful to Wilbur H. Duncan, James Allison, and Tom Patrick of Georgia; Al Schotz of Alabama; Ron Wheeland and Jack Herring of Mississippi; and Alan Weakley of North Carolina. Particularly special gratitude is extended to Robert B. McCartney of Woodlanders Nursery in Aiken, South Carolina, and Angus K. Gholson of Chattahoochee, Florida. Through their knowledge and assistance, many native plants were made available to my sight for the first time, ultimately resulting in a great number of the images drawn in this book. I also have the late Robert K. Godfrey to thank, for memorable days filled with enlightening experiences among the flora of northern Florida. And finally, it is only through the prodigious help and persistence of the University of Georgia Press that a mass of keys and illustrations have been transformed into this publication.

Woody Plants of the Southeastern United States

Introduction

Purpose of This Book

The southeastern United States is a region with great diversity in plant species and plant community habitats. Even the woody segment of the flora includes more than eight hundred types of plants. Toward recognition of this woody flora, many publications have been written for the region, though the majority have been designed for use during the growing season in this temperate climate. Few are designed for those times of the year when the deciduous plants are dormant, when characters of identification seem to disappear with the fallen leaves. In actuality, most of the woody plants are as easily identified by twig, bud, and bark characters as by leaves, if one is familiar with the traits to observe in winter.

Provided herein are dichotomous keys, illustrations, and condensed identification information to aid in identifying woody plants of the Southeast in winter. I have tried to include all species known to occur in the region, with accompanying drawings of those species that have definitive winter characters. Because twigs are not always diagnostic enough to separate species in winter, not every species is illustrated. Some examples of genera where twigs cannot be used to identify all the included species are *Amorpha, Crataegus,* and *Vaccinium.* In these and other genera where species identification is difficult to impossible during the dormant months, other means of identification become necessary. The use of flowers, fruit, or summer leaf characters is accordingly mandated within this book to identify to the species level, though such referral to nonwinter characters has been included only as a last resort.

All drawings are my own artwork, using freshly collected specimens. Although the concept of this book has been influenced by publications such as Trelease (1925), Petrides (1958), Harlow (1966), and Preston and Wright (1988), I developed the style and content of the keys and illustrations within the book in my own instructional courses and independent study over a span of eighteen years.

How to Use This Book

Identification of woody plants using any type of guidebook is a task requiring some level of knowledge in botanical terms. The more species included within a book, the more complicated and extensive the task becomes, and the more terms are necessary. This book is no exception. An understanding of the basic terms presented in these introductory pages will facilitate the use of keys that follow. The detailed introduction, with important words in italics, may also be used in lieu of the glossary.

Terminology and Taxonomy

Botanical terminology utilizes many words from Greek and Latin origins to specifically describe plant features. I have tried to limit the use of technical botanical terminology in the keys and descriptions for easier use by novice botanists as well as those not inclined toward any botanical experience. Standard botanical texts such as Fernald (1950), Steyermark (1925), and Radford, Ahles, and Bell (1968) provide more detailed glossaries of botanical terms.

One of the first concepts a novice botanist must grasp is the science of plant classification and the application of botanical names, which is the field of *taxonomy.* Plants are classified as to their assumed relationships to each other, from *species* to *genus* to *family* and beyond. There may be difficulties in validation of these segregate units of classification in some plants and among taxonomists over time, especially at the basic level of species. Specific botanical names, which are usually derived from words of Latin or Greek origin, consist of two words, the genus and the species. This specific epithet, along with the author of the name, is popularly known as the Latin, or scientific, name. The term *taxa* (singular, *taxon*) designates described plant entities and may be used as an alternative to *species* when there is sufficient disagreement among taxonomists as to whether the plant is a true species or not. If not widely accepted as a distinct species, the plant may be considered a *variety* or *forma,* but all such units of identity are still taxa.

Changes in taxonomic treatment of a plant may be due to disagreement concerning its distinctiveness, reclassification based on new information, or other nomenclatural (name) issues. Alternate names used for plants due to taxonomic changes are commonplace in published works, and the Latin names of species may differ considerably with author and time of publication. In cases where a plant has had more than one Latin name assigned to it, the "correct" name is determined by current taxonomic studies and the alternate names are *synonyms.* The taxonomy and names used in this book generally follow Kartesz (1994), though I departed from that reference in a few instances. Only the most commonly encountered synonyms are included in this book. Readers interested in detailed taxonomic treatments of native plants and more complete lists of synonyms should consult other publications, such as the above-mentioned reference.

Using Keys for Identification

A *key* is simply a method of elimination. Choices are commonly given as paired statements, and after reading each pair of statements (called a *couplet*), the reader chooses the statement that applies to the specimen in hand. The applying statement (called a *lead*) will direct the reader to another couplet, or to a direct identification. Both leads normally begin with the same word or are numbered alike, to aid in their recognition. The process may be lengthy if many species are involved, and the branching pattern of choices accounts for the term for this type of identification process, the *dichotomous key.*

The dichotomous key has two popular formats, *indented* and *bracketed.* In the indented key, the two leads of each couplet are indented similarly from the left margin, with all consecutive couplets included and indented systematically underneath the corresponding lead. See the species key to *Hypericum* or *Prunus,* as an example. The bracketed key uses space more efficiently by keeping the two leads of a couplet together, and only alternating couplets are indented. But directional numbers after the leads become necessary. Both types of keys are used within this book: the bracketed mainly in the preliminary section, and indented keys in the species guide section.

This book includes such a great number of species that a preliminary divisional key is provided to segregate several large groupings of plants, this being the Master Key to Diagnostic Keys on page 15. This master key should be used first to isolate an unknown specimen into the groups shown as dichotomous keys A through H. Once a specimen seems identified through use of one of the main keys A to H, a genus name is usually given to direct the user to the alphabetical portion of the book. The genus is located within this portion, and where there are several species involved in such a genus, another key is offered to separate the species. Illustrations are provided where diagnostic value is evident in twigs, leaves, or fruit that may be present in winter.

With extensive use of the keys, there will no doubt be instances when a specimen will not fit the descriptive elements provided. This may be due to natural variation, absence of characters needed, or errors in the key or in the use of the key, or the specimen may represent a species not included in the book. Working backward within the keys to discover possible erroneous diversions will be simpler if the sequence of couplet numbers is written down, especially in the case of long keys.

Scope of This Book

I have tried to include all native species of woody plants occurring within the defined southeastern U.S. region, as well as those exotic species of woody plants that are known to have naturalized within the region. This latter group is always expanding, since the discovery of naturalized exotics continues to increase. Depending on the various taxonomic viewpoints that can be employed, the region supports about 650 to 870 species of native woody plants, and at least 85 naturalized woody exotic species. This book includes identification traits for 695 native species and varieties, and 189 exotic species and varieties. These 884 plant taxa are supplemented by mention of 123 additional varieties that are not presented in the identification process, due to their insignificance in the scope of this book.

A *woody plant* is here defined as one with visible dormant parts remaining alive in winter and with buds for shoot growth present above the surface of the ground. The presence of true wood (secondary xylem) is not always apparent nor significant in the case of diminutive shrubs and groundcovers, so the presence of aboveground buds seems a more suitable criterion for including these plants within this treatment. Certain species, however, may be more woody in some areas than in others, particularly in the southern portions of their range.

A *native* woody plant is here considered to be one that occurred within the southeastern region prior to colonization by Europeans. Native Americans likely contributed to movement of many plants well before European colonization, but such movement of plants involved few woody species that were not native to some portion of the Southeast. During European settlement, many new plants were introduced into the region, and that trend continues today. A vast number of plant species and horticultural selections are now widely cultivated. A small percentage of these exotics have been found to *naturalize,* or grow without aid of human cultivation. Naturalized exotics vary in rates of naturalization, depending on habitat, climate, and dispersal agents. Some exotics aggressively overgrow and outcompete native plants in certain habitats; some are only occasionally seen where the situations are conductive to their survival; some spread only by roots from original cultivated specimens; and others may seem naturalized but are only remnants from original plantings. The naturalized exotics included within this book are those known, through my own experience and available resources, to be found in at least two separate locations in the Southeast.

The Southeastern Region

The southeastern region for which this guide is written is shown in Figure A. This region is considered on geographical rather than political terms. The temperate climate and vegetation that in my opinion best typify the general southeastern region lie between 29 to 39 degrees north latitude and 75 to 96 degrees west longitude. This region of coverage excludes the bulk of peninsular Florida, since the transition to subtropical vegetation in that area is of a different floral character.

The physiographic provinces of the southeastern region shown in Figure A are mentioned throughout the text, in remarks about plant range and habitat. The natural ranges of many plants follow these provincial zones, due to geological and climatic differences that are peculiar to each. General characteristics of each province are summarized here.

The Coastal Plain

The coastal plain covers an extensive portion of the Southeast, being the province lowest in elevation (sea level to about 152m [500ft]), mildest in winter climate, and with limited topographic relief. Once a seafloor, the soils are characteristically sandy, or composed of clay or silt deposits from eroding peripheral provinces. In some areas, upland ridges or sandhills are derived from old dunes or beaches, or deposited by rivers and winds. These deep sands have rapid water drainage, harsh sunlight conditions, and a high incidence of fire. Many plants native to these sandhills possess special adaptations to resist the effects of heat and dessication by

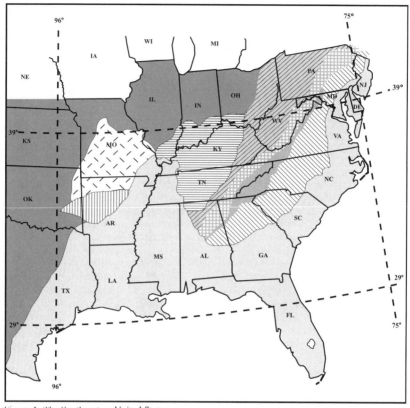

Figure A. The Southeastern United States

PHYSIOGRAPHIC REGIONS OF THE SOUTHEASTERN UNITED STATES

Region of coverage of this book lies within 29° to 39° N latitude and 75° to 96° W longitude

sunlight reflecting from the bleached sands. Other uplands of the coastal plain may be composed of clays, such as the alkaline uplands of the Black Belt and Jackson Prairies of Alabama and Mississippi, where vegetation reflects adaptation to high pH levels. Fertile loess deposits, such as those near the Mississippi River in western Tennessee, support a high diversity of hardwood forests. The lowland zones of the coastal plain have variable water tables, owing to soil substrate and degree of drainage. Where flooding is frequent or drainage is slow, swamps, bogs, marshes, pocosins, bays, and wet depressions occur. Pocosins and shrub bays have acidic, peat-derived soils that are poor in fertility. Pine flatwoods and savannas are flat habitats that may be seasonally wet, but overall are fairly dry, and the pine canopy component is maintained by the incidence of fire.

The coastal plain may be separated into two zones, the inner (or upper) coastal plain, and the outer (or lower). The inner coastal plain roughly corresponds to the limits of river swamps that carry loads of silt or alluvium with flooding, from which the terms *brownwater* or *redwater* swamp are derived. The outer coastal plain has as its upper limits the extent of the blackwater swamps, where dark, tannin-stained floodwaters carry little silt and the elevations above sea level are slight. Near the coastal regions, salt spray and saltwater intrusion into marshes and maritime forests influence vegetation patterns. Shell mounds, or middens, near coastal marshes were created by native people over many generations, and the calcareous remnant shells may support unusual plant communities.

The *Mississippi Embayment* area is a portion of the coastal plain that extends from southern Illinois to the Gulf Coast, and from eastern Texas to southern Alabama. The Mississippi River is the central watercourse of this valley, which has received heavy sediment loads from areas to the north.

The Piedmont

The Piedmont, a plateau of rolling hills and occasional small mountain ranges that are actually eroded foothills of the Appalachians, lies on the eastern side of the Blue Ridge from New Jersey to Alabama. Underlying rock is mostly metamorphic and concealed by clay-loam soils, but granitic domes and outcrops occur in some areas. There are few sedimentary or calcareous deposits. Elevations range from around 457m (1500ft) adjacent to the Blue Ridge or on summits of Piedmont mountain ranges, down to 91m (300ft) at the fall line, where streams drop into the sandy sedimentary soils of the coastal plain.

The Appalachian Mountains

The Appalachian Mountain Province of the southeastern region is itself subdivided into three distinctive and separate provinces. The easternmost escarpment of the mountains, extending from Maryland to Alabama, is the *Blue Ridge*. This province contains the most extensive topographic relief, with elevations from about 305m (1000ft) to a maximum of 2038m (6686ft) at the highest peak in the eastern United States. Within this province, lower elevations of 305 to 762m (1000–2500ft) host varied hardwood forests in moist sites, oak and oak-pine forests in disturbed and drier soils. The middle elevations of 762m to 1220m (2500–4000ft) receive higher annual precipitation than lower zones and include diverse hardwood forests of high quality, especially in deeper soils of coves and high valleys. The high elevations of 1220m to more than 1830m (4000–6000ft) have oak forests transitional to a mixed hardwood forest, ultimately to a boreal character above 1524m (5000ft). This boreal zone of the Blue Ridge hosts coniferous forests similar to those much farther to the north, due to the habitats induced by a short growing season, cooler climate, and high annual precipitation. The rocks of the Blue Ridge are predominately metamorphic in origin, with overlying soils of acidic nature.

The *Ridge and Valley* Province lies to the west of the Blue Ridge. This section of the Appalachians contains mostly sedimentary rocks that have been folded by great pressure and weathered into ridges and valleys oriented from the northeast to the southwest.

The *Cumberland Plateau* lies to the west of the Ridge and Valley Province. A broad, uplifted, and eroded plateau of sedimentary rocks, it transitions into the Cumberland and Allegheny Mountains of eastern Kentucky and West Virginia. Outcroppings of limestone are common, and soils overlying such substrates are often calcareous, or high in pH.

The Interior Plateau

The interior plateau of central Tennessee and western Kentucky involves several geologically distinct subprovinces, but overall it is a region of rich and varied forests, with both calcareous and circumneutral soil conditions.

The Ozark Plateau and Ouachita Mountains

The Ozark Plateau was uplifted by underlying magma and eroded into a diverse landscape of ridges and narrow valleys, with highest elevations to about 762m (2500ft). The sedimentary rocks are commonly exposed within the region due to the eroding action of streams, with limestone outcroppings common. Just to the south lie the Ouachita Mountains, an older range allied to the time of formation of the Appalachians, but separated from that formation by the Mississippi Embayment. Elevations of about 823m (2700ft) occur, and exposed rocks are mostly sedimentary in origin.

The Plains and Prairie

The Plains and Prairie Province enters the Southeast only in Missouri, eastern Kansas, Oklahoma, and Texas, though prairie remnants and communities exist well eastward, to Alabama. Soils typically are alkaline, with calcium carbonate or calcite sometimes being close to the surface. The extensive herbaceous plant community is usually dependent on the incidence of fire for its maintenance, which prunes back woody encroachment.

Vegetation and Habitats

As with all naturally occurring vegetation, woody plants are adapted to various soils and habitats that govern their distribution and abundance. Where tolerance is limited, ranges are also frequently limited, or populations are found only within specific conditions. Following are some common terms used in this book to describe habitat and range of woody plants.

Mesic soils or habitats remain relatively moist throughout the year. Derived from the Greek word *mesos,* meaning "middle," a mesic condition is intermediate between the extreme of dryness, which is termed *xeric,* and the extreme of wetness, *hydric.*

Neutral soils are those with a pH range of 7; if around 7 they are *circumneutral.* Many nutrients and plant absorption capabilities to make use of nutrients are at an optimum range in these types of soils, so diversity of plant species is often higher in such sites. By contrast, lower pH soils are *acidic,* and as the range drops below about 4.5 pH, soils are considered very acidic and only acid-tolerant plants may thrive. *Calcareous* soils have higher pH ranges, with an abundance of calcium. These soils often lie over limestone or amphibolite rock formations where magnesium is available, or over calcite where deficiencies of magnesium may occur. High pH soils can pose particular problems with iron and other mineral uptake in many plants. *Calcophiles* are plants adapted to high pH habitats.

An *endemic* plant is one that is known to occur only in a limited geographic area. With the majority of plants, wider ranges are the norm. A population or area of occurrence of a species far removed from the bulk of its range is considered *disjunct.* In some species, ranges transcend continents such as in the *circumboreal* or *circumpolar* species of the north. These plants have spread around the pole in northern regions that were once connected, though today are separated by oceans.

Morphological Terms Used in This Guide

Deciduous species are those that lose foliage during a dormant period, which typically begins in the autumn in the southeastern region. Bare branches and twigs of the deciduous species are a primary focus throughout this book, since the majority of the region's plants are in this category. The rate of leaf abscission is not always consistent within deciduous species that have wide latitudinal ranges. In the southern or coastal portions of their range, longer growing seasons and milder climates permit later leaf fall, so these plants may be termed *late-deciduous* or *subevergreen* if some leaves persist to spring. If foliage remains on the twigs beyond the appearance of the next spring leaves, the plant is considered a true *evergreen.* The evergreen and subevergreen woody plants are included in Key A, the persistent foliage providing a convenient means of identification.

The *habit* of a plant is an important trait to observe in the field and is used to describe its growth appearance. A *tree* is here considered to be a woody plant with a well-developed main trunk at least 10cm diameter at 1.75m (about 5.5ft) above the ground. A *shrub* has smaller dimensions at maturity and usually multiple stems. This categorization is a generality, since many specimens or species approximate both of these habit descriptions, depending on habitat, range, or genetic variation. A woody plant that climbs on other plants or objects and has no rigid stem capable of supporting its own weight is termed a *vine*. A woody plant reclining on the ground or spreading over it may be termed a *groundcover,* or it may be a vine without available supports.

A climbing vine is *twining* if its stem twists around other objects. Many vines use *tendrils* or *aerial rootlets* to cling to surfaces. *Prostrate, creeping,* or *sprawling* plants are those inclined to do so along the ground, whereas a *clambering* plant may recline over other plants or grow into them for partial support and may appear more vinelike.

Many woody plants spread by subterranean stems or roots that send up multiple stems and induce a clumping or colony-building habit. Populations derived by sprouting of one or more individuals may form a *colony,* though if derived only from one individual, the colony is *clonal* in nature. Where stems are produced from roots, the plant is *suckering*. A *rhizomatous* plant is one whose aboveground stems are produced from an underground stem. When the sprouts occur at the tip of a horizontal stem above- or underground, the plant is *stoloniferous.*

Leaf Terminology

A simple leaf is composed of a *blade* attached to the twig by a *petiole.* The petiole usually falls with the leaf at abscission time. When the petiole is very short or absent, the leaf is *sessile.* In some plants, *stipules* may be produced on the petiole base or on the twig at the nodes. Stipules are small, leaflike structures normally seen in spring, but they may persist beyond that season or become modified into spines.

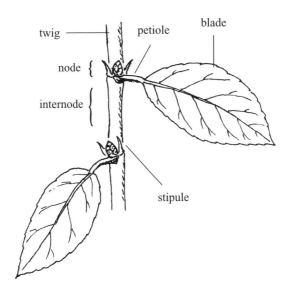

Figure B. A simple leaf

8 · *Introduction*

A *midrib* is the large central vein in a leaf, transitioning to the petiole below the blade. *Main veins* branch from the midrib into a successively smaller network in cases where the leaf is said to have *net venation*. In *parallel venation,* many straight, unbranched veins run side by side from the base of the blade to the tip (as in monocots). In variations of the net venation, several main veins may originate from a common point near the blade base in *palmate venation,* or arise along the midrib in *pinnate venation*. If main veins bend or curve along the edge of the blade, such venation is termed *arcuate*.

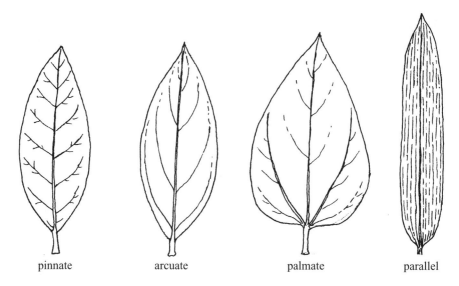

Figure C. Leaf venation

A leaf is *simple* if it has only one blade. *Compound* leaves have completely divided blades; each portion is called a *leaflet*. If all leaflets originate from a common point at the end of the petiole, it is termed *palmately* compound (or *trifoliate*, if there are only three leaflets). When leaflets are attached along a central stem (*rachis*), the leaf is *pinnately* compound, with the petiole below the lowest pair of leaflets. A *bipinnate* leaf has the equivalent pinnate leaflets divided again into smaller leaflets, with these side rachi along the main central rachis called *pinnae*.

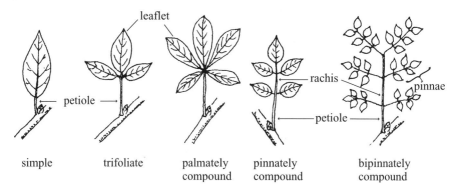

Figure D. Leaf types

Leaf arrangement on the twig is *alternate* where one leaf is borne at a distinct distance from another above and below. *Opposite* leaves occur in pairs directly opposite each other at the same point on the twig, and if three or more leaves occur at the same point, the arrangement is *whorled*. If twig growth is slow, as in spur shoots, alternate leaves may be closely clustered and appear whorled. Actively growing or elongated terminal shoots should be inspected for ascertaining true leaf arrangement.

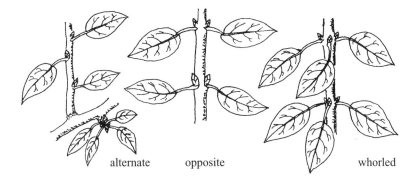

Figure E. Leaf arrangement

Leaves may be *lobed*, with sinuses between lobes, and lobes may be palmate or pinnate, depending on the related venation. The margins of leaves may be *entire* (untoothed) or *serrated* (toothed). The type of serrations may be *remote* (barely discernible or scarce), *finely serrate, coarsely serrate,* or *doubly serrate* (small teeth on larger teeth). If teeth are rounded, they are

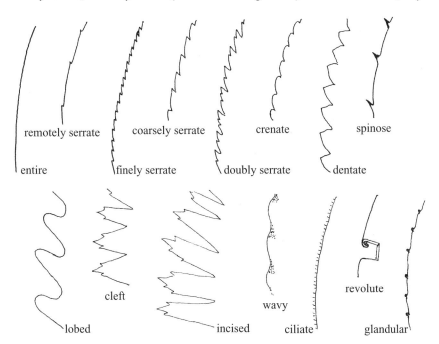

Figure F. Leaf margins

10 · Introduction

termed *crenate*; if large and outward projecting they are *dentate*. Teeth are *glandular* if a small swelling (gland) is seen at the tooth apex or in its axil. Margins may also be *wavy, ciliate* (with fine marginal hairs), *spinose* (with spine-tipped teeth), or *revolute* (curled under).

Many botanical terms describe the shape of leaf blades and their tips and bases. Where leaves are present in winter, or where leaf shape, tips, or bases are important diagnostic characters between difficult species, these may occasionally be mentioned in the text. Otherwise these terms are not used. Please consult a summer guide to investigate these details more closely.

Twig Terminology

The current or most recent woody growth of the ends of a stem are the *twigs*. The previous season's growth, more than a year old, is the *branchlet*, sometimes called a second-year stem. A *branch* is three or more years of age. In winter, twigs normally bear the most conspicuous buds, unless the buds are sunken into the twig. Compared with the older branchlet, twigs may also have a different color or varying degree of hairiness (pubescence), and small bud scale scars show where the previous year's bud opened and grew onward as the twig.

A *spur shoot* is short, knobby, and slow growing, and it usually produces clustered leaves or flowers. It usually is seen alongside a branchlet or branch and is tipped with a bud or a spine.

A *node* of the twig is the location where vascular tissue exits and enters the twig, and bears the leaves. Between nodes are the *internodes*, spaces where no leaves are borne.

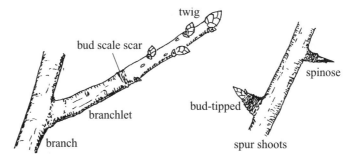

Figure G. Twig terminology

Where a leaf falls from a twig (abscission), a *leaf scar* is left behind. Leaf scars tend to have distinctive average shapes for each species. They can be *raised* (protruding) or *sunken* (craterlike), and their arrangement (alternate, opposite, whorled) corresponds to the leaf arrangement. Within the leaf scar, the vascular bundles can usually be seen as dots, bumps, or other

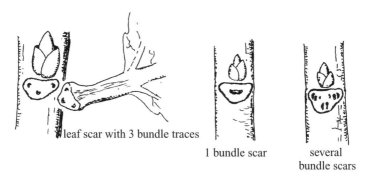

Figure H. Leaf scars

abnormalities in color or texture, or as clusters of dots. These *bundle scars*, or *bundle traces*, are important aids to winter identification of woody plants and particularly useful in separating out groups of plants in dichotomous keys. The use of an 8× or 10× lens is necessary in many cases of identification using bundle scars since some leaf scars are tiny. Determination of the number of bundle scars per leaf scar is a prominent question early in the keys, but in species where bundle scars are often obscure, reference to Key D is suggested. Bundle scars may be easier to see if a sharp blade is used to slice a thin layer off the face of the leaf scar or off the base of the petiole in persistent leaves.

In species identification, *buds* are among the most important features of twigs. Buds may be sunken or submerged in the twig in a few species but usually are visible above the leaf scars. A *naked* bud has the first set of leaves folded or wrinkled and visible as the outside of the bud; a dense covering of hairs may obscure these. *Bud scales* are modified leaves that protect embryonic leaves within the bud. A *single bud scale* may cover the bud in a small number of genera. When two bud scales meet but do not overlap at the edges they are termed *valvate*. Overlapping bud scales are *imbricate* and can number from two to many per bud.

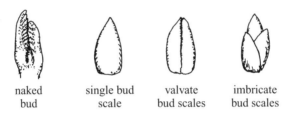

Figure I. Buds

In the majority of woody plants, the position of most buds on a twig is *axillary*, which means buds are borne within the axils of leaves or are visible just above the leaf scar if leaves have abscised. Where the twig has ended growth by forming a bud on the very tip, a *terminal bud* is present. A terminal bud is often larger than the axillary buds, or *lateral buds*. When the growing twig tip withers or falls away, one of the lateral buds ends up being close to the winter twig tip and substitutes as a terminal bud for extension of the next year's growth. Such a case is called a *false terminal bud,* or *pseudoterminal bud,* and a telltale *branch scar* or branch stub is

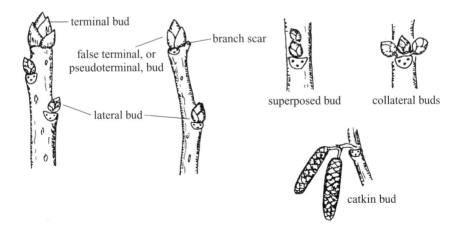

Figure J. Bud positioning

12 · Introduction

usually visible where the twig tip has fallen opposite the last leaf scar. Lateral buds can also be *superposed,* where two or more are vertically positioned above a leaf scar, or *collateral,* where side by side at the same node. Such additional buds at a single node are called *accessory* buds; sometimes these may be flower buds that vary in appearance or size from lateral vegetative buds. A *catkin bud* is a type of flower bud that is elongate and multiscaled in winter, developing in spring into the type of inflorescence known as a catkin.

Stipule scars are present on twigs of species where stipules are produced. These scars are usually seen near the upper corner of the leaf scar. Small stipules leave tiny, slitlike scars, while larger stipules leave more conspicuous scars, sometimes larger on one side of the leaf scar than the other. In a few species, the stipule encircles the twig or nearly so, leaving ringlike scars at the node.

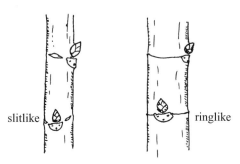

Figure K. Stipule scars

Lenticels are dotlike or elongate pores in the epidermis, or outer bark, of twigs and other stems. Depending on species, these may be conspicuously raised or bumplike, inconspicuous, or lacking entirely. In some plants (as in many birches and cherries), the lenticels are horizontally elongated on larger stems, appearing as streaks or lines.

The *pith* occupies the central portion of a twig, and is composed of soft, thin-walled cells that are not functional in sap transfer. It is easily compacted when cut by a dull blade; to see its actual anatomical characteristics in sectional cuts it is important to use a very sharp cutting edge. The shape of the pith in cross section can be circular or nearly so (*terete*), to lobed or angular. An *angled* pith is often five-sided or five-lobed. When pith is cut lengthwise, in a slice that splits the twig, several important variations in its morphology can be seen. If the pith seems

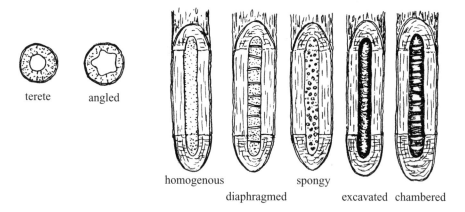

Figure L. Pith

solid or unchanging, it is *homogenous*. If the pith seems homogenous but with horizontal plates or thickened walls scattered throughout, it is *diaphragmed*. If the pith seems perforated with holes, like a sponge, it is *spongy*. Pith that seems to be missing, with a hollow central cavity in the twig, is called *excavated*. A pith made up of consecutive horizontal walls in an otherwise excavated condition is *chambered*.

A twig is said to be *armed* if spines are present. Spines are described in several ways, depending on morphological origin. A true *thorn* is an outgrowth of wood, rigid and not easily broken from the twig (as in *Maclura,* Figure 301). *Prickles* are borne on the epidermal, or outer, bark and are easily detached from the twig if pressed to one side (as in *Rosa,* Figure 462). In some plants, stipules may be modified into spines (as in *Robinia,* Figure 459), or spines may terminate spur shoots or twig tips (as in *Pyracantha,* Figure 387). *Bristles* are stiff trichomes ("hairs") that do not become woody or sharp.

Other features of twigs mentioned within this book include *tendrils* (as in *Parthenocissus,* Figure 342) and *aerial rootlets* (as in *Campsis,* Figure 83), which are appendages used by climbing vines. A *scurfy* covering is a granular surface that is actually made up of short, massed hairs or masses of *peltate scales* (as in *Croton,* Figure 146). A peltate scale is a rounded or disk-shaped appendage attached by a stalk, often with a gland in the center. It is actually a modified hair, best visible with a 10× lens. A *bloom* is a whitish covering of wax that may be present on twigs, fruit, or leaves. The term *glaucous* is used to describe pale or whitened surfaces, often due to a bloom.

A twig is said to be *angled* if it is not rounded or terete in cross section (as in *Bignonia,* Figure 67). A *lined* twig bears fine and slightly raised lines, usually below the leaf scars (as in *Physocarpus,* Figure 351). If these lines are sufficiently prominent, they give the twig a *ridged* appearance (as in *Populus,* Figure 372).

Any surface that is smooth and free of hairs is said to be *glabrous*. Plant hairs are actually *trichomes,* cellular appendages that take on many variations and with many botanical terms used to describe them. Overall, the term *pubescent* is used to denote the presence of hairs. Short, dense, or woolly pubescence can be referred to as *tomentose* or *lanate,* or in other terms depending on length and straightness of the hairs. Sparse, elongate hairs are *hirsute,* or *hispid* if stiff and bristly. Hairs that branch from a central stem, resembling starlike or shrublike shapes when viewed with magnification, are *stellate*. If the surface is covered with many small bumps, it is *verrucose* (as in some *Vaccinium*); do not confuse this with *lenticellate,* which means bearing visible and distinct lenticels.

Bark

The bark of woody plants is a lifeless tissue designed to protect the living vascular tissue beneath. It is composed mostly of cork cells, and the outer appearance can vary widely, depending on the content of suberin, pigments, waxes, and other ingredients; the thickness, size, and weathering rate of the cork cells; and the underlying distribution of the cork cambium. Most woody plants have conspicuous bark colors and patterns of weathering that can be recognized through experience but are difficult to describe. Noting the appearance of bark on mature and juvenile portions of trees and shrubs while in the field is important, as these can be valuable in identification. In general, bark is described in this book as one of three categories: *smooth, papery,* or *rough*. Smooth bark is thin and relatively unbroken but can be *striated* (striped or lined with a contrasting color) or *mottled* (with differential color patterns). Papery bark is exfoliating, or shed in thin, papery or curly layers. Rough bark is thick or hardened, and may be *warty* (raised excrescences), *scaly* (stiff plates that are free and curling on edges or ends), or *furrowed*. A furrowed bark may be divided into ridges that extend longitudinally with the trunk or in *plates* that are flat-topped and not so conspicuously longitudinally elongated. Particularly thick, short, or squarish plates account for a *blocky* bark texture. The odors or colors of inner bark can also be of great value in identification, as inner bark is actually vascular phloem tissue, where most of the food reserves of the woody plant is transported and where sap odors are most conspicuous.

Flowers and Fruits

Few flowers are seen in winter months in woody plants, so references to these are minimal. Several difficult taxa, however, cannot be separated without the use of flowers, so some information is offered in these cases. A plant with only *staminate* flowers produces pollen only and is considered male. By contrast, a female plant produces *pistillate* flowers that will form fruits following pollination. If a plant has both types of flowers, or flowers with both sets of staminate and pistillate parts, it is *monoecious*. A *dioecious* plant has only one type of flower and is primarily either pistillate or staminate. Some plants may actually have both pistillate and staminate flowers, but one set may not be fertile or completely formed; thus the plant will be *functionally dioecious*. The arrangement of flowers into an *inflorescence* is also of value in identification, as the type of inflorescence is also mirrored to some degree in the maturation of fruits.

Fruits or their remnants are often present on woody plants in winter or may be present on the ground under mature plants. These are valuable articles in identifying species, sometimes greatly simplifying the process. Fruits are generally considered in two categories, *fleshy* and *dry*. A fleshy fruit has a succulent or moist flesh surrounding the seed. This flesh is a derivative of the ovary of a flower or from additional floral structures. An *aril*, for example, is a fleshy covering derived from the stalk of the ovule.

A dry fruit may have a leathery or hardened covering for the seed, which either splits when the fruit is mature (*dehiscent*) or does not (*indehiscent*). Fruits may be *compound* where derived from separate pistillate units or carpels in a flower (*aggregate*), or where derived from a close cluster of separate flowers in an inflorescence (*multiple*). A *simple* fruit is derived from one pistil in one flower. Some common types of fruits in woody plants are

Achene. A small, hard fruit from a simple pistil, with a thin seed coat, as in *Brunnichia*.

Berry. A fleshy fruit, with several immersed seed, as in *Vaccinium*.

Capsule. A dry fruit, usually with several chambers or locules, from a compound pistil, often dehiscent along lines or sutures, as in *Rhododendron*.

Cone. A dry type of fruit with layered and overlapping scales, most often associated with the gymnosperms, where seed are matured ovules held between bracts, not in ovaries, as in *Pinus*.

Drupe. A fleshy fruit, usually 1-seeded, with a stony seed coat (endocarp), such as in *Prunus*.

Follicle. A dry, 1-chambered fruit derived from a simple pistil, dehiscent on one side, sometimes referred to as a pod, as in *Zanthoxylum*.

Hip. An aggregate of achenes, with a partially fleshy receptacle forming the outer wall of the fruit, mostly peculiar to the genus *Rosa*.

Legume. A dry, 1-chambered fruit derived from a simple pistil, dehiscent on 2 sides, peculiar to the family Fabaceae, the legume or pea family, as in *Robinia*.

Nut. A dry fruit, usually 1-seeded and hard-coated, enclosed within a husk or involucre, as in *Carya*.

Nutlet. A small nut or nutlike fruit held within a thin involucre or associated with a bract, as in *Carpinus*.

Pome. A fleshy fruit, derived from an ovary with fused carpels and a floral cup or hypanthium contributing to the flesh and peel of the fruit, as in *Malus*.

Samara. A winged achene-like fruit, as in *Acer*.

Strobile. A conelike fruit derived from a spike or catkinlike inflorescence, composed of nutlets growing between protective layers of bracts, as in *Betula*.

Master Keys to Genera

Master Key to Diagnostic Keys

1. Leaves evergreen or persistent over most of winter season, or if leaves absent, a succulent-stemmed species lacking green leaves (such as cacti) .... Key A
1. Leaves deciduous and absent during winter season; not a succulent 2
   2. Leaf scars opposite or whorled (2 or 3 scars per node) 3
   2. Leaf scars alternate (1 scar per node) or closely clustered 4
3. True terminal bud present .... Key B
3. True terminal bud lacking .... Key C
   4. Leaf scar obscured, or bundle scar count indistinguishable .... Key D
   4. Leaf scar with 1 to several distinguishable bundle scars 5
5. Leaf scar with 1 bundle scar (or 2 in *Ginkgo*), or many small bundle traces closely crowded as to appear as a single scar 6
5. Leaf scar with 3 or more separate bundle scars 7
   6. True terminal bud present .... Key E
   6. True terminal bud lacking .... Key F
7. Leaf scar with 3 bundle scars .... Key G
7. Leaf scar with 4 or more bundle scars .... Key H

Key A

(Leaves persistent over part or all of winter)

1. Leaves reduced or appear to be lacking; stems green and succulent	2
1. Leaves visible; stems not succulent	3
2. Spines present, or stems flattened (cacti)	*Opuntia* (See also *Escobaria*)
2. Spines lacking; stems round, segmented	*Sarcocornia*
3. Leaves narrow, needlelike (acicular or linear, or 3mm or less wide and over 4 times as long)	4
3. Leaves not needlelike, or width over 4mm	25
4. Leaves (needles) bundled at the base, mostly over 4cm long	*Pinus*
4. Leaves singly attached to twig, under 3cm long	5
5. Leaves gray, scurfy; an epiphyte growing on trees	*Tillandsia*
5. Leaves green or not scurfy; rooted plants	6
6. Leaves noticeably broadest at the sessile base, tapering to a narrow tip; mostly 1cm long or less	7
6. Leaves broadest near the middle, or with a short petiole	11
7. Leaves opposite or whorled	8
7. Leaves alternate, though sometimes closely clustered	9
8. Leaves spinose-tipped and white-lined; a conifer	*Juniperus*
8. Leaves not as above; a dicot	*Hypericum*
9. Leaves taper from above the middle to a short point	*Pyxidanthera*
9. Leaves taper from the base to an elongate tip	10
10. Stems not jointed; no stipules	*Hudsonia*
10. Stems jointed; membranaceous stipules	*Paronychia*
11. A conifer; leaves persistent for 2–3 years; trees (or shrubs in *Taxus*)	12
11. A flowering plant, nonconiferous; leaves not as above; all shrubs	17
12. Needles angular, perceptively 3- or 4-sided when rolled between fingers	*Picea*
12. Needles flat	13
13. Needles sharply tipped	14
13. Needles blunt or without sharp tips	16
14. Needles dull green or not glossy, mostly under 2.5cm long; small trees or shrubs (if spur shoots present, see *Cedrus*)	*Taxus*
14. Needles dark glossy green, mostly over 3cm long; tall trees	15
15. Odor of bruised foliage strong, rather rank	*Torreya*
15. Odor of bruised foliage weakly pine-scented	*Cunninghamia*
16. Needles sessile	*Abies*
16. Needles on short stalks	*Tsuga*
17. Leaves alternate or whorled	18
17. Leaves opposite	23 (if sap milky, see *Cynanchum*)
18. Leaves whorled, or alternate only near twig tips	19
18. Leaves alternate along length of all twigs	21
19. Leaves under 6mm long; low, creeping shrubs	*Erica*
19. Leaves mostly over 8mm long; upright or sprawling shrubs	20
20. Leaves strongly revolute; buds apparent	*Ceratiola*
20. Leaves not revolute; buds sunken	*Lycium carolinianum*
21. Leaves under 1cm long; a small, prostrate shrub	*Pyxidanthera*
21. Leaves mostly over 1cm long; an upright shrub	22
22. Leaves thick, stiff; terminal bud present	*Ilex myrtifolium*
22. Leaves thin, flexible; no terminal bud	*Baccharis angustifolia*
23. Nodes near shoot ends with silvery, membranaceous bracts (stipules)	*Paronychia*
23. Nodes not as above	24

24. Crushed foliage with mintlike odor; fruit horizontal	*Conradina*
24. Not as above; capsular fruit in upright clusters	*Hypericum*
25. Leaves palmately lobed or cleft and 25cm or more in length (includes petiole)	26
25. Leaves not palmately lobed or cleft, or less than 25cm long	29
26. Leaf venation netted; twigs green	*Manihot*
26. Leaf venation parallel; no twigs (palm family)	27
27. Leaf petioles serrated	*Serenoa*
27. Leaf petioles entire	28
28. Leaf sheaths with needlelike spines	*Rhapidophyllum*
28. Leaf sheaths spineless	*Sabal*
29. Leaves strap-shaped or bayonet-like and over 20cm long; twigs lacking	*Yucca*
29. Leaves not as above; twigs present	30
30. Leaves compound	31
30. Leaves simple	46
31. Leaves or twigs with spines or prickles	32
31. Leaves and twigs lacking spines and prickles	38
32. Spines occur on leaf margins only, not on twigs	*Mahonia*
32. Spines occur on leaf rachis or twig	33
33. Leaves trifoliate; leaflet margins entire	*Erythrina*
33. Leaves not trifoliate, or leaflet margins toothed	34
34. Leaflets under 1cm wide; spines at twig nodes only	35
34. Leaflets over 1cm wide; prickles scattered on rachis or twig	36
35. Pinnae more than 2 per leaf	*Acacia*
35. Pinnae 2 per leaf, appear as 2 pinnate leaves	*Parkingsonia*
36. Bruised parts very aromatic; upright shrubs or trees	*Zanthoxylum*
36. Bruised parts weakly or not at all aromatic; suckering shrubs or with vinelike habit	37
37. Twigs angular; pith angled in cross section	*Rubus*
37. Twigs and pith rounded in cross section	*Rosa*
38. Leaves with 2 leaflets and a tendril between; climbing vines	39
38. Leaves with 3 or more leaflets; shrubs or trees	40
39. Tendril forked and twining irregularly; a common native vine	*Bignonia*
39. Tendril divides into 3 short, clawlike forks; uncommonly naturalized exotic	*Macfadyena*
40. Leaves trifoliate	41
40. Leaves pinnate or bipinnate	42
41. Leaflets with a few teeth near apex; leaves alternate	*Sibbaldiopsis*
41. Leaflets not toothed; leaves opposite	*Jasminum*
42. Leaves bipinnate	*Nandina*
42. Leaves pinnate	43
43. Leaves thick, very leathery; plant stem short, mostly subterranean	*Zamia*
43. Leaves not as above; obvious aboveground stems present	44
44. Leaves less than 5cm long	*Pentaphylloides*
44. Leaves mostly 7cm or longer	45
45. Leaflets 5cm or more in length; tip acute to acuminate	*Sapindus*
45. Leaflets shorter than 5cm; tip rounded or obtuse	*Sophora* (if leaflets toothed, see *Schinus*)
46. Leaves tiny and scalelike, appressed closely to twig	47
46. Leaves spreading, not as above	52
47. Crushed foliage lacking strongly aromatic oils; not a gymnosperm	48
47. Crushed foliage strongly aromatic; a gymnosperm	50
48. Tall shrubs or trees, usually near water	*Tamarix*
48. Low or matlike shrubs	49
49. Leaves opposite	*Calluna*
49. Leaves alternate	*Hudsonia*
50. Twigs with foliage rounded in cross section, branched irregularly	*Juniperus*

50. Twigs with foliage flattened, branching in flat sprays	51
51. Leaves on branchlets mostly 2mm or less	*Chamaecyparis*
51. Leaves on branchlets mostly 3mm or more	*Thuja*
52. Leaves grasslike, with parallel venation (bamboos)	53
52. Leaves not as above; venation netted or obscure	55
53. Stems (culms) flattened on one side, above node	*Phyllostachys*
53. Stems (culms) rounded, not conspicuously flattened on one side	54
54. Branches solitary at each node; culms usually green (if sheaths are persistent, see *Sasa*)	*Pseudosasa*
54. Branches 3 or more at each node; culms usually brownish or olive-green	*Arundinaria*
55. Leaves obviously opposite or whorled	56
55. Leaves alternately arranged or closely clustered	88
56. Plant vinelike, sprawling, or a very low shrub	57
56. Plant an upright shrub or tree (if growing as a parasite on tree limbs, see *Phoradendron*)	77
57. Leaves succulent; shrubs of coastal dunes and brackish marshes (if vinelike, with milky sap, see *Cynanchum*; if leaves whitened beneath, see *Avicennia*)	58
57. Leaves not succulent; habitat not brackish	59
58. Leaves mostly less than 1.5cm long, opposite; in wet soils	*Batis*
58. Leaves over 1.5cm; some leaves alternate; often on dunes	*Iva*
59. Leaves minty scented when crushed	60
59. Leaves not minty-scented when crushed	62
60. Leaves on lower stem widened at base, with a few conspicuous long hairs	*Stachydeoma*
60. Leaves tapered to base, widest near middle; long hairs lacking	61
61. Leaf margins strongly revolute; leaf undersides glaucous; leaf length over 3 times the width	*Conradina*
61. Leaf margins not as above; green beneath; leaf width ⅓ or more of length	*Calamintha*
62. Leaf margin toothed	63
62. Leaf margin entire	69
63. Habit usually less than 30cm high; stems sparingly branched	64
63. Habit taller or well-branched	65
64. Leaf tips blunt, teeth large, rounded; multiple stems arising from a rhizome (if leaves under 2cm, see *Linnaea*)	*Pachysandra*
64. Leaf tips pointed, teeth sharp; plant clumplike or with a single stem	*Chimaphila*
65. Leaves very thick, stiff, mostly 2cm long or less	*Paxistima*
65. Leaves thin, or over 2cm long	66
66. Stems green; terminal buds conspicuous	*Euonymus*
66. Stems gray; terminal buds lacking	67
67. Leaves widest near middle, bases acute; 3 bundle scars; stiffly erect shrubs	*Iva frutescens*
67. Leaves widest near base; 1 bundle scar; rambling or vinelike habit	68
68. Buds scaled; leaf teeth small, sharp, incurved	*Sageretia*
68. Buds hairy, scales not apparent; leaf teeth conspicuous, blunt	*Lantana*
69. A climbing vine	70
69. A low shrub, possibly sprawling but not climbing	71
70. Twigs glabrous; leaves long-pointed	*Gelsemium*
70. Twigs hairy; leaves short pointed	*Lonicera japonica*
71. Twigs distinctly ridged or winged below nodes; leaves sessile, opposite; buds not evident; clusters of small leaves at nodes and twig tip	*Hypericum*
71. Twigs and leaves not with above combination of characteristics	72
72. Leaves less than 1cm long	*Erica*
72. Leaves mostly over 1cm long	73
73. Leaves 2cm long or longer	74
73. Leaves less than 2cm long	75

74. Leaves whorled, dull green; petioles distinct; erect shrub — *Kalmia*
74. Leaves opposite, glossy, nearly sessile; groundcover — *Vinca*
75. Plant appressed to ground; twigs green, supple — *Mitchella*
75. Plant with stiffly erect twigs or stems — 76
  76. Leaves ovate or elliptic; margins entire; clusters of small capsules often persist — *Leiophyllum*
  76. Leaves mostly oblanceolate; margins often with a few teeth; clusters of capsules lacking — *Paxistima*
77. Plants of seacoast, in tidal or brackish soils — 78
77. Plants more inland, not as above — 80
  78. Leaves glaucous beneath; only of tidal shores in FL — *Avicennia*
  78. Leaves not glaucous; widespread along coast (if terminal buds present, go to 85) — 79
79. Fruit heads spinose, persist over winter on twig tips — *Borrichia*
79. Fruit heads not as above — *Iva*
  80. Buds naked, scurfy; twigs strongly 4-sided or ridged — *Buddleya*
  80. Buds not as above; twigs terete, or if slightly 4-sided then terminal bud much larger than laterals — 81
81. Buds partly or mostly concealed by small leaves, or persistent brown capsules usually present — 82
81. Buds apparent; fruit not as above — 83
  82. Leaves petiolate, whorled near twig base; capsules rounded — *Kalmia*
  82. Leaves sessile or subsessile, opposite; capsules pointed — *Hypericum*
83. Leaves remotely toothed or with a short bristle on tip (mucronate) — 84
83. Leaves entire on margin, tip not mucronate; clusters of blue drupes often present — 86
  84. Leaves mostly over 2.5cm wide, dull green — *Lonicera*
  84. Leaves mostly 2cm or less wide, glossy on top surface — 85
85. Buds reddish, scurfy; leaves with small, obscure teeth — *Viburnum obovatum*
85. Buds not as above; leaf teeth prominent — *Abelia*
  86. Leaves mostly 10–15cm long; axillary clusters of fruit or flower buds often present; terminal bud with 2 valvate scales — *Osmanthus*
  86. Leaves and twigs not with above combination of characteristics — 87
87. Leaves sessile or nearly so, rarely over 1.5cm wide, blunt or rounded at tip — *Forestiera segregata*
87. Leaves short-petiolate, mostly 2cm wide or more, tip pointed — *Ligustrum*
  88. Cluster or raceme of small capsular fruits usually persistent; 1 bundle scar — 89
  88. Cluster of fruits not as above, but single bundle scar may be present — 105
89. Leaf margins toothed (may be finely so) — 90
89. Leaf margins entire or distinctly revolute — 95
  90. Habit usually under 20cm; single-stemmed or nearly so — *Chimaphila*
  90. Habit larger; usually multistemmed or well-branched — 91
91. Terminal clusters of flower buds usually present in winter — 92
91. Flower bud clusters axillary or lacking — 93
  92. Leaf tips mucronate; leaflike bracts present along flower bud racemes — *Chamaedaphne*
  92. Leaf tips not mucronate; racemes not as above — *Pieris*
93. Leaves bluntly tipped (rounded or obtuse) — *Zenobia*
93. Leaves sharply tipped (acuminate) — 94
  94. Pith solid — *Leucothoe*
  94. Pith chambered — *Agarista*
95. Capsules globular; apex flat or blunt (except for spikelike style) — 96
95. Capsules elongated, pointed at apex — 102
  96. Twigs and leaves distinctly hirsute — 97
  96. Twigs or leaves glabrous to sparsely hairy — 98
97. Leaves under 1cm long; plants under 30cm height; capsules 5-parted — *Kalmia hirsuta*

97. Leaves over 1cm long; plants taller than above; capsules 7-parted — *Befaria*
98. Leaves strongly revolute and glaucous below; a short shrub of wet sites in the North — *Andromeda* (if leaves slightly glaucous but not strongly revolute, see *Kalmia*)
98. Leaves not as above; distributed mostly in the South — 99
99. Capsule with 5 pale suture lines, its tip abruptly flattened — *Lyonia*
99. Capsule not as above — 100
100. Leaves thin; twigs lined or winged below nodes; tiny blackened stipules often persist; bark smooth or mottled — *Lagerstroemia*
100. Leaves thick or firm; twigs not as above; no stipules; bark furrowed or with small scales — 101
101. Sprawling or vinelike habit; some leaves bearing small teeth — *Pieris*
101. Short to tall shrubs (occasionally rhizomatous but not vinelike); all leaves entire — *Kalmia*
102. Leaves mostly 15mm long or less; capsules 3-locular — *Leiophyllum*
102. Leaves mostly 2cm long or more; fruit not as above — 103
103. Buds and leaves crowded at twig tip; terminal buds with more than 6 imbricate scales — *Rhododendron*
103. Buds not crowded, or with fewer than 6 scales, or terminal bud lacking — 104
104. Terminal bud present; fruit 3mm long or less; leaf scars triangular — *Cyrilla*
104. Terminal bud lacking; fruit over 3mm; leaf scars shield-shaped — *Lyonia*
105. Habit vinelike, or a shrub 1m or less in stature — 106
105. Habit a taller shrub or tree — 141
106. Habit a climbing vine — 107
106. Habit nonclimbing; low, shrubby, or sprawling — 111
107. Prickles, spines, or stiff spur shoots present — 108
107. Prickles, spines, and stiff spur shoots lacking — 109
108. Tendrils present on persistent petioles; prickles present on stem at nodes and internodes — *Smilax*
108. Tendrils lacking; spine-tipped side shoots often present — *Sageretia*
109. Leaves with sessile or clasping base — *Aster carolinianus*
109. Leaves with distinct petioles — 110
110. Leaves thick, glossy; climbing by aerial rootlets on twigs and branchlets — *Hedera* (if sap milky, see *Ficus*)
110. Leaves thin, dull or light green; climbing by tendrils — *Brunnichia*
111. Leaves palmately lobed — *Ribes*
111. Leaves not palmately lobed — 112
112. Twigs and petioles with tiny brown dots or peltate scales — 113
112. Twigs and petioles lack peltate scales — 114
113. Leaves mostly close-clustered at twig tips; low shrubs — *Croton*
113. Leaves not as above; tall or clambering shrubs with spur shoots — *Elaeagnus*
114. Leaves mostly 2.5cm or wider, very hairy on surfaces and petioles — 115
114. Leaves narrower or not hairy as above — 117
115. Stems appressed to the ground, rarely erect — *Epigaea*
115. Stems not as above; in sandy soil of coastal plain — 116
116. Leaves lanceolate, or much longer than wide — *Lupinus*
116. Leaves ovate — *Smilax pumila*
117. Crushed leaves with distinct wintergreen odor — *Gaultheria*
117. Crushed leaves lack wintergreen odor — 118
118. Inner bark and crushed leaves with strong peppery odor; buds naked, hairy — *Asimina*
118. Not with above combination of characteristics — 119
119. Leaves thin, narrowly spatulate; stems distinctly jointed — *Polygonella*
119. Leaves or stems not as above — 120
120. Plant with few-branched, erect stems only 5–20cm in height — 121
120. Plant well-branched or taller than above — 123
121. Leaves finely toothed on margin — *Licania*

121. Leaves with only a few large teeth or points on margin, or entire 122
  122. Leaf blade not much longer than broad; teeth blunt *Pachysandra*
  122. Leaf blade much longer than broad; teeth pointed when present *Quercus minima*
123. Buds poorly visible or inconspicuous; fruiting twigs usually dying back in winter 124
123. Buds plainly visible; twigs woody to tip, not as above 128
  124. Leaves occasionally opposite or subopposite; plants of coastal brackish soils *Iva*
  124. Leaves all alternate or closely clustered; plants of inland sandy soils or swamps 125
125. Leaves less than 2cm long *Ceanothus microphyllus*
125. Leaves over 2cm long 126
  126. Leaf tip blunt or rounded; native to sandy FL soils *Garberia*
  126. Leaf tip mostly acute or pointed; plant not confined to FL 127
127. Leaves tapered to base; plants of xeric, sandy soils *Chrysoma*
127. Leaves auriculate or partially clasping at base; plants of moist to wet sites *Aster carolinianus*
  128. Bundle scars 3 or more, or spines on twigs 129
  128. Bundle scar single 133
129. Bundle scars numerous; twigs spineless; fruit an acorn *Quercus*
129. Bundle scars 3, or spines on twigs; fruit not an acorn 130
  130. Stipules persistent; no spines on twigs *Sebastiania*
  130. Stipules deciduous; spines or 3 bundle scars present 131
131. Fruits fleshy, red or orange, usually present in winter; a naturalized exotic *Pyracantha*
131. Fruits not as above; native species in sandy coastal plain soils 132
  132. Spines often present on twigs; leaves entire on margin *Sideroxylon*
  132. Spines lacking; leaves often with a few teeth beyond middle *Myrica*
133. Terminal bud present, larger than laterals; shrubs or trees with smooth, grayish bark 134
133. Terminal bud lacking, or low or creeping shrubs with shreddy or flaky older bark 135
  134. Leaves 4 or more times longer than broad, finely toothed; sparsely branched shrubs of wet sites *Stillingia*
  134. Leaves not as above, remotely toothed or crenate; well-branched shrubs or trees *Ilex*
135. Twigs and leaves hirsute 136
135. Twigs and leaves pubescent to glabrous but not hirsute 137
  136. Leaves under 2cm long; terminal buds present *Kalmia hirsuta*
  136. Leaves mostly over 2cm; terminal bud lacking *Gaylussacia mosieri*
137. Plant stems mostly elongated and creeping over the ground 138
137. Plant stems erect, shrubby or well-branched 140
  138. Leaves clustered near twig tips, often cupped *Arctostaphylos*
  138. Leaves clustered at twig base or well spaced, mostly flat 139
139. Buds rounded; leaves numerous, may be opposite near twig base; twigs finely lined *Leiophyllum*
139. Buds pointed; few leaves on branchlets, all alternate; twigs green or red, unlined *Vaccinium*
140. Resin dots usually present on leaf surfaces; 2d-year stems brown to reddish *Gaylussacia*
140. Resin dots lacking; 2d-year stems greenish (but may be red on one side) *Vaccinium*
141. Leaves distinctly silvery below, punctate; twigs covered by peltate scales *Elaeagnus*
141. Leaves and twigs not as above 142
  142. Terminal buds large, single-scaled; ringlike stipule scars encircle twigs at nodes *Magnolia*
  142. Not with above combination of characteristics 143
143. Petioles winged; twigs green, often with spines *Citrus*
143. Petioles not winged; twigs not as above 144
  144. Terminal bud lacking 145
  144. Terminal bud present 149
145. Crushed leaves spicy scented, or with tiny yellowish resin dots; 3 bundle scars *Myrica*
145. Not with above combination of characteristics 146

146. Stipules usually persistent; 3 bundle scars — *Sebastiania*
146. Stipules not persistent; bundle scars single or obscure — 147
147. Spines or spur shoots usually present — *Sideroxylon*
147. Spines or spur shoots lacking — 148
   148. Twig tips usually die back in winter; leaf tips rounded; downy seed head remnants often persist in winter — *Garberia* (if some teeth are on leaf margin and twigs lined, see *Baccharis*)
   148. Twigs and leaves not as above; fruit a berry — *Vaccinium*
149. Buds naked, dark brown, hairy; several bundle scars — *Asimina*
149. Buds scaled, or bundle scars 1 or 3 — 150
   150. Crushed leaves and twigs distinctly spicy-scented — 151
   150. Crushed parts inodorous or not spicy-scented — 155
151. Twigs or leaf undersides hairy — 152
151. Twigs and leaves glabrous — 153
   152. Leaves thin, or with a few teeth on margin; 3 bundle scars — *Myrica*
   152. Leaves thick, margins entire; 1 bundle scar — *Persea*
153. Leaves with anise- or root beer–like odor when crushed; twigs with large pith; bark smooth — *Illicium*
153. Leaves with camphorlike odor, or odor similar to bay leaves; pith not large; bark furrowed — 154
   154. Leaves slightly glaucous beneath, with bay leaf odor when crushed; twigs angular — *Persea*
   154. Leaves green beneath, with glands in lower vein axils and camphorlike odor when crushed; twigs round — *Cinnamomum*
155. Leaves finely and sharply toothed; bark thickly furrowed; tall, slender trees — *Gordonia*
155. Not with above combination of characteristics — 156
   156. Bundle scars obscure; spines or spinose spur shoots present — 157
   156. Bundle scars visible; not as above — 158
157. Leaf margins finely toothed; clusters of orange fruit often persist into winter — *Pyracantha*
157. Leaf margins entire; no orange fruit produced — *Sideroxylon*
   158. Bundle scars 3 or more — 159
   158. Bundle scar single — 161
159. Leaves with tiny resinous dots (punctate); buds blunt or rounded — *Myrica*
159. Leaves not punctate; buds pointed, at least the terminal buds — 160
   160. Bundle scars 3; twigs green or reddish, may be odorous when scraped — *Prunus caroliniana*
   160. Bundle scars numerous; twigs gray to brown, inodorous — *Quercus*
161. Crushed leaves with sweet odor/taste; pith chambered — *Symplocus*
161. Not as above; pith solid — 162
   162. Bark gray and smooth; leaves with some teeth on margin, or twigs slightly hairy — *Ilex*
   162. Bark shallowly furrowed or scaly on mature trunk; leaves entire and twigs glabrous — 163
     163. Leaf underside bears tiny black dots; apex minutely notched (emarginate); venation obscure — *Cliftonia*
     163. Leaf without dots; apex not emarginated; venation prominent — *Cyrilla*

Key B

(Opposite leaves; terminal bud present)

1. Leaf scar with 1 bundle scar, or many minute bundle traces closely crowded so as to appear as a single scar or U-shaped line   2
1. Leaf scar with separated bundle scars (more than 1), or bundle scars indistinguishable   12
  2. Terminal bud with 6 or more imbricate, glabrous, acute or sharply tipped bud scales   3
  2. Terminal bud naked, or with 2 to 4 valvate or imbricate bud scales; bud scales acute and hairy, or blunt or truncated and scurfy   7
3. Twigs green, 4-angled and ridged longitudinally; pith spongy; bud scales greenish or pinkish   *Euonymus*
3. Not with above combination of characteristics   4
  4. Pith chambered or excavated; twigs lined   *Forsythia*
  4. Pith solid; twigs grayish and unlined   5
5. Twigs medium sized; terminal buds normally over 3mm long   *Chionanthus*
5. Twigs slender; terminal buds about 3mm long or less   6
  6. Lateral buds usually superposed   *Forestiera*
  6. Lateral buds usually single per node   *Ligustrum*
7. Bud scales long-tipped (acuminate)   8
7. Bud scales not acuminate; surface scurfy or densely hairy   10
  8. Terminal bud with 2 valvate scales   *Pinckneya*
  8. Terminal bud with more than 2 imbricate scales   9
9. Lateral buds superposed   *Forestiera*
9. Lateral buds not superposed   *Ligustrum* (if twigs green, see *Euonymus*)
  10. Terminal buds scaled; lateral buds usually single per node; trees   *Fraxinus*
  10. Terminal buds naked or scales inconspicuous; laterals often superposed; shrubs   11
11. Terminal bud naked, scurfy, often slightly stalked   *Callicarpa*
11. Terminal bud dark red-brown, hairy, sessile   *Clerodendrum*
  12. Bundle scars 3   13
  12. Bundle scars more than 3, or scars indistinguishable   27
13. Habit a vine   14
13. Habit a shrub or tree (also includes diminutive subshrubs)   15
  14. Buds appearing naked or with 2 valvate bud scales   *Decumaria*
  14. Buds with 4 or more imbricate scales   *Lonicera*
15. Terminal bud with 2 valvate bud scales, single-scaled, or naked   16
15. Terminal bud with 4 or more imbricate bud scales   20
  16. Twigs grayish or brown   17
  16. Twigs reddish or greenish   19
17. Terminal buds much larger than subtending leaf scar (longer than width of scar, at least)   *Viburnum*
17. Terminal buds very small, about as long as width of subtending leaf scar or shorter   18
  18. Buds partially sunken   *Philadelphus*
  18. Buds not sunken   *Hydrangea*
19. Leaf scars raised, darkened, joined by a fine line or ledge; widespread species across Southeast   *Cornus*
19. Leaf scars not as above; trees of Appalachian Mountains in Southeast   *Acer*
  20. Lowermost bundle scar often partially divided; twigs stout, lenticellate   *Aesculus*
  20. Bundle scars and twigs not as above   21
21. Twigs often green or reddish, or brown and lenticellate; trees   *Acer*
21. Twigs grayish or brown, not conspicuously lenticellate; shrubs   22
  22. Twigs hairy, or with rows of hairs or stellate trichomes   23
  22. Twigs smooth and hairless   26

23. Twigs bearing hairs in longitudinal lines or rows	*Weigela*
23. Twigs bearing hairs scattered over surface	24
24. Twigs and most bud scales with coarse, stellate hairs	*Deutzia*
24. Twigs and bud scales not as above	25
25. Bark exfoliating; dried capsular fruits usually persist	*Philadelphus*
25. Bark smooth; no capsular fruits borne (fruit fleshy)	*Viburnum*
(if scraped bark rankly odorous, see also *Rhamnus*)	
26. Terminal bud length less than twice the width of subtending leaf scar	*Hydrangea*
26. Terminal bud length greater than twice the width of subtending leaf scar	*Viburnum*
(if buds globular, see *Lonicera*)	
27. Leaf scar with several noticeable bundle scars	28
27. Leaf scar mostly hidden by dried petiole bases, or bundle scars poorly distinguishable	
	Lonicera
28. Bark exfoliating on twigs or branchlets; bud scales incompletely valvate, narrowing to bulbous tips	*Hydrangea quercifolia*
28. Bark not exfoliating; bud scales imbricate, not as above	29
29. Bud scales 6 or more, glabrous, tightly imbricate	*Aesculus*
29. Bud scales not as above	30
30. Buds scurfy; leaf scars flat or widely notched on top margin; trees	*Fraxinus*
30. Buds dark red-brown, hairy; leaf scar with a narrow notch or cleft on top margin; shrubs	
	Clerodendrum

Key C

(Opposite or whorled leaves; no terminal bud)

1. Spines usually present on twigs or branchlets	2
1. Spines lacking	4
2. Buds naked, hairy, smaller than subtending leaf scar; twig tips often dying back over winter	*Lantana*
2. Buds scaled, larger than subtending leaf scars; twigs not as above	3
3. Twigs slender, lined; buds collateral; a vinelike, clambering shrub	*Sageretia*
3. Twigs moderately slender, round and smooth; an upright shrub	*Rhamnus*
4. Bundle scar single, or many small bundle traces so closely crowded they appear as a single large scar	5
4. Bundle scars 3 or more, or bundle scars indistinguishable	19
5. Buds prominent, with several imbricate, slightly keeled scales	6
5. Buds reduced or partially obscured, naked or with very small imbricate or valvate scales	9
6. Twigs lined or ridged below corners of leaf scar, or leaf scars connected by a ridge or line	7
6. Twigs unlined at leaf scars	8
7. Twigs with longitudinal lines from leaf scars; tall shrubs with rigid twigs; pith white, solid	
	Syringa
7. Twigs with horizontal lines connecting leaf scars; arching shrubs with slender twigs; pith tiny, excavated between nodes	*Symphoricarpos*
8. Twigs and buds greenish	*Buckleya*
8. Twigs and buds grayish to reddish-brown	*Nestronia*
(if stipules or their scars are present, see *Rhamnus*)	
9. Habit a climbing or twining vine	10
9. Habit shrubby	12
10. Tendrils lacking; slender, twining vine; sap milky	*Trachelospermum*
10. Tendrils present (on persistent petioles); high-climbing vines without milky sap	11
11. Tendril forked and twining irregularly; a common native vine	*Bignonia*
11. Tendril divides into 3 short, clawlike forks; an uncommonly naturalized exotic	*Macfadyena*
12. Buds partially sunken, less than ½ the size of the subtending and prominent leaf scar	13

12. Buds not sunken, larger than above description 15
13. Buds hairy; pith interrupted through nodes; tomato-like sap odor; an exotic shrub of dry to moist sites *Lantana*
13. Buds scaled or sunken; pith solid or hollow; no odor like above; a native shrub of wet sites 14
14. Leaf scars often whorled; pith solid; lenticels large *Cephalanthus*
14. Leaf scars opposite or subopposite; pith excavated; twigs often dying back in winter *Decodon*
15. Twigs velvety; leaf scar U-shaped; buds often superposed *Vitex*
    (if scurfy scales present, see *Callicarpa*)
15. Not with above combination of characteristics 16
16. Buds with several imbricate scales; buds may be collateral 17
16. Buds not as above 18
17. Buds often collateral; elongate stipules often persistent; bark smooth *Sageretia*
17. Buds usually single; no stipules; outer bark shreddy *Symphoricarpos*
18. Buds scurfy or hairy, naked or with valvate scales *Buddleya*
18. Buds not as above, may be concealed in base of persistent small leaves *Hypericum*
19. Bundle scars 3 20
19. Bundle scars more than 3, or scars indistinguishable 33
20. Bud scales 4 or fewer and tightly adhering, or buds hidden 21
20. Bud scales usually 6 or more; scales not tightly adherent, at least near apex 27
21. Buds purplish or deep red, smooth and glabrous, appearing 1-scaled 22
21. Buds not as above 23
22. Twigs very slender; some nodes alternate *Salix purpurea*
22. Twigs rigid, not slender; all nodes opposite *Viburnum*
23. Buds sunken and barely visible behind leaf scar; twigs dark brown *Philadelphus*
    (if inner bark spicy aromatic, see *Calycanthus*)
23. Buds clearly visible; twigs not as above 24
24. Buds naked or valvate-scaled; twigs slightly lined, ridged, or lenticellate, gray to brown 25
24. Bud scales imbricate; twigs round, unlined, green or reddish 26
25. Buds prominent, often 2 at end of twig *Viburnum*
25. Buds reduced; twigs tend to die back in winter *Iva*
26. Twigs glabrous; buds conspicuous; small stipule scars do not connect leaf scars *Staphylea*
26. Twigs slender, hairy at tip; buds tiny; leaf scars connected by a line or ledge *Abelia*
27. Pith excavated, wholly or partially 28
27. Pith solid (homogeneous) or diaphragmed 29
28. Buds broad, rather globular; twigs glabrous to finely hairy; fruits berrylike *Lonicera*
28. Buds not as above; twigs scabrous; dried capsular fruits often persist *Deutsia*
29. Petiole base remnants often persistent, or habit a vine *Lonicera*
29. Petiole base remnants lacking, and a shrub 30
30. Leaf scars not connected by lines or ridges; fruit fleshy, not persistent in winter *Rhamnus*
30. Leaf scars connected by a line or ridge; dry fruits or their remnants often persistent 31
31. Pith large, light brown; twigs hairy or with lines of hairs connecting nodes *Diervilla*
31. Pith small, or white; twigs not as above 32
32. Pith small; buds projecting upward (toward twig apex) *Rhodotypos*
32. Pith large; buds outward projecting *Hydrangea paniculata*
33. Bundle scars 4 or more 34
33. Bundle scars indistinguishable 41
34. Habit a climbing vine; twigs with tendrils or aerial rootlets 35
34. Habit a shrub or tree; twigs not as above 36
35. Twigs slender, green or reddish, with tendrils *Bignonia*
    (if tendrils clawlike, see *Macfadyena*)
35. Twigs not slender, with aerial rootlets, gray or light brown *Campsis*

36. Twigs robust; lenticels conspicuous; pith large or excavated   37
36. Twigs not as above; pith solid, white, not overly enlarged   39
37. Buds conspicuous; leaf scars half-moon or crescent-shaped; usually 5 bundle scars; shrubs   *Sambucus*
37. Buds very small; leaf scars large, concave, with a circular bundle scar pattern; trees   38
  38. Leaf scars opposite; pith chambered or excavated, solid only at nodes   *Paulownia*
  38. Leaf scars mostly 3 per node (whorled); pith homogenous   *Catalpa*
39. Twigs hirsute or hairy; some leaf scars may be subopposite or alternate; trees   *Broussonetia*
39. Twigs glabrous or only faintly hairy; all leaves opposite; shrubs   40
  40. Twigs dark brown; buds small, dark, hairy, appear naked   *Calycanthus*
  40. Twigs green; buds prominent; bud scales imbricate, greenish or red-tinted   *Staphylea*
41. Habit a vine   42
41. Habit a shrub   44
  42. Leaf scars conspicuous or raised; buds elongated   *Lonicera*
  42. Leaf scar concave or indistinct; buds small, blunt or rounded   43
43. Buds sunken partially into round, grayish twigs   *Clematis*
43. Buds visible, with truncated outer valvate scales; twigs reddish, angled   *Bignonia*
  44. Twigs green, distinctly squared in cross section; dried petiole remnant covers leaf scar   *Jasminum*
  44. Twigs not as above   45
45. Buds hairy or scurfy, naked or valvate-scaled; twigs partially 4-angled   *Buddleya*
45. Buds and twigs without above combination of characteristics   46
  46. Twigs grayish; buds flattened; shrubs of brackish soils   *Borrichia*
  46. Twigs brownish; buds not as above; shrubs of nonbrackish sites   47
47. Buds partially or wholly concealed by persistent small leaves or their bases; twigs longitudinally winged or lined below leaf scars   *Hypericum*
47. Buds visible; leaf scars often with persistent remnants of petioles; twigs not as above   48
  48. Twigs very slender; branches arching, flexible   *Symphoricarpos*
  48. Twigs slender to moderately stout; tall, stiffly erect shrubs   *Lonicera*

Key D

(Leaves alternate; bundle scar count obscured)

1. Leaf scar partially or wholly covered by dried petiole remnants; nodes covered by scales, or internodes between buds with many peglike leaf attachment points   2
1. Leaf scar clearly visible at nodes; not as above   11
  2. Nodes covered by scales, or internodes with lines and peglike bumps; trees or tall shrubs   3
  2. Nodes not as above; vines or weakly upright shrubs   5
3. Twigs long, slender, tips often dying back in winter   *Tamarix*
3. Twigs not as above; scaled buds present on twig tip   4
  4. Twigs lined; spur shoots usually present; a northern conifer   *Larix*
  4. Twigs not as above; a southern conifer   *Taxodium*
5. Spines or prickles present   6
5. Spines and prickles lacking   8
  6. Spines present only at nodes, elongate and often branched   *Berberis*
  6. Spines or prickles scattered on stems, none branched   7
7. Buds with 3 or more imbricate bud scales   *Rubus*
7. Buds with no scales visible, or appearing single-scaled   *Smilax*
  8. Habit a shrub; leafy or elongate stipules persistent, at least near twig apex   9
  8. Habit a vine; stipules not commonly persistent (but stipule scars may be present)   10
9. Bark shreddy; white, cobwebby hairs on twig apex   *Pentaphylloides*
9. Bark smooth to furrowed; no hairs as above   *Indigofera*

10. Tendrils lacking; curved, crescent-shaped scars on each side of node; a low-climbing or clumplike vine *Aster carolinianus*
10. Tendrils present; node scars not as above; a high-climbing vine *Brunnichia* (if leaf scars large, see *Ampelopsis*)
11. Spines or thorns present on most twigs, or buds sunken and essentially not visible 12
11. Spines or thorns lacking; buds visible 28
   12. Buds sunken, not discernible 13
   12. Buds visible, though may be small 15
13. Leaf scars very small, clustered on swollen nodes; spines terminal on twig *Lycium*
13. Leaf scar single per node; spines, when present, paired at node 14
   14. Twigs greenish; hairs cover area of sunken buds *Sophora*
   14. Twigs brown; sunken buds mostly within/under leaf scar *Robinia*
15. Twigs green 16
15. Twigs grayish to brown 17
   16. Twigs heavily ridged; thorns flattened, with single leaf scar at lower margin *Poncirus*
   16. Twigs not as above; leaf scar on upper margin of thorn base *Parkingsonia*
17. Spines paired or branched 18
17. Spines solitary and unbranched 20
   18. Spines slender, flexible, mostly 3- or more branched *Berberis*
   18. Spines rigid, 2-branched or paired 19
19. Spines paired, 1 on each side of node *Zizyphus*
19. Spines pale or whitened, 1 per node but usually 2-branched *Acacia*
   20. Spines terminate spur shoots or short branch tips (look for leaf scars on spines) 21
   20. Spines (thorns) all nodal 25
21. Twigs and buds with silvery or brown peltate scales *Elaeagnus*
21. Twigs and buds lack peltate scales 22
   22. Inner bark with rank odor when bruised, yellowish-colored *Rhamnus*
   22. Inner bark with bitter-almond odor or odorless, not yellowish 23
23. Bark of young trunks or branchlets smooth but with horizontally elongated lenticels; scraped bark often with bitter-almond odor *Prunus*
23. Bark of young stems may be smooth but not lenticellate as above; scraped bark without bitter-almond odor 24
   24. Buds blunt, about as wide as long or wider *Sideroxylon*
   24. Buds pointed, longer than wide *Pyracantha* (if twigs angular or lined, see *Pyrus*)
25. Spines weak or easily broken; sap in freshly cut twigs clear 26
25. Spines stout, not easily broken; sap in freshly cut twigs milky 27
   26. Spines paired at nodes *Zizyphus*
   26. Spines single per node *Prosopis*
27. Stipules and stipule scars lacking; shrubs or small trees *Sideroxylon*
27. Stipules or their scars present; medium to large trees 28
   28. Terminal bud often appearing present; basal scales slightly hairy; twigs red-brown, with epidermal slits; thorns straight; bark scaling (exfoliating) into loose strips *Cudrania*
   28. Terminal bud lacking; bud scales glabrous; twigs gray; thorns curved slightly or strongly; bark furrowed *Maclura*
29. Twigs deep green, heavily ridged or lined below leaf scars; buds rounded with greenish, rounded bud scales 30
29. Twigs and buds not as above 31
   30. Collateral buds present *Baccharis*
   30. Buds single per node *Cytissus*
31. Twig apex with white, cobwebby hairs; dried stipules connect around node *Pentaphylloides*
31. Twigs not as above; no connecting stipules 32
   32. Lateral buds with 4 or fewer bud scales visible 33

32. Lateral buds with 5 or more bud scales visible   43
33. Habit a vine; twigs long and twining   34
33. Habit a shrub or tree; twigs not twining   35
   34. Leaf scar with knoblike swelling to each side; bud scales loose and rather ragged at tips   *Wisteria*
   34. Leaf scar not as above; bud scales smooth and tightly adherent to entire bud   *Berchemia*
35. Buds pointed, longer than wide; bundle traces are a raised, fibrous mass in center of leaf scar; brittle, winglike ridges below leaf scar corners   *Lagerstroemia*
35. Buds not as above; bundle traces otherwise; no winglike ridges   36
   36. Buds often superposed   *Amorpha*
   36. Buds not superposed (but may be collateral)   37
37. Leaf scars crescent-shaped or U-shaped, ½ as wide as twig circumference or wider   *Iva*
37. Leaf scars not as above   38
   38. Twigs covered by long, stiff trichomes; collateral buds often present   *Broussonetia*
   38. Twigs glabrous or finely hairy; collateral buds rarely present   39
39. Lateral bud partially hidden behind raised ledge above leaf scar   *Sesbania*
39. Lateral bud not as above   40
   40. Endmost bud at twig tip short, rounded on tip; twigs moderately stout; trees   *Cudrania*
   40. Endmost bud pointed; twigs very slender; shrubs   41
41. Stipules or their scars obvious; twigs lightly lined below leaf scars   42
41. Stipules lacking; twigs unlined, or dotted, usually greenish at least on 1 side   *Vaccinium*
   42. Twigs reddish; bud scales 2 or 3, obscure; sap milky   *Sebastiania*
   42. Twigs brown or gray; bud scales distinct, imbricate; sap not milky   *Spiraea*
43. Habit a vine   *Akebia*
43. Habit a shrub or tree   44
   44. Habit a low shrub; buds small, mostly 3mm long or less   45
   44. Habit a tall shrub or tree; buds often longer than above   47
45. Twigs rather short, or with spur shoots   *Chaenomeles*
   (if bitter-almond odor present in scraped twigs, see *Prunus*)
45. Twigs elongate but dying back at tips; often slightly lined   46
   46. Collateral buds or elongate, persistent stipules usually present; considerable winter dieback occurring on twigs   *Lespedeza*
   46. Not as above; capsular fruits often persisting on twig tips   *Spiraea*
47. Bud scales loosely imbricate, gray-hairy   48
47. Bud scales not gray-hairy   49
   48. Twigs reddish or partly green; buds often clustered near twig tip, or collateral buds present on some lateral nodes   *Prunus persica*
   48. Twigs grayish; buds single per node   *Pyrus calleryan*a
49. Lumplike swelling present at base of some lateral buds, or collateral buds present; no spur shoots   *Planera*
49. Lumplike swelling not present; collateral buds lacking; spur shoots present   50
   50. Buds blunt or rounded; bud scales entire, but often with a recurved tip   *Ginkgo*
   50. Buds pointed; bud scale margin hairy, toothed, or ragged   51
51. Young stems often with horizontally elongated lenticels; inner bark greenish, with bitter-almond odor; buds often present at twig tip   *Prunus*
51. Young stems not as above; inner bark yellowish, with different rank odor; twig tips dying back, or branch scar prominent   *Rhamnus*

Key E

(Alternate leaves; 1 bundle scar; terminal bud present)

1. Terminal bud with 5 or more imbricate scales and several smaller lateral buds crowded at its base; small capsular fruits persist through winter   2
1. Terminal bud without above combination of characteristics   3
  2. Bud scales pale-margined or loosely imbricate; capsules short, on much longer stalks; outer bark papery, exfoliating   *Menziesia*
  2. Bud scales dark-margined or very tightly imbricate; capsules elongated, longer than their stalks; bark not papery   *Rhododendron*
3. Internodes with many small, peglike attachment points where leaves (needles) were borne   *Larix*
3. Internodes not as above   4
  4. Terminal bud with 4 or more grayish or reddish bud scales; tiny blackened stipules often persist; spur shoots commonly present   *Ilex*
  4. Buds and twigs without above combination of characteristics   5
5. Twigs and buds covered with silvery or brownish peltate scales   *Elaeagnus*
5. Twigs and buds not as above   6
  6. Terminal bud scales distinctly hairy or scurfy   7
  6. Terminal bud scales glabrous or nearly so, but may be minutely bumpy (verrucose)   20
7. Twig apex with white, cobwebby hairs; persistent stipules connecting around nodes   *Pentaphylloides*
7. Twigs not as above   8
  8. Pith chambered   9
  8. Pith solid (homogenous) or spongy   10
9. Bark of twigs, branchlets, and trunk smooth   *Symplocus*
9. Bark of twigs or branchlets slightly shreddy or with loose threads; trunk bark rough; *Halesia*
  10. Spines or spiny spur shoots present   11
  10. Spines and spiny spur shoots lacking (nonspiny spur shoots may be present)   13
11. Outermost bud scales much shorter than bud   *Prunus*
11. Outermost bud scales about as long as entire bud   12
  12. Stipule scars present   *Cudrania*
  12. Stipule scars lacking   *Sideroxylon*
13. Stipule scars large, dark; buds scurfy, stellate-pubescent   *Fothergilla*
13. Stipule scars and buds not as above   14
  14. Bundle scar distinctly raised within leaf scar; twigs brown or grayish   15
  14. Bundle scar not distinctly raised; twigs green or reddish   18
15. Lateral buds smaller than subtending leaf scars, naked or obscurely valvate-scaled   *Clethra*
15. Lateral buds distinctly larger than subtending leaf scars, with 2 or 3 imbricate bud scales   16
  16. Twigs slender, finely lined below leaf scars; multistemmed, arching shrubs   *Spiraea*
  16. Twigs not as above; trees or very large, stiffly erect shrubs   17
17. Lateral buds hairy; twigs often slightly flattened at nodes   *Stewartia*
17. Lateral buds glabrous; twigs rounded at nodes   *Halesia*
  18. Dried stipules or their scars present   19
  18. Stipules lacking   21
19. Outer, imbricate bud scales only partially covering the conspicuously hairy inner bud   *Ceanothus*
19. Buds not as above   20
  20. Buds often collateral but rarely superposed; largest buds with 5 or more visible bud scales   *Prunus*
  20. Buds often superposed but rarely collateral; largest buds mostly with 3 to 4 bud scales   *Ilex*

21. Pith brown; leaf scars much larger than lateral buds *Franklinia*
21. Pith white; leaf scars smaller than lateral buds; scraped twigs spicy-scented
Lindera melissifolia
  22. Buds superposed, or flower buds stalked, globular 23
  22. Buds not superposed; no stalked flower buds 25
23. Twigs not aromatic when scraped; buds often superposed; no larger flower buds present
Ilex
23. Twigs spicy-scented when scraped; stalked, globular flower buds usually present 24
  24. Buds superposed, greenish; twigs greenish-gray *Lindera*
  24. Buds usually red, single per node except for collateral flower buds; twigs slender, reddish, zigzag *Litsea*
25. Twigs covered by long trichomes (hirsute); buds naked or covered by granular pubescence
Befaria
25. Twigs not hirsute; buds scaled (at least laterals) 26
  26. Scraped twigs spicy-scented; terminal bud scales imbricate, tightly adherent *Sassafras* (if there are many bud scales but no spicy scent, see *Rhododendron*; if thorns present, see *Cudrania*)
  26. Scraped twigs not as above; terminal bud scales loose at tips, outer ones extending beyond the length of inner bud scales 27
27. Pith white, often chambered in 2d-year stems 28
27. Pith green, homogenous 29
  28. Lateral buds wider than long, or bluntly pointed; pith chambered *Symplocus*
  28. Lateral buds longer than wide, acute or sharply pointed; pith often chambered only in branchlets *Halesia*
29. Twigs distinctly lined or ridged below leaf scars; a clambering, vinelike shrub *Lycium*
29. Twigs not lined or ridged; habit not vinelike 30
  30. Lateral buds mostly larger than leaf scars 31
  30. Lateral buds mostly smaller than leaf scars 32
31. Twigs with short, glandular hairs; a short shrub of coastal plain *Kalmia*
31. Twigs glabrous; a northern shrub of moist mountain uplands *Ilex* (= *Nemopanthus*)
  32. Twigs with raised ridge or line extending down from lower tip of leaf scar (giving slight angular aspect to twig) *Cyrilla*
  32. Twig rounded, not ridged; leaf scar rounded at lower tip *Elliottia*

Key F

(Alternate leaves; 1 bundle scar; no terminal bud)

1. Buds or leaf scars clustered (buds superposed or collateral in 2s or more, or leaf scars in close clusters on nodes) 2
1. Buds solitary, over single leaf scar 12
  2. Buds superposed or collateral; leaf scars single per node 3
  2. Buds single per node or poorly visible; leaf scars usually 3 or more per node 8
3. Buds naked, or with scurfy pubescence or peltate scales 4
3. Buds obviously scaled 5
  4. Buds and twigs with peltate scales; collateral buds usually present *Elaeagnus*
  4. Buds naked or scurfy; superposed buds usually present *Styrax*
5. Buds usually superposed (but collateral flower buds may occur); scraped twigs spicy-scented, mostly unlined and smooth *Lindera*
5. Buds collateral; twigs not spicy scented, usually having lines or ridges 6
  6. Twigs green; stipules lacking *Baccharis*
  6. Twigs grayish; elongated stipules often persistent, or their scars obvious 7
7. Branchlets bearing stiff side shoots; leaf scars often subopposite on twigs *Sageretia*
7. Branchlets and twigs not as above; twigs long, slender, wandlike *Andrachne*

8. Nodal thorns or spines present, or twig tips or spur shoots spiny 9
  8. Thorns and spines lacking 11
 9. Habit clambering or vinelike; twig tips or spur shoots commonly spine-tipped *Lycium*
 9. Habit shrubby or treelike; spines at nodes only 10
   10. Spines usually branched, located below buds; shrubs *Berberis*
   10. Spines (thorns) not branched, lateral to buds; trees *Maclura*
 11. Buds clearly visible, with 1 to 3 accompanying basal leaf scars; stipules often persistent
    Ceanothus
 11. Buds poorly evident; leaf scars 4 or more per node; stipules lacking *Hibiscus*
   12. Habit a vine; twigs twining 13
   12. Habit a shrub or tree; twigs not twining 15
 13. Twigs green, round and smooth; buds sharply pointed, appressed to twig; pith green
    Berchemia
 13. Twigs gray or brownish, ridged or lenticellate; buds blunt or outward projecting; pith not
    green 14
   14. Scraped twigs rankly scented; pith white, solid (homogenous) *Celastrus*
   14. Scraped twigs odorless; pith excavated except at nodes *Solanum*
 15. Twigs deep green and heavily ridged, or spiny 16
 15. Twigs not deep green, not spiny 22
   16. Twigs green 17
   16. Twigs gray or brown 18
 17. Large, flattened thorns present *Poncirus*
 17. Thorns lacking; twigs heavily ridged or lined *Cytissus*
   18. Spines nodal, with bud to 1 side of base; buds rounded or blunt; sap milky 19
   18. Spines terminating spur shoots; buds pointed; sap clear 20
 19. Twigs and buds hairless; twigs moderately stout; stipules or their scars present *Maclura*
 19. Twigs or buds often hairy; twigs slender; stipules lacking *Sideroxylon*
   20. Buds and twigs silvery-dotted or with peltate brown scales *Elaeagnus*
   20. Buds and twigs lacking peltate scales 21
 21. Inner bark with yellowish color *Rhamnus*
 21. Inner bark greenish *Prunus*
   22. Twigs jointed in appearance; bud scales tiny but numerous (6 or more), fingerlike or with
     elongate tips *Stillingia*
   22. Twigs and bud scales not as above 23
 23. Endmost bud and twig tip bent toward 1 side; pith usually chambered or diaphragmed in
    2d-year twigs; trees with gray, warty bark *Celtis*
 23. Not with above combination of characteristics 24
   24. Twigs round (terete), stiff and brittle (rather inflexible); small to large trees 25
   24. Twigs slightly flattened, lined, or slender and flexible; shrubs 28
 25. Twigs distinctly hairy; sap milky; at least some leaf scars subopposite or nearly so
    Broussonetia
 25. Not with above combination of characteristics 26
   26. Twigs reddish on 1 side, aromatic when scraped; buds rounded, red *Oxydendrum*
   26. Twigs brownish or gray; not as above 27
 27. Buds appressed to twig, covered mostly by 2 bud scales *Diospyros*
 27. Buds not as above *Halesia*
   28. Buds completely covered by 2 tight bud scales 29
   28. Buds with more than 2 scales, or incompletely covered by 2 outer bud scales 30
 29. Bud scales valvate; buds appressed to twig; upper margin of leaf scar mostly straight
    Vaccinium
 29. Bud scales imbricate; buds not appressed; upper margin of leaf scar convex or crowned
    Lyonia
   30. Twig apex with white, cobwebby hairs; stipules connecting around nodes
    Pentaphylloides

30. Twigs not as above 31
31. Stipule scars present, or stipules may be persistent at nodes nearest twig tip; if stipule scars are tiny, then twigs usually lined or ridged below leaf scars 32
31. Stipules and stipule scars lacking; twigs not lined 37
  32. Twigs very slender, lined; buds and twig size distinctly reduced toward apex of twigs; most buds and leaf scars 2mm wide or less; stipules deciduous, their scars tiny 33
  32. Twigs not as above; stipules often persistent on nodes near apex of twig 34
33. Pith white; buds 2- to 4-scaled .... *Andrachne*
33. Pith brown; buds usually with more than 4 scales, or buds very hairy .... *Spiraea*
  34. End bud larger than laterals; persistent stipules longer than buds, project upward; bud scales few, only partially covering the very hairy bud surface beneath ...... *Ceanothus*
  34. End bud about equal in size to laterals, or smaller; stipules not as above 35
35. Leaf scar less than ½ the width of bud; persistent stipules often elongate, curling; inner bark relatively inodorous .... *Indigofera*
35. Leaf scar about as wide as bud; stipules short or straight; inner bark usually with strong odor 36
  36. End bud smaller than laterals of lower nodes; buds rounded or slightly hairy, often superposed .... *Amorpha*
  36. All buds about equal size; buds pointed, glabrous, sometimes collateral but not superposed .... *Rhamnus*
37. End bud much larger than other laterals, distinctly hairy; inner bark spicy-scented *Lindera*
37. End bud not as above; inner bark inodorous 38
  38. Buds mostly with 5 or fewer visible scales; fruits fleshy (berries), not persistent in winter 39
  38. Buds mostly with 6 or more visible scales; fruits dry (capsules), usually persistent in winter 40
39. Resin globules visible with lens on inner bud scales; older stem bark mostly smooth; branchlets gray, brownish, or reddish .... *Gaylussacia*
39. Resin globules lacking; older stem bark usually scaly; branchlets often verrucose and green, at least on one side .... *Vaccinium*
  40. Capsules much longer than broad; bud scales minutely serrate along margin .... *Lyonia mariana*
  40. Capsules wider than long; bud scales entire on margin 41
41. Buds wider than twig diameter; pith spongy; lower leaf scar margin wrinkled *Zenobia*
41. Buds usually not as wide as twig; pith homogenous; leaf scar margins smooth; terminal racemes of flower buds often present in winter .... *Leucothoe*

Key G

(Alternate leaves; 3 bundle scars)

1. Thorns, spines, or prickles present on most twigs or branchlets 2
1. Thorns, spines, and prickles lacking 24
  2. Spines terminating short twigs or spur shoots 3
  2. Spines nodal or internodal, not as above 10
3. Inner bark yellowish, with rank odor; some buds may be subopposite ...... *Rhamnus*
3. Inner bark not yellowish; odor not rank (if slightly so, see *Prunus*); buds alternate ... 4
  4. Spinose spur shoots usually flanked by a bud on each basal side 5
  4. Spinose spur shoots with single bud on one basal side, rarely with 2 6
5. Bud scales blunt or rounded, tightly imbricate, hairless; a low, dense, naturalized exotic shrub .... *Chaenomeles*
5. Bud scales abruptly pointed into a projecting tip, often with hairs; a tall, rare native shrub .... *Mespilus*

6. Terminal or endmost bud much larger than other buds, with reddish scales; a tree with scaly bark and stout, stiff spur shoots *Malus*
6. Not with above combination of characteristics 7
7. Twigs lined below leaf scars; most buds over 6mm long, with grayish, very hairy scales *Pyrus*
7. Twigs not lined; buds smaller, with glabrous or slightly hairy scales 8
  8. Bark of branchlets or young trunks smooth but with horizontally elongated lenticels; scraped bark may have bitter-almond odor *Prunus*
  8. Bark of young stems smooth but not lenticellate; no odor as above 9
9. Buds bluntly pointed or rounded, about as wide as long or wider; fruit black when ripe, rarely persistent; native shrub or small tree *Sideroxylon*
9. Buds distinctly pointed, longer than wide; fruit orange, often persistent; exotic, sparingly naturalized shrub *Pyracantha*
  10. Spines paired at nodes, or easily broken from twig surface if forced to one side 11
  10. Spines solitary at nodes (but may be forked) and not easily broken (a thorn) 16
11. Petiole base persistent, covering leaf scar; twigs 5-angled, or covered by a pale bloom *Rubus*
11. Petiole base not persistent; twigs round, lacking bloom 12
  12. Spines mostly solitary at nodes, directly under leaf scar; twig tips often dying back in winter *Erythrina*
  12. Spines paired at nodes, or scattered along internodes 13
13. Buds sunken within or near top of leaf scar, not visible; terminal bud obviously lacking *Robinia*
13. Buds visible, not sunken; terminal bud may appear present 14
  14. Bud scales imbricate, 4 or more visible per bud; leaf scar narrow; twigs slender *Rosa*
  14. Bud scales indistinct, or appear valvate; leaf scar taller; twigs moderately stout 15
15. Buds woolly or with bumpy surface; scales indistinct; inner bark aromatic; trunk bark smooth or warty, often with prickles *Zanthoxylum*
15. Buds with a few elongate, smooth, distinct scales; inner bark not aromatic; trunk bark scaly or roughly furrowed *Zizyphus*
  16. Thorns arising directly above node and bud; buds partially sunken within twig *Gleditsia*
  16. Thorns or spines arising beside or below nodal buds; buds fully visible 17
17. Spines arising directly below leaf scars 18
17. Spines arising to 1 side of node, slightly above leaf scar 21
  18. Leaf scars obscure, or minute and on thorn bases; trees or large shrubs often over 2m tall 19
  18. Leaf scars obvious, or not on thorn bases; shrubs mostly under 1.5m tall 20
19. Twigs dark reddish-brown; leaf scars and buds clustered above base of thorns *Acacia*
19. Twigs greenish; leaf scars on upper surface of thorn base *Parkingsonia*
  20. Buds often collateral or several leaf scars present; inner bark yellowish *Berberis*
  20. Buds single at nodes; inner bark greenish *Ribes* (if twigs green, see *Erythrina*)
21. Pith diaphragmed, or excavated in sprouts; width of leaf scar ½ or less of twig diameter; thorns of twigs gray or light brown *Sideroxylon*
21. Pith homogenous; width of leaf scar ⅔ or more of twig diameter; thorns of twigs darkened or reddish 22
  22. Thorns lacking buds or leaf scars except at base, near twig union *Crataegus*
  22. Thorns (spur shoots) with leaf scars or buds along length, usually with 2 distinct and equal size buds at base, on each side 23
23. Bud scales tightly imbricate, the tips blunt or rounded; a low, dense, sparingly naturalized exotic shrub *Chaenomeles*
23. Bud scales pointed, plump, the tips free; a tall, rare native shrub *Mespilus*
  24. Terminal bud lacking 25
  24. Terminal bud present 89

25. Habit a vine; twigs and stems climbing or trailing 26
25. Habit a shrub or tree 31
  26. Buds scaled; stipules or stipule scars visible 27
  26. Buds naked, sunken, or not scaled; stipule scars lacking 29
27. Twigs hairy; bud scales rather loose, hairy; stipule scars large . . . . *Pueraria*
27. Twigs smooth; bud scales tightly imbricate, smooth; stipule scars tiny 28
  28. Twigs brownish and lenticellate; buds short-ovoid . . . . *Akebia*
    (if buds elongate, see *Schisandra*)
  28. Twigs greenish, or reddish on one side; buds pointed, flattened . . . . . *Berchemia*
29. Buds superposed, partially sunken into a silvery-silky raised area within the narrow, horseshoe-shaped leaf scar . . . . *Aristolochia*
29. Buds not as above; leaf scar not horseshoe-shaped 30
  30. Leaf scars large, semicircular, concave; twigs lustrous, reddish on one side
    . . . . *Menispermum*
  30. Leaf scars rather bean-shaped, with raised ridges on top margin; twigs dull green
    . . . . *Cocculus*
31. Buds reduced (less than ½ the size of leaf scar); rounded or domelike on medium to robust twigs or sunken partially or entirely within twig 32
31. Buds very evident (over ½ the size of leaf scar); pointed, or if rounded then with 4 or more imbricate scales; dried stipules may persist 42
  32. Branch scars obscure; leaf scars small, often with smaller accessory leaf scars . . . *Hibiscus*
  32. Branch scars prominent; leaf scars large, triangular to partly 3-lobed 33
33. Bud scales visible on small, rounded buds 34
33. Bud scales not evident; buds sunken or hidden by hairs 38
  34. Twigs and buds dark reddish-brown; some side buds with 4 or more visible scales . . . *Cercis*
  34. Twigs gray, greenish, or light brown; buds rarely with more than 3 visible scales . . . 35
  35. Buds often superposed; small stipule projections may persist, though easily broken; bundle scars resemble rusty-colored circular patches 36
  35. Buds not superposed; stipules lacking; bundle scars not as above 37
    36. Twigs glabrous; a tree . . . . *Albizzia*
    36. Twigs hairy; a shrub . . . . *Amorpha*
37. Pith green; buds glabrous . . . . *Sapindus*
37. Pith white; buds brown with short hairs . . . . . . *Melia* (if end bud pointed, see *Nyssa ogeche*)
  38. Pith large, pinkish; twigs robust; buds sunken, but obviously superposed . . . *Gymnocladus*
  38. Pith and twigs not as above 39
39. Buds single per node, covered by hairs, visible within U-shaped leaf scar; scraped twigs with lemonlike or weakly skunklike odor . . . . *Ptelia*
39. Buds superposed, partially sunken, or obscure; no odor as above in scraped twigs . . . 40
  40. Bud scales tiny, glabrous; buds partially sunken . . . . *Gleditsia*
  40. Bud scales not visible; buds completely sunken or nearly so, or covered by hairs . . . 41
41. Buds sunken within leaf scar; twigs brown . . . . *Robinia*
41. Buds mostly concealed by hairs; twigs green . . . . *Sophora*
  (if hairy buds visible above leaf scar, see *Amorpha*)
    42. Buds single-scaled (*note*: be careful here not to confuse a raised lateral ridge on the bud scale with a true seam, as between 2 valvate scales) . . . . *Salix*
    42. Buds with 2 or more bud scales (if scales obscured by hairs or buds appear naked, go to 87) 43
43. Buds with 2 or 3 bud scales visible 44
43. Buds with 4 or more bud scales visible 52
  44. Stipules small, triangular, darkened, persistent, or stipular projections present . . 45
  44. Stipules and stipular projections not persistent, but stipule scars may be present . . 47
45. Twigs glabrous; buds usually smaller than subtending leaf scar, in outline . . . *Sebastiania*
45. Twigs hairy near apex or around buds, or buds obviously larger than subtending leaf scar 46
  46. Habit a tree; mature bark furrowed; sap milky . . . . *Sapium*

46. Habit a shrub; bark smooth or lenticellate; sap clear *Amorpha*
(if persistent stipules are longer than the lateral buds, see *Ceanothus*)
47. Stipule scars distinct, crescent-shaped on greener side of twig; leaf scars elliptical; outer bark tough and stringy *Tilia*
47. Stipule scars small and slitlike, or inconspicuous; not as above 48
    48. Branch stub present, nearly as long or longer than end bud; buds often superposed or collateral 49
    48. Branch scar small, stub lacking or not as above; buds single per node; catkin buds may be present 50
49. Habit a tree; buds reddish; bark scaly *Cercis*
49. Habit a shrub; buds gray to brown; bark smooth or lenticellate *Amorpha*
    50. Buds stalked or reddish; bud scale margins slightly obscured; inner bark yellowish *Alnus*
    50. Buds sessile, not reddish; bud scales easily discernible 51
51. Habit a tree; buds distinctly longer than wide *Betula*
51. Habit a shrub; buds about as long as wide *Corylus*
    52. Buds generally with 4 or 5 visible scales 53
    52. Buds generally with 6 or more visible scales 74
53. Glandular or hispid trichomes present; outer bark exfoliating *Rubus*
53. Not with above combination of characteristics 54
    54. Stipules or stipular projections usually persist at 1 or both sides of leaf scar, especially on nodes closer to twig apex 55
    54. Stipules or stipular projections not persistent 61
55. Lateral buds partially hidden by raised ledge above leaf scar *Sesbania*
55. Lateral buds conspicuous, not as above 56
    56. Twigs vivid green; collateral buds often present *Kerria*
    56. Twigs gray to brownish; buds single or superposed 57
57. Endmost bud usually as large or larger than leaf buds of lower nodes 58
57. Endmost bud smaller than other buds, or twig tips dried or withered 59
    58. Twigs conspicuously lined; bud scales loose, free at tips or edges; inner bark not odorous *Physocarpus*
    58. Twigs not lined, or barely so; bud scales tight; inner bark rankly odorous *Rhamnus*
59. Buds blunt or rounded, occasionally superposed *Amorpha*
59. Buds pointed, rarely superposed 60
    60. Twigs lined; stipules close to buds and projecting upward *Neviusia*
    60. Twigs unlined; stipules curling away from twig, brittle and easily broken *Indigofera*
61. Habit a shrub with conspicuously lenticellate bark; scraped or crushed parts spicy-scented 62
61. Habit a tree, or a shrub without lenticellate bark and spicy scent 65
    62. Buds often superposed; flower buds (if present) globular, stalked, collateral to a leaf bud *Lindera*
    62. Buds single or rarely collateral; flower buds not as above 63
63. Catkin buds lacking; shrubs of sandy soils in coastal plain *Myrica*
63. Catkin buds usually present; shrubs of mountains or Piedmont 64
    64. Catkin buds hairy; twigs hairy; a shrub of well-drained sites *Comptonia*
    64. Catkin buds mostly glabrous; twigs essentially hairless; a shrub of damp or boggy mountain sites *Myrica gale*
65. Collateral buds often present, especially on basal nodes of twig 66
65. Collateral buds lacking 68
    66. Buds pointed; twigs zigzag, minutely hairy *Planera*
    66. Buds blunt or rounded; twigs essentially straight 67
67. Inner bark yellowish, rankly scented; twigs gray *Rhamnus*
67. Inner bark greenish, essentially inodorous or with bitter-almond scent; twigs reddish or greenish *Prunus*
    68. Buds as wide as long or wider; outer bud scales about as long as entire bud *Sideroxylon*

68. Buds longer than wide; outer bud scales much shorter than bud length   69
69. Buds closely appressed to twig, endmost one bent toward branch scar; pith partially chambered or only so at nodes   *Celtis*
69. Buds not as above; pith homogenous   70
  70. Twigs slightly lined below leaf scars; bud scales loosely imbricate   *Pyrus calleryana*
  70. Twigs unlined; bud scales tight   71
  71. Habit a shrub; some stiff, bristlelike trichomes near buds   *Corylus*
  71. Habit a tree; twigs glabrous, or hairs soft or minute   72
    72. Buds more than 2-ranked; spur shoots usually present; bark smooth, papery, or in scaly plates on old trunks   *Betula*
    72. Buds 2-ranked; spur shoots lacking; bark furrowed or scaly   73
    73. Leaf scars with sunken face; bark scaly; small trees of wet coastal plain sites   *Planera*
    73. Leaf scars not as above; bark furrowed or with scaly ridges; medium to large trees widespread in various sites   *Ulmus*
      74. Collateral buds usually present, or buds with a small lump on 1 side, near base   75
      74. Buds single at nodes (if buds stalked or superposed and reddish, see *Cercis*)   76
      75. Bud base with small lump or swelling on 1 side; face of leaf scar darkened; twigs minutely hairy   *Planera*
      75. Bud base not as above; leaf scar light reddish-brown; twigs glabrous   *Ulmus parvifolia* (if leaf scars prominently raised, see *Prunus*)
        76. Catkin buds usually present at or near twig apex   77
        76. Catkin buds lacking   81
        77. Catkin buds covered with pale hairs; rank odor present in scraped bark   *Leitneria*
        77. Catkin buds not conspicuously hairy; rank odor lacking   78
          78. Buds elongated or pointed, longer than wide; scraped twigs odorless; trees   79
          78. Buds short, blunt, about as long as wide; spicy aroma in scraped bark; low shrubs   80
          79. Bud scales finely lined, darker along margins; bark lenticellate, scaly or furrowed   *Ostrya*
          79. Bud scales not lined, light-colored along margins; bark smooth, gray   *Carpinus*
            80. Lenticels prominent on twigs and branchlets; rare species of wet sites   *Myrica gale*
            80. Lenticels not as above; twigs usually hairy; widespread shrub of well-drained sites   *Comptonia*
        81. Buds rounded or spherical   82
        81. Buds acute or pointed   83
          82. Scraped twigs spicy scented; shrubs or small trees with smooth, gray bark   *Myrica*
          82. Scraped twigs inodorous; tall trees with furrowed bark   *Ulmus pumila*
          83. Bud scales covered by hairs; leaf scar narrow, darkened   *Pyrus calleryana* (if leaf scar is light brown, see *Mespilus*)
          83. Bud scales glabrous or with only a few hairs; leaf scar as deep as wide, or nearly so   84
            84. Stipule scars small, slitlike, or dried stipules persist; a tree (if a shrub, go to 55)   85
            84. Stipule scars not apparent; a shrub, or a small tree in *Prunus*   87
            85. Buds scales lighter-colored on margins; buds slightly angular in cross section; bark of trunk smooth, gray   *Carpinus*
            85. Bud scales darker on margins; buds not angular; trunk bark furrowed or scaly   86
              86. Basal or lighter-colored part of bud scales greenish; buds not 2-ranked on twig; horizontally elongated lenticels present on branchlets or young stems   *Ostrya*
              86. Basal part of bud scales brownish; buds 2-ranked on twig; no horizontally elongated lenticels   *Ulmus*
              87. Bud scales partly green, entire on margins   *Pyrularia*
              87. Bud scales reddish-brown, slightly toothed or ciliate on margins   88
                88. Inner bark yellowish or with strong odor; some nodes often subopposite   *Rhamnus*
                88. Inner bark not as above, or with bitter-almond odor; nodes all alternate or clustered   *Prunus*
89. Pith chambered   90
89. Pith not chambered   92

90. Lateral buds superposed; twigs robust; mature bark thick, furrowed; large trees *Juglans*
90. Lateral buds single per node; twigs slender; bark smooth or thin; shrubs or small trees 91
91. Stipules elongated, persist at nodes near twig apex; scraped twigs rankly aromatic; fruit berrylike *Rhamnus*
91. Stipules lacking or not as above; scraped twigs inodorous; fruit capsular, often persist over winter *Itea*
92. Terminal buds naked, or with 3 or fewer bud scales 93
92. Terminal buds with 4 or more bud scales 100
93. Twigs and buds glabrous 94
93. Twigs or buds hairy or scurfy, or covered with peltate scales 96
94. Twigs greenish; leaf scars mostly subopposite *Cornus alternifolia*
94. Twigs reddish-brown; leaf scars distinctly alternate 95
95. Twigs roughened by numerous raised lenticels; buds slightly gummy; catkin buds may be present; bud scale margins often indistinct *Alnus crispa*
95. Twigs not as above; buds not gummy; catkin buds lacking; bud scale margins obvious *Aronia melanocarpa*
96. Twigs and buds dotted with silvery or brown peltate scales; sap milky *Croton*
96. Twigs and buds not dotted; sap clear 97
97. Stipule scars lacking; buds sessile, not stalked *Nyssa ogeche*
97. Stipule scars present; buds usually stalked, or hairy 98
98. Buds with long hairs; stipules elongate, persistent; small shrubs *Ceanothus* (if inner bark rankly odorous, see *Frangula*)
98. Buds short-hairy, scurfy, or glabrous; stipules not as above; large shrubs 99
99. Buds naked or scurfy; accessory buds or flower buds/flower remnants often present *Hamamelis* (if buds solitary and stipule scars large and dark, see *Fothergilla*)
99. Buds with 1 to 3 tight bud scales; buds single per node; catkin buds usually present *Alnus*
100. Stipules slender, persistent at nodes near apex of twig 101
100. Stipules not as above, or plant with occasional stipules not keying by above characteristics 103
101. Lateral buds partially hidden by a raised ledge above leaf scar *Sesbania*
101. Lateral buds fully visible 102
102. Buds dome-shaped; collateral buds usually present; twigs unlined *Chaenomeles*
102. Buds oblong and pointed, single per node, rather loosely scaled; twigs lined *Physocarpus* (if bud scales tight and twigs unlined, see *Ostrya*, which actually has no terminal bud)
103. Trees with horizontally elongated lenticels on smooth stem parts; end bud nearly the same size as laterals; small stipule scars present 104
103. Trees without combination of above characters, or a shrub 105
104. Scraped twigs with bitter-almond odor; terminal bud scales usually 6 or more *Prunus*
104. Scraped twigs lacking bitter-almond odor; terminal bud scales usually number 4 or 5 *Betula*
105. Bud scales of terminal bud hairy 106
105. Bud scales of terminal or endmost bud hairless, or nearly so, to naked eye 112
106. Buds about as long as wide, or sessile catkin buds present 107
106. Buds distinctly longer than wide 111
107. Inner bark of scraped twigs yellow-green or bright green, rankly odorous; lenticels large *Leitneria*
107. Inner bark and lenticels not as above 108
108. Lateral bud wider than leaf scar; terminal bud slightly larger than laterals *Sideroxylon*
108. Lateral bud as wide as or narrower than leaf scar; terminal bud much larger than laterals 109
109. Bud scales thick or rather fleshy, with dark, projecting tips *Mespilus*
109. Bud scales thin, not as above 110
110. Twigs greenish or olive-green; pith large; mature bark furrowed *Nyssa ogeche*

110. Twigs reddish or gray-brown; pith small; mature bark scaly   *Malus*
111. Twigs finely hairy; leaf scars darkened; stem bark scaly or thinly furrowed
    Pyrus calleryana
111. Twigs glabrous; leaf scars not darkened; stems smooth or furrowed on large or old trunks
    Amelanchier
    112. Leaf scars crescent-shaped (over twice as wide as tall)   113
    112. Leaf scars not as above   120
113. Terminal bud over twice as long as wide   114
113. Terminal bud about twice as long as wide, or shorter   116
    114. Habit a low shrub (height usually 1m or less); bark exfoliating   *Ribes*
    114. Habit a tall shrub or tree; bark smooth or furrowed   115
115. Bud scales smooth, reddish, somewhat keeled or with an edge on the back   *Aronia*
115. Bud scales hairy or greenish, not as above   *Amelanchier*
    116. Habit a shrub; twigs slender   117
    116. Habit a tree; twigs moderately stout; thick, stiff spur shoots usually present   119
117. Lateral buds mostly with 2 or 3 bud scales   *Aronia*
117. Lateral buds with 4 or more scales; at least the lower scales with blackened, elongate tips
    118
    118. Twig hairy at apex or bases of buds; a rare native shrub   *Mespilus*
    118. Twigs glabrous; a sparingly naturalized exotic   *Exochorda*
119. Bud scales reddish; terminal buds with white hairs at tip or base   *Malus*
    (if buds rounded, see *Crataegus*)
119. Bud scales brownish, slightly keeled at apex; buds mostly glabrous   *Pyrus communis*
    120. Habit a tree; bark roughened and furrowed, or scaly   121
    120. Habit a shrub; bark smooth, though often lenticellate (if a vine, see *Schisandra*)   124
121. Stipule scars present, or buds lustrous   122
121. Stipule scars lacking   123
    122. Bundle scars round, dark with a pale border; lowermost bud scales of lateral bud meet near middle of top margin of leaf scar   *Liquidambar*
    122. Bundle scars not as above; lowermost bundle scar elongate, almost divided; lowermost bud scale of lateral bud spans width of bud base directly atop leaf scar   *Populus*
123. Inner bark aromatic; outer terminal bud scales nearly as long as entire bud, with loose tips; pith solid (homogenous)   *Cotinus*
123. Inner bark inodorous; bud scales tightly imbricate, not as above; pith diaphragmed   *Nyssa*
    124. Buds large (6–20mm long), with partly green, loosely imbricate scales   *Pyrularia*
    124. Buds not as above   125
125. Buds often collateral or superposed   *Lindera*
125. Buds single per node   126
    126. Buds and twig tips with tiny but conspicuous resinous dots; catkin buds lacking; plants native to coastal plain   *Myrica*
    126. Buds and twigs lacking dots; catkin buds often present; plants native mostly to mountains   127
127. Stipule scars present; buds longer than wide, with 3 or 4 bud scales   *Alnus crispa*
127. Stipule scars lacking; buds short, with 4 or more visible scales   128
    128. Bud scales usually 4 or 5; gray hairs cover twigs; widespread shrubs of dry or well-drained sites   *Comptonia*
    128. Bud scales usually 6 or more; some tiny white hairs on twigs, or glabrous; rare shrubs of wet sites   *Myrica gale*

Key H

(Alternate leaves; 4 or more bundle scars)

1. Sap hazardous! Vines or low shrubs with terminal buds naked or few-scaled, light brown, hairy, and lateral buds single per node, or a tall shrub/small tree of wet sites with terminal buds covered by 3 to 5 reddish, valvate scales   *Toxicodendron*
1. Sap not as above, safe for further handling; without above combination of characteristics, but if terminal buds present and naked, at least some superposed lateral buds commonly occur   2
  2. Terminal bud usually flanked by 2 or more buds of nearly equal size, all crowded on twig tips, each bud with at least 4 small, imbricate bud scales; fruit an acorn; pith angled in cross section   *Quercus*
  2. Not with above combination of characters   3
3. Terminal bud present, usually much larger than laterals (except in *Fagus*)   3
3. Terminal bud lacking; most buds of similar size and shape   19
  4. Buds naked, hairy, brown; peppery odor in scraped bark   *Asimina*
  4. Not as above   5
5. Leaf scar very wide, encompassing ½ or more of twig circumference   6
5. Leaf scar not as wide as above   8
  6. Twigs spiny (prickles)   *Aralia*
  6. Twigs without prickles or spines   7
7. Scraped twigs with rank odor; inner bark bright yellow   *Xanthorhiza*
7. Scraped twigs with cherrylike odor; inner bark green   *Sorbus*
  8. Stipule scars ringlike, encircling ½ or more of twig at each node; sap clear   9
  8. Stipule scars not ringlike, or sap milky (if any spines occur, go to 36)   11
9. Buds with 8 or more brown, imbricate bud scales; inodorous inner bark   *Fagus*
9. Buds with 1 or 2 scales; aromatic inner bark   10
  10. Terminal buds covered by 2 valvate bud scales   *Liriodendron*
  (if scales imbricate, go to 17)
  10. Terminal buds covered by a single bud scale, this scale with a thickened hinge or line on one side of bud, near base   *Magnolia*
11. Pith chambered   *Juglans*
11. Pith solid (homogenous) or diaphragmed   12
  12. Pith angular or 4- or 5-lobed in cross section (if a shrub with catkin buds, see *Alnus*)   13
  12. Pith rounded in cross section   14
13. Lowermost scale of lateral bud spans width of bud base directly atop leaf scar; stipule scars small but present; bundle scars in 3 distinct groups; terminal bud with 4 or more tightly imbricate bud scales   *Populus*
13. Lowermost bud scale not as above; stipule scars lacking; bundle scars rather scattered; if terminal bud has more than 3 scales visible, outer ones loosely imbricate   *Carya*
  14. Leaf scars crescent-shaped; stipule scars tiny or lacking; bark roughened   15
  14. Leaf scars oval to shield-shaped or cordate; stipule scars large, distinct; bark smooth   17
15. Twigs with a few elongated lenticels; cherrylike odor in scraped inner bark   *Sorbus*
15. Twigs roughened by many small, rounded lenticels; inner bark odor not cherrylike   16
  16. Pith dense or hardened; terminal bud scales truncate on tips, or keeled   *Pistacia*
  16. Pith soft; terminal bud scales rather loose or free at tips, not as above   *Cotinus*
17. Terminal bud conical, long-tipped, 2-scaled   *Ficus*
17. Terminal bud 3- or more-scaled   18
  18. Terminal bud glabrous, with greenish bud scales   *Vernicia*
  18. Terminal bud squat or rounded, the scales covered by reddish-brown hair   *Firmiana*
19. Habit a vine; twigs long and sinewy, or with tendrils   20
19. Habit a shrub or tree; twigs not as above   28

20. Vines climbing by tendrils, which are located on nodes 21
20. Vines climbing by twining; tendrils lacking 24
21. Scraped twigs odorless; twig lenticels not conspicuous; twigs ridged or finely lined *Vitis*
21. Scraped twigs unpleasantly aromatic; lenticels of twig corky, raised, conspicuous 22
   22. Tendrils unforked; inner bark rankly odorous; twigs with large, corky, orange lenticels *Cissus*
   22. Tendrils forked; inner bark and lenticels not as above 23
23. Tendrils forked or branched more than once; often 4 visible bud scales *Parthenocissus*
23. Tendrils forked once; rarely more than 3 bud scales visible, or buds sunken *Ampelopsis*
   24. Buds scaled; leaf scars in outline nearly equal in size to bud, or smaller 25
   24. Buds inconspicuous, hairy, no scales visible; leaf scars much larger than buds 26
25. Buds sharply pointed, with 2 or 3 smooth bud scales; twigs lack conspicuous lenticels *Berchemia* (if stipules persist, see *Wisteria*)
25. Buds blunt, bud scales 5 or more; twigs prominently lenticellate *Akebia*
   26. Leaf scars egg-shaped, not notched on top margin *Calycocarpum*
   26. Leaf scars notched on top margin 27
27. Twigs glabrous, lustrous, reddish on at least 1 side *Menispermum*
27. Twigs dull green, often hairy *Cocculus*
   28. Catkin buds usually present 29
   28. Catkin buds lacking 31
29. Lateral buds with indistinct scales, hidden behind raised ledge above leaf scar *Rhus*
29. Lateral buds fully visible, with 3 or more bud scales 30
   30. Buds sessile, with 3 or more distinct brown scales *Corylus*
   30. Buds stalked, with obscure scales, or reddish *Alnus*
31. Buds naked, hairy 32
31. Buds scaled, or if appearing scaleless then surface bumpy, wavy, or granular 34
   32. Pith large, yellow-brown; twigs robust or hairy *Rhus*
   32. Pith not large, green or white; twigs slender or moderately stout, glabrous 33
33. Twigs light brown; outer bark tough and stringy; a shrub *Dirca*
33. Twigs dark brown; twigs rather brittle; a tree *Cladrastis*
   34. Buds covered by 1 wrinkled bud scale; leaf scar surrounds bud; stipule scar ringlike, encircles twig *Platanus*
   34. Not with above combination of characteristics 35
35 Thorns or spines present 36
35. Thorns or spines lacking 40
   36. Spines easily broken from epidermis when bent to 1 side (prickles) *Zanthoxylum*
   36. Spines woody (thorns), not easily broken 37
37. Spines forked at base; twigs greenish; fresh sap not milky *Parkingsonia*
37. Spines or thorns not forked; twigs gray to brownish; sap of freshly cut twigs milky 38
   38. Stipules lacking; shrubs or small trees *Sideroxylon*
   38. Stipules or their scars present; medium to large trees 39
39. Terminal bud may appear present, its lower scales hairy; thorns mostly straight; trunk bark scaly, with thin, curling plates *Cudrania*
39. Terminal bud lacking; scales hairless; thorns curved slightly or strongly; trunk bark furrowed, with thick ridges or noncurling plates *Maclura*
   40. Buds with 2 or 3 visible bud scales 41
   40. Buds with 4 or more bud scales 45
41. Twigs robust; pith large; buds much smaller than leaf scars *Ailanthus*
41. Twigs slender; pith small; buds equal to or exceed size of leaf scars 42
   42. Pith 5-angled or 5-lobed in cross section *Castanea*
   42. Pith rounded (terete) in cross section 43
43. Habit a shrub; twigs slender *Corylus*
43. Habit a tree; twigs moderately stout 44

44. Twigs with long, stiff hairs; collateral buds often present; leaf scars often subopposite; sap milky *Broussonetia*
44. Twigs glabrous or finely hairy; no collateral buds; leaf scars all alternate; sap clear *Tilia*
45. Inner bark with rank odor and yellowish color; narrow stipules often persist *Rhamnus*
45. Inner bark inodorous, greenish; persistent stipules lacking 46
  46. Stipule scars ringlike, encircling at least ½ of twig; bark smooth, gray *Fagus*
  46. Stipule scars small, slitlike; bark furrowed, scaly, or brown 47
47. Habit a shrub; twigs with scattered, brittle hairs *Corylus*
47. Habit a tree; twigs glabrous or with close, short pubescence 48
  48. Buds distinctly longer than wide, except in spherical flower buds that may occur at lower twig nodes *Ulmus*
  48. Buds about as wide as long, or wider *Morus*

Descriptive Text and Keys to Species

Abelia × grandiflora (Andre) Rehd.   GLOSSY ABELIA   Family Rosaceae   *Figures 1 and 2*

A shrub sparingly naturalized in the SE, native to Asia. Usually encountered around home sites or urban areas where it may spread or persist from cultivation. The small, opposite leaves are glossy above, with a few large teeth on the margin, and may persist into winter on twig tips. The slender twigs may die back after severe freezes. The 3 tiny bundle scars may be difficult to see or may appear as 1, due to swollen margins of leaf scars.

Abies P. Mill.   FIR   Family Pinaceae

Two species are native in the se. US, identified primarily by cone structure. Separation of these 2 e. firs is also possible by using 10× or more magnification to count the number of stomatal rows to each side of the midpoint of the underside of the leaves; 4–8 rows in balsam fir and 8–12 in Fraser fir. All firs bear upright, barrel-shaped cones that release seed by disintegrating, the cone scales falling with the seed and leaving behind the central axis as a spikelike, upright stem. The cones are borne only on the uppermost branches, near the trunk. Leaves are flat, white-lined beneath (these "lines" are the rows of stomata), aromatic, straight and arranged mostly in 1 plane on lower branches, more spirally arranged and upturned on upper branches. The twigs bear circular scars where the nearly stalkless leaves are removed. Buds large, blunt, resinous. Bark gray, thin, and smooth to slightly scaly, often with bumplike bark "blisters" that contain clear resin. These are conifers native to the highest elevations in the Appalachians.

1. Cone scale bracts well exserted from between cone scales; trees native to high mountain elevations of TN, NC, and s. VA   1. *A. fraseri*
1. Cone scale bracts little or not exserted from between cone scales; trees native to WV and n. VA mountains and northward   2. *A. balsamea*

1. *Abies fraseri* (Pursh) Poir.   FRASER FIR   *Figure 3*

This conifer is the native fir of TN, NC, and s. VA, of high elevations in the Blue Ridge. The relatively pure stands of the highest peaks have been decimated by an introduced insect, the balsam woolly adelgid, in the 1970s. Regeneration after the wholesale death of most mature trees has occurred, but continued attacks of the trees are expected as the insect population again rebounds with the thick stands of maturing young fir. This fir is also extensively cultivated in the Christmas tree industry in the mountains of NC.

2. *Abies balsamea* (L.) P. Mill.   BALSAM FIR

The balsam fir is separated into 2 varieties, 1 with no exserted bracts between the cone scales (var. *balsamea* L.) and the other with slightly exserted bracts (var. *phanerolepis* Fernald). The nonexserted form is not in the SE, being prevalent in n. and w. portions of the species' range, mainly in Canada. The exserted forms are more prevalent in the ne. US and southward into n. VA, with the length of bracts increasing with the southward range. Since Fraser fir is quite similar but with more extensively exserted bracts and ranges northward to s. VA, it is thought by some to represent a disjunct phase of the balsam fir and may not deserve a separate species status. An alternate opinion is that var. *phanerolepis* may be of hybrid origin between Fraser and balsam fir.

Acacia P. Mill. ACACIA Family Fabaceae

Three woody or semiwoody species are native to the SE. The 2 woody species are extremely southern in their distribution, and the semiwoody prairie acacia can be expected only in parts of AR and MO. All acacias have obscure bundle scars and lack terminal buds. Twigs of the 2 woody species bear forked white thorns at many nodes. Fruit is a legume.

1. Twigs semiwoody, dying back near to the ground each winter; nodal spines lacking; prickles recurved or lacking 1. *A. angustissima* var. *hirta*
1. Twigs woody, rarely with dieback; spines nodal, forked
   2. Fruits curved; spines usually 1–4cm long 2. *A. minuta*
   2. Fruits mostly straight; spines rarely over 1cm 3. *A. farnesiana*

1. *Acacia angustissima* var. *hirta* (Nutt.) B. L. Robinson PRAIRIE ACACIA

A semiwoody shrub of grasslands, prairies, and dry or rocky woods of TX and AR, north to s. MO and NE. Height rarely over 1m and colonizing by woody rhizomes. Twigs hirsute, without stipular spines. Legume flat, 4–8cm long, often "pinched" between seed. Leaves bipinnate, with 8–14 pairs of pinnae.

2. *Acacia minuta* (M. E. Jones) Beauchamp SMALL'S ACACIA *Figure 4*

Multistemmed shrub or small tree, TX eastward along the coast to w. FL. Legume oval to nearly round in x-section, 3–6cm, tapered on both ends but with a slightly curved beak at apex. Leaves even bipinnate, with 1–4 pairs of pinnae, each pinna with about 30 leaflets; veins indistinct; petioles glabrous. Flower stalks glabrous, to 1cm long; flowers usually appear before or with leaf emergence in spring. The validity of this species is arguable; it may be merely a western variation of *A. farnesiana*, as differences are minor and the ranges overlap. (*A. smallii* Isely)

3. *Acacia farnesiana* (L.) Willd. SWEET ACACIA

A shrub or small tree, native to tropical Amer. but apparently naturalizing in several s. coastal areas of the SE. Most common in s. FL, sporadic up the e. Atlantic Coast to se. GA, and westward along the Gulf Coast possibly to TX. Legume similar to above species but apex obtuse to merely acute. When plant is in leaf, veins are visible in the leaflets, and petioles usually pubescent. Flower stalks pubescent, 15–20mm; flowers usually appear from dormant buds after leaves have appeared.

Acer L. MAPLE Family Aceraceae

The maple genus here includes 9 native species and 1 exotic for the SE, with a few additional important varieties or subspecies. The silver maple and boxelder are by nature bottomland species but are so widely planted they may appear more common in urban and suburban areas than in the wild. In the SE, the striped and mountain maples are restricted to the Appalachians. Sugar maple is not characteristic of s. forests, but 2 s. types do occur in the Piedmont and coastal plain from VA into FL and TX. These are the Florida and chalk maples, considered separate species here but sometimes considered subspecies or varieties of sugar maple by others. Florida and chalk maples are not reliably distinguishable by winter twigs alone. The black maple shares a similar range with sugar maple and is considered only a subspecies or variety by some botanists. Twigs of all maples have opposite leaf scars, 3 bundle scars, and scaled buds. Terminal buds are present. Fruit is a samara (winged seed), normally borne paired and clustered.

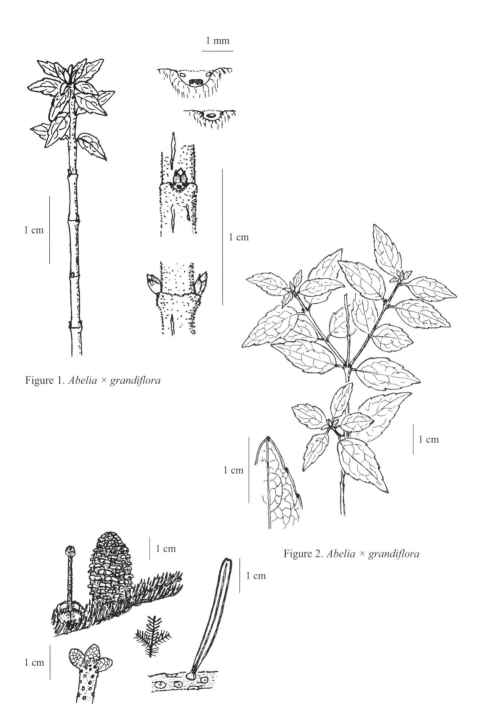

Figure 1. *Abelia × grandiflora*

Figure 2. *Abelia × grandiflora*

Figure 3. *Abies fraseri*

46 · Acer

1. Terminal bud with 2 valvate scales
    2. Twigs finely pubescent; terminal bud sessile to slightly stalked   1. *A. spicatum*
    2. Twigs glabrous; terminal bud distinctly stalked   2. *A. pensylvanicum*
1. Terminal bud with 4 or more imbricate scales
    3. Terminal bud sharply pointed, with 8 or more brownish, acute scales
        4. Twigs rarely over 2mm diam.; terminal buds usually 4mm or less in length; bark gray to glaucous; coastal plain and Piedmont trees (twigs not definitive further)
            5. A tall tree, or straight-trunked if short; leaves glaucous beneath   3. *A. barbatum*
            5. A short tree, stems multiple or crooked; leaves green beneath   4. *A. leucoderme*
        4. Twigs often 3mm or more diam.; terminal buds usually 5mm or more in length; bark gray, brown, or darkened; trees of Appalachians or areas north and west
            6. Twigs brown; terminal buds with glabrous scales or few hairs; bark grayish or light brown on mature trunks   5. *A. saccharum*
            6. Twigs yellow-brown to gray; terminal buds often hairy on tip or with a glaucous bloom on some bud scales; bark dark brown to blackish on mature trunks   6. *A. nigrum*
    3. Terminal bud rather bluntly pointed with 4–6 reddish or greenish scales
        7. Tip of terminal bud hairy between last pair of outer scales; trunk bark finely furrowed
            8. Twigs green or reddish, may have a glaucous bloom; twigs 5–6mm diam.; sap not milky   7. *A. negundo*
            8. Twigs reddish-brown, usually 6–8mm diam.; sap milky   8. *A. platanoides*
        7. Tip of terminal bud not as above; all buds reddish or green on shady side; trunk bark smooth when young, becoming scaly
            9. Twigs with rank odor when scraped   9. *A. saccharinum*
            9. Twigs without rank odor   10. *A. rubrum*

1. *Acer spicatum* Lam.   MOUNTAIN MAPLE   *Figure 5*

A n. species, limited to higher elevations of the Appalachians in the SE. Often grows as a shrub or very small tree in the boreal zone of these mountains. Valvate bud scales and finely pubescent to velvety twigs are characteristic. Fruit clusters droop from twig tips in autumn, having been developed from an erect, spikelike flower cluster in spring. Bark pale brown, papery to scaly, and often lichen-covered.

2. *Acer pensylvanicum* L.   STRIPED MAPLE   *Figure 6*

An Appalachian species in the SE, though growing at much lower elevations northward. This is an understory tree, with a trunk diam. rarely exceeding 3dm. Fruit clusters pendent in autumn. The green bark of twigs and juvenile stems often becomes striped with white lines in the fissures of expanding outer bark. Older bark gray, thin, and smooth.

3. *Acer barbatum* Michx.   FLORIDA MAPLE   *Figure 7*

A tree of moist soils across the Piedmont and coastal plain. The leaves are glaucous and hairy on the underside but smaller than typical sugar maple. Although not attaining the size of sugar maple, this species does attain a straight, tall stature quite unlike the next. Fruits mature in autumn. The bark is usually gray, furrowed, or in plates. [*A. saccharum* ssp. *floridanum* Chapm.; *A. floridanum* (Chapm.) Pax]

4. *Acer leucoderme* Small   CHALK MAPLE

A small tree, often with multiple or crooked stems. Not common overall but may be locally abundant on slopes and hilly woods in the Piedmont region. The leaves are green and hairy on the undersides. The bark is often pale brown to whitish, except on old trunks. Twigs are similar to the Florida maple, but often have the small, dried leaves persisting on the twigs of juvenile wood. Fruits mature in autumn. [*A. saccharum* ssp. *leucoderme* (Small) Desmarais]

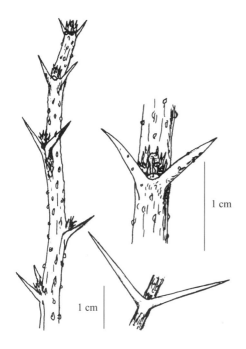

Figure 4. *Acacia minuta*

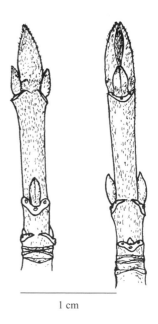

Figure 5. *Acer spicatum*

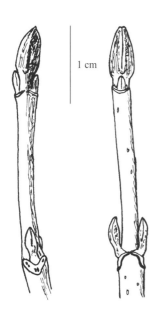

Figure 6. *Acer pensylvanicum*

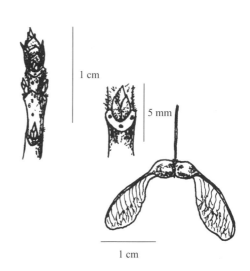

Figure 7. *Acer barbatum*

48 · *Acer*

5. *Acer saccharum* L.   SUGAR MAPLE   *Figure 8*

Predominately a n. and central forest species, limited in the SE primarily to the Appalachians and areas west of there. It is also a favorite tree for ornamental and shade tree purposes, planted widely over many areas of the SE, though faring better out of the coastal plain. Sap leaking out of bird pecks in the trunk bark often permits mildews to stain the bark black, resembling fire charring. Fruits mature in autumn.

6. *Acer nigrum* Michx. f.   BLACK MAPLE   *Figure 9*

Rare in the SE; most often seen in central forest regions and northward. The leaves are larger and less toothed on the lobes than in sugar maple and are often lightly pubescent beneath. The bark is usually darker brown to black, and twigs are stouter than in sugar maple and more yellow-brown. The leaf petioles often have persistent stipules, especially on robust twigs. Fruits mature in autumn. [*A. saccharum* ssp. *nigrum* (Michx. f.) Desmarais]

7. *Acer negundo* L.   BOXELDER   *Figure 10*

Twigs are green and glabrous in typical boxelder, but in w. sections of the SE some fine pubescence may be noticed on twigs of var. *texanum* Pax. Other varieties extend across the continent to CA, making this species the most widely dispersed native maple. This is a streamside and bottomland tree, but it has seen wide use as a street and lawn tree due to its rapid growth rate. The fruits hang in elongate clusters in autumn; trunk bark brownish and furrowed.

8. *Acer platanoides* L.   NORWAY MAPLE   *Figure 11*

This tree most closely resembles our native sugar maple in foliage, though other features are distinctly different. It is an exotic species from Europe that has been widely planted as a quick-growing shade tree in the NE, where it has become a commonly naturalized species. In the SE it is occasionally naturalized, chiefly near urban areas or home sites where the large seed has dropped from planted individuals. The bark is tightly furrowed, and freshly cut parts exude a milky sap. Fruits mature in autumn.

9. *Acer saccharinum* L.   SILVER MAPLE   *Figure 12*

A native bottomland tree most similar to red maple in twig characters. The whitened undersurfaces of the leaves is most characteristic and can be apparent even in fallen winter leaves. The rank odor of the inner bark, 1 of the major separating winter characters between this maple and the red maple, may be difficult to discern in many individuals over the winter months; it is most conspicuous in late winter or spring. Trunk bark gray, scaly. A common habit of the trees in urban areas and when growing in the open is a crown with long, up-reaching branches. The natural habit of the tree to develop a long trunk is rarely allowed in urban cultivation, the trees often being topped or pruned excessively. Fruits green, mature in spring.

10. *Acer rubrum* L.   RED MAPLE   *Figure 13*

Our most widely distributed tree in terms of habitat, found from boreal forests of the Appalachians to lowland swamps of the coast. The winter features cannot be used alone to distinguish 3 important s. forms of the red maple; these being the "typical" red maple, the swamp red maple, and the trilobed red maple. The swamp red maple, var. *drummondii* (Hook & Arn. ex Nutt) Sarg., is found in the outer and mostly southernmost sections of the coastal plain and has dense, pale pubescence on the leaf undersides and larger fruit (often 3–5cm) than in typical red maple. The trilobed red maple, var. *trilobum* Torr. & Gray ex K. Koch, has as its major distinction 3-lobed leaves that are cuneate to rounded at the base. It is commonly found in moist to

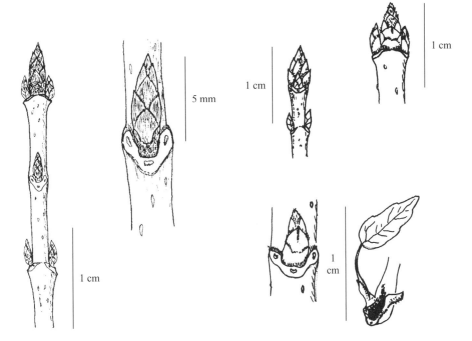

Figure 8. *Acer saccharum*

Figure 9. *Acer nigrum*

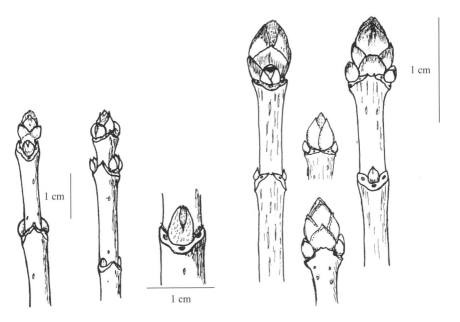

Figure 10. *Acer negundo*

Figure 11. *Acer platanoides*

wet sites in the coastal plain and inland to bottomlands in the Appalachians; the fruit is normally less than 3cm long, as in the "typical" red maple, var. *rubrum,* that is widespread throughout the SE, from mountain uplands to the coastal plain. Leaves are usually sharply toothed and with a cordate base in the typical variety, with the red fruits maturing in spring.

Aesculus L.   BUCKEYE   Family Hippocastanaceae

Five species are native to the SE, with a few additional varieties noted here. One exotic, the horse-chestnut, may occasionally be encountered, but is rare as an escape from cultivation. Twigs of all show large terminal buds, except where fruit has been borne at twig tips and the resulting 2 paired buds have a flat fruit stem scar between them. The leaf scars are opposite and have several bundle scars (sometimes in 3 groups). The fruit is actually a capsule holding 1–6 seed, rather than being a "nut" as is popularly believed. The seed contain a bitter, poisonous alkaloid that renders them inedible for humans and most animals. The same alkaloid is found in the leaves, thus these plants are rarely browsed.

1. Bud scales blackish, glued together by resinous secretions   1. *A. hippocastanum*
1. Bud scales not so dark, not sticky or resinous beneath or between
   2. Terminal bud scales 4–8, grayish brown or covered by a glaucous bloom or wax; large, suckering shrubs   2. *A. parviflora*
   2. Terminal bud scales 10 or more, brown; trees or nonsuckering shrubs
      3. Bud scales of terminal bud slightly keeled; twigs usually dark brown or dark greenish-brown; scraped bark with rank odor; fruit husks bear weak prickles   3. *A. glabra*
      3. Bud scales of terminal bud not keeled; twigs usually light to medium brown or gray, only slightly rank when scraped; fruit husks smooth   *from here on, twigs are not definitive for species*
         4. A tall tree of Appalachians and some areas west   4. *A. flava*
         4. A small tree or large shrub of the Piedmont or coastal plain
            5. Native to Piedmont; seed dark; fruit husk thick   5. *A. sylvatica*
            5. Native to coastal plain; seed pale reddish-brown; fruit husk thin   6. *A. pavia*

1. *Aesculus hippocastanum* L.   HORSE-CHESTNUT   Figure 14

A European tree that rarely naturalizes in the SE. Seedlings near cultivated trees are occasionally encountered where squirrels have aided in "planting." Mature trees bear dark gray to blackish, scaly bark and large, darkened buds, though seedlings may offer up for winter identification only reddish buds with a varnished appearance. Fruit husks are prickly, and seed are somewhat flattened and with a very large hilium, or "eye."

2. *Aesculus parviflora* Walt.   BOTTLEBRUSH BUCKEYE   Figure 15

Native to the coastal plain from SC to AL in rich soils under hardwoods in ravines, slopes, and streamsides. A popular ornamental shrub. The whitish waxed bud scales and slender, erect fruit stems are distinctive. Fruit husks thin; seed pale brown to orange. Usually suckers and forms multiple stems or widening colonies, especially when free of competition.

3. *Aesculus glabra* Willd.   OHIO BUCKEYE   Figure 16

A variable species, shrublike to tree-sized. Range is mostly west of Appalachians, with largest individuals seen in KY, TN, and AL. Occurs near limestone in nw. GA. Shrubby forms such as var. *arguta* (Buck.) B. L. Robinson are known from TX. The bark is light gray, warty to scaly. Presence of rank odor in bruised foliage and twigs is often diagnostic.

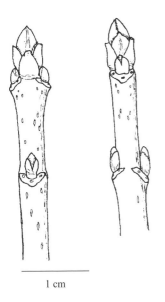

Figure 12. *Acer saccharinum*

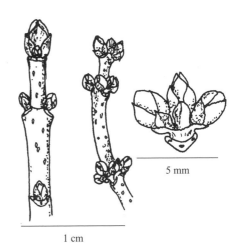

Figure 13. *Acer rubrum*

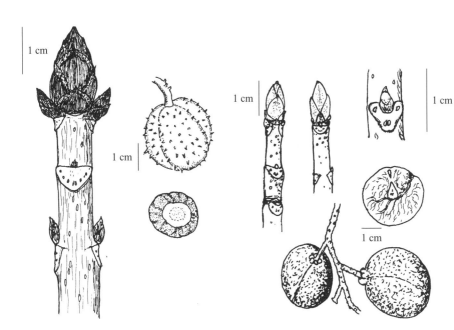

Figure 14. *Aesculus hippocastanum*

Figure 15. *Aesculus parviflora*

4. *Aesculus flava* Ait.   YELLOW BUCKEYE   *Figure 17*

A tree, to 30m tall in rich Appalachian soils. Ranges from the mountains of WV to GA, west to central TN, KY. Attains largest sizes in the Blue Ridge in coves and cool slopes of middle to high elevations. Bark gray and smooth when young, becoming scaly or with large gray to brown plates on older trunks. Fruit husks are thick, smooth. (*A. octandra* Marsh.)

5. *Aesculus sylvatica* Bartr.   PAINTED BUCKEYE

A shrub or small tree most common in rich woods and bottomlands of Piedmont forests from VA to AL, but occasional in inner coastal plain of VA and NC. Difficult to separate by twigs alone from the next species when the 2 occur together. Seed dark brown.

6. *Aesculus pavia* L.   RED BUCKEYE

A shrub or small tree, common across most of the coastal plain from NC to TX, north to MO, and sporadically in the Piedmont of NC to AL. Seed normally a light brown or orange-brown color. Twigs may be grayish. Flowers yellow to red. Several varieties have been described based on flower color and pubescence characters, such as var. *flavescens* (Sarg.) Correll in TX. A popular ornamental because of its red flowers.

Agarista populifolia (Lam.) Judd   FLORIDA-LEUCOTHOE   Family Ericaceae   *Figure 18*

A large shrub native to the coastal plain region, SC to FL. An evergreen ericad with light green, lanceolate, finely toothed leaves and rounded capsules that often persist on the twigs in winter.

Ailanthus altissima (P. Mill.) Swingle   AILANTHUS   Family Simaroubaceae   *Figure 19*

An Asian exotic tree that is thoroughly naturalized across the SE. The stout twigs have a large pith, large leaf scars, and small, rounded buds; there is no terminal bud. A characteristic odor suggesting buttered popcorn exudes from handled foliage in summer and vaguely from scraped twigs in winter. Bark thin, gray-brown. The fruit is a samara.

Akebia quinata (Houtt.) Dcne.   AKEBIA   Family Lardizabalanaceae   *Figure 20*

One species in the genus is known to be naturalized in the SE. This is a climbing Asian vine with brown, lenticellate twigs and many-scaled buds. Bundle scars often are obscure. The flowers may be white to maroon; fruit large, berrylike, splitting open when mature, rarely seen.

Albizia julibrissin Durazzini   MIMOSA   Family Fabaceae   *Figure 21*

An Asian tree commonly encountered as a naturalized species in the SE. The trunk of this small tree is frequently short; multiple stems may be produced. The twigs usually have superposed buds, orange lenticels, and stipular projections but no terminal bud. Legume is flat, thin. Bark gray, thin, smooth.

Alnus P. Mill.   ALDER   Family Betulaceae

Four species are native to the SE, and 1 exotic may occasionally be encountered (*A. glutinosa* European alder). All bear catkin buds near or at twig tips and few-scaled leaf buds that are frequently stalked except in one species. Both pistillate and staminate flower catkins occur on the same twig but as separate flower buds. The pistillate catkin buds are much smaller than the staminate. Fruit conelike, woody, and persistent. The bark is thin, grayish, overlying the yellowish inner bark that turns orange-red after exposure to air.

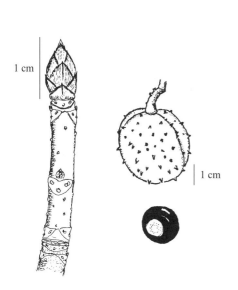

Figure 16. *Aesculus glabra*

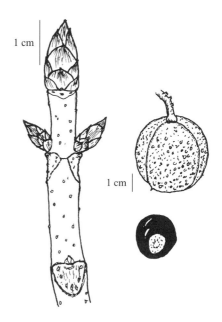

Figure 17. *Aesculus flava*

Figure 18. *Agarista populifolia*

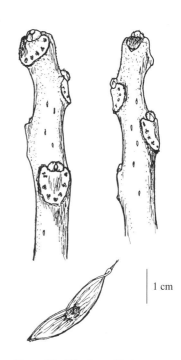

Figure 19. *Ailanthus altissima*

1. Female catkin buds not visible; male catkin buds sessile, not pendent; leaf buds not stalked, with reddish, glossy, uneven scales; a shrub of high mountain openings, only on Roan Mountain in NC-TN   1. *A. viridis* ssp. *crispa*
1. Female catkin buds visible; male catkin buds stalked and often pendent; leaf buds stalked
   2. Flowering in autumn, mostly in leaf axils; cones 2 or 3 per cluster   2. *A. maritima*
   2. Flowering in spring, on twig tips; cones 3–10 per cluster
      3. Habit treelike, with single long trunk; bark scaly; cones long-stalked   3. *A. glutinosa*
      3. Habit shrubby, with multiple stems; bark not scaly; cones short-stalked
         4. Female catkin buds pendent, often lower than male buds; trunk bark glossy, reddish-brown, with prominent, horizontally elongated lenticels   4. *A. incana* ssp. *rugosa*
         4. Female catkin buds not as above; trunk bark dull, gray, no prominent lenticels; our most common species   5. *A. serrulata*

1. *Alnus viridis* ssp. *crispa* (Ait.) Turrill   GREEN ALDER   *Figure 22*

A shrubby alder of high elevations on Roan Mountain, in NC/TN. This species reappears in PA and ranges northward, common across much of Canada. Three other subspecies of this alder occur in boreal regions of N. Amer. and Europe.

2. *Alnus maritima* (Marsh.) Muhl. ex Nutt.   SEASIDE ALDER   *Figure 23*

A tall shrub of unusual distribution, long known from coastal MD-DE and in OK. Recently discovered in boggy uplands in n. GA. The only alder blooming in late summer or autumn. The fruits mature over winter and release seed in the next year.

3. *Alnus glutinosa* (L.) Gaertn.   EUROPEAN ALDER   *Figure 24*

An exotic tree that naturalizes near cultivated specimens, especially so in n. states of the SE. The straight trunk and narrow crown are unlike any native e. alder. Bark becomes dark gray, scaly. Multitudes of fruit often hang from the twigs in winter.

4. *Alnus incana* ssp. *rugosa* (DuRoi) Clausen   SPECKLED ALDER

A shrubby n. species of upland swamps and boggy streamsides, ranging into WV, MD, and VA in the SE. This alder bears conspicuously lenticellate and often shiny trunk bark.

5. *Alnus serrulata* (Ait.) Willd.   SMOOTH ALDER   *Figure 25*

A large, usually multistemmed shrub of moist soils, very common across the SE. One of the first woody plants to flower in late winter or early spring.

Amelanchier Medic.   SERVICEBERRY   Family Rosaceae

Seven species are native to the SE. Two additional varieties can be recognized for 1 of our species. The twigs are not reliably definitive as to species, so other characters must be used in identification. All flower in early spring, and the soft pomes mature by June in all but high mountain elevations. Bark is gray, smooth, thin, becoming furrowed on old trunks of the more arborescent species.

1. Leaves within buds not folded together lengthwise; rare in SE, known only in high mountain bogs of WV   1. *A. bartramiana*
1. Leaves within buds folded flatly together, lengthwise; widespread across several se. states
   2. Buds slenderly elongate, longer ones 4 or more times longer than wide; bud scales hairy only on margins or near tip of bud; tall shrubs or trees, not forming colonies
      3. Stems clumped and multiple; flower racemes erect; hypanthium tomentose externally
         2. *A. canadensis*

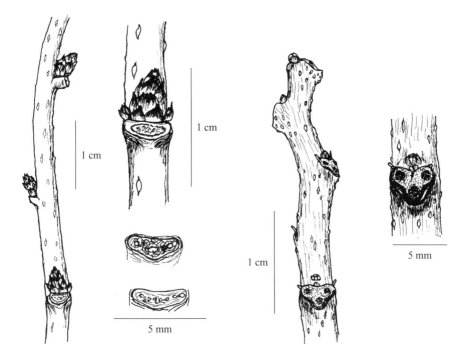

Figure 20. *Akebia quinata*

Figure 21. *Albizia julibrissin*

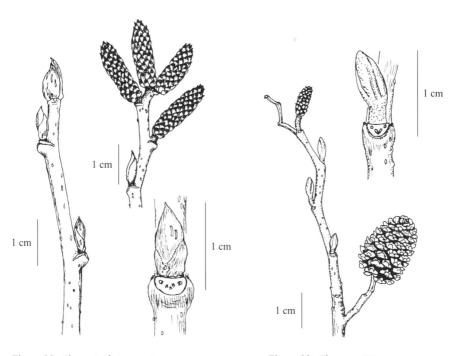

Figure 22. *Alnus viridis* ssp. *crispa*

Figure 23. *Alnus maritima*

3. Stems rarely clumped; flower racemes droop; hypanthium glabrous externally
   4. Hairs commonly on bud scale margins; young leaves densely hairy   3. *A. arborea*
   4. Hairs rarely on bud scales; young leaves glabrous or nearly so   4. *A. laevis*
2. Buds not as above, less than 4 times longer than wide; bud scales hairy over majority of surface; often in colonies or with several sprouts from roots
   5. Height usually over 1m; lateral leaf veins prominent, lead to large teeth; of mountains
                                                                              5. *A. sanguinea*
   5. Height usually under 1m; leaf veins not as above, teeth small; native to Piedmont and coastal plain
      6. Flowers before leaves finish unfolding from bud; ovary summit glabrous; all pedicels 2–5mm   6. *A. obovalis*
      6. Flowers when leaves nearly half-grown; ovary summit tomentose; lower pedicels 6–15mm   7. *A. stolonifera*

1. *Amelanchier bartramiana* (Tausch) M. Roemer   BARTRAM SERVICEBERRY

A clump-forming species up to 2m high, with rather narrow leaves and only 1–3 flowers per inflorescence. A n. species of moist woods and bogs, reaching the SE in a few high mountain bogs of WV.

2. *Amelanchier canadensis* (L.) Medic.   THICKET SERVICEBERRY

A tall shrub with multiple stems in a clump, up to 8m tall. Mostly native to the coastal plain, from VA to MS in the SE, but ranging north to Newfoundland. The summit of the ovary in the flower is glabrous.

3. *Amelanchier arborea* (Michx. f.) Fern.   DOWNY SERVICEBERRY   *Figure 26*

The most common and widespread species in the SE, often reaching stature of a small tree. The heavy pubescence of young leaves and other parts seems to be the chief distinguishing trait between this and the next species, and this becomes of poorer value as leaves naturally become more glabrous later in the season. Within *A. arborea,* the var. *alabamensis* (Britt.) G. N. Jones is known from the coastal plain of AL and scattered locales eastward, with rather obscurely toothed leaves and a pubescent ovary summit. The granite serviceberry, var. *austromontana* (Ashe) Ahles, is known from a few rocky, mountainous areas of NC to GA; it often has smaller (2–3cm) leaves with widely cuneate bases.

4. *Amelanchier laevis* Wieg.   ALLEGHENY SERVICEBERRY

A small tree, limited to the Appalachians in the SE. Of largest size and abundance in middle to higher elevations of the Blue Ridge. Young leaves are reddish-tinted and glabrous. The succulent, sweet fruit is purplish-red to nearly black, 10–15mm. [*A. arborea* var. *laevis* (Wieg.) Ahles]

5. *Amelanchier sanguinea* (Pursh) DC.   ROUNDLEAF SERVICEBERRY

A shrubby species of the Blue Ridge of VA and NC. Rare and localized among rocky woods and outcrops in the middle elevations, where it may occur as a multistemmed shrub, to 3m high, or a colony of stems over rock and under 1m. The young leaves are glossy and red-tinted above, white-tomentose below. More widely distributed north of the SE.

6. *Amelanchier obovalis* (Michx.) Ashe   COASTAL SERVICEBERRY   *Figure 27*

A stoloniferous shrub, forming colonies of slender erect stems, usually less than 1m high. Most often confused with the next species, due to similarities in habit and foliage. Flower racemes

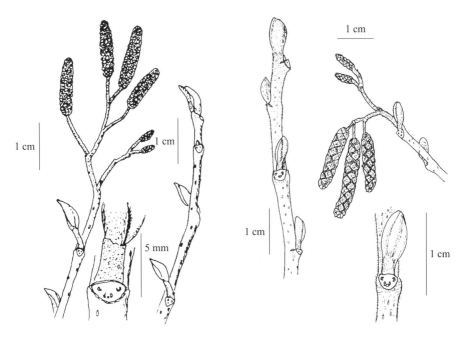

Figure 24. *Alnus glutinosa*   Figure 25. *Alnus serrulata*

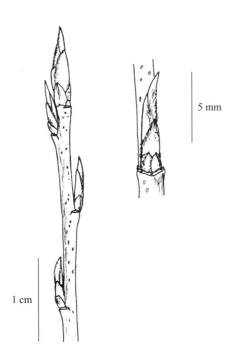

Figure 26. *Amelanchier arborea*

erect, with very short stalks throughout (2–5mm). This is a coastal plain species occurring from NJ to SC in moist, sandy, or peaty soils.

7. *Amelanchier stolonifera* Wieg.   RUNNING SERVICEBERRY

Stoloniferous and colonial, to 1m high. The flower racemes are erect as in the preceding species, but individual flower stalks are longer, especially lower ones (6–15mm). Occurs sporadically in sandy woods of the coastal plain from NC to GA and further inland amid dry, rocky woods across the Piedmont to the Appalachians of NC, TN. [*A. spicata* auc. non. (Lam.) K. Koch.]

Amorpha L.   INDIGO-BUSH   Family Fabaceae

There are 11 woody species native to the SE, plus 1 significant variety. Only 1 species can be considered common and widespread across the region, the others rather limited of range or habitat. All are shrubs that bear small, 1-seeded legumes. The twigs lack terminal buds and often show superposed buds and small stipular projections. Sometimes there are terminal, spike-like remnants of the fruit-bearing stems in winter. A somewhat kerosene-like odor may be detectible in the scraped inner bark of some species. Winter twig characteristics alone cannot reliably separate all the species in this genus; many of the species must be distinguished by examining flowers or fruits.

1. Plant height rarely over 1m, usually 5dm or less; leaf petiole mostly shorter than width of lowest leaflet
    2. Twigs and legume densely pubescent
        3. Twigs usually die back to near base of plant in winter; end of leaflets with swollen vein tip   1. *A. herbacea*
        3. Twigs not as above; leaflet with tiny bristle; prairie species of w. parts of SE
                                                                                                 2. *A. canescens*
    2. Twigs and legume glabrous or very sparsely pubescent
        4. Leaflets mostly 1cm long or less; racemes 5cm or less   3. *A. georgiana* var. *georgiana*
        4. Leaflets over 1cm long; racemes 10cm or more long   4. *A. georgiana* var. *confusa*
1. Plant height usually over 1m; petiole longer than width of lowest leaflet
    5. Calyx lobes distinct, acute, about as long as calyx tube, or longer
        6. Habit sparsely branched; buds distinctly hairy; bud scales not distinct   5. *A. paniculata*
        6. Habit shrubby, well-branched; buds sparingly pubescent; scales distinct   6. *A. schwerinii*
    5. Calyx lobes small, blunt to rounded, less than ½ as long as the tube
        7. Glands on calyx few or inconspicuous; leaves lustrous, very aromatic when bruised, blacken with drying   7. *A. nitens*
        7. Glands on calyx conspicuous; leaves not as above
            8. Leaflet stalks not glandular; leaflet tips usually with a gland or point, rarely notched
                9. Calyx lobes less than 0.6mm; leaflets often tipped with a bulbous gland   8. *A. glabra*
                9. Calyx lobes usually over 0.6mm; leaflets tipped with bristle or point   9. *A. fruticosa*
            8. Leaflet stalks glandular; leaflet tips usually notched; native to TX, OK, AR
                10. Shrub of Ouachita Mountains in AR, OK; bud scales distinct   10. *A. ouachitensis*
                10. Shrubs of TX and adjacent OK; buds tiny or scales not conspicuous
                    11. Flowers bluish; stipules glabrous; of coastal plain in OK, ne. TX   11. *A. laevigata*
                    11. Flowers purplish; stipules pubescent; of limey hills in central TX
                                                                                                 12. *A. roemeriana*

1. *Amorpha herbacea* Walt.   DWARF INDIGO-BUSH

A semiwoody shrub, to 1m tall, often dying back to near ground level each winter. Most common in the coastal plain from NC to FL, but occasionally in Piedmont. Noticeably short-pubescent throughout. The bluish flowers open May–July.

2. *Amorpha canescens* Pursh   DOWNY INDIGO-BUSH

A 1–2m shrub of prairies and sparsely wooded hills, widespread in Midwest, entering the SE in MO, OK, AR, TX. Conspicuously white-pubescent, though the form *glabrata* (Gray) Fassett may be nearly hairless. Foliage often blue-green. Blooms May–July, flowers bluish.

3. *Amorpha georgiana* Wilbur   GEORGIA INDIGO-BUSH

A short shrub rarely over 5dm tall, of sandy pinelands from NC to GA, mostly inner coastal plain. The leaflets are smaller, and flower racemes shorter and less-branched than in var. *confusa*. Flowers purplish, in Apr–May.

4. *Amorpha georgiana* var. *confusa* Wilbur   SAVANNA INDIGO-BUSH

A short shrub of moist pine savannas in the outer coastal plain of a few NC and SC counties. Leaflets usually 12–20mm long, flower clusters 10–25cm long and branched. Flowers blue, in June–July.

5. *Amorpha paniculata* T. & G.   SWAMP INDIGO-BUSH   *Figure 28*

A rather leggy shrub, to 3m tall, with elongate, tomentose twigs and large paniculate flower clusters to 40cm long. Rare; found in low, moist to wet woodlands in the coastal plain of LA, AR, e. TX. Leaflets 3–8cm long, softly pubescent and prominently veined. Flowers violet, but yellowish stamens more conspicuous, May–June.

6. *Amorpha schwerinii* Schneid.   PIEDMONT INDIGO-BUSH

This 1–2m shrub seems to prefer rocky soils of the Piedmont and a few adjacent stations, from NC to MS. Rare and local in distribution. The calyx lobes are 2–3.5mm long. Leaflet tips often bulbous-glandular, and twigs slightly pubescent. Flowers purplish, Apr–June.

7. *Amorpha nitens* Boynton   SHINING INDIGO-BUSH

A shrub 1–2.5m tall, widespread but uncommon from SC to LA in the coastal plain and inland to s. IL. Occurs in moist pinelands, streambanks, rocky slopes, from low elevations to at least 733m (2405ft) (Mt. Cheaha AL). A kerosene-like odor in bruised foliage is conspicuous in this species. Leaves usually glossy, turning black when dried. Twigs glabrous. Flowers deep purplish, Apr–June. The legume is curved, glabrous, with no glandular dots.

8. *Amorpha glabra* Desf. ex Poir.   MOUNTAIN INDIGO-BUSH

Similar to the next species, and perhaps only a variation of it. Overall a glabrous species. Endemic to the s. Appalachians of NC, TN, AL, GA, SC. Scattered in rocky woods and dry to moist slopes, especially along wooded mountain roadsides. Flowers purplish, May–July.

9. *Amorpha fruticosa* L.   COMMON INDIGO-BUSH   *Figure 29*

The most common and widely distributed species of the genus, essentially throughout the SE. Variation in leaflet size, shape, and pubescence has led to several varieties being described, none mentioned here. Most often seen in moist soils. Flowers purplish, Apr–June.

10. *Amorpha ouachitensis* Wilbur   OUACHITA INDIGO-BUSH   *Figure 30*

Endemic to the Ouachita Mountains of AR, OK, where it is rare and localized in rocky woods of ridges and slopes. The twigs are sparingly pubescent to glabrous. Flowers purplish, Apr–June. Legume to 1cm, conspicuously gland-dotted.

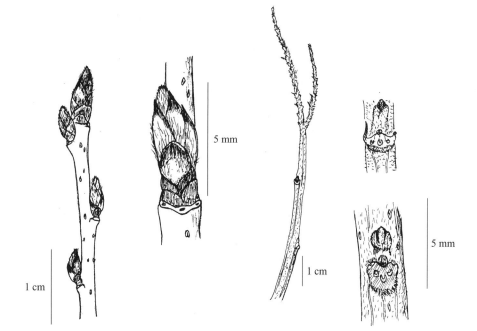

Figure 27. *Amelanchier obovalis*

Figure 28. *Amorpha paniculata*

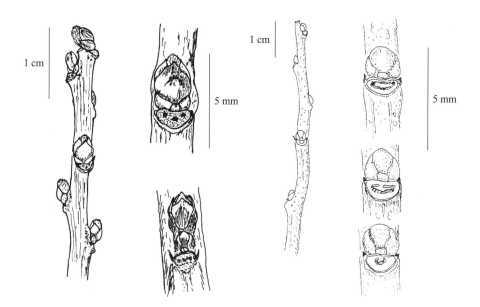

Figure 29. *Amorpha fruticosa*

Figure 30. *Amorpha ouachitensis*

11. *Amorpha laevigata* Nutt.   SMOOTH INDIGO-BUSH

A rare and localized shrub of ne. TX and adjacent OK, in gravelly or dry soils of open woodlands and prairies. The very glandular to warty leaflet stalks, petiole, and calyx are glabrous. Flowers bluish, in 15–30cm racemes in June. Possibly a form of the central TX species *A. texana* Buckl., which is slightly pubescent and bears 10–20cm racemes in May.

12. *Amorpha roemeriana* Scheele   ROEMER INDIGO-BUSH

A shrub of streamsides through limey hills of central TX. Flowers purplish, May–June.

Ampelopsis Michx.   AMPELOPSIS   Family Vitaceae

Of this genus, 2 are native; 1 is a naturalized exotic species. All are climbing vines with prominently lenticellate twigs and forked tendrils. The fruit is a berry, maturing in late summer or autumn.

1. Twigs bear sparse hairs, especially youngest, smallest twigs; petioles hairy
                                1. *A. brevipedunculata*
1. Twigs glabrous
 2. Buds visible; leaf scars usually rounded to elliptical; mature fruit shiny black 2. *A. arborea*
 2. Buds sunken; leaf scars U-shaped to shield-shaped; mature fruit bluish   3. *A. cordata*

1. *Ampelopsis brevipedunculata* (Maxim.) Trautv.   PORCELAIN-BERRY

An Asian species most closely resembling the native *A. cordata,* but leaves distinctly 3-lobed, and all parts pubescent. Birds feed on the dotted, bright blue fruit; this is the chief method of dissemination for this rapidly spreading vine.

2. *Ampelopsis arborea* (L.) Koehne   PEPPERVINE   *Figure 31*

Common throughout the coastal plain from VA to TX, occasionally more inland, and up the Mississippi embayment to IL. Prefers moist soils. Leaves are bipinnate. The shiny black fruits are succulent but not palatable to humans.

3. *Ampelopsis cordata* Michx.   AMERICAN AMPELOPSIS   *Figure 32*

A high-climbing native vine of bottomlands and moist, fertile woodlands. Range is similar to *A. arborea*. The simple leaves are grapelike in shape but not lobed. The fruit is a blue berry, not edible by humans. The snakelike trunks of old vines often hang bare of foliage from high attachment points in the tree canopy, with brown bark that is deeply furrowed.

Andrachne phyllanthoides (Nutt.) Coult.   MAIDENBUSH   Family Euphorbiaceae *Figure 33*

A rare shrub of rocky soils, in sections of AR and MO. The slender twigs are lined and without terminal buds. Leaf scars bear 1 bundle scar. The fruit is a capsule. The milky sap is most noticeable in summer.

Andromeda polifolia var. *glaucophylla* (Link) DC.   BOG-ROSEMARY   Family Ericaceae
                                      Figure 34

Widespread in the NE, but reaching the SE only in WV, where this low shrub occurs in high mountain bogs. The N. Amer. plants are considered only a variety of the Eurasian species. The evergreen leaves are small, with revolute margins and whitened undersides.

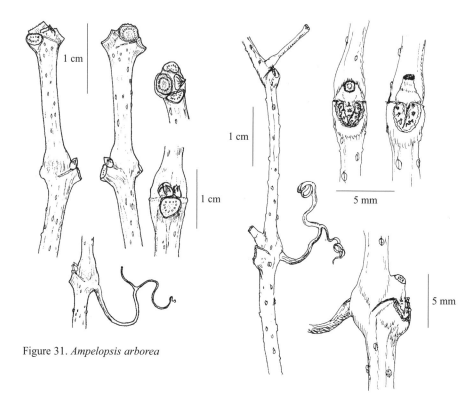

Figure 31. *Ampelopsis arborea*

Figure 32. *Ampelopsis cordata*

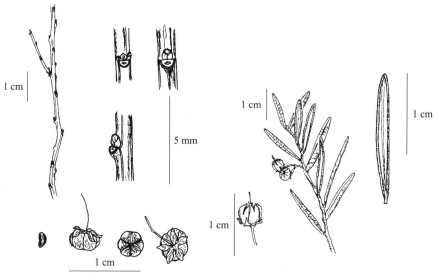

Figure 33. *Andrachne phyllanthoides*

Figure 34. *Andromeda polifolia* var. *glaucophylla*

Aralia spinosa L.   DEVIL'S WALKING STICK   Family Araliaceae   *Figure 35*

A coarsely branched small tree, native across a wide range of soils throughout the SE. The large, prickly twigs have conical terminal buds and leaf scars that surround much of the girth of the twig. Flowers in summer; berries ripen in Sept in large, terminal masses and promptly fall or are taken by birds.

Arctostaphylos uva-ursi (L.) Spreng.   COMMON BEARBERRY   Family Ericaceae *Figure 36*

A native, creeping shrub of the northernmost sections of the SE (VA, WV), more common to the north of this region and there circumboreal in range, across Canada to AK, and in Eurasia. In rocky or sandy soils. The small, ovate leaves may take on a reddish hue in the winter months. The rounded, fleshy fruits often persist into winter.

Aristolochia L.   BIRTHWORT   Family Aristolochiaceae

Two woody species in this genus are native to the SE. Both are twining vines bearing green twigs and horseshoe-shaped leaf scars. The superposed buds are partially sunken. Fruit is an elongate capsule that splits along several seams to release flattened, rather heart-shaped seed individually separated by partitions.

1. Twigs glabrous   1. *A. macrophylla*
1. Twigs velvety-pubescent   2. *A. tomentosa*

1. *Aristolochia macrophylla* Lam.   DUTCHMAN'S-PIPE   *Figure 37*

A high-climbing vine of moist woodlands in the Appalachian region. The older stems bear thin gray bark that is very slightly furrowed. The stems are tough, coiled or ropelike, often hanging from the tree canopy or twined around trunks of small trees. (*A. durior* Hill)

2. *Aristolochia tomentosa* Sims   WOOLLY PIPEVINE

A vine of calcareous slopes and rich woods, especially in bottomlands and streamsides in most se. areas west and southwest of the Blue Ridge, and extending eastward to SC along upper sections of some coastal plain rivers.

Aronia Medic.   CHOKEBERRY   Family Rosaceae

Two native species of this genus are widely recognized as native in the SE. One apparent hybrid between these species reproduces and perpetuates independently as a new species. These are shrubs of a wide variety of habitats. Twigs resemble those of some cherries in appearance, with a strong candied-cherry odor in scraped bark. Dried fruits (berrylike pomes) may persist into early winter. These shrubs may produce colonies by multiple stems from the root system. Due to showy white flowers and colorful fruit, they are occasionally planted as ornamentals. Taxonomy and nomenclature of the chokeberries has a complicated history. These plants have been alternately included in the genera of *Aronia, Photinia, Pyrus,* and *Sorbus* by various botanists and taxonomists. The similarity of floral structure in all these genera is the chief reason for the confusion. Accordingly, there is an extensive synonymy of names for the chokeberries, a result of differing opinion in generic, species, and varietal rank. Only a few of these synonyms are included here for each species; for more complete synonymy, other technical floral treatments should be referenced.

1. Twigs glabrous; bud scales often 4 or less on terminal bud, 1 or 2 on lateral buds
                                                                                                       1. *A. melanocarpa*

Figure 35. *Aralia spinosa*

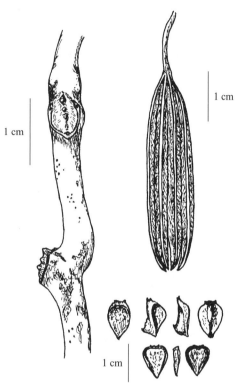

Figure 37. *Aristolochia macrophylla*

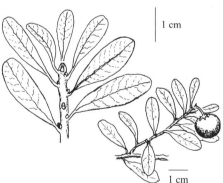

Figure 36. *Arctostaphylos uva-ursi*

1. Twigs with some hairs, especially near tip; bud scales more numerous
   2. Lateral bud scales mostly 2 or 3   2. *A. arbutifolia*
   2. Lateral bud scales mostly 4 or 5   3. *A. prunifolia*

1. *Aronia melanocarpa* (Michx.) Ell.   BLACK CHOKEBERRY   *Figure 38*

A tall shrub, most common in the Appalachian region of VA to AL, where it prefers higher elevations in moist or rocky sites. Also ranges sporadically eastward in VA, NC, and westward across TN. Fully ripened fruits glossy black and juicy. [*Photinia melanocarpa* (Michx.) Robertson & Phipps; *Pyrus melanocarpa* (Michx.) Willd.; *Sorbus melanocarpa* (Michx.) Heynh.]

2. *Aronia arbutifolia* (L.) Pers.   RED CHOKEBERRY   *Figure 39*

The most common of the 3 chokeberries, ranging across most of the SE. Common in wet soils of lowlands and in rocky seeps in uplands. The fruits are red, rather dry or mealy, with pubescent stems. [*Photinia pyrifolia* (Lam.) Robertson & Phipps; *Pyrus arbutifolia* (L.) L. f.; *Sorbus arbutifolia* (L.) Heynh.]

3. *Aronia prunifolia* (Marsh.) Rehd.   PURPLE CHOKEBERRY   *Figure 40*

Intermediate in characters of the above 2, and presumed to be a hybrid. However, since it reproduces true in much of its range, it may be considered a "valid" species. Most commonly encountered in the Appalachian area and adjacent provinces, similar to the range of *A. melanocarpa*. Prefers moist to wet sites. Fruits purplish, juicy when fully ripe. [*Photinia floribunda* (Lindl.) Robertson & Phipps; *Pyrus floribunda* Lindl; *Sorbus arbutifolia* var. *atropurpurea* (Britt.) Schneid; *Aronia atropurpurea* Britt.]

Arundinaria Michx.   CANE   Family Poaceae

This genus of bamboos is represented in the SE arguably by 2 native species or divided into subspecies of a single variable species. The unclear taxonomy is caused by variation of habit and sporadic flowering that make identification difficult. Both types described here may occur as 1m tall plants on less-optimum sites. One trait used to separate the 2 is based on the presence of air canals in the outer zone of the rhizomes, seen in cross section with 10× magnification. Both types of canes spread by underground rhizomes, with the aboveground culms round in cross section in the internodes. Canebrakes are thickets of cane along rivers and bottomlands, historically covering extensive areas but very much reduced by white settlement during conversion of lands to agricultural uses.

1. Rhizomes without air canals; branches from culms often under 3dm long; culm midstem sheaths shorter than associated internode; upper surface of leaves usually glabrous
                                                                                             1. *A. gigantea*
1. Rhizomes with a row of air canals; culm branches usually over 3dm; culm midstem sheaths longer than internode; leaf upper surfaces pubescent   2. *A. gigantea* ssp. *tecta*

1. *Arundinaria gigantea* (Walt.) Muhl.   LARGE CANE   *Figure 41*

Widespread across the SE, favoring rich lowlands where it may form extensive thickets (canebrakes). Heights of 6–8m are seen in optimum sites, as short as 1m on less-optimum sites. Branches of the culms are formed before the culms reach full height, according to McClure (1973). The flowering and subsequent death of this cane is reported to be at 30- to 40-year intervals.

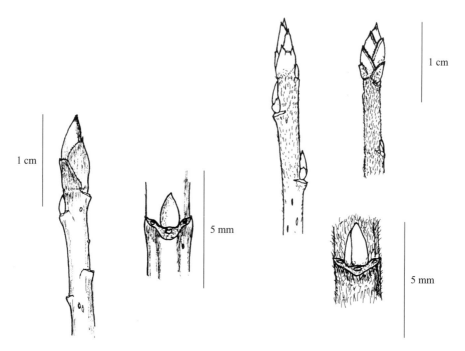

Figure 38. *Aronia melanocarpa*

Figure 39. *Aronia arbutifolia*

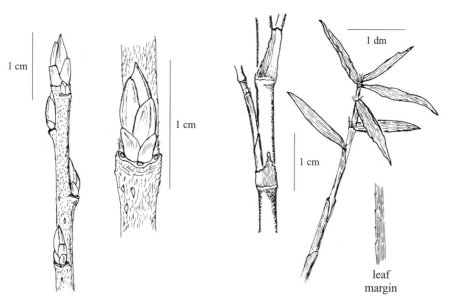

Figure 40. *Aronia prunifolia*

Figure 41. *Arundinaria gigantea*

2. *Arundinaria gigantea* ssp. *tecta* (Walt.) McClure   SMALL CANE

Mostly limited to the coastal plain of VA to AL, where it may form canebrakes in low, damp soils. Heights of 1–3m are the norm. Culm branches often appear in spring on culms that have previously grown to maximum height, according to McClure (1973). This cane has been reported to flower in 3- to 4-year intervals, with time of subsequent death not certain.

Asimina Adans.   PAWPAW   Family Annonaceae

Eight species are native to the SE, but 1 of these grows in central FL, out of range of this book. The other 7, along with a few notable varieties and hybrids, are not easily described or identified using winter characters like twigs alone. All have terminal buds, hairy buds, superposed buds, and 5 or more raised bundle scars. Most have a strong peppery odor in the scraped inner bark. Flower buds, usually on lowermost nodes of the twig, are globular and larger than other laterals. The 2 common and more widely known species, *A. triloba* and *A. parviflora*, bear thinner leaves than the others. The 5 shrubby pawpaws of the Deep South (chiefly of s. GA and n. FL) bear leaves that may be semipersistent into winter, or at least some of the thick, revolute-margined leaves can be located underneath the plants. These s. pawpaws occur in sandy soils in or near pine forests, and in such areas they may be more obvious when fires have reduced competition and promoted their resprouting. Flowers of most s. species are white. Fruits of all pawpaws are large berries, edible but with variable palatability. Seed dark brown, suggestive of a bean but not edible.

1. Shrub or small tree of moist soils in hardwood forests; leaves thin, usually acuminate on tip
   2. Shrub; flower buds borne on twigs of plants of 1.5m high or less   1. *A. parviflora*
   2. Tree; flower buds borne on twigs of plants mostly over 2m height   2. *A. triloba*
1. Shrubs of sandy soils in pine forests of s. sections of coastal plain; leaves thick, tip blunt
   3. Shrubs usually under 1.5m tall; diam. of last twig internode usually 1.5–2mm
     4. Shrub height rarely over 3dm tall; leaf petiole usually 5–8mm   3. *A. pygmaea*
     4. Shrub height usually 3–15dm; leaf petiole usually 2–5mm
       5. Twigs with raised lenticels; leaf length 4× width or more; of drier pinelands
                                                                                     4. *A. longifolia*
       5. Twigs without raised lenticels; leaves not as above; of moist pine flatwoods
                                                                                      5. *A. reticulata*
   3. Shrubs often over 1.5m tall; diam. of last twig internode usually 2.5–3mm
     6. Twigs with pale hairs; few or no raised lenticels   6. *A. incana*
     6. Twigs with reddish hairs; pale, raised lenticels present   7. *A. obovata*

1. *Asimina parviflora* (Michx.) Dunal   SMALL-FLOWER PAWPAW   *Figure 42*

A shrub usually between 1–3m in height, the taller plants tending to have broad, shrubby crowns or multiple stems. Occurring in rich soils of bottomlands and slopes, mostly in hardwood forests. May be seen on occasion in sandier soils near the coast. Widespread across the SE, but most common in the coastal plain and Piedmont; ranges inland to sections of the Appalachians. The small flowers are 1–2cm wide, appearing before or with the leaves in early spring, with light maroon petals and an odor reminiscent of aged raw beef. Flies are the principal pollinators.

2. *Asimina triloba* (L.) Dunal   PAWPAW   *Figure 43*

A small tree, 3–12m tall, usually with a single, straight trunk. Fairly common and widespread across the entire SE in rich, moist soils. Most common in bottomlands and hardwood swamp forests. The roots sprout prolifically; often a thicket or colony of smaller plants will occur around older specimens. The flowers are 2–3cm wide, with maroon petals and a slight meaty odor, similar to the above species. Fruit 5–15cm long, ripe in Aug–Sept. Cultivar selections for superior fruit production are known.

68 · *Asimina*

3. *Asimina pygmaea* (Bartr.) Dunal.   PYGMY PAWPAW   *Figure 44*

A low shrub of pine flatwoods and savannas from se. GA to central FL. Usually 2–3dm tall, with few-branched arching stems that bear maroon-colored flowers with petals to 3cm long. The odor, like other pawpaws with maroon-colored flowers, is not pleasant. Leaves thick, 4–8cm long, with revolute margins.

4. *Asimina longifolia* Kral   NARROW-LEAF PAWPAW   *Figure 45*

Shrubs mostly 1–1.5m tall, of sandy pinelands in n. FL and se. GA, AL. The elongate leaves are rather linear, thick, with revolute margins. Two varieties are described: var. *longifolia,* with leaves most linear, widest near middle, and flowers white; var. *spathulata* Kral, with leaves more spatulate, widest near tip, and flowers pink to light maroon. The latter variety is mostly found in the w. parts of the species' range. Flower petals to 8cm long. (*A. angustifolia* A. Gray)

5. *Asimina reticulata* Shuttlew. ex Chapm.   FLATWOODS PAWPAW   *Figure 46*

A shrub 3–15dm tall, mostly occurring in the moist flatwoods of central FL, barely reaching ne. FL and the range of coverage of this book. Usually under slash pine or among saw palmetto in slash or longleaf flatwoods. Leaves thick, oblong, to 8cm long, with revolute margins. Flower petals white, to 7cm long.

6. *Asimina incana* (Bartr.) Exell.   FLAG PAWPAW, WOOLY PAWPAW   *Figure 47*

A shrub 1–1.5m tall or a little taller when vigorously resprouted after fire. Occurs on sandy pinelands from central FL to se. GA. Leaves thick, 5–8cm long, flat and only slightly revolute on margin. Pale or whitish hairs adorn leaf undersides, twigs, and buds. Flowers showy, with white petals to 7cm long; appear before or during the advent of spring leaves, having been developed from buds set on twigs of previous season. (*A. speciosa* Nash)

7. *Asimina obovata* (Willd.) Nash   BIG-FLOWER PAWPAW   *Figure 48*

A large shrub, 2–4m tall, mainly of sandy pinelands and scrub of central FL, barely reaching ne. FL and the range of coverage of this book. Leaves thick, revolute on margin, mostly 5–10cm long, with reddish hairs below. The showy white flowers are similar in size to *A. incana* but appear near the ends of the elongated spring shoots. (*A. grandiflora* W. Bartr.)

Aster carolinianus Walt.   CLIMBING ASTER   Family Asteraceae   *Figure 49*

One woody species of this large, mostly herbaceous genus is native to the SE. The vinelike climbing aster is found in swamps of the coastal plain from NC to TX. Flowers are pinkish to lavender, appearing in autumn and occasionally continue into winter. The leaves are likewise persistent for some time and have clasping bases. Bare twigs show few features in leaf scars or bundle scars and may die back considerably by spring, depending on extent of freezing weather.

Avicennia germinans (L.) L.   BLACK MANGROVE   Family Avicenniaceae   *Figure 50*

This native tree is better known in coastal areas of s. FL and in certain tropical regions, but it ranges to a few scattered Gulf Coast sites such as Cedar Key in n. FL and sporadically to TX. It is reduced by colder climate to a shrubby stature in its n. range, usually 2m or shorter, often appearing as masses of short stems in tidal mud deposits. Fingerlike projections grow from the roots for aeration (pneumatophores). The evergreen leaves are whitened beneath. Fruit is a single-seeded, pubescent pod.

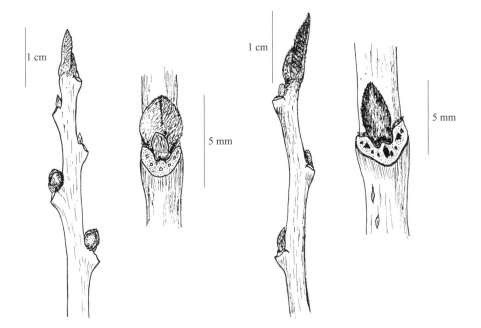

Figure 42. *Asimina parviflora*

Figure 43. *Asimina triloba*

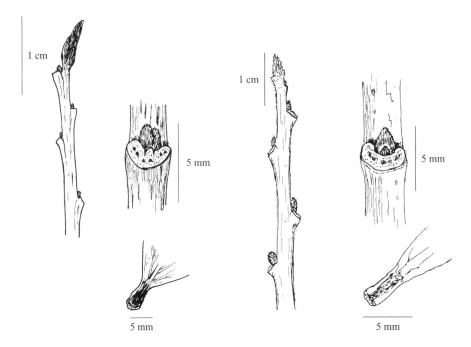

Figure 44. *Asimina pygmaea*

Figure 45. *Asimina longifolia*

Baccharis L. SALTBUSH Family Asteraceae *Figure 51*

There are 4 woody species native to the coastal plain of the SE, not identifiable by twigs alone. Fortunately, remnant leaves can often be found in winter, and near the coast these shrubs are more tardily deciduous. Greenish, lined twigs and collateral buds are common to all species. The opening fruit heads with their accompanying white "down" (pappus bristles) are conspicuous in autumn, when they may be mistaken at a distance for flowers. Most of our species also occur in the West Indies or Bahamas.

1. Leaves linear, 3mm wide or less — 1. *B. angustifolia*
1. Leaves wider
  2. Heads mostly sessile, in clusters along twig in axils of leaves; leaves only sparsely dotted — 2. *B. glomeruliflora*
  2. Heads mostly stalked, near twig tips and on long stems from leaf axils; leaves with many resinous dots
    3. Leaves often toothed; a widespread and common species — 3. *B. halimifolia*
    3. Leaves mostly entire, rarely toothed; a rare species — 4. *B. dioica*

1. *Baccharis angustifolia* Michx. NARROW-LEAF GROUNDSEL-BUSH *Figure 52*

A coastal plain shrub, ranging from NC to peninsular and gulf coastal FL. Most common in vicinity of brackish and salt marshes, interdune swales, and other sites very near the sea.

2. *Baccharis glomeruliflora* Persoon SOUTHERN GROUNDSEL-BUSH *Figure 53*

Range and habitat similar to above species, though this shrub is often overlooked due to its less-conspicuous leaf shape. Leaves tend to be thick and persist over winter months more than other *Baccharis* species.

3. *Baccharis halimifolia* L. GROUNDSEL-TREE *Figure 54*

The most common and widespread of the genus in the SE, presumed native to the coastal plain from MA to TX, but seemingly in the process of spreading inland. Not uncommon in waste areas, roadsides, railways, and fields throughout the Piedmont, occasionally into the Appalachian region. A large shrub with brownish-gray, deeply furrowed or fibrous bark, and resinous-dotted leaves and twigs. Leaves variously shaped, margins entire or bearing several large teeth.

4. *Baccharis dioica* Vahl GULF GROUNDSEL-BUSH *Figure 55*

A shrub of brackish soils near the coast, usually in or near marshes and dune swales. Rare in the se. region covered by this book; only known from Dauphin Island AL southward, it is found in the keys of s. FL and in the West Indies. Leaves 1–3cm, resin-dotted as in *B. halimifolia* but margins usually entire.

Batis maritima L. SALTWORT Family Bataceae *Figure 56*

One species in the SE, inhabiting salty coastal soils near tidewater areas. A low, creeping shrub with succulent, salty-tasting leaves. Found from SC to TX along brackish shores, southward to tropical Amer. and Hawaii. The axillary, lumplike fruits are dispersed via water currents.

Befaria racemosa Vent. TARFLOWER Family Ericaceae *Figures 57 and 58*

An evergreen shrub of coastal plain pinewoods from se. GA to s. FL. The twigs bear whitish, bristlelike hairs; leaves ovate and veins obscure. White, sticky flowers are fragrant and borne at the top of new growth. Fruit a 7-parted capsule.

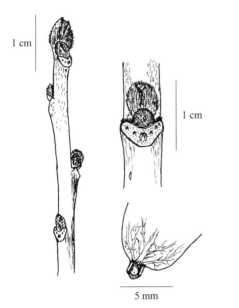

Figure 46. *Asimina reticulata*

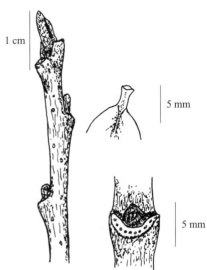

Figure 47. *Asimina incana*

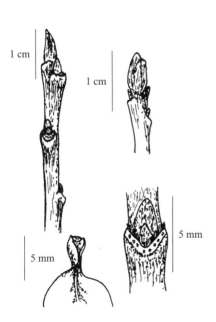

Figure 48. *Asimina obovata*

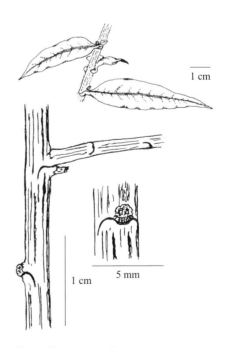

Figure 49. *Aster carolinianus*

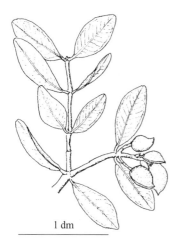

Figure 50. *Avicennia germinans*

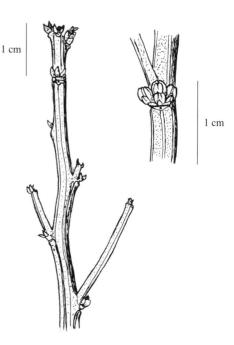

Figure 51. *Baccharis*

Figure 52. *Baccharis angustifolia*

Figure 53. *Baccharis glomeruliflora*

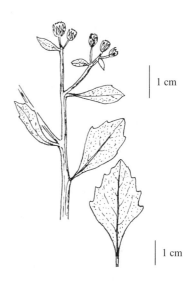

Figure 54. *Baccharis halimifolia*

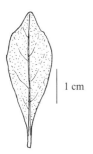

Figure 55. *Baccharis dioica*

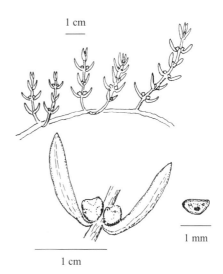

Figure 56. *Batis maritima*

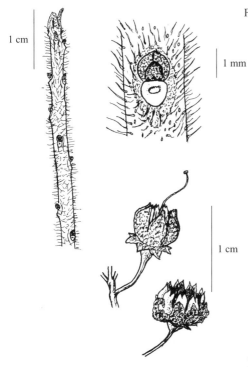

Figure 57. *Befaria racemosa*

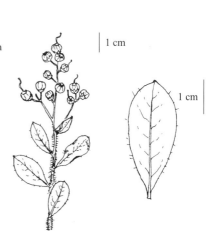

Figure 58. *Befaria racemosa*

Berberis L.   BARBERRY   Family Berberidaceae

Three species of these spiny shrubs are treated here; the single native species is not common. Twigs of all have clustered axillary buds flanked by several leaf scars and spines beneath. Inner bark is yellow. Flowers yellow; fruit a red berry.

1. Spines mostly unbranched 1. *B. thunbergii*
1. Spines mostly branched
   2. Twigs usually reddish; spines on older stems usually curve downward; peglike petiole bases rarely persist around bud base 2. *B. canadensis*
   2. Twigs usually grayish; spines on older stems not curved, though maybe drooping; peglike petiole bases persist around bud base 3. *B. vulgaris*

1. *Berberis thunbergii* DC.   JAPANESE BARBERRY   Figure 59

An exotic shrub that has naturalized from VA to GA, mostly in Appalachian and Piedmont regions. The most commonly encountered barberry in open woods, forest edges, and pastures. Leaves are untoothed.

2. *Berberis canadensis* P. Mill.   AMERICAN BARBERRY   Figure 60

The only native species of barberry, very sparsely distributed in rocky soils of Appalachian uplands and peripheral regions, from WV to GA and westward to IN and MO. Overall rare and localized today, this short shrub was eradicated from many areas in the past, as it is an alternate host for wheat stem rust. Leaves may bear a few teeth on margins.

3. *Berberis vulgaris* L.   COMMON BARBERRY   Figure 61

A European shrub rarely encountered as a naturalized exotic in the SE. More widely cultivated and naturalized across uplands in the SE in the past, but wheat rust eradication efforts were also concentrated on this species, an alternate host for the fungus.

Berchemia scandens (Hill) K. Koch   SUPPLEJACK   Family Rhamnaceae   Figure 62

A native twining vine with thin, smooth bark. Common in the Piedmont and coastal plain from VA to TX, and northward to IL. Prefers moist hardwood forest, bottomlands, and swamps. The twigs are reddish or green, tough, with raised leaf scars showing 1–3 obscure bundle scars. Larger stems grayish green, coiled snakelike around tree trunks or hanging from high in the canopy. Sapsuckers often peck holes in these larger stems. The fruit is an oblong drupe.

Betula L.   BIRCH   Family Betulaceae

Seven native birches occur in the SE. The only 1 ranging over the entire SE is river birch (*B. nigra*). In the Appalachians, the black birch (*B. lenta*), and the yellow birch (*B. alleghe-niensis*) are relatively common. Catkin buds (staminate) are present on twig tips of mature trees. Fruit a strobile, a conelike cluster of tiny seed and layers of bracts, the structure disintegrating after maturity.

1. Scraped twigs aromatic (wintergreen fragrance); lenticels on twigs not raised or rough to touch
   2. Bark smooth, gray to dark on young stems; twigs and buds glabrous
      3. Leaves oblong-ovate, mostly over 6cm 1. *B. lenta*
      3. Leaves oval, mostly under 6cm 2. *B. uber*
   2. Bark papery, gray to yellowish on young stems; twigs and buds slightly hairy
                                                                           3. *B. alleghaniensis*

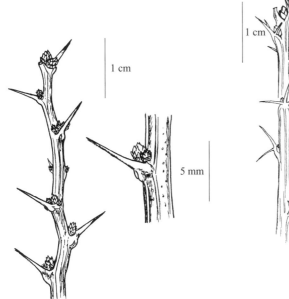

Figure 59. *Berberis thunbergii*

Figure 60. *Berberis canadensis*

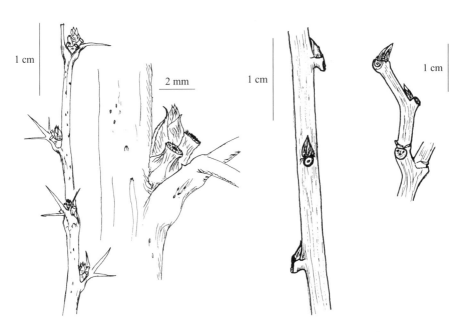

Figure 61. *Berberis vulgaris*

Figure 62. *Berchemia scandens*

1. Scraped twigs without wintergreen odor; lenticels on twigs raised or rough, or twigs roughened by hairs
    4. Bark pinkish or red-brown, papery on young stems; twigs and buds softly and thickly pubescent   4. *B. nigra*
    4. Bark smooth and lenticellate on young stems, becoming whitened or pinkish, exfoliating only in horizontal strips; bud scales mostly glabrous but resinous within; twigs glabrous to lightly pubescent
       5. Mature trunks smooth and grayish white, not papery; twigs gray   5. *B. populifolia*
       5. Mature trunk bark papery or exfoliating; twigs brownish or dark red-brown
          6. Twigs glabrous, drooping on mature trees; rarely naturalized exotic   6. *B. pendula*
          6. Twigs minutely to noticeably pubescent, not drooping; rare native trees of high elevations
             7. Bark chalky white; twigs pubescent   7. *B. papyrifera*
             7. Bark pinkish-white; twigs nearly glabrous   8. *B. papyrifera* var. *cordifolia*

1. *Betula lenta* L. SWEET BIRCH   Figure 63

A moderately large tree and perhaps the most common birch of the s. Appalachians. Occurs as far south as n. GA. Prefers moist soils in lower and middle elevation coves, streamsides, and cool slopes. The dark, smooth bark of young stems becomes scaly with age, with broad, curling gray plates.

2. *Betula uber* (Ashe) Fern. VIRGINIA ROUND-LEAF BIRCH

Similar in most respects to *B. lenta* and possibly representing only a genetic variant of that species. Leaves are much smaller, almost rounded in outline. Endemic to Smyth County VA and very rare. Germinated seedlings of *B. uber* predominately appear as "reversions" to typical *B. lenta*. Of smaller mature size than *B. lenta,* as evidenced from the few natural trees that exist.

3. *Betula alleghaniensis* Britt. YELLOW BIRCH   Figure 64

A large tree with age, in higher Appalachian elevations of NC and TN northward. Common in the Blue Ridge boreal zone and along cool streams and slopes where it may mingle with *B. lenta*. The gray to yellowish papery bark of young stems thickens with age and becomes dark grayish to brownish and roughly scaly on old trunks, often obscured by lichens or moss.

4. *Betula nigra* L. RIVER BIRCH   Figure 65

A widespread tree over the SE along streams, alluvial floodplains, and disturbed lowlands near water. Young stems conspicuously shreddy and papery with grayish outer layers of bark revealing pinkish inner layers, becoming overall grayish and thickly scaly with age. Seed drop earlier than other birches, scattered on the wind and waters by early summer. Rare in middle and higher mountain elevations; intolerant of shade.

5. *Betula populifolia* Marsh. GRAY BIRCH

A small native tree, occurring mostly north of the se. region. A few rare populations appearing natural occur in stony mountain soils of VA and WV. The bark is a grayish white, with pronounced black triangular patches where the branches join the trunk. The bark is not papery or peeling. Leaves bear a prolonged tip and wide base.

6. *Betula pendula* EUROPEAN WHITE BIRCH

A native of Europe, commonly planted in the Appalachian region and VA Piedmont. On rare occasions naturalizing by seed near cultivated specimens. Mature trees develop pendulous branches. Bark slightly papery on mature trunks.

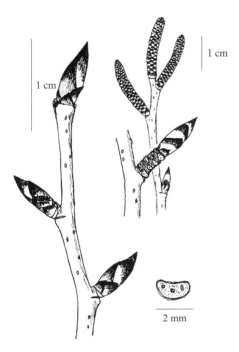

Figure 63. *Betula lenta*

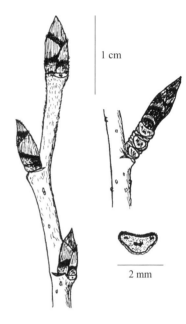

Figure 64. *Betula alleghaniensis*

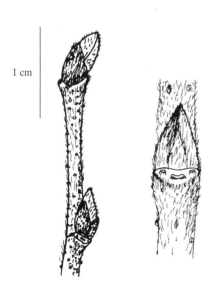

Figure 65. *Betula nigra*

7. *Betula papyrifera* Marsh. PAPER BIRCH *Figure 66*

This n. tree, so common across much of N. Amer., essentially does not reach the SE, although it has been reported from the n. VA mountains. Leaves tend to be tapered to rounded at the base, twigs slightly pubescent, and bark chalky white and exfoliating. This birch is occasionally planted as an ornamental in the Blue Ridge region of the SE.

8. *Betula papyrifera* var. *cordifolia* (Regel) Fern. MOUNTAIN PAPER BIRCH

Another n. birch that appears in a few disjunct populations in the high mountains of WV, VA, NC, and TN. A variety of *B. papyrifera* or a separate species. The cordate-based leaves, less pubescent twigs, and pinkish-white bark are the primary morphological traits used for separation, though some genetic differences have also been reported. (*B. cordifolia* Regel)

Bignonia capreolata L. CROSSVINE Family Bignoniaceae *Figure 67*

A native vine ranging throughout the SE, except absent from the middle and higher elevations of the Blue Ridge. Most common in moist forests of the coastal plain in floodplains and rich slopes. Climbs by use of tendrils, 1 borne between the paired leaflets of each leaf. Leaflets may persist into winter, or at least the petiole is persistent. No terminal bud; laterals with outer pair of bud scales thickened and truncated. Twigs 4-ridged. Fruit a flat capsule.

Borrichia frutescens (L.) DC. SEA-OXEYE Family Asteraceae *Figures 68 and 69*

A coastal shrub, ranging from VA to TX in the SE. Found along tidal flats and brackish soils of marshes, often forming extensive colonies or thickets under 1.5m in height. The gray twigs may die back in winter near the tips. Buds and bundle scars sometimes obscure. A raised ridge often forms a point between the leaf scars. Fruit is a spinose head terminating some twigs, persistent over winter. The opposite leaves are gray-green, thickened, sometimes persistent.

Broussonetia papyrifera (L.) L'Her ex Vent PAPER-MULBERRY Family Moraceae
Figure 70

An exotic tree with smooth, yellow-brown bark and bristly hairy twigs. Often found around urban areas, forming thickets by root sprouts. Native to Asia; originally cultivated in the SE for silkworm production. Most of the plants in the SE are male, producing only staminate flowers. Pistillate plants forming the rounded, reddish fruits are rarely seen. Naturalized mostly in lower elevations of the Blue Ridge eastward and southward; common in the Piedmont and coastal plain of VA to TX.

Brunnichia ovata (Walt.) Shinners LADIES' EARDROPS Family Polygonaceae *Figure 71*

A native vine of coastal plain bottomlands that climbs by tendrils at ends of side shoots just below the ringed, jointlike nodes. Twigs are lined and often die back from the tips in winter. Leaves have wide, truncate bases. The fruit is a pendent brownish achene in a twisted or ribbed sac. This vine may climb high into the canopy.

Buckleya distichophylla (Nutt.) Torr. PIRATEBUSH Family Santalaceae *Figure 72*

A rare native shrub that is parasitic on the roots of other woody plants, chiefly hemlock trees. Habitat is typically rocky soils in scattered areas of the Blue Ridge of VA, NC, TN. The olive-green twigs have no terminal buds. Fruit is an oblong, greenish to yellow drupe that is inedible for people. Bark thin, greenish or brown, with prominent white lenticels.

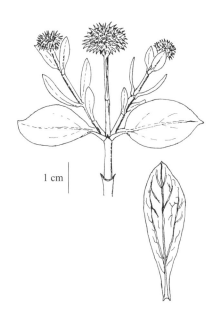

Figure 66. *Betula papyrifera*

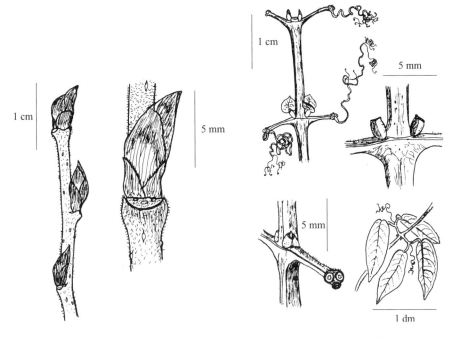

Figure 67. *Bignonia capreolata*

Figure 68. *Borrichia frutescens*

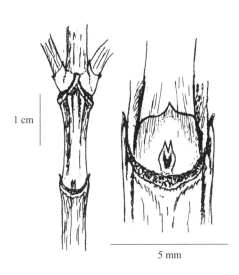

Figure 69. *Borrichia frutescens*

Buddleja L. BUTTERFLY-BUSH   Family Verbenaceae

Two species of these Asian shrubs are treated here as sparingly naturalized exotics of the SE mostly near sources of cultivation. The twigs often die back in winter, and capsular fruits are usually present.

1. Buds white-woolly; leaf scars connected at nodes by a ledge	1. *B. davidii*
1. Buds yellow-brown or grayish and scurfy; leaf scars not as above	2. *B. lindleyana*

1. *Buddleja davidii* Franch.   BUTTERFLY-BUSH   Figure 73

An exotic shrub that is widely cultivated. Occasionally naturalizes by seed and may spread by roots near cultivated specimens. Twigs usually reddish, hairy, often with juvenile leaves that are white-woolly beneath. Capsules elongate.

2. *Buddleja lindleyana* Fort. ex Lindl.   LINDLEY BUTTERFLY-BUSH   Figure 74

A less commonly cultivated species in the SE but known to naturalize in some areas by seed. Twigs gray, slightly scurfy but not hairy, often with shreddy outer bark. Leaves nearly glabrous beneath. Capsules ovoid.

Calamintha P. Mill.   BASIL   Family Labiatae

Low, slender-branched, brittle shrubs of the coastal plain, with strong minty fragrance to inner bark and crushed leaves. At least some dried leaves, if not green foliage, can usually be found on the twigs during winter. The 3 species are best identified using leaves. Persistent fruits are axillary to leaves, appearing as ribbed, 1cm or less calyx tubes that contain the tiny nutlets inside. Bark of stems is shreddy or exfoliating.

1. Leaves oblong-elliptic, entire, and revolute on margin	1. *C. coccinea*
1. Leaves ovate or obovate, at least some with a few large teeth	
2. Leaves widest at or below middle; petiole 2–5mm long	2. *C. georgiana*
2. Leaves widest beyond middle; sessile, no petiole	3. *C. dentata*

1. *Calamintha coccinea* (Nutt. ex Hook.) Benth.   SCARLET BASIL   Figure 75

A slender, spindly-branched shrub, to 1m high, native to sandy pinelands of the coastal plain from se. GA through n. FL to s. AL and se. MS. Flowers red, appear in summer.

2. *Calamintha georgiana* (Harper) Shinners   GEORGIA BASIL   Figure 76

A low, densely branched shrub 1–4dm high. Found in sandy or rocky woodlands of the coastal plain from NC to LA. Sporadic and localized overall; most abundant in FL panhandle to MS. Flowers pinkish, appear in summer.

3. *Calamintha dentata* Chapm.   TOOTHED BASIL   Figure 77

A low, 1–4dm shrub with slender but erect branches. Endemic to central panhandle region of FL, in sandy pinelands. Flowers pinkish, appear in summer.

Callicarpa L.   BEAUTYBERRY   Family Verbenaceae

Of these shrubs, 1 native species and 1 sparingly naturalized exotic occur in the SE. Both have similar twigs, with naked buds and a close covering of peltate scales. The inner bark has a rank odor. The lavender-colored fruits may persist into early winter; these are drupes borne in clusters at the nodes. Twigs may die back after severe freezes.

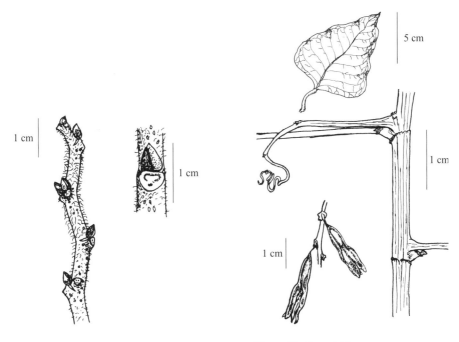

Figure 70. *Broussonetia papyrifera*

Figure 71. *Brunnichia ovata*

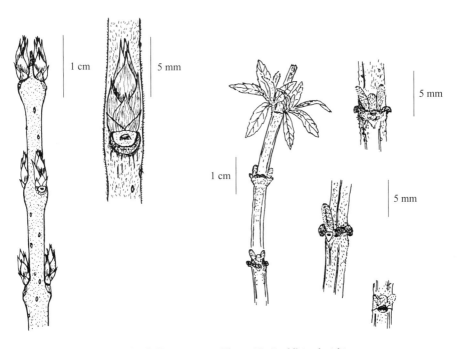

Figure 72. *Buckleya distichophylla*

Figure 73. *Buddleja davidii*

1. Twigs usually 3–4mm diam. at internodes; fruit peduncles 2–6mm long   1. *C. americana*
1. Twigs usually 2–2.5mm diam. at internodes; fruit peduncles 10–20mm long 2. *C. dichotoma*

1. *Callicarpa americana* L.   BEAUTYBERRY   Figure 78

A coarse shrub 1–2m tall, widespread across the SE but absent from middle and higher elevations of the Blue Ridge. Occurs in a wide diversity of sunny to semishaded habitats, from rock outcrops to boggy thickets and sandy pinelands. Rare in very shady hardwood forest understories. Leaves pubescent and scurfy beneath. Fruits pinkish-lavender to magenta.

2. *Callicarpa dichotoma* (Lour.) K. Koch   PURPLE BEAUTYBERRY   Figure 79

An exotic shrub that is known to naturalize in a few areas of VA and NC. Cultivation of this Asian species is occasional across the SE. Similar in appearance to the native species, but leaves are nearly glabrous beneath; the fruits more purplish and borne as a group on a longer peduncle; and twigs are more slender. White-fruited variations of both species are known.

Calluna vulgaris (L.) Hull   SCOTS HEATHER   Family Ericaceae   Figure 80

An exotic, evergreen shrub that is occasionally found naturalizing near areas of its cultivation, mostly north of NC. Overall rarely naturalized in the SE. The scalelike leaves somewhat resemble those of a juniper, but are 4-ranked, about 1mm long, slightly triangular in cross section. Bases of old leaves persist on branchlets, seemingly with bases split into 2 "tails." Flowers are white to pinkish.

Calycanthus floridus L.   SWEETSHRUB   Family Calycanthaceae   Figure 81

A single species is native, but with 2 varieties. Overall a shrub of moist woodlands of the Appalachians and areas south and east of these mountains. The dark brown twigs have swollen nodes and bundle scars that may appear as 3 but more often with smaller trace bundles among the larger 3. Fruit is a large follicle with bean-sized, dark brown seed rattling inside (poisonous if eaten). The var. *floridus* L. has pubescent twigs and leaves, and is mostly found in the Piedmont and coastal plain from VA to FL and MS, though occasionally in the Appalachians. The var. *glaucous* (Willd.) T. & G. [var. *laevigatus* (Willd.) T. & G.] is essentially glabrous and is typically found in Appalachian areas from n. AL and GA northward, though occasional in moist woodlands as far south as n. FL or to the coastal plain of SC, NC.

Calycocarpum lyonii (Pursh) Gray   CUPSEED   Family Menispermaceae   Figure 82

A native, twining vine that often dies back to near the ground in areas where winters are cold. Ranges chiefly west of the Blue Ridge, from KY to IL, MO, south to e. TX and east to sw. GA, w. FL. Typically in rich and alluvial soils of hardwood forests, though occasionally in calcareous uplands. The twigs are green, finely lined, with large leaf scars. The 2cm fruit is rarely seen, but hangs in clusters on the pistillate (female) plants; seed concave on 1 side; all parts inedible.

Campsis radicans (L.) Seem. ex Bureau   TRUMPET-CREEPER   Family Bignoniaceae
                                                                             Figure 83

A native vine climbing by fibrous aerial rootlets. Most commonly seen ascending to the canopy in bottomlands and swamps, though widely dispersed across nearly the entire SE and to be expected in any habitat except middle and higher Blue Ridge elevations. Opposite leaf scars are joined at the nodes by a ridge or line. Fruit is a plump, elongate capsule filled with winged seed. Produces large, orange to reddish trumpet-shaped flowers in summer.

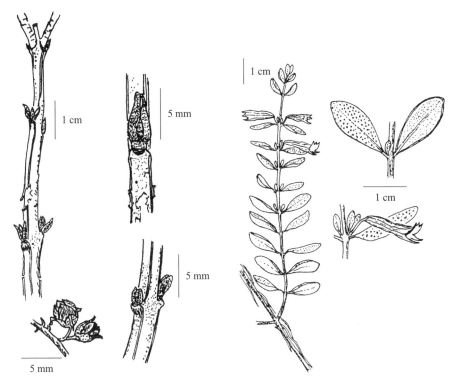

Figure 74. *Buddleja lindleyana*

Figure 75. *Calamintha coccinea*

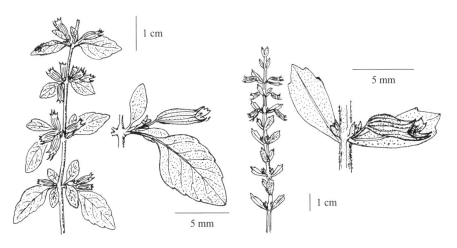

Figure 76. *Calamintha georgiana*

Figure 77. *Calamintha dentata*

Carpinus caroliniana Walt.   AMERICAN HORNBEAM   Family Betulaceae   *Figure 84*

A small native tree, widespread and common over most of the SE. The gray bark is thin and smooth, though growth of wood underneath is not symmetrical, and the resulting bulges and fluted areas give a sinewy appearance to the trunk. The twigs are slender, lenticellate, with no terminal bud. Bud scales several, each dark with a lighter margin. Larger buds that hold staminate catkins within can be found on some twigs. Most commonly found in bottomlands and along streams. Fruit is a nutlet attached to the base of a lobed and toothed bract, several borne together as a pendent cluster in autumn. The var. *virginiana* (Marsh.) Furlow is more common in interior areas of the SE, from VA to n. GA, and west to AR and northward; it tends to have sharper teeth and more pointed tips on the leaves and fruit bracts.

Carya Nutt.   HICKORY   Family Juglandaceae

Thirteen species are treated here for the SE, all native trees. Twigs bear terminal buds that are larger than laterals, the buds appearing naked, imbricate-scaled, or nearly valvate-scaled. Bundle scars are scattered or in 3 or 4 general groups. Pith homogenous, angled in cross section. The fruit is a nut, borne in a 4-parted husk. Identification by twigs alone is difficult with some species or individuals; if fallen fruit remnants or leaf fragments can be found, identification is easier.

1. Bud scales of terminal bud scurfy, nearly valvate, or bud appearing naked; shell of nut very thin, easily cracked
    2. Buds dark gray or brown; twigs hairy or slightly so near base of terminal bud; trunk bark scaly
        3. Buds dark gray; twigs hairy; pith light brown; trunk bark gray to brownish, furrowed or slightly scaly; nut elongate, round in cross section, smooth   1. *C. illinoinensis*
        3. Buds brown or reddish-brown; twigs hairy only at apex; pith dark brown; trunk bark light gray, scaly; nut short, flattened, finely grooved over surface   2. *C. aquatica*
    2. Buds yellowish or pale silvery-brown or buff; twigs glabrous; trunk bark not scaly
    3. *C. cordiformis*
1. Bud scales imbricate, not scurfy but may bear silvery, reddish, or yellow dotlike scales or glands (peltate scales)
    4. Buds and adjacent twig surfaces densely and finely dotted with silvery scales; fruit husk silver-dotted, sutures distinctly winged; nut dark or mottled, unridged and round in cross section   4. *C. myristiciformis*
    4. Buds not as above; fruit not as above; nut ridged, or not symmetrically round in cross section
        5. Outer, darker bud scales falling away early or easily from terminal buds, leaving exposed the pale, tightly adherent inner scales that are smooth or minutely silky, or covered with glandular dots
            6. Glandular dots lacking; twigs stout, usually over 6mm diam.; terminal buds often over 1cm long   5. *C. alba*
            6. Glandular dots present; twigs and buds smaller than above
                7. Lateral buds usually retain outer, darker scales; yellow dots on terminal buds usually present; leaves pale beneath; rachis with tufts of hair at leaflet junction   6. *C. pallida*
                7. Lateral buds usually lack outer darker scales, visible scales tight and margins fairly obscure; reddish dots may occur on buds, no yellow dots; leaves lighter colored but not pale beneath; rachis glabrous to hairy but not with conspicuous tufts as above
                    8. Fruit husk commonly 3–4mm thick; trees of areas west of Mississippi River
                    7. *C. texana*
                    8. Fruit husk usually 2–3mm thick; trees of dry, sandy soils of peninsular FL
                    8. *C. floridana*
        5. Outer, darker bud scales remain, these usually with long, loose tips or undulating, rather brittle margins

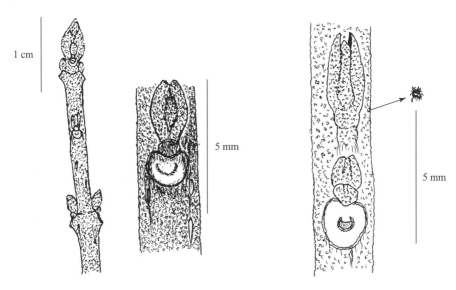

Figure 78. *Callicarpa americana*　　　　　　　Figure 79. *Callicarpa dichotoma*

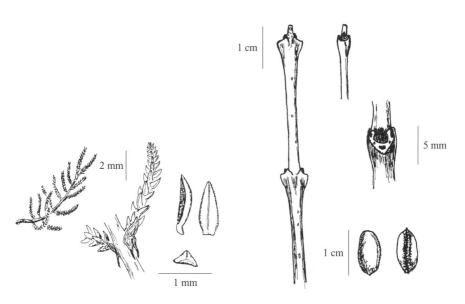

Figure 80. *Calluna vulgaris*　　　　　　　Figure 81. *Calycanthus floridus*

9. Inner bud scales or apex of buds hairy; fruit husks 5mm or thicker, falling free from nut in 4 parts; nuts heavily angled or ridged
   10. Twigs mostly 4–6mm diam.; terminal bud gradually tapered beyond base, nearly glabrous   9. *C. carolinae-septentrionalis*
   10. Twigs mostly 6mm or more diam.; terminal bud abruptly tapered from about the middle, usually pubescent
      11. Twigs reddish-brown or olive-brown; fruit nearly globular, less than 4.5cm long
            10. *C. ovata*
      11. Twigs sandy, buff, or orange-brown; fruit slightly oblong, usually 5cm or longer
            11. *C. laciniosa*
9. Inner bud scales or bud apex glabrous; fruit husk 4mm or thinner (if thicker, then husk persisting around nut); nuts not strongly ridged or angular   the "pignut complex"
   12. Fruit husks usually retain nut or split only halfway to base of fruit; fruit pear-shaped with a knoblike base (stipe); bark furrowed to slightly scaly on old trunks
            12. *C. glabra*
   12. Fruit husks splitting to base of fruit along at least one suture; fruit lacks stipe, though may bear a stalk; sutures often slightly winged near apex; mature bark scaly
            13. *C. ovalis*

1. *Carya illinoinensis* (Wang.) K. Koch   PECAN   Figure 85

Presumed native to the Mississippi River vicinity and nearby regions west, this tree has been so extensively cultivated across much of the SE that naturalized trees are today encountered over all but the Appalachian sections of the region. The fruit husks are 2–4mm thick, winged on the sutures, but being rather fleshy they shrivel quickly after falling. Nuts oblong, thin shelled, not angled, round in cross section. Bark brownish, in scaly plates.

2. *Carya aquatica* (Michx. f.) Nutt.   WATER HICKORY   Figure 86

A lowland tree, mainly of floodplains in the coastal plain from VA to TX and north to IL. The winged fruit husks are 1–2mm thick. The small, brown, slightly flattened nuts have grooved surfaces with a thin shell and bitter meat. Bark of mature trees scaly, often with long, free strips or curling plates.

3. *Carya cordiformis* (Wang.) K. Koch   BITTERNUT HICKORY   Figure 87

A large, common tree across most of the SE, scarce or missing only along the Gulf Coast and in FL. The winged fruit husks are 2–3mm thick. Nuts are slightly flattened, pale, somewhat heart-shaped, with a thin shell and bitter meat. Bark grayish to brown, tight and smooth to shallowly furrowed, rarely slightly scaly. Found in floodplains, valley forests, rich coves, or on slopes and hills. Ascends nearly to the boreal zones of s. Appalachian elevations.

4. *Carya myristiciformis* (Michx. f.) Nutt.   NUTMEG HICKORY   Figure 88

A tall tree, scattered sporadically in its range from TX to NC, mostly in the coastal plain, in calcareous soils or over marl clay substrates. An overall majority of the trees are concentrated in areas of LA, AR, and MS. The fruit husks are winged, and 1–2mm thick; nut shell fairly thick; meat not bitter. Bark gray, smooth at first, becoming scaly with long, loose strips or plates.

5. *Carya alba* (L.) Nutt. ex Ell.   MOCKERNUT HICKORY   Figure 89

Widespread and common across the SE in a wide variety of habitats. Fruit husks 5–10mm thick; nut angled, pointed on both ends; shell thick; meat sweet. The stout, hairy twigs, pale terminal buds, and gray, furrowed bark usually distinguish this tree easily in winter. [*C. tomentosa* (Poir.) Nutt.]

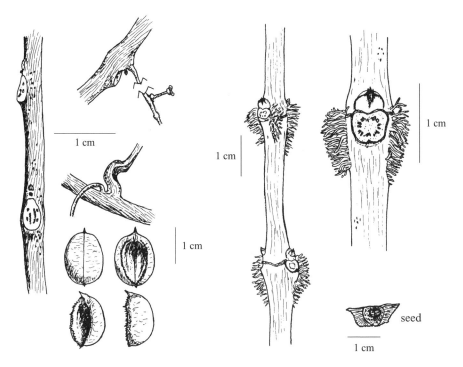

Figure 82. *Calycocarpum lyonii*

Figure 83. *Campsis radicans*

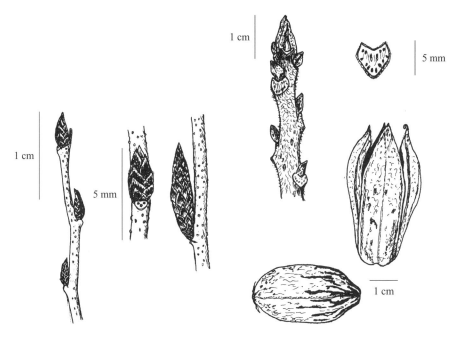

Figure 84. *Carpinus caroliniana*

Figure 85. *Carya illinoinensis*

6. *Carya pallida* (Ashe) Engl. & Graebn.   SAND HICKORY   *Figure 90*

Common across most of the SE east of the Mississippi River. Most commonly seen in sandy or rocky soils. The fallen hairy petioles of leaves can be an aid to identification in winter, coupled with the rather slender twigs and thin fruit husks. Yellow dots are usually present on the terminal buds. Fruit husks 2–4mm thick, usually yellow-dotted, split to the fruit base or only halfway. Nut slightly angled, with a thick shell and sweet meat. Bark is gray to nearly black and furrowed.

7. *Carya texana* Buckl.   BLACK HICKORY   *Figure 91*

This tree is essentially limited to areas west of the Mississippi River on lowlands as well as rocky uplands from LA to central TX, north to KS, IL. Most conspicuous in e. TX and the Ozark region. Reddish scales dot the leaf undersides and some bud scales, especially early in the season. Fruit husks are 2–5mm thick. The nut is slightly angled or not at all angled; shell thick; meat usually sweet. Bark gray to black, furrowed.

8. *Carya floridana* Sarg.   SCRUB HICKORY   *Figure 92*

A small tree endemic to central FL, barely reaching the range of coverage of this book. Found in sandy soils of the scrub community where it may be dwarfed on the deeper sands to form only a shrub. Fruit husks 2–3mm thick; nut usually not angled; shell fairly thick; meat sweet. Bark gray, thinly furrowed.

9. *Carya carolinae-septentrionalis* (Ashe) Engl. & Graebn.   SOUTHERN SHAGBARK HICKORY   *Figure 93*

A tall tree, sometimes considered only a variety of the more widely ranging shagbark hickory. Ranges from central NC to central MS, north to TN. Typically over granitic or calcareous rock in the Piedmont and in bottomlands or floodplains. The leaflets are narrow and nearly glabrous, and the twigs more slender, glabrous, and with narrower, darker-colored bud scales than in the next species. Fruit husks 5–10mm thick. Bark gray, with long scales loose at the ends, as in the next species. [*C. ovata* var. *australis* (Ashe) Little]

10. *Carya ovata* (P. Mill.) K. Koch   SHAGBARK HICKORY   *Figure 94*

Widely scattered across the SE but rare in the coastal plain. Most common on uplands of areas west of the Blue Ridge over calcareous rock or in basic or circumneutral soils. Also occurs in bottomlands and hills of the Piedmont. Twigs stout, slightly hairy, with large, loosely scaled buds. Fruit husks 5–15mm thick; nut angled; shell fairly thick; meat sweet and often considered the best flavored of the genus. Bark gray, scaling into long, curling strips or plates.

11. *Carya laciniosa* (Michx. f.) G. Don   SHELLBARK HICKORY   *Figure 95*

A large tree, most typically found in alluvial bottomlands and rich lowland soils near streams. Occurrence in the SE is sporadic, mostly limited to areas west of the Blue Ridge with scattered pockets extending to central MS and AL, s. TX, n. GA, and to the coastal plain of NC. Twigs are robust, with terminal buds to 23mm. Fruit husks are 8–14mm thick, finely hairy on the surface when in a green state. Nut shell thick; meat sweet. Fruits are the largest of the genus; in some trees near the Mississippi River, fruit lengths of 8cm are known. Bark gray to brown, furrowed to slightly scaly, with some loose plates or strips.

12. *Carya glabra* (P. Mill.) Sweet   PIGNUT HICKORY   *Figure 96*

A variable species, and with all variations considered it is found throughout the entire se. region covered by this book. The "typical form" of the species bears rather slender, glabrous

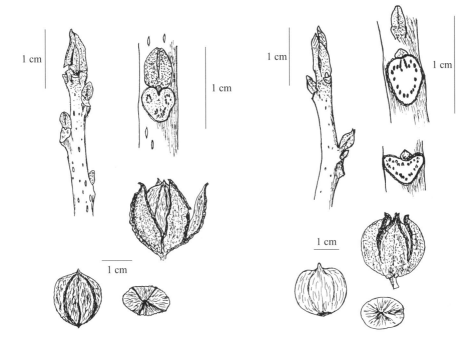

Figure 86. *Carya aquatica*

Figure 87. *Carya cordiformis*

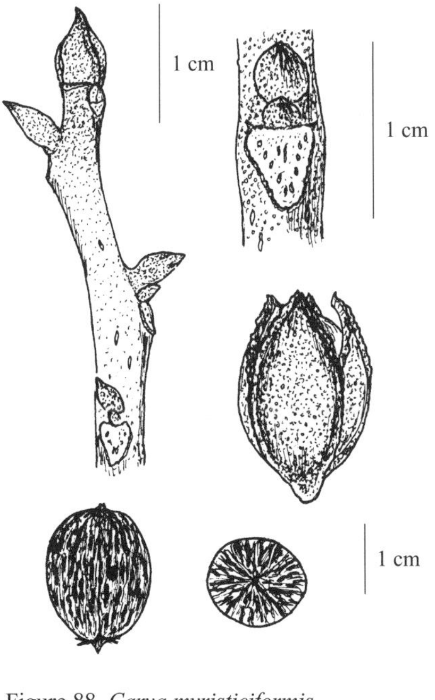

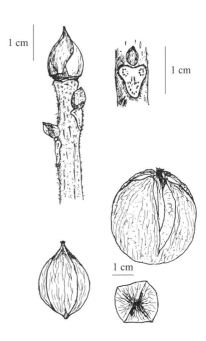

Figure 88. *Carya myristiciformis*

Figure 89. *Carya alba*

twigs 3–5mm diam. at the internodes, glabrous leaves with 5 leaflets, pear-shaped fruits 2–3cm long with thin (1–2mm) husks that do not split to the fruit base, often retaining the somewhat flattened nut, and roughly furrowed bark. There are coastal plain forms (var. *megacarpa* [Sarg.] Sarg. and var. *leiodermis* Sarg.) that have thicker twigs (5–7mm diam.), larger leaves with 5–9 leaflets, larger fruits (3–5cm) with thicker husks (3–5mm), less-flattened nuts, and thinner, smoother bark. The var. *leiodermis* of LA and TX additionally has pubescent spring foliage. These coastal pignuts often grow to large sizes in the maritime and hardwood hammock forests of GA and the Gulf states. Additional inland varieties of pignut include types with differences in pubescence, fruit size, husk thickness and dehiscence, and nut shape. One of the better-known variant forms of pignut is the red hickory or sweet pignut, treated here as a separate species.

13. *Carya ovalis* (Wang.) Sarg.   RED HICKORY   Figure 97

A large tree that is arguably distinct from pignut, due to intermediate trees known. The major identifying characters used to separate this hickory from other pignuts are leaves more commonly with 7 leaflets, reddish petioles, stouter twigs (6–7mm at internodes), fruits not pear-shaped, all or some fruit husks split to base, husks slightly winged on sutures, nut pointed on both ends and weakly angled, and bark more scaly, sometimes to the extent of resembling shagbark hickory. The red hickory ranges across much of the SE, mostly inland and north of the lower coastal plain, and seems most common and typical in the Appalachian region.

Castanea P. Mill.   CHESTNUT   Family Fagaceae

Three native species and 1 exotic are considered here. The more arborescent types (chestnuts) bear 3 nuts per fruit, even though 1 or more of these can be undeveloped due to poor pollination. The chinquapins are small trees or shrubs and bear only 1 nut per fruit. All species have in common alternate leaf scars with many bundle traces, stipule scars, angular pith, few-scaled buds, and lack of terminal buds. The edible nuts are borne in a spiny husk. The chestnut blight, a fungal disease introduced accidentally in the early 1900s, has reduced 2 of our native species from a former existence as conspicuous trees to occasional, resprouting understory species. Twigs alone are not always sufficiently diagnostic for winter identification, though the types with similar twig characters often occupy different habitats or regions.

1. Twigs glabrous, with conspicuous lenticels
    2. Fruits split along 4 sutures, enclosing 3 nuts; leaf petioles mostly 15–30mm long; widespread east of the Mississippi River   1. *C. dentata*
    2. Fruits split along 2 sutures, enclosing 1 nut; leaf petioles mostly under 15mm long; restricted west of Mississippi River   2. *C. ozarkensis*
1. Twigs pubescent, or lenticels few and/or inconspicuous
    3. Twigs reddish-brown; fruits 2–2.5cm wide, split mainly on 2 sutures, enclosing 1 nut
                                                                                                3. *C. pumila*
    3. Twigs grayish; fruits over 4cm wide, split along 4 sutures, enclosing 3 nuts
                                                                                            4. *C. mollissima*

1. *Castanea dentata* (Marsh.) Borkh.   AMERICAN CHESTNUT   Figure 98

Once a large and common tree in the Appalachians, ranging less commonly in adjacent provinces of the SE, and rare or absent in much of the coastal plain. By the late 1930s essentially all mature trees were killed by the introduced fungal chestnut blight. Sprouts from trees continually killed back to the roots are fairly common in the Appalachians, with occasional fruiting specimens seen, especially in middle and higher elevations of the Blue Ridge. Leaves are light green and glabrous, with distinct triangular teeth. Bark smooth and gray, becoming furrowed and thicker with age. The nut bears a long style on the apex.

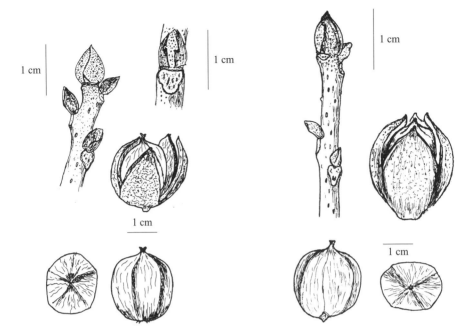

Figure 90. *Carya pallida*

Figure 91. *Carya texana*

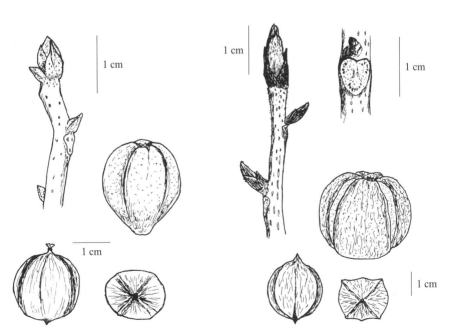

Figure 92. *Carya floridana*

Figure 93. *Carya carolinae-septentrionalis*

2. *Castanea ozarkensis* Ashe   OZARK CHINQUAPIN

Once a tree, to 20m tall, mostly of AR and adjacent states near the Ozark region. Reduced by the chestnut blight to resprouts from roots. Leaves are nearly glabrous. This species has been treated as a variety of *C. pumila,* but it differs from that species in less pubescent leaves and twigs, longer fruit spines, longer petioles, and more arborescent habit.

3. *Castanea pumila* (L.) P. Mill.   ALLEGHENY CHINQUAPIN   Figure 99

A shrub or small tree occurring throughout most of the SE in thinly wooded forests and disturbed lands. Hairs usually are conspicuous on twigs and leaf undersides, at least early in the season. The short trunk often leans, or multiple stems occur. This species is susceptible to the chestnut blight, though not so quickly and virulently attacked as the other 2 native species. Fruit husk spines and leaf petioles are usually less than 1cm long. Bark gray, smooth at first, furrowed with age. Coastal plain (NC to MS) forms of this species may bear sparsely spaced spines on the fruit husk (*C. pumila* var. *ashei* Sudw.), bear twigs and leaves becoming nearly glabrous late in the season (*C. alnifolia* var. *floridana* Sarg.), or be rhizomatous and form colonies 1.5m or less in height (*C. alnifolia* var. *alnifolia* Nutt.). These variations from sandy soils are likely influenced by ecological factors, as intermediates exist that provide a morphological transition to the "typical" inland form of *C. pumila*.

4. *Castanea mollissima* Blume   CHINESE CHESTNUT   Figure 100

An Asian tree that has been widely planted across the SE and occasionally naturalizes near sources of cultivation. Resistant but not immune to the chestnut blight. Leaves of mature trees are gray-velvety beneath. Like other exotic chestnuts, the trunk is usually rather short or well divided into a crown broader than high. Fields or forest openings with ample light and space promote survival of this chestnut rather than competitive forest understories. Hybrids between this species and other chestnuts, including *C. dentata,* have also been widely planted, so indeterminate specimens may be encountered. Other exotic chestnuts may also be encountered in the SE, but *C. mollissima* is the most frequently naturalized species.

Catalpa Scop.   CATALPA   Family Bignoniaceae

There are 2 native species, both of which have naturalized far beyond their original ranges by aid of cultivation. Twigs alone cannot be used to separate the 2, but minor differences in the fruit and seed are diagnostic in winter, if they are available. Both species have large, heart-shaped leaves with long petioles, stout twigs with craterlike, whorled leaf scars, and no terminal bud. The fruit is an elongate capsule filled with 2 rows of flat, winged seed, divided by an inner wall. The seed are freed when the capsule splits in half longitudinally. Showy white flowers appear in late spring.

1. Capsules 15–40cm long, 6–10mm diam., each half (valve) thin-walled and flattening after seed dispersal; seed 4–6mm wide; wing tapered into a pointed tuft of hair 1. *C. bignonioides*
1. Capsules 23–60cm long, 10–20mm diam., valves thick-walled and remaining semiround after seed dispersal; seed 8–9mm wide; wing rounded, with a ragged fringe of hairs
2. *C. speciosa*

1. *Catalpa bignonioides* Walt.   SOUTHERN CATALPA   Figure 101a

A tree native to the Gulf coastal plain from w. GA and FL to central MS. Through extensive cultivation, it is now naturalized over essentially the entire SE, typically in moist soils. Often a short and wide-crowned tree, to 15m. Bark brown, scaly. Leaves acute or acuminate on apex, strongly scented when bruised, to 30cm long, including a 10–15cm petiole. Leaf scars elliptic

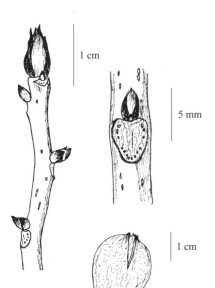

Figure 94. *Carya ovata*

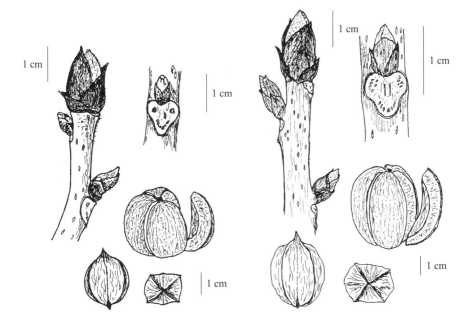

Figure 95. *Carya laciniosa*

Figure 96. *Carya glabra*

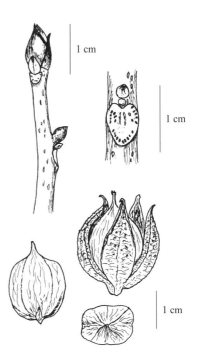

Figure 97. *Carya ovalis*

to orbicular. Flower clusters (panicles) 20–30cm, with numerous 2–4cm flowers. This species is more commonly encountered in the Piedmont and coastal plain regions than the next.

2. *Catalpa speciosa* (Warder) Warder ex Engelm.   NORTHERN CATALPA   *Figure 101b*

Native to the upper Mississippi River region, from w. TN and north to s. IL, IN. Extensively cultivated to the point of being sporadically naturalized throughout the SE, though most commonly so in n. sections of the area. Usually seen in bottomlands, along streams, and in other moist, rich soils. May reach heights of 30m. Bark ridged or deeply fissured; not as scaly as in *C. bignonioides*. Leaves are more long-pointed, less strongly scented, and reach 35cm long, including the rounded petiole that more often gives orbicular leaf scars on twigs than in *C. bignonioides*. Typically a taller, straighter tree than the other species, though open-grown specimens may develop a broad crown.

Ceanothus L.   REDROOT   Family Rhamnaceae

Three species occur in the SE, all low shrubs with slender, stipulate twigs and lingering fruit remnants in winter. Many more species of this genus occur in the w. US. The dark, elongate stipules may persist into winter. Terminal buds present or lacking, hairy, usually scaleless. The small white flowers are borne in spring to early summer. Fruit 3-lobed and capsule-like, with the persistent floral tube providing a disklike or conelike base. Roots reddish-colored.

1. Leaves persistent, less than 1cm long   1. *C. microphyllus*
1. Leaves deciduous, 2cm or more in length
   2. Lateral buds often with elongate basal scales; tips of fruit with ridge or crest on each lobe
        2. *C. americanus*
   2. Lateral buds usually not as above; fruit tips without crest   3. *C. herbaceous*

1. *Ceanothus microphyllus* Michx.   LITTLELEAF CEANOTHUS   *Figure 102*

A low evergreen shrub, 2–8dm tall, with greenish-yellow stems and tiny leaves. Occurs in sandy pinelands of the coastal plain in FL and adjacent s. GA, AL. Leaves 2–8mm long. Fruits 3–5mm wide, the tips smooth and without a crest.

2. *Ceanothus americanus* L.   NEW JERSEY TEA   *Figure 103*

A common shrub distributed throughout the SE on well-drained soils. Most often found as a 5–10dm shrub in dry woods, rocky sites, openings, and road banks. Twigs yellow-green to reddish, pubescent to glabrous. In some areas often dying back nearly to the ground in winter; a deep or large root system promotes vigorous resprouting. Dried leaves were used as a substitute for English tea (though caffeine free). Plants with smaller leaves and typically in drier or sandy habitats, especially in the coastal plain, may be treated as var. *intermedius* (Push) K. Koch.

3. *Ceanothus herbaceus* Raf.   REDROOT   *Figure 104*

A rare shrub in the SE, known from rocky riverbeds in WV and VA. Reported from AL and GA, but mostly a prairie species. It typically occurs among grasses and on rocky slopes of plains and prairies, TX and AR northward. Twigs pubescent and often dying back in winter.

Cedrus Trew   CEDAR   Family Pinaceae   *Figure 105*

Exotic conifers with evergreen, sharp-tipped leaves (needles) that are triangular to somewhat 4-sided in cross section. Short spur shoots are present on branchlets. Barrel-shaped cones borne

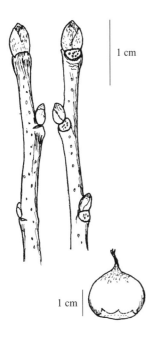

Figure 98. *Castanea dentata*

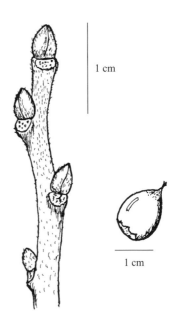

Figure 99. *Castanea pumila*

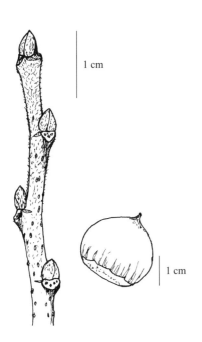

Figure 100. *Castanea mollissima*

Figure 101. a: *Catalpa bignonioides;* b: *C. speciosa*

96 · Cedrus

upright on outer branches, disintegrating after seed maturity. Three species are commonly planted in the SE; naturalization is rare. The following key will only be useful for typical forms of the species, not cultivars.

1. Leaves mostly 30–50mm, blue-green color; cones rounded at apex   1. *C. deodara*
1. Leaves usually less than 30mm long; cones more often flat or concave at apex
    2. Leaves mostly 20–30mm, long, dark green; twigs glabrous or nearly so   2. *C. libani*
    2. Leaves mostly 15–25mm, long, blue-green; twigs pubescent   3. *C. atlantica*

1. *Cedrus deodara* (Roxb. ex D. Don) G. Don f.   DEODAR CEDAR

A commonly cultivated species in the SE, to 50m tall. Shoots and branch tips tend to droop on mature trees. Twigs very pubescent. Cones 7–10cm long. Native to Himalayas.

2. *Cedrus libani* A. Rich.   CEDAR-OF-LEBANON

A dark or bright green crown to 40m height, with broad, horizontally layered appearance to the branches is characteristic. Branches not drooping as in *C. deodara*. Cones 8–10cm. Native to Asia Minor.

3. *Cedrus atlantica* (Endl.) Carr.   ATLAS CEDAR

The bluish tint of foliage is distinctive of this species. The crown of mature trees is pyramidal but rather open or sparse, to 40m. Cones 5–7cm. Native to North Africa. [*C. libani* ssp. *atlantica* (Endl.) Battand. & Trabut]

Celastrus L.   BITTERSWEET   Family Celastraceae

One native and 1 exotic species are treated here. Both are twining vines with no terminal buds and 1 bundle scar in the leaf scar. Plants are dioecious (either male or female), the pistillate plants bearing clusters of rounded capsules that are 3-valved, splitting to reveal an inner 3-parted fleshy portion consisting of a red aril covering pale, smooth seed. Old stem bark grayish or brown, shallowly furrowed. Roots orange-colored.

1. Twigs grayish, unlined, lenticels inconspicuous; bud scales not keeled or spinose-tipped
                                                                                                                          1. *C. scandens*
1. Twigs brown, often ridged or lined, lenticels conspicuous; outer bud scales keeled and spinose-tipped   2. *C. orbiculatus*

1. *Celastrus scandens* L.   AMERICAN BITTERSWEET   *Figure 106*

A native vine with a broad range from the e. US and Canada to WY and NM. In the SE it is sporadic from VA to n. GA, and west to TX, mostly in upland woods and thickets in basic to calcareous soils. Genetically threatened through cross-pollination by the expanding naturalization of *C. orbiculatus*. The fruit clusters are borne on the ends of the twigs, the outer capsule orange-colored.

2. *Celastrus orbiculatus* Thunb.   ORIENTAL BITTERSWEET   *Figure 107*

An Asian vine most extensively naturalized in the SE from VA to n. GA, especially in areas of the Blue Ridge and adjacent Piedmont. The spread of this invasive species is slower in areas of calcareous soils, but the availability of fruit to birds, the primary dispersal agent, continues to increase. Commonly seen along roadsides, urban areas, streamsides, and fence lines. The fruit clusters are borne in axils of leaves, the outer capsule yellow.

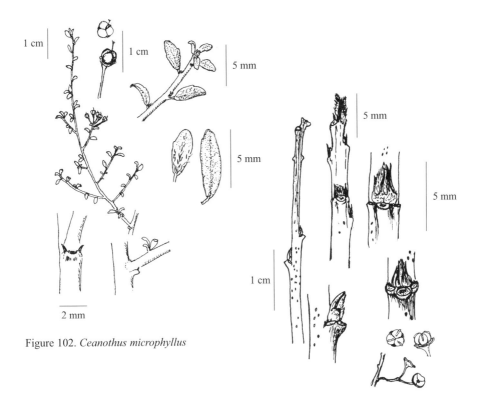

Figure 102. *Ceanothus microphyllus*

Figure 103. *Ceanothus americanus*

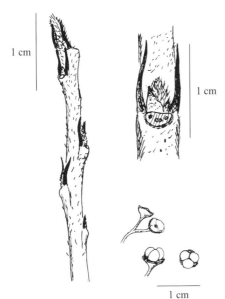

Figure 104. *Ceanothus herbaceus*

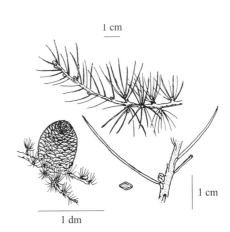

Figure 105. *Cedrus*

Celtis L.   HACKBERRY   Family Ulmaceae   *Figure 108*

Three native species are treated here for the SE. None are distinguishable by twigs alone, and on juvenile specimens there are no good characters for identification any time of year. Leaf variations in texture, margin, and shape are influenced by juvenility and environmental factors of the site, as well as genetic variation across the wide geographical ranges. The closely appressed lateral buds and zigzag twigs are characteristic of all species. The bundle scars are tiny and obscure, 3 per leaf scar but often appearing as 1. Bark of mature trunk gray, smooth, becoming roughly warty. Fruit a thin-skinned drupe borne in leaf axils in autumn. The following key is provisionally offered as a means of separating the typical forms of the 3 species. If fallen leaves or fruits can be found, these are useful in winter identification.

1. Trees over 8m tall; leaf length more than twice the width
    2. Leaves usually sparingly toothed to nearly entire, glabrous or nearly so, prolonged into a narrow acuminate to falcate tip; fruit 5–8mm long, nearly always lacking the peglike style at tip, and skin not puckered against seed in dry fruit; seed 4–7mm long   1. *C. laevigata*
    2. Leaves usually coarsely toothed, scabrous above, simply long-acuminate at tip; fruit 8–11mm, commonly with a peglike style on tip; skin puckered against seed in dry fruit; seed 7–9mm   2. *C. occidentalis*
1. Trees rarely over 6m, often contorted or shrubby; leaf length less than twice the width
     3. *C. tenuifolia*

1. *Celtis laevigata* Willd.   SUGARBERRY

A tree, to 30m on optimum sites, distributed over most of the SE except the Blue Ridge region. Typically seen in rich lowland soils such as bottomlands, but found in a wide range of habitats, including rocky, dry limestone hills. This is the *Celtis* that is usually encountered as a tree in the Atlantic and Gulf coastal plains. Leaves are rather yellow-green and glabrous on both surfaces (except scabrous on juvenile plants), tapering from the base into a prolonged tip. Bark gray, smooth to warty.

2. *Celtis occidentalis* L.   HACKBERRY

Usually a tree, ranging mostly north of the range of *C. laevigata*, but overlap occurs from near coastal NC across the Piedmont to GA and west to n. OK. Considerable variation in foliage and habit occurs in this species from east to west. Most commonly seen in moist, rich soils, but tolerant of dry, rocky uplands as well. Leaves are usually dark green and scabrous above, yellow-green and hairy beneath; long-acuminate on the tip but not conspicuously narrowing from the base, as in above species. Bark more commonly roughened by warty ridges or knobs, with narrow, smooth sections between.

3. *Celtis tenuifolia* Nutt.   GEORGIA HACKBERRY

A shrub, or small tree at best, usually with gnarly branches and short, crooked twigs. Found in dry woods or calcareous, rocky uplands over most of the SE, but scarce in s. GA to e. LA and north in the Mississippi embayment region to w. TN. Leaves usually scabrous and darker green above, sparingly toothed near the apex or nearly entire, the tip often short-acuminate. Fruit 5–8mm long, the skin not adhering and puckering to the seed when dried; seed 4–7mm.

Cephalanthus occidentalis L.   BUTTONBUSH   Family Rubiaceae   *Figure 109*

A native shrub of moist to wet soils. Common across most of the SE, though scarce in the Blue Ridge. The gray twigs have prominent lenticels, partially sunken buds, and opposite or whorled leaf scars. There is no terminal bud, and often the twig tips die back in winter. Stipules or their remnants often persist, and the single bundle scar appears U-shaped. The fruit is a

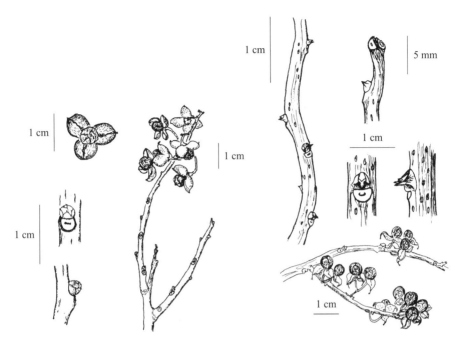

Figure 106. *Celastrus scandens*

Figure 107. *Celastrus orbiculatus*

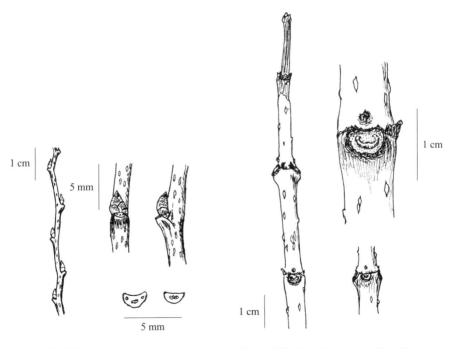

Figure 108. *Celtis*

Figure 109. *Cephalanthus occidentalis*

99

globular head of seed (nutlets), superficially similar to a small sycamore fruit. Globular heads of white flowers appear in summer.

Ceratiola ericoides Michx.   ROSEMARY   Family Ericaceae   *Figure 110*

This evergreen shrub is native to upland deep sands of the coastal plain from SC to s. MS. Sometimes locally abundant in the drier sands of hills and ridges under longleaf pine or scrub oak. Usually 1–2m in height, with bushy, rounded habit and shreddy bark. Leaves needlelike, 6–14mm long, appearing whorled on upright, hairy twigs. Flowers open in autumn in leaf axils. Pistillate plants produce yellowish drupes about 3mm diam. by Nov.

Cercis canadensis L.   REDBUD   Family Fabaceae   *Figure 111*

A small native tree well known and widely cultivated for its early spring flowers. Common across most of the SE, except uncommon in parts of the outer coastal plain and middle to higher Blue Ridge elevations. Twigs are zigzag with closely appressed buds (flower buds larger and slightly stalked) and prominent lenticels. A terminal bud is lacking. Bark scaly. Flowers pink to magenta, appearing before leaves in early spring. Fruit a legume, usually 6–10cm long.

Chamaecyparis thyoides (L.) B.S.P.   ATLANTIC WHITE-CEDAR   Family Cupressaceae
Figure 112

This native coniferous tree inhabits damp to wet peaty or acidic sandy soils in the coastal plain. Scattered from MD to central MS in depressions, swamps, and along streams. May be locally abundant in some areas, but absent over wide areas between populations. The 1–2mm, appressed, scalelike leaves usually have a gland on the back. Cones globose, 5–8mm, woody, brown, with about 3 pairs of flattened, disklike scales. Bark brownish, fibrous, shallowly furrowed into narrow ridges, sometimes twisting around trunk. The twigs and branchlets grow in flat sprays but tend not to be held in consistently flat planes to the extent of the genus *Thuja*.

Chamaedaphne calyculata var. *angustifolia* (Ait.) Rehd.   LEATHERLEAF   Family Ericaceae   *Figure 113*

A rhizomatous native shrub, generally 1m high or less, with uncommon and scattered distribution in the coastal plain of NC and adjacent parts of SC. Also known from a boggy site in the Blue Ridge but more common and typical of regions to the north. Found in acidic, peaty bogs and wet depressions. The other varieties are found in boreal regions across N. Amer. and Eurasia. The evergreen leaves are 2–5cm long, scurfy beneath, and reduced in size toward twig tips. The capsular fruit is 5-lobed. [*Cassandra calyculata* (L.) D. Don]

Chimaphila Pursh   PIPSISSEWA   Family Ericaceae

Two species occur in the SE, both diminutive shrubs less than 2dm tall. The evergreen leaves are alternate but often closely clustered. Creeping underground rhizomes may send up several of the slender stems. Fruit is a capsule borne on elongated terminal stems, 5-parted and about 5mm diam.

1. Leaves with whitish vein markings, a few large teeth, and widest near the rounded base
    1. *C. maculata*
1. Leaves dark green, with small serrations beyond the narrowed base
    2. *C. umbellata* ssp. *cisatlantica*

1. *Chimaphila maculata* (L.) Pursh   PIPSISSEWA, SPOTTED WINTERGREEN   *Figure 114*

A common species across most of the SE, in dry, acidic woodlands. There is no wintergreen odor, despite the name.

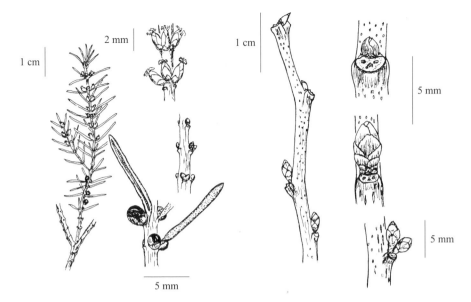

Figure 110. *Ceratiola ericoides*

Figure 111. *Cercis canadensis*

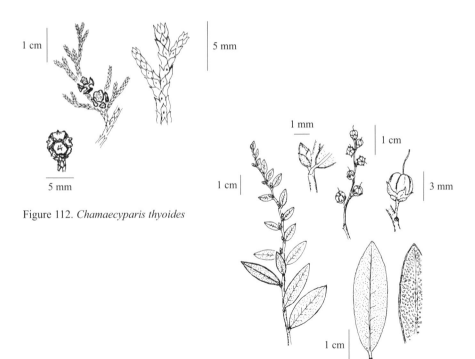

Figure 112. *Chamaecyparis thyoides*

Figure 113. *Chamaedaphne calyculata* var. *angustifolia*

2. Chimaphila umbellata ssp. cisatlantica (Blake) Hulten   PRINCE'S-PINE   *Figure 115*

Common in the NE, sparingly extending into the SE in dry, acidic woodlands of the Appalachians of WV, VA, and to central VA, NC. Two other varieties occur in the n. hemisphere, across N. Amer. and Eurasia.

Chionanthus virginicus L.   FRINGE-TREE   Family Oleaceae   *Figure 116*

A native shrub or small tree of wide distribution across the entire SE, with conspicuous leaf variation in foliage texture and surface appearance. Where found in xeric or sandy regions of the coastal plain, the thicker foliage, larger fruit, and more glabrous aspect can appear quite different from the inland forms that may have thin, dull, pubescent leaves. Grows on a wide variety of sites; rock outcrops, glades, dry woods, pine forests, mountain slopes, bottomlands, streamsides, swamps. Twigs have terminal buds, large leaf scars with 1 bundle scar, and prominent lenticels. The showy white flowers in spring have promoted extensive cultivation across the South. Plants are dioecious; the pistillate ones producing oval, purplish drupes to 2cm long. An additional species, *C. pygmaeus* Small, described from peninsular FL is out of range of this book, and it is sufficiently similar to other dry-site forms of *virginicus* that it may only deserve subspecies or varietal status.

Choenomeles speciosa (Sweet) Nakai   FLOWERING QUINCE   Family Rosaceae *Figure 117*

An exotic shrub that occasionally naturalizes near areas of its cultivation or along roadsides. Seen in the Blue Ridge region and adjacent provinces as a 1–2m bushy shrub. Twigs often bear spines on spur shoots or nodes. Buds rounded, tight-scaled. Fruit greenish-yellow, rounded, with dense, sour flesh; borne on branchlets by very short stems in autumn. The reddish flowers (occasionally white) appear in early spring.

Chrysoma pauciflosculosa (Michx.) Greene   SHRUB-GOLDENROD   Family Asteraceae
Figure 118

A native shrub of dry sands in the coastal plain from s. NC to n. FL and west to s. MS. Leaves evergreen, or at least some leaves can be found on the twigs in winter. Leaf surfaces gray-green, finely reticulate, the leaves alternate but closely clustered on the short branches that arise near the main trunk. Twigs below these leaves show a scalelike pattern of leaf scars over their surface. The yellow flowers are held high above the main branches by a long (to 1m), more sparsely leaved stem, blooming in late summer or autumn. The yellow flowers mature into a head of achenes in winter.

Cinnamomum camphora (L.) J. Presl.   CAMPHOR-TREE   Family Lauraceae   *Figure 119*

An evergreen, exotic tree that has sporadically naturalized in the outer coastal plain from NC to s. FL and along the Gulf Coast to LA. More commonly seen in FL, where it has been extensively cultivated as an ornamental and shade tree. Leaves are glabrous, long-petioled, spicy-aromatic when crushed, with a pale, thickened margin and glands in the main vein axils beneath. Twigs green, glabrous, aromatic; terminal flower bud much larger than laterals. Bark brownish, fissured.

Cissus trifoliata (L.) L.   MARINE-IVY   Family Vitaceae   *Figure 120*

A native vine climbing by tendrils arising from the twigs. Peculiar in distribution, occurring near the Gulf Coast of FL, west to TX and north to MO, se. KS; also from TX to AZ. Uncommon; mostly in rocky soils and maritime habitats. Twigs bear large orange lenticels, small buds, and a strong, unpleasant odor in the inner bark. Leaves are trifoliate, thick and

Figure 114. *Chimaphila maculata*

Figure 115. *Chimaphila umbellata* ssp. *cisatlantica*

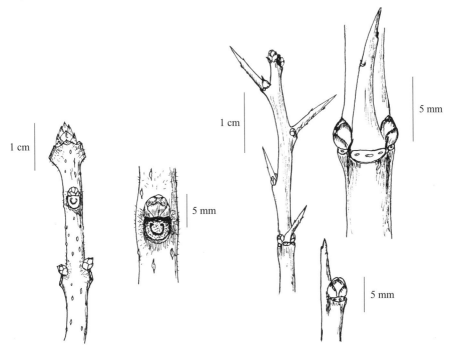

Figure 116. *Chionanthus virginicus*

Figure 117. *Choenomeles speciosa*

fleshy, rankly fragrant when crushed. Fruit is a dark berry, to 8mm diam., inedible. (*Cissus incisa* auct. non DesMoulins)

Citrus aurantium L.   SOUR ORANGE   Family Rutaceae   *Figure 121*

This is the most commonly naturalized *Citrus* in the SE, reaching the size of a small tree where winter freezes are not too severe. Most often encountered in FL. The evergreen leaves have a winged petiole; twigs are green and often spinose. The fruit is yellow to orange in color, to 9cm wide, with a thick, often wrinkled skin, flattened or impressed at the end; flesh juicy, very acidic, sour to bitter to taste.

Cladrastis kentukea (Dum.-Cours.) Rudd   YELLOWWOOD   Family Fabaceae   *Figure 122*

A small tree of sporadic occurrence in the Blue Ridge region and westward to IL, AR, OK, and to s. AL, MS. Most commonly seen in coves or on rich slopes with circumneutral soils under mixed hardwood forests or over limestone. The glabrous dark brown twigs lack terminal buds; lateral buds naked, dark brown, nearly surrounded by the leaf scars. Wood and inner bark of trunk yellow. Bark gray, smooth. Legume 5–8cm long; seed light olive-brown, bean-shaped, 5–6mm long. [*C. lutea* (Michx. f.) K. Koch]

Clematis L.   CLEMATIS   Family Ranunculaceae

Of the many se. species in this genus, 3 are here included as woody species; the others are more commonly herbaceous. All are vines climbing by twining leaf petioles. The opposite, compound leaves may lose the leaflets in winter, but twining petioles that hold the vines to another plant usually remain. Often the stems die back considerably in winter, only the basal sections being woody. Fruits are heads of achenes with long, curling, plumelike tails (styles). Bark of larger stems brown or gray, grooved to shreddy. Of the species considered, leaves and flowers are needed for identification.

1. Leaf margins entire, not toothed   1. *C. terniflora*
1. Leaf margins with a few large teeth, or with pointed lobes
  2. Leaves with 3 leaflets; common species sprawling over shrubs and small trees
     2. *C. virginiana*
  2. Leaves with 5–7 leaflets; uncommon and often high-climbing   3. *C. catesbyana*

1. *Clematis terniflora* DC.   YAM-LEAF CLEMATIS

An exotic vine from Japan that may be semievergreen in the s. sections of the SE. Naturalized widely across the region but most commonly in the coastal plain and Piedmont, especially near urban areas and disturbed habitats.

2. *Clematis virginiana* L.   VIRGIN'S-BOWER   *Figure 123*

A sprawling vine, woody near the base with most seasonal growth dying back in winter. A widespread species in upland provinces from VA to FL, west to LA, north to KS, less commonly seen in the coastal plain.

3. *Clematis catesbyana* Pursh   WOODBINE

Often a high-climbing vine, but sprawling when no tall trees are available to climb. Overall range is similar to that of *C. virginiana,* but most common in the coastal plain. Occurs in calcareous or rich, basic soil in regions interior of the coastal plain, but overall uncommon in uplands.

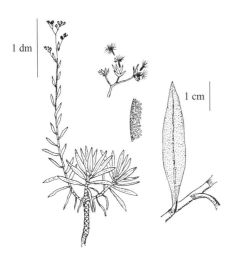

Figure 118. *Chrysoma pauciflosculosa*

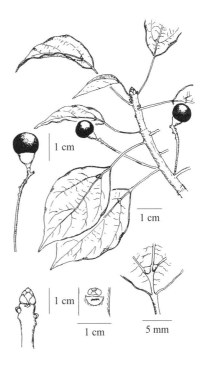

Figure 119. *Cinnamomum camphora*

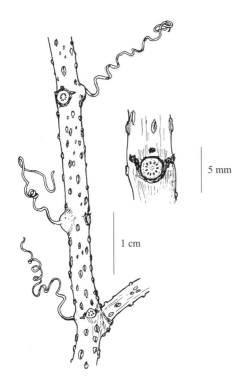

Figure 120. *Cissus trifoliata*

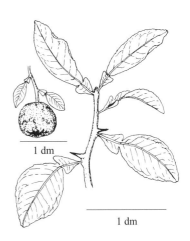

Figure 121. *Citrus aurantium*

Clerodendrum L. GLORYBOWER   Family Verbenaceae

Two species of this exotic genus are treated here, as occasional escapes from cultivation. The plants spread via root sprouts and seed, and may be encountered in disturbed sites near urban or residential areas. Twigs have raised, opposite leaf scars, often superposed buds, and unpleasantly aromatic inner bark. Fruit is a bluish drupe, flanked by a showy red calyx, borne in terminal clusters in autumn.

1. Twigs pubescent, rounded at internodes; leaves tapered at base, not toothed
1. *C. trichotomum*
1. Twigs glabrous, 4-angled; leaves cordate, toothed
2. *C. japonicum*

1. *Clerodendrum trichotomum* Thunb.   HARLEQUIN GLORYBOWER   *Figure 124*

The most commonly encountered species in the SE. A shrub often planted for the sweetly aromatic summer flowers. May spread by the roots in loose soils.

2. *Clerodendrum japonicum* (Thunb.) Sweet   JAPANESE GLORYBOWER

A shrub, less commonly naturalized; reported as an escape in a few coastal plain sites in GA, FL.

Clethra L.   PEPPERBUSH   Family Clethraceae

Two species are native to the SE, both shrubs. Twigs pubescent, with 1 bundle scar in each leaf scar; terminal buds present, very hairy, naked. Fruit a capsule, borne in elongate racemes. Bark reddish-brown, shreddy. Twigs are not useful for separating the species in winter, however the natural ranges of the plants do not overlap.

1. Leaves acuminate at tip; tall shrubs of Appalachian region   1. *C. acuminata*
1. Leaves acute at tip; short shrubs of coastal plain   2. *C. alnifolia*

1. *Clethra acuminata* Michx.   MOUNTAIN PEPPERBUSH   *Figure 125*

A tall, slender shrub, to 4m tall, of moist to wet soils in the Appalachian region of WV and VA to n. GA, west to TN, KY. Most often seen in rocky outcrops, seepage slopes, along streams, and in high elevation woodlands and balds.

2. *Clethra alnifolia* L.   SWEET PEPPERBUSH   *Figure 126*

A short shrub, rarely to 3m tall, often forming colonies through stolons or root sprouts. Found in the coastal plain throughout the SE, from MD to TX. Common in moist, sandy or peaty soils of pocosins, bays, bogs, barrens, swamp forests, and depressions in pine forests.

Cliftonia monophylla (Lam.) Britt. ex Sarg.   BUCKWHEAT-TREE   Family Cyrillaceae
Figure 127

An evergreen shrub or small tree of the outer coastal plain, from s. SC to FL and west to LA. Forms thick stands in FL to MS, these locally referred to as black titi swamps. Common in damp sands and peaty soils of bays, bogs, flatwoods, and streamsides. The leaves have obscure venation and tiny dots over the undersurface; the tip is emarginate (notched). Leaf scars show a single bundle scar. The fruit is 3- or 4-winged, resembling a buckwheat seed; borne in racemes. Bark dark, shallowly furrowed.

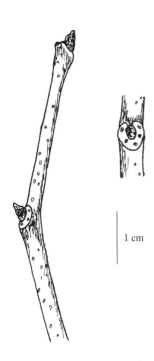

Figure 122. *Cladrastis kentukea*

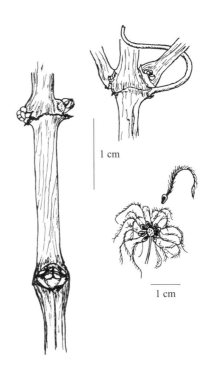

Figure 123. *Clematis virginiana*

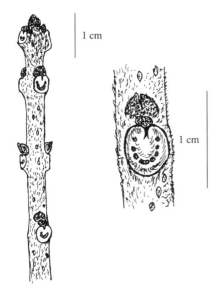

Figure 124. *Clerodendrum trichotomum*

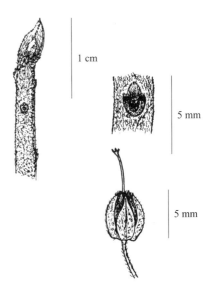

Figure 125. *Clethra acuminata*

Cocculus carolinus (L.) DC.   SNAILSEED   Family Menispermaceae   *Figure 128*

A native twining vine distributed throughout most of the SE, though scarce or absent from parts of the Appalachians and in VA. Usually seen in sunny forest openings, disturbed areas, and roadside thickets, especially in the coastal plain. Stems are green, with small, naked buds. The 3 bundle scars are small and difficult to discern. The fruit is a red drupe, containing a flattened, rough-edged seed that is suggestive of a snail's shell.

Comptonia peregrina (L.) Coult.   SWEET-FERN   Family Myricaceae   *Figure 129*

A native shrub, usually 1m or less in height, deriving its common name from the spicy aroma of the sap and the deeply incised leaves that are suggestive of a fern leaf pinnae. Common in the North, ranging into the SE along the Appalachians to SC. Occasional in the Piedmont of NC, VA. Found in dry, sandy woodlands, road banks, rocky sites, often where natural fires aid in reducing competition. The hairy twigs bear catkin buds in winter. Fruit resembles a brown, bristly burr. Colonies are formed by sprouts from the spreading root system.

Conradina Gray   CONRADINA, ROSEMARY   Family Labiatae

Five species are native to the SE in areas north of central peninsular FL; several more species are known from that state, south of the range of this book. All 5 species mentioned here are low, aromatic, brittle-stemmed shrubs that have persistent leaves in winter, so twigs are not used to identify them. Leaves are opposite and fascicled, with strongly revolute margins, obscure veins, dotted upper surfaces, and mostly pubescent lower surfaces. The minty odor and white, pink, or lavender flowers are distinctive. Fruits are tiny nutlets borne in the elongated, persistent calyx tube.

The common name of "rosemary" for these shrubs is derived from their very similar appearance and odor to true rosemary, *Rosmarinus* L., of the Mediterranean region.

1. Leaves dull green or gray-green above
    2. Leaves gray-green and very pubescent above, 2mm or less wide, mostly less than 15mm long   1. *C. canescens*
    2. Leaves dull green, glabrous or nearly so above, over 2mm wide and over 15mm long
            2. *C. etonia*
1. Leaves glossy above, bright or dark green
    3. Calyx tube glabrous or very finely hairy; plants native to Florida
        4. Leaves nearly all under 15mm long; flower cymes short-stalked, with 4 or fewer flowers.
            3. *C. glabra*
        4. Leaves often over 15mm; flower cymes distinctly stalked, mostly 5- to 8-flowered
            4. *C. grandiflora*
    3. Calyx tube distinctly hairy; plant native to Cumberland Plateau region   5. *C. verticillata*

1. *Conradina canescens* Gray   SHORTLEAF ROSEMARY   *Figure 130*

A low, bushy or sprawling shrub with a gray-green appearance, very pubescent. Found in sandy pinelands, scrub, and dunes in w. FL, s. AL, se. MS. Leaves 5–15mm long, grayish, hairy, very narrow due to margins rolled under almost to the midrib. (*C. brevifolia* Shinners)

2. *Conradina etonia* Kral & McCartney   ETONIA ROSEMARY   *Figure 131*

A slender shrub, to 1.5m, very rare; known only from deep, white sands in scrub habitat near Etonia River in central FL. Leaves 15–30mm long, 3–9mm wide; upper surface dull green, glabrous or with short, downy hairs; margin rolled under but not to near midrib; lateral veins visible beneath.

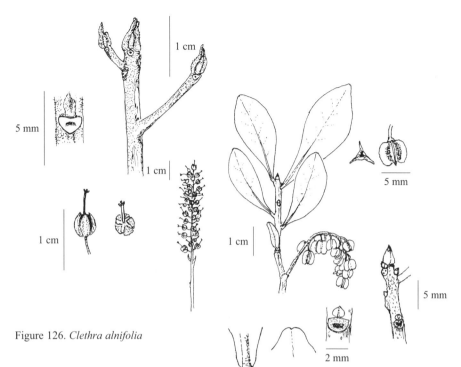

Figure 126. *Clethra alnifolia*

Figure 127. *Cliftonia monophylla*

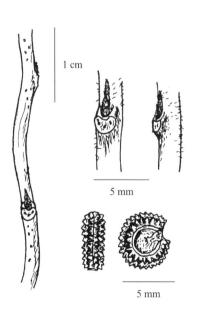

Figure 128. *Cocculus carolinus*

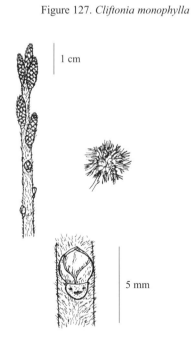

Figure 129. *Comptonia peregrina*

3. *Conradina glabra* Shinners   APALACHICOLA ROSEMARY   *Figure 132*

A low shrub, rare and locally distributed in sandy pinelands and ravine edges of the Florida panhandle, near the Apalachicola River. Leaves 10–15mm long; upper surface bright green, glabrous or nearly so; margins rolled under halfway to midrib.

4. *Conradina grandiflora* Small   LARGE-FLOWERED ROSEMARY   *Figure 133*

A slender shrub, to 1m tall. Barely reaching the area of coverage of this book on the east coast of FL. Occurs in sandy pinelands and scrub of peninsular FL from the Miami area north to vicinities of Daytona Beach and Orlando. Leaves are narrowly spatulate, 10–25mm long, bright green and lustrous above, the margins rolled under to near the midrib; no side veins are obvious underneath. Flowers may appear nearly year-round in stalked clusters. The calyx tube is finely pubescent.

5. *Conradina verticillata* Jennison   CUMBERLAND ROSEMARY   *Figure 134*

A rare sprawling shrub of gravelly or sandy soils near major streams, known only from a few sites in the Cumberland Plateau region of KY and TN. Leaves 12–25mm long; upper surface glossy green, glabrous; very narrow, margin rolled under nearly to midrib.

Cornus L.   DOGWOOD   Family Cornaceae

Eleven native species of dogwood are treated here for the SE. Only about 4 of these can be easily identified by twigs alone in winter. Branchlet, leaf, and fruit details are needed for identification of the rest. All have terminal buds, 3 bundle scars, and only 1 differs in not having opposite leaves and naked or 2-scaled buds. Fruit of all dogwoods is a drupe, red in 2 species but bluish or white in the others.

1. Leaf scars alternate or subopposite; terminal buds imbricate-scaled   1. *C. alternifolia*
1. Leaf scars opposite; terminal buds naked or valvate-scaled
   2. Shrub diminutive, rarely over 20cm high   2. *C. canadensis*
   2. Shrub taller, or treelike
      3. Twigs with stalked terminal flower buds; lateral buds not evident; jointed appearance of twigs   3. *C. florida*
      3. Twigs not as above
         4. Buds and twigs smooth and glabrous; large, brown or purplish spots evident on the green twigs   4. *C. rugosa*
         4. Buds or twigs bear some hairs, no spots; twigs more commonly red or reddish on 1 side; includes 7 dogwoods that cannot be separated by twigs alone:
            5. Pith of branchlets brown
               6. Leaves rough (scabrate) on top surface, woolly-hairy below; habit tall, slender   5. *C. drummondii*
               6. Leaves smooth to slightly pubescent above, not scabrate; hairs appressed below; habit shrubbier
                  7. Leaf undersides with whitish hairs; leaves under 4cm wide; cymes elongate, fruit white   6. *C. racemosa*
                  7. Leaf undersides with brown or reddish hairs; leaves often over 4cm wide; cymes rounded, fruit blue   7. *C. amomum*
            5. Pith of branchlets white
               8. Leaves rough (scabrate) on top surface, obviously pubescent beneath
                  9. Twigs rarely scabrous; leaf undersurface hairs whitish; fruit white, 6mm or less; w. species   8. *C. drummondii*
                  9. Twigs commonly scabrous; leaf undersurface hairs gray or brownish; drupes blue, to 8mm; e. species   9. *C. asperifolia*

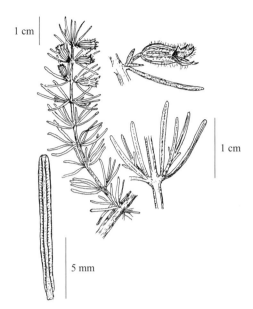

Figure 130. *Conradina canescens*

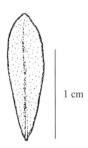

Figure 131. *Conradina etonia*

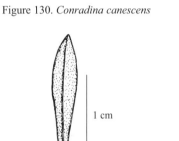

Figure 132. *Conradina glabra*

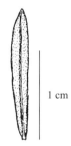

Figure 133. *Conradina grandiflora*

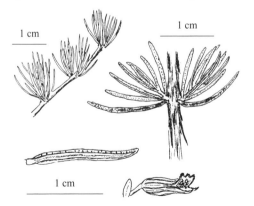

Figure 134. *Conradina verticillata*

    8. Leaves smooth, not scabrous above, obscurely pubescent to glabrous beneath
        10. Leaves light green beneath; cyme round-topped; fruit blue; habit tall, slender
            10. *C. foemina*
        10. Leaves whitened beneath; cyme flat-topped; fruit white; habit shrubby
            11. *C. sericea*

1. *Cornus alternifolia* L. f.   ALTERNATE-LEAF DOGWOOD   Figure 135

A small tree, to 8m, common in the Appalachian region but sporadic across the Piedmont and coastal plain from VA to FL and west to MS, AR. The bark is thin, smooth, green at first, becoming gray. Inner bark not pleasantly aromatic. Fruit blue-black on red stalks.

2. *Cornus canadensis* L.   BUNCHBERRY

A dwarf shrub, to 3dm tall and spreading by rhizomes that send up unbranched stems that usually die back after flowering and fruiting. A n. species barely reaching the SE in WV. Not tolerant of the higher soil temperatures over the course of the growing season in the SE.

3. *Cornus florida* L.   FLOWERING DOGWOOD   Figure 136

A small tree widespread over most of the SE and well known for its showy spring flower bracts. Common in the understory of upland hardwood and oak-pine forests. Plants in more humid habitats suffer from attack by the dogwood anthracnose, a fungal disease. Bark fairly thin, brown, blocky.

3. *Cornus rugosa* Lam.   ROUND-LEAF DOGWOOD

A shrub, to 3m tall, barely reaching the SE in rocky uplands and mountains of WV, VA. The smooth, green twigs with purplish spots are distinctive of this n. species of dogwood.

4. *Cornus drummondii* C. A. Meyer   ROUGH-LEAF DOGWOOD

A tall shrub or small tree, to 10m. Primarily a midwestern species that is limited in the SE to states west of the Appalachians and AL. Habitat is variable, from rocky, dry, calcareous soils to moist bottomlands. Pith of branchlets may be white or brownish. Fruits usually 5–6mm diam. The form *priceae* (Small) Rickett of KY and TN may have 3–4mm fruit and very prolonged leaf tips.

5. *Cornus racemosa* Lam.   GRAY DOGWOOD   Figure 137

A shrub, to 3m tall, with flower and fruit clusters (cymes) longer than they are broad. A n. species, reaching the SE in MD and west across VA, WV, KY, MO. Occasionally recorded from more s. stations, but sometimes cultivated, and naturalization may be involved. Often sprouts from the roots, forming thickets.

6. *Cornus amomum* P. Mill.   SILKY DOGWOOD   Figure 138

A shrub, to 3m, usually clumplike and multistemmed. Distributed across much of the SE, though less common in the coastal plain than in areas inland. Typically in moist to wet soils, especially streambanks and swampy thickets. The var. *obliqua* (Raf.) J. S. Wilson has leaves that are paler beneath and have more prolonged tips and more tapered bases than are typical for the species; this variety found mostly from WV, KY to AR, OK.

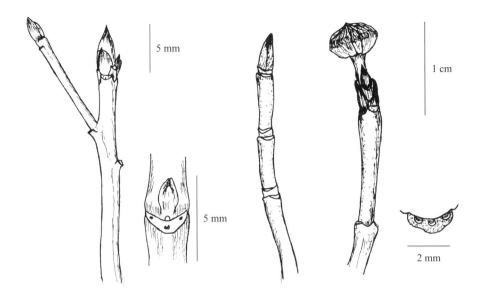

Figure 135. *Cornus alternifolia*

Figure 136. *Cornus florida*

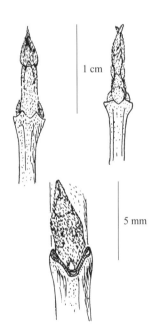

Figure 137. *Cornus racemosa*

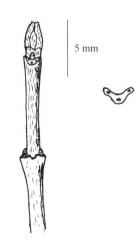

Figure 138. *Cornus amomum*

7. *Cornus asperifolia* Michx. COASTAL ROUGH-LEAF DOGWOOD

A slender shrub, to 3m, most similar to *C. drummondii* but an e. species found in the outer coastal plain from NC to FL and west to AL. Of scattered distribution in moist to wet bottomlands, floodplains, swamps.

8. *Cornus foemina* P. Mill. SWAMP DOGWOOD

A slender shrub or small tree, to 5m, widespread across most of the SE, but uncommon out of the coastal plain and absent from much of the Appalachian region. Typical of moist soils of lowlands such as swamps, streamsides, wetlands, bottomlands. (*C. stricta* Lam.)

9. *Cornus sericea* L. RED-OSIER DOGWOOD

A stoloniferous shrub forming thickets to 2m high. A n. species ranging southward to WV, in wet soils of bogs, lakeshores, and swampy streamsides. The stems are bright red, inducing extensive cultivation as a "winter color" plant. Locally naturalized on occasion, in some parts of the upper SE, on account of nearby cultivation. (*C. stolonifera* Michx.)

Corylus L. HAZELNUT Family Betulaceae

Two native species occur in the SE, both shrubs. Twigs lack terminal buds; catkin buds present on twigs of mature stems; pith angled in cross section. Fruit is a nut enclosed in a leafy husk.

1. Catkin buds stalked; lateral buds blunt, often with 4 or more bud scales; fruit husks ragged at apex   1. *C. americana*
1. Catkin buds sessile; lateral buds pointed, usually 3- to 4-scaled; fruit husks drawn out into an elongate beak   2. *C. cornuta*

1. *Corylus americana* Walt. AMERICAN HAZELNUT *Figure 139*

A clumplike shrub, to 3.5m tall, with glandular hairs on new growth. Widespread across most of the SE, except uncommon in the coastal plain from NC to GA, and absent from FL and the Gulf coastal plain. Often seen in moist soils near streams, bottomlands, or rocky uplands with basic or calcareous soils. The tips of the nuts are visible within husks of mature fruit.

2. *Corylus cornuta* Marsh. BEAKED HAZELNUT *Figure 140*

A low shrub, usually about 1m tall or less, forming colonies by root sprouts. Distributed across the n. half of the SE, mostly from the VA and NC Piedmont west to MO. Occurs south in the Appalachian region to GA. Often seen in dry woodlands and rocky slopes. The elongate husks are densely covered by short, stiff hairs that easily detach and penetrate/irritate skin if fruits are handled. This species ranges to w. Canada and CA, where it grades into var. *californica* (A. De Candolle) Sharp.

Cotinus obovatus Raf. AMERICAN SMOKETREE Family Anacardiaceae *Figure 141*

This small native tree is not common in the SE but may be locally abundant over limestone in certain regions. It is known from calcareous, rocky uplands from nw. GA to OK, with major range segments in TN, AR, MO. Also occurs in central TX. The lenticellate twigs bear terminal buds with several scales and an aromatic inner bark. The bundle scars may appear as 3, or in 3 groups. The seed is borne in a papery sac on female trees only (plants dioecious). The name refers to the billowy inflorescences, with many hairy and threadlike bracts. Bark gray, scaly.

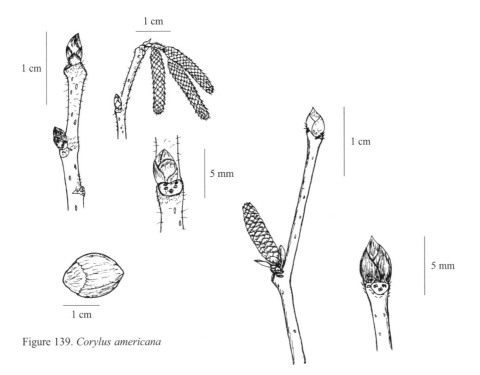

Figure 139. *Corylus americana*

Figure 140. *Corylus cornuta*

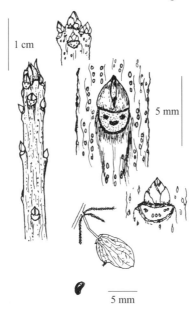

Figure 141. *Cotinus obovatus*

Crataegus L. HAWTHORN Family Rosaceae *Figures 142, 143, and 144*

The complex of hawthorn species in the SE may, depending on taxonomic viewpoint, number from about 23 to more than 400 species. In this text 31 are treated as species, though several of these may warrant varietal status or synonymy after closer study. Conversely, some entities not treated here as distinct species may deserve that rank after similar study. Many hawthorns described as species between 1899 and 1910 are poorly understood, and many perhaps represent clinal variation, hybridization, or aberrant clonal populations of widely variable species. Regardless of the taxonomic difficulties of species recognition, these. hawthorns are relatively easy to assign to intergeneric series. The species discussed after the key are thus grouped in their respective series, which are presented alphabetically.

Hawthorns are small deciduous trees or sometimes shrubs with nodal thorns and small pomes (5–25mm) as fruit. All have blunt, imbricate-scaled buds in winter and white flowers in spring. No species can be reliably identified in winter by twigs alone, though collectively as a genus they are easily recognized. Identification of many species requires careful study of flowers, leaves, and fruit. In winter identification, important features to note include bark, growth habit, average thorn appearance, and presence of remnant fruit. Fallen or dried leaves or fruits are valuable clues worth searching for, under or on specimens in the field. Brittle leaves and fruits can be soaked in water to reclaim flexibility. The calyx lobes and number of remnant stamens are important fruit characters, and in 2 cases the seed are diagnostic. Winter twigs that may bear hairs occur in the species *ashei, berberifolia, flava, harbisonii, marshallii, mollis, triflora*, and *uniflora*. Others may have hairs on twigs early in the growing season. Hawthorns should be expected in a wide range of habitats, most particularly in disturbed lands, grazed pastures, and fencerows, and under sparse forest canopies. The following key to species necessitates use of leaves and/or fruit.

1. Leaves tapered at base, acute to cuneate (a few from elongating shoots may have rounded bases); petioles less than 1cm long, or winged to within 1cm of base 2
1. Leaves wide at base, mostly rounded, flattened, or cordate (a few may have an abruptly tapered junction into petiole); petioles over 1cm, excluding winged portion near blade that may be 5mm or less 25
2. Bark of mature and older trunks gray, reddish-mottled (reddish or red-brown inner bark revealed by curling outer gray scales) 3
2. Bark of mature trunks scaly or furrowed, not mottled 7
3. Leaves spatulate, often 3-lobed at tip; few or no thorns (ser. *Microcarpae*) 17. *C. spathulata*
3. Leaves not spatulate, unlobed or lobed along sides; thorns usually present 4
4. Petiole over 10mm long; fruits mature Aug–Oct (ser. *Virides*) 31. *C. viridis*
4. Petiole 2–8mm, or winged to base; fruits in May–June (ser. *Aestivales*) 5
5. Leaves mostly elliptic, 5–7cm long; bark scales mostly light gray; native to areas west of FL 2. *C. opaca*
5. Leaves mostly obovate, 2–5cm long; bark scales often olive-brown or brownish-gray; mostly FL and eastward in range 6
6. Leaves dull above, with reddish hairs below *C. rufula* (1a)
6. Leaves glossy above, glabrous or with grayish or reddish hairs only in vein axils 1. *C. aestivalis*
7. Fruit with an elevated calyx, the calyx lobes borne on perimeter of a collar or neck on end of fruit 8
7. Fruit calyx sessile 10
8. Leaves often 3cm or less in length (ser. *Parvifoliae*) 19. *C. uniflora*
8. Leaves over 3cm 9
9. Twigs slender, purplish; leaves of longer twigs deeply lobed; fruit mostly subglobose; bark brown and deeply fissured on mature trunks; usually 20 stamens
(ser. *Pulcherrimae*) 21. *C. pulcherrima*
9. Twigs red-brown or yellow-brown; leaves mostly shallowly lobed; fruit longer than it is broad; bark gray to brownish, scaly, or fissured only on lower parts of old trunks; usually 10–15 stamens (ser. *Intricatae*) 13. *C. intricata*

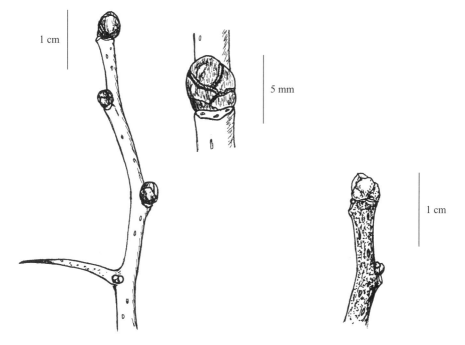

Figure 142. *Crataegus*

Figure 143. *Crataegus*

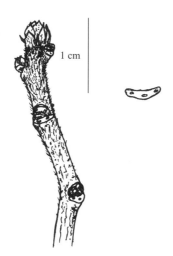

Figure 144. *Crataegus*

10. Bark deeply furrowed or blocky, not scaly, or leaves mostly under 3cm long   11
10. Bark scaly, at least on part of trunk; leaves over 3cm long   13
11. Leaves with few or no glands on petiole/blade margin; calyx lobes persistent on fruit, deeply toothed and foliaceous; spines very slender, straight
(ser. *Parvifoliae*) 19. *C. uniflora*
11. Leaves glandular on margin of petiole/blade junction; calyx lobes not as above; spines stout (ser. *Flavae*)   12
  12. Twigs slender, geniculate or weeping; principal leaf veins 3 or 4 per side
12. *C. lacrimata* group
  12. Twigs not as above; leaf veins 5 or 6 per side   11. *C. flava* group
13. Leaves conspicuously glandular along margin of petiole/blade junction, pubescent on both surfaces (use lens for old leaves); calyx lobes deeply glandular-serrate (ser. *Triflorae*)   14
13. Leaves not glandular, or if barely glandular then leaves pubescent only below, to glabrous; calyx lobes not as above   16
  14. Leaves thin; twigs usually hairy; 30 or more stamens   28. *C. triflora*
  14. Leaves firm or stiff; twigs usually glabrous in winter; 20 stamens   15
15. Largest leaves mostly under 8cm long, base distinctly cuneate, margin rarely lobed; twigs slender; fruit soft, succulent; calyx lobes usually persistent; habit shrubby or multistemmed
29. *C. ashei*
15. Largest leaves often 8–10cm, base short-cuneate, margin often shallowly lobed; twigs rather stout; fruit dense, not succulent; calyx lobes semipersistent; habit usually treelike
30. *C. harbisonii*
  16. Principal lateral veins of leaves run to sinuses as well as lobe points (if unlobed, teeth crenate)   17
  16. Principal lateral veins of leaves run only to lobe points   18
17. Leaves hairy on veins below; spines to 4cm; fruit red, ripe May–June
(ser. *Aestivales*) 2. *C. opaca*
17. Leaves glabrous; spines under 2cm; fruit blue-black, ripe Aug–Sept
(ser. *Brevispinae*) 4. *C. brachyacantha*
  18. Fruit soft, flesh succulent; seed hollowed, pitted, or channeled on inner side (ser. *Macracanthae*)   19
  18. Fruit not soft, flesh dry or mealy; seed plane on inner side   20
19. Leaves usually glabrous, veins impressed on top surface, prominent beneath; fruit subglobose   15. *C. succulenta*
19. Leaves pubescent beneath, veins not conspicuously impressed above; fruit usually oblong
16. *C. calpodendron*
  20. Leaves rhombic to obovate, often as wide as they are long; calyx lobes broadly triangular
(ser. *Rotundifoliae*) 24. *C. margaretta*
  20. Leaves mostly obovate, ovate, or elliptic, longer than wide; calyx lobes elongate   21
21. Leaf veins sunken on upper surface of leaves, prominent beneath (ser. *Punctatae*)   22
21. Leaf veins obscure or at least not as above   23
  22. Leaves usually glabrous; petioles not glandular; calyx lobes glabrous   22. *C. punctata*
  22. Leaves usually pubescent beneath; petioles with a few glands; calyx lobes pubescent
23. *C. collina*
23. Leaves often 3cm wide or more, pubescent beneath   (ser. *Molles*) 18. *C. mollis* forms
23. Leaves mostly less than 3cm wide or glabrous (ser. *Crus-galli*)   24
  24. Leaves 4–7cm long; twigs and leaves glabrous   7. *C. crus-galli*
  24. Leaves mostly less than 4cm long; twigs or leaf undersides hairy   8. *C. berberifolia*
25. Bark of mature and older trunks gray, reddish-mottled (reddish or red-brown inner bark revealed by curling outer gray scales)   26
25. Bark not mottled   27
  26. Twigs hairy; leaves deeply incised-lobed, hairy beneath; fruit ellipsoid
(ser. *Apiifoliae*) 3. *C. marshallii*
  26. Twigs glabrous; leaves not as above; fruit subglobose   (ser. *Virides*) 31. *C. viridis*

27. Leaves conspicuously glandular along margin of petiole/blade junction, pubescent on both surfaces (use lens for old leaves); calyx lobes deeply glandular-serrate 28
27. Leaves not glandular, or if barely glandular then leaves pubescent only below, or glabrous; calyx lobes not as above 30
  28. Calyx tube elevated on fruit; thorns slender; usually 10 stamens
                                                                                             C. biltmoreana (13a) (ser. Intricatae)
  28. Calyx sessile; thorns stout; stamens 20 or more (ser. Triflorae) 29
29. Leaves thin; fruit soft, succulent; habit usually shrubby or multistemmed; often 30 or more stamens 28. C. triflora
29. Leaves thick; fruit hard or dense; habit single-stemmed or treelike; 20 stamens
                                                                                                            29. C. harbisonii
  30. Larger leaves 5–12cm long but most under 8cm; fruit usually 1cm or less diam. 31
  30. Larger leaves 7–16cm, many over 8cm; fruit over 1cm long or thick 39
31. Fruit with an elevated calyx 32
31. Fruit with sessile calyx 34
  32. Habit usually shrubby; stamens 10; petiole often winged for half its length from blade
                                                                             (ser. Intricatae) 14. C. boyntonii
  32. Habit rarely shrubby; stamens 20; petiole rarely winged more than ⅓ length from blade 33
33. Bark brown, deeply fissured on mature trunks; twigs slender, purplish
                                                                                (ser. Pulcherrimae) 21. C. pulcherrima
33. Bark gray, scaly; twigs moderately stout, red-brown (ser. Pruinosae) 20. C. pruinosa
  34. Main lateral leaf veins run to sinuses as well as lobe points; fruit 2–6mm diam., wider than long (ser. Cordatae) 6. C. phaenopyrum
  34. Main lateral leaf veins run only to lobe points; fruit 7–8mm long, longer than wide 35
35. Leaves rhombic to obovate, rounded at base only on some twigs; widest beyond middle of leaf blade 36
35. Leaves mostly ovate to deltoid, base rounded, truncate, or cordate on most twigs; widest below midpoint of blade 37
  36. Leaf veins impressed and reticulate above; marginal teeth acute; seed channeled on inner side (ser. Macracanthae) 15. C. succulenta
  36. Leaves not as above; marginal teeth obtuse; seed plane on inner side
                                                                               (ser. Rotundifoliae) 24. C. margaretta
37. Stamens 15–20 (ser. Tenuifoliae) 27. C. schuettii
37. Stamens 5–10 38
  38. Fruit soft, fleshy, red, usually obovoid or oblong; leaves glabrous, lobe tips acuminate and slightly reflexed (ser. Tenuifoliae) 26. C. macrosperma
  38. Fruit hard, dry or mealy, green or red-blushed, subglobose; leaves with sparse hairs or tufts in vein axils, lobe tips mostly acute, not reflexed (ser. Silvicolae) 25. C. iracunda
39. Petioles, fruit stalks, and veins of leaf undersides distinctly hairy; leaf teeth mostly acute
                                                                                   (ser. Molles) 18. C. mollis
39. Petioles, fruit stalks glabrous or nearly so; leaf undersides sparsely hairy to glabrous; leaf teeth mostly acuminate 40
  40. Calyx lobes evenly glandular-serrate; leaves longer than broad, base abruptly narrowed to truncate; stamens 5–10 (ser. Coccineae) 5. C. coccinea
  40. Calyx lobes irregularly serrate; leaves wide as long or wider, base rounded to subcordate; 20 stamens (ser. Dilatatae) 41
41. Petioles often hairy near blade; leaves slightly hairy; trunk usually long, straight; fruit rarely angular 9. C. dilatata
41. Petioles glabrous; leaves glabrous; trunk usually short or with multiple stems; fruit slightly angular 10. C. coccinioides

Ser. *Aestivales* (Sarg.) Schneider: Two species, both called "mayhaws" and both typically found in wet sites in freshwater lowlands of the outer coastal plain. Fruits are shiny, red, ripening from May–June, with sour, succulent flesh. Bark scaly, mottled in the 2 easternmost species, less so in *C. opaca*.

1. *Crataegus aestivalis* (Walt.) Torr. & Gray. To 10m, ranging from s. VA to n. FL, se. AL. Leaves mostly obovate, lustrous above. The similar entity *C. rufula* Sarg., known as "rufous mayhaw," is found in w. FL, sw. GA, se. AL. Its leaves are usually dull green above and overall pubescent below, with dense reddish hair tufts in vein axils; may only be a variety of *C. aestivalis*.

2. *Crataegus opaca* Hook & Arn. To 10m, from nw. FL to e. TX, s. AR. Leaves mostly elliptic, dull green above, often with hairs or hair tufts in vein axils beneath; margins obscurely crenate-toothed. Called "western mayhaw" or "riverflat haw."

Ser. *Apiifoliae* (Loud.) Rehd.: A single species.

3. *Crataegus marshallii* Eggl. To 8m, typically found in moist soils of bottomlands and streamsides of the coastal plain and Piedmont from se. VA to e. TX and north to se. MO. Occasional in more upland sites. Bark scaly, mottled; the trunk slender, single or multiple. Leaves thin, deeply incised-lobed; the main side veins run to sinuses as well as lobe points. Fruits ellipsoid, 5–9mm long; flesh soft. Twigs usually hairy; thorns 1–3cm or lacking. Called "parsley haw."

Ser. *Brevispinae* (Beadle ex Schneider) Rehd.: A single species.

4. *Crataegus brachyacantha* Sarg. & Engelm. To 12m, in moist soils of LA to e. TX north to sw. AR. Leaves glossy, with conspicuous vein reticulation, crenate teeth, and winged petioles; when lobed, the leaves have main side veins running to sinuses as well as lobe points. Fruit nearly black but with a waxy bloom that renders them blue-colored, to 13mm diam.; flesh not sweet. Thorns 9–20mm or lacking, occasionally terminating spur shoots (unusual for hawthorns). Bark gray, very scaly. Called "blueberry haw."

Ser. *Coccineae* (Loud.) Rehd.: A single species in the SE.

5. *Crataegus coccinea* L. A small tree, to 8m in the SE, usually with a long, single trunk and oblong, open crown. Common in the NE, reaching the SE in the Appalachians and adjacent Piedmont regions of WV, VA, and n. NC. Leaves rather large, 6–15cm, shallowly lobed, slightly pubescent beneath. Fruit usually longer than it is broad, 10–18mm long; flesh soft; calyx lobes evenly serrate above base. Bark gray, scaly. Called "scarlet haw."

Ser. *Cordatae* Beadle ex Eggleston: A single species.

6. *Crataegus phaenopyrum* (L. f.) Medic. A small tree, to 8m, with a preference for rich soils of uplands from PA to n. FL, west to AR, MO, IL. Widely cultivated, and naturalizing in areas where it may not have been native. Leaves lobed, often rather maplelike, glossy above, with main side veins running to sinuses and lobes. Fruit 4–6mm diam., red, lustrous, persist into winter. Twigs glabrous. Bark gray, scaly. Commonly called "Washington hawthorn." [*C. cordata* (Mill.) Ait., *C. youngii* Sarg.]

Ser. *Crus-galli* Loud.: More than 80 species for the SE have been described from this group; only 2 are considered here. Both have oblanceolate to obovate leaves, mostly toothed and unlobed, though some may be shallowly lobed on sprouts or elongating twigs.

7. *Crataegus crus-galli* L. A small tree, to 10m, with a broad crown of stiff, wide-spreading branches. Common across the SE in a wide variety of habitats but mostly in rich or moist soils. Leaves, fruit stems, and twigs glabrous. Fruit 8–12mm diam.; flesh hard or dry, rarely sweet. Thorns 3–10cm on twigs, to 15cm on trunk. Called "cockspur hawthorn." The variety *pyra-*

canthifolia Ait. has leaves only 1–1.5cm wide. The thornless var. *inermis* is more commonly cultivated.

8. *Crataegus berberifolia* T. & G. A shrub or small tree, to 8m, with a broad crown. Native to the coastal plain from NC to e. TX and north to w. KY, s. IL. Most common from s. AL to e. TX, and typical of moist lowlands in LA and TX. Grows on dry uplands and calcareous soils in other parts of the range. Leaves, fruit stems, and young twigs hairy. Fruit 6–10mm diam. The originally described species of *C. berberifolia* in TX and LA may be considered a different species from the more widely dispersed *C. engelmanii* Sarg., but the latter is treated here as a synonym. Called "barberry haw."

Ser. *Dilatatae* (Sarg.) Palmer: Two species in the SE, both rare and of limited range. Leaves of both are lobed, as wide or wider than they are long, with a truncate to subcordate base. Fruits red, usually 10–23mm wide; flesh thick, succulent, sweet; calyx lobes deeply serrate, their bases enlarged and reddened.

9. *Crataegus dilatata* Sarg. A small tree, to 7m, with a tall, straight trunk, open crown, and wide-spreading branches. Reported from a few locations in KY and OH in rich or calcareous soils. Leaves usually pubescent. More common in the NE. Called "broadleaf haw."

10. *Crataegus coccinioides* Ashe. A small tree, to 6m, with a dense, wide crown and short trunk. Leaves glabrous, often glandular on the petiole. Fruit often wider than long, 13–23mm, slightly angular. Native to dry or calcareous hills and streamsides from s. IL to se. KS, OK, and n. AR. Called "Kansas hawthorn."

Ser. *Flavae* (Loud.) Rehd.: Two species are treated here as representative of 2 extremes of variation in this taxonomically difficult group. About 90 species have been described, but most of these are poorly understood and often inseparable. Nearly all entities in this group favor xeric woodlands and deep, sandy soils. Petioles are distinctly glandular. Bark is furrowed or blocky, thicker than other hawthorns, giving more protection in the fire-prone habitats these plants occupy.

11. *Crataegus flava* Ait. This species originally was described from a cultivated specimen in Europe, presumed native of the coastal plain from NC to FL. Exact matches in the wild are rare, but similar entities occur sporadically west to s. AL and north to the Piedmont and Blue Ridge of AL, NC. Leaves mostly widest near middle, ovate or broadly elliptical, sometimes obovate. Twigs tend to point upward, or at least are not slender and drooping. Fruit pyriform, yellow; stamens 10; anthers purplish. Several named species that are sufficiently different from the originally described species of *C. flava* may be worthy of recognition, but none are separable in winter and are here considered in synonymy. Of these, fruit is generally 10–13mm diam., yellow to orange-red; flesh dense to mealy, soft after maturity. Called "yellow haw." Ranging to the Appalachians and with only 10 stamens are *C. aprica* Beadle and *C. alleghaniensis* Beadle, the former with more globose fruit and yellow anthers; the latter with subpyriform fruit and purple anthers. Similarly ranging to the s. Appalachians and with 12–20 stamens and purple anthers are *C. extraria* Beadle, with ellipsoid fruit, and *C. ignava* Beadle, with subglobose fruit.

12. *Crataegus lacrimata* Small. Range is similar to above species, but much more common in the sandy coastal plain soils. Conspicuous and plentiful on sand ridges from NC to AL. Leaves mostly obovate to spatulate, widest beyond middle (though juvenile twigs may have nearly circular leaves). Twigs slender, zigzag (geniculate), drooping or weeping on mature plants. Fruit yellow to red, 10–20mm wide; flesh dry to soft and sweet, with considerable variation among types. The original *C. lacrimata* was described from w. FL only, but there are so many similar, closely related entities across the SE that all are here grouped under the older name of *C. lacrimata,* for simplification; they are not separable in winter. All are "weeping hawthorns."

Including *dispar* Beadle, *floridana* Sarg., *impar* Beadle, *lanata* Beadle, *lepida* Beadle, *meridiana* Beadle, *michauxii* Pers., *munda* Beadle, *ravenellii* Sarg., *senta* Beadle, this group of hawthorns has over 80 species earlier described, mostly by Beadle.

Ser. *Intricatae* (Sarg.) Rehd.: A complex and poorly understood series of the SE, with many similar entities that cannot be determined in winter. The 2 species treated here are interpreted broadly, inclusive of about 33 species previously named for the region. Most members of this group have glabrous, glandular leaves (especially when young) and hard or dry fruit with an elevated calyx. These are upland hawthorns, mostly in rocky woods, hills, pastures, and cutover lands.

13. *Crataegus intricata* Lange. Mostly shrubs, to 3m. Widespread across the SE, though most commonly seen in the Appalachian region and rocky uplands west to MO, AR. Wide variation in leaf shape and fruit, though leaves are mostly tapered at base (sometimes rounded on juvenile growth), with a slightly winged and glandular petiole. Thorns are slender and straight or, more rarely, stout and slightly curved. Fruit greenish to red-blushed or rarely red or yellow; calyx lobes glandular; anthers usually 10, rarely 20. Among several forms that may warrant species status, but are here considered in synonymy, are *C. biltmoreana* Beadle (13a) (leaves and stems of fruit hairy; calyx persistent, conspicuously gland-toothed), *C. straminea* Beadle (fruit oblong or pyriform, yellow, short-stemmed; leaves mostly elliptic), *C. rubella* Beadle (leaves mostly ovate or elliptic, tip and base acute, margins weakly lobed; fruit pyriform, red), *C. communis* Beadle (leaves oval; fruit globose), and *C. sargentii* Beadle (leaves ovate, often shallowly incisely lobed; fruit globose with 20 stamens, of nw. GA, adjacent AL, TN).

14. *Crataegus boyntonii* Beadle. Shrubby, or a small tree, to 8m, mostly in Appalachian and Piedmont regions from WV, VA to AL. Leaves glabrous, rather stiff, mostly broadly ovate, lobed, the base short-cuneate to truncate. Fruit longer than it is broad, angled, green to red-blushed; flesh hard and dry; calyx lobes mostly lack glands; stamens 10. A similar entity is *C. buckleyi* Beadle, of the s. Appalachians and foothills, with mostly ovate leaves and globose fruit with glandular calyx lobes.

Ser. *Macracanthae* (Loud.) Rehd.: Two species in the SE, both uncommon and more typical of regions to the north. The seed have deep channels or pitlike depressions on their inner side and are 2 or 3 per fruit; fruit flesh soft after full maturity, succulent though thin; fruits usually borne upright, persisting into winter.

15. *Crataegus succulenta* Schrad. ex Link. A shrub or small tree, to 6m, often with multiple stems in the SE or crown wider than height. Known to extend southward to the Appalachians of NC, west in uplands to MO. Leaves firm and glabrous when mature; upper surface dark green and lustrous with conspicuous impressed veins. Fruit subglobose, 6–10mm wide, red. Twigs usually dark reddish-brown; thorns lustrous, 4–9cm. Called "fleshy haw." Includes *C. neofluvialis* Ashe.

16. *Crataegus calpodendron* (Ehrh.) Medic. A slender shrub or a small tree, to 6m, often with multiple and leaning stems. Sporadic in the Appalachian region and uplands west to MO; also seen occasionally in the Piedmont of NC to AL in calcareous soils or on rich slopes and streamsides. Rare in the coastal plain of GA to MS. Leaves thin, dull green, pubescent, veins not as conspicuous as in above species; petiole winged well below blade. Fruit oblong or rarely subglobose, to 12mm, orange-red. Twigs often hairy, yellow-brown; thorns grayish, 3–6cm. [*C. tomentosa* L, *C. chapmanii* (Beadle) Ashe]

Ser. *Microcarpae* (Loud.) Rehd.: A single species.

17. *Crataegus spathulata* Michx. A shrub or small tree, to 8m, mostly in the coastal plain from s. VA to FL, west to TX, north to s. MO. Prefers moist soils, but tolerant of drier uplands. Uncommon in the Piedmont from SC to AL. Trunk slender, often leaning or multiple; bark

scaly, mottled. Leaves usually less than 3cm wide, mostly spatulate, petiole winged to near base. Fruit rarely over 7mm diam. Thorns 2–5cm, often lacking on mature growth. Called "littlehip hawthorn."

Ser. *Molles* (Beadle ex Sarg.) Rehd.: A single species is treated here for the SE, inclusive of about 33 species previously described for the region.

18. *Crataegus mollis* Scheele. A small tree, to 12m, typically in moist lowlands such as bottomlands and stream valleys and in calcareous soils of more upland sites. Sporadic in n. VA, WV mostly west of Appalachians and w. AL to central TX and northward. Most common in TX and Mississippi embayment area in the SE. Leaves pubescent beneath and on petiole, 5–12cm long, rarely glandular, usually lobed (barely lobed in *C. meridionalis* Sarg.), mostly widest below middle of blade, base rounded. Fruit 13–25mm; flesh variable from dry and dense to soft and sweet. Bark of older trunks usually brownish, furrowed; scaly when young. Variant forms with leaves more often short-cuneate at base and widest at middle of blade include *C. texana* Buckley of e. TX and *C. meridionalis* of s. AL, MS. The entity *C. viburnifolia* Sarg. of e. TX has yellow fruit and cream-colored anthers. All are known as "downy hawthorn."

Ser. *Parvifoliae* (Loud.) Rehd.: A single species.

19. *Crataegus uniflora* Muenchh. Mostly a shrubby species, occasionally a small tree, inhabiting dry woods and rocky or sandy soils throughout the SE. Leaves mostly obovate, often under 3cm long, glossy above, pubescent below; petiole 2–5mm. Fruit solitary to 3 per twig, 7–12mm diam.; flesh hard or dry; calyx lobes persistent, foliaceous, deeply serrate. Twigs usually hairy; thorns 1–6cm, very slender and straight. Called "one-flowered hawthorn."

Ser. *Pruinosae* (Beadle ex Sarg.) Rehd.: A single species treated here for the SE, inclusive of about 38 named species.

20. *Crataegus pruinosa* (Wendl. f) K. Koch. A tall shrub or small tree, to 10m, with an open crown. A common n. species, ranging southward in the Appalachians to AL and west in uplands to MO. A s. form (*C. gattingeri* Ashe) extends into the Piedmont and coastal plain from VA to LA. Leaves lobed, widest near the base, occasionally more short-cuneate and widest near the middle, glabrous (except hairy in var. *virella* [Ashe] Kruschke). Fruit subglobose, green to red, often with a waxy bloom (pruinose); calyx elevated; flesh hard or dry; stamens usually 20. Called "frosted hawthorn."

Ser. *Pulcherrimae* (Beadle ex Palmer) Robertson: A single species is treated here as a broadly defined entity that includes at least 15 named taxa in the SE. Some of these taxa may deserve separate species status, but none can be separated in winter. Nearly all have leaves with evenly proportioned lobes, shallow to deeply incised, with straight, parallel side veins not conspicuously reticulate between. Twigs and branchlets tend to be slender, purplish-brown in color, and glabrous. The fruit is 7–14mm diam., with hard or dry flesh and elevated calyx. Bark of mature trunks is conspicuously brown, fissured, not scaly. This group is glandular and glabrous in spring, closely related to ser. *Intricatae*.

21. *Crataegus pulcherrima* Ashe. A small tree, to 8m, with a slender trunk or multiple, often leaning stems. An understory species of rich hardwood and hardwood-pine forests in the coastal plain, mostly in sw. GA, n. FL west to LA and sporadically north to nw. central GA and adjacent AL. Various forms may be locally common while absent over wide intervening areas. The entities *C. ancisa* Beadle and *C. mendosa* Beadle are known to occur in the n. parts of the range to the foothills of the Appalachians. Other named entities included in synonymy here are *alma* Beadle, *austrina* Beadle, *contrita* Beadle, *illustris* Beadle, *opima* Beadle, *robur* Beadle. The specific name *pulcherrima* means "beautiful."

Ser. *Punctatae* (Loud.) Rehd.: Two species are treated here for the SE. Both have mostly obovate leaves with conspicuously impressed veins on the top surface and toothed, relatively weakly lobed margins.

22. *Crataegus punctata* Jacq. A small tree, to 12m, most common north of the se. region but ranging down the Appalachians to GA, mostly at higher elevations in the Blue Ridge. Also known from calcareous regions from MO to AR, TX. Some TX forms may deserve separate species status. Leaves typically glabrous, not glandular except when young. Fruit 2–22mm diam., red or rarely yellow, dotted; flesh thick, dense or mealy, not sweet. Thorns 3–8cm, often branched on trunk. Crown is wide and flattened. Called "dotted hawthorn."

23. *Crataegus collina* Chapm. A small tree, to 8m, scattered from sw. VA to central GA, west to e. TX and areas north. Occurs on hills and uplands in acidic or calcareous soils, and in moist lowlands. Widespread but rarely seen in high numbers in any 1 location. Crown tends to be oblong, open; trunk fairly straight; sometimes shrubby on dry sites. Leaves usually pubescent beneath and glandular on petiole. Fruit 8–12mm wide, greenish or red; flesh hard and dry, or mealy, not sweet. Calyx lobes pubescent. One of the earliest hawthorns to flower in spring. Known as "hill thorn" in some areas.

Ser. *Rotundifoliae* (Eggleston) Rehd.: A single species included here for the SE. Members of this series are mostly n. or w. hawthorns not reaching the se. region.

24. *Crataegus margaretta* Ashe. A shrub or small tree, to 7m, very rare in the SE; reported from a few sites in MO, KY, WV in rocky soil. Leaves variable in shape but overall widest beyond middle of blade, shallowly lobed, with crenate serrations; glabrous and with slightly impressed veins above. Fruit 9–12mm, dull red; flesh thin, mealy.

Ser. *Silvicolae* (Beadle ex Sarg.) Palmer: A single species.

25. *Crataegus iracunda* Beadle. A tall shrub or small tree, to 10m, with an open, irregular crown. Sporadic but widespread over much of the SE. Known from moist lowlands and rich upland soils from VA to LA. Leaves mostly rounded or truncate at the base, as wide as long or wider, lobed, pubescent to nearly glabrous. Fruit 8–14mm long, greenish to red-blushed; flesh hard, dry; calyx small, not elevated; 10 stamens. Twigs hairy when young, glabrous or nearly so when mature.

Ser. *Tenuifoliae* (Beadle ex Sarg.) Rehd.: Two species are treated here for the SE, neither considered synonymous with the n. species *C. flabellata* (Bosch.) K. Koch, which is often done in other texts.

26. *Crataegus macrosperma* Ashe. A shrub or small tree, to 7m, fairly common in the Appalachian region from WV, VA to n. AL, and sporadically west to IL, AR. Trunk usually single, angular or sinewy, with oval crown. Leaves thin, glabrous (sparsely hairy above when young). Fruit longer than broad, 12–20mm long, usually dull or pruinose; flesh soft, sweet; 10 stamens.

27. *Crataegus schuettii* Ashe. Similar to above species, but mostly restricted to the Appalachians from WV, VA to n. GA. Leaves may retain a few sparse hairs on top surface after maturity and are more deeply lobed on juvenile growth. Fruit 10–14mm long, bright red; flesh thin, mealy to succulent; 20 stamens. (*C. basilica* Beadle)

Ser. *Triflorae* (Beadle) Rehd.: Three species in the SE, 2 from the closely related ser. *Bracteatae* (Palmer) Rehd. that are here considered too similar to warrant separation. All have conspicuously glandular leaves and calyx lobes, especially in spring, pubescent leaves, deeply serrate calyx lobes, and fairly large fruit, 12–23mm diam.

28. *Crataegus triflora* Chapm. Shrubby, usually multistemmed, but may reach 6m. Found sporadically from nw. GA to s. AL, west to e. LA in calcareous prairies, over limestone, and in rich, forested slopes and ravines. Leaves thin, dull green, 2–7cm long, slightly lobed, widest at middle or below. Fruit red; flesh soft, succulent; calyx lobes large, persistent. Stamens most often 30 or more. Called "three-flower haw." Includes *C. austromontana* Beadle, a type described from n. AL with 10 stamens.

29. *Crataegus ashei* Beadle. Shrub or small tree, usually multistemmed, to 6m. Rare and localized in calcareous prairies and clay soils of s. AL, s. and central MS, to e. LA, mostly associated with Black Belt and Jackson Prairie soils. Leaves 5–7cm, thicker than in above species, more singly toothed, rarely lobed, slightly glossy above, widest mostly just beyond middle of blade. Fruit red; flesh soft, succulent; calyx lobes usually persistent. Called "Ashe hawthorn."

30. *Crataegus harbisonii* Beadle. A small tree, to 8m, usually with a single trunk. Very rare; currently known only from the Nashville basin, though recorded from 4 other sites in TN. In calcareous or rich soils under thin hardwood forest canopies. Leaves 5–10cm, glossy above, thick, coarsely toothed, and very shallowly lobed, widest near middle of blade. Fruit orange-red; flesh dense, hard or mealy, not sweet; calyx lobes persistent or not. Called "Harbison hawthorn."

Ser. *Virides* (Beadle ex Palmer) Rehd.: A single species is considered here, inclusive of about 33 named entities for the area.

31. *Crataegus viridis* L. A small tree, to 12m, of the coastal plain and Piedmont from VA to FL, west to TX and north to IL. Usually in moist lowlands, but also seen in upland sites in calcareous soils. Often planted as an ornamental, especially the cultivar 'Winter King'. Trunk often long and sturdy, with scaly, mottled bark. Leaves variable, usually tapered at base. Fruit 5–8mm, often persistent into winter. All parts glabrous, though more variation in pubescence and fruit size and color is seen in TX. Called "green hawthorn." One horticulturally important entity, *C. nitida* (Engelm.) Sarg., is suspected to be of hybrid origin between *C. viridis* and *C. crus-galli*. It is native and reproducing along the Mississippi River lowlands from OH to AR; leaves thicker and glossier than in *C. viridis;* fruit 8–10mm diam. Known in cultivation as "glossy hawthorn."

Croton L.   CROTON   Family Euphorbiaceae

Three native species of these shrubs are woody enough to warrant inclusion here. All have naked terminal buds, 3 small bundle scars, and a reddish sap that oozes from cut tissues in the growing season. A granular, dotlike surface to the leaves and twigs comes from the presence of stellate trichomes and peltate scales that are silvery with a brownish center. Twigs are not very useful to separate species, but usually some leaves are persistent. Twigs may die back in winter freezes. Fruit is 3-lobed, capsular, with 3 grayish seed expelled from within. An additional species, *C. michauxii* Webster (*C. linearis* Jacq.), is predominately herbaceous, of sandy pinelands of the coastal plain from SC to TX but may survive the winter intact aboveground in areas of warm climate, such as FL. It has slender, unbranched stems, sessile leaves, and 1-seeded fruit.

1. Leaves mostly ovate; a sprawling shrub of dunes near the beach 1. *C. punctatus*
1. Leaves more elliptic to oblong; upright shrubs of other habitats
 2. Leaves over 2cm wide, often truncate at base; a branching shrub over 3dm tall
    2. *C. alabamensis*
 2. Leaves 15mm or less wide, usually obtuse or acute at base; slender, herblike habit under 3dm 3. *C. argyranthemus*

1. *Croton punctatus* Jacq.   BEACH-TEA   Figure 145

A shrub of beach dunes from NC to TX, also southward onto shores of tropical Amer. Leaf width about half its length; leaves 2–6cm long, 1–4cm wide.

126 · Croton

2. *Croton alabamensis* E. A Smith ex Chapm.   ALABAMA CROTON   *Figure 146*

A rare shrub of rocky streamsides and calcareous soils of n. and central AL and adjacent TN. Occurs on shales and limestones of the border of the Cumberland Plateau and coastal plain. Sprawling but with some stems rising erect to 1.5m.

3. *Croton argyranthemus* Michx.   SILVER CROTON   *Figure 147*

A small, herblike shrub that survives the winter aboveground in its FL range west to s. AL, in the range of coverage of this book. Occurs in sandy pinelands and drier pine flatwoods. The blunt-tipped leaves are silvery beneath, with brown dots, and with a mostly brown-peltate midrib and petiole.

Cudrania tricuspidata (Carr.) Bureau ex Lavallee   CHINESE CHE   Family Moraceae
Figure 148

An exotic tree that has become sparingly naturalized, usually by root sprouts near areas of its cultivation. It most closely resembles the native Osage-orange (*Maclura pomifera*), with which it is closely related. The twigs bear straight, sharp spines above or beside the buds. Bark is scaly, with the vertically elongated scales curling at the tips, becoming thicker and more ridged with age, but retaining the scaly character that is different from the more fissured bark of *Maclura*. Inner bark yellowish to light orange. Bundle scars are obscure. Fruit is a berrylike multiple, 2–4cm diam., with milky sap when young but turning red, sweet, and succulent when fully mature and fallen from the branches.

Cunninghamia lanceolata (Lamb.) Hook.   CHINA-FIR   Family Cupressaceae   *Figure 149*

An exotic conifer widely cultivated in the SE, rarely naturalizing. Included here since it is so often encountered as a remnant of cultivation and suspected of being naturalized. Leaves are needlelike, flat, 3–5cm long, spinose-tipped, sessile, and with 2 broad stomatal bands (whitish and wax-covered) that cover majority of width of leaf undersurface. Terminal buds small and slender on juvenile twigs; larger, conical, and often surrounded by additional buds on mature wood; flower buds spherical. Cones subglobose, 3–4cm, with pointed, serrated scales. Bark brown, fibrous, furrowed. A tall tree, to 20m, with dense crown and drooping branchlets and twigs. Often litters the ground with fallen twigs.

Cynanchum L.   MARSH MILKVINE   Family Asclepiadaceae

Twining native vines with milky sap, mostly herbaceous and dying back to ground level in the n. parts of their range, but the gray-green stems sometimes persist over winter in the s. sections. Leaves opposite, linear, usually drooping or reflexed. Flowers in umbels; fruit a follicle, splitting to release downy-winged flat seed, much in the fashion of the true milkweeds (genus *Asclepias* L.). Both species typically found along salt marshes and coastal shores.

1. Leaves sessile; fruit enlarged near base and to 7mm wide   1. *C. angustifolium*
1. Leaves short-stalked; fruit narrow, to 3mm wide   2. *C. scoparium*

1. *Cynanchum angustifolium* Pers.   SAND-VINE   *Figure 150*

Distributed from coastal NC to TX, along salt marshes and maritime forest edges near brackish shores or dunes. Leaves are sessile, 4–8cm long, 1–5mm wide. Follicles are noticeably widened at base, mostly lance-ovoid or tapering from the enlarged base, 4–7cm long, 5–7mm broad.

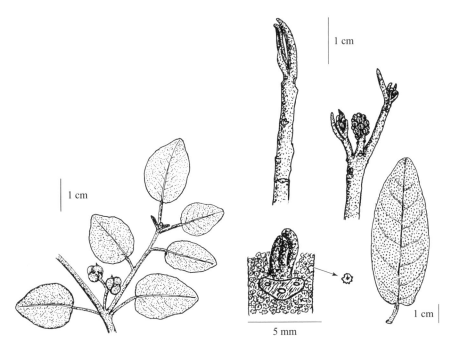

Figure 145. *Croton punctatus*

Figure 146. *Croton alabamensis*

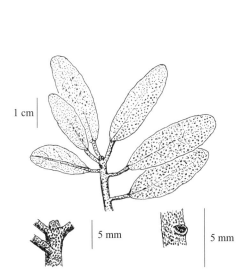

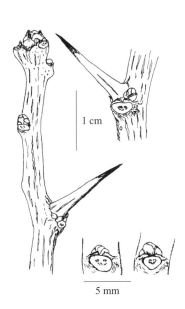

Figure 147. *Croton argyranthemus*

Figure 148. *Cudrania tricuspidata*

2. Cynanchum scoparium Nutt.   MARSH MILKVINE

Distributed from se. SC to FL in marsh borders and hammocks near the coast. Also in scrub in FL. Leaves bear a short petiole; blades 2–5cm long, 2–5mm wide. Follicles are narrow, not so swollen at base as above species, rather cylindric, and 3–5cm long, 1–3mm broad.

Cyrilla racemiflora L.   TITI   Family Cyrillaceae   *Figures 151 and 152a*

A shrub or small tree of the coastal plain from VA to FL and west to TX. Ranges into the Piedmont in moist lowlands from NC to GA. Common in flatwoods depressions, along watercourses, bays, pocosins, shrubby bogs, and similar low areas where moisture is available. May form dense stands in areas of GA and FL, called "titi swamps." The glabrous leaves are deciduous to subevergreen, depending on the latitude of range and climatic influence, mostly 5–8cm long, 12–25mm wide, with finely reticulate venation, glossy upper surfaces, and short petioles. Twigs have distinctive triangular leaf scars with a single raised bundle scar; terminal buds are present, with tapering outer scales that project beyond the inner portions of the bud. Fruits are small and bottlelike, about 2mm long, borne in racemes 6–15cm long. Bark brownish, shallowly fissured. Grades into another distinctive form in FL, in which the leaves are much smaller, mostly 5–10mm wide; this has been called *C. parvifolia* Raf. (Figure 152b)

Cytissus scoparius (L.) Link   SCOTCH-BROOM   Family Fabaceae   *Figure 153*

An exotic shrub, to 2m tall, with numerous deep green twigs so heavily ridged and flexible that no other plant is similar. The leaf scars are tiny and obscure. Planted as an ornamental for the yellow flowers in spring as well as for the green of stems in winter. Naturalizing near sources of cultivation, sometimes forming dense stands. A serious invasive pest species in the nw. US. Fruit is a flattened legume to 5cm long. In the SE usually seen naturalized near urban areas and waste places.

Decodon verticillatus (L.) Ell.   WATER-WILLOW   Family Lythraceae   *Figure 154*

A native aquatic shrub distributed across nearly all the SE but rarely seen in the Appalachian region. Mostly in the coastal plain of the SE, near freshwater shores, especially ponds, lakes, swamps. The stems above water rarely more than 0.5m high but arching and rooting in mud. Some above-water stems die back in winter. Below-water stems have thick, spongy bark. Forms colonies by the spreading, rooting branches. Leaf scars opposite, buds partially sunken and obscure. The dried capsules may persist into winter.

Decumaria barbara L.   CLIMBING HYDRANGEA   Family Hydrangeaceae   *Figure 155*

A native vine, climbing by aerial rootlets on the twigs. Found in Piedmont and coastal plain forests from VA to FL and LA, and north to TN. Climbs the trunks of trees or sprawls over rocks and other objects on the ground in rich woodlands or swamps. Leaves may be late-deciduous; upper surfaces glossy, dark green. The twigs have a jointed appearance and bear small, rusty-colored terminal buds. The clusters of capsules may persist in winter.

Deutzia Thunb.   DEUTZIA   Family Hydrangeaceae

All exotic flowering shrubs that occasionally naturalize near areas of their cultivation. Three species have been noticed as naturalized in the SE. All have clusters of stellate hairs on leaves and twigs, opposite leaf scars, and excavated (hollow) pith in the branchlets. Fruit is an elongate capsule.

1. Twigs usually without terminal bud; bark brown, peeling; leaves with large, crenate-dentate teeth   1. *D. scabra*

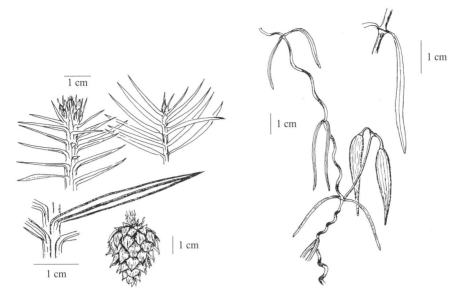

Figure 149. *Cunninghamia lanceolata*

Figure 150. *Cynanchum angustifolium*

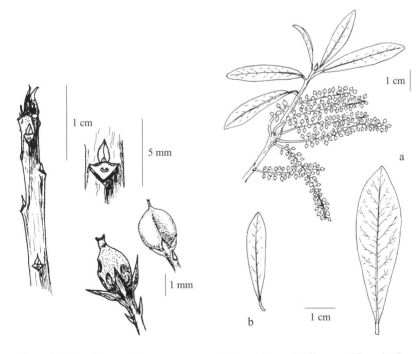

Figure 151. *Cyrilla racemiflora*

Figure 152. a: *Cyrilla racemiflora;* b: *C. parvifolia*

130 · Deutzia

1. Twigs often with terminal bud; bark gray, close; leaves finely serrate
   2. Leaves almost glabrous beneath, base broadly tapered or rounded   2. *D. gracilis*
   2. Leaves hairy below, base narrowed   3. *D. parviflora*

1. *Deutzia scabra* Thunb.   ROUGHLEAF DEUTZIA   *Figure 156*

A tall shrub, to 3m, with narrow panicles of 1.5–2cm white flowers in June. Leaves stellate-pubescent on top and bottom, rounded at base. Native to China and Japan.

2. *Deutzia gracilis* Sieb. & Zucc.   SLENDER DEUTZIA   *Figure 157*

A tall shrub, to 2m, with upright panicles or racemes of 1.5–2cm white flowers. Branches arching. Leaves nearly glabrous beneath. Native to Japan.

3. *Deutzia parviflora* Bunge   SMALLFLOWER DEUTZIA

A tall shrub, to 2m, with corymbs of white flowers each about 1cm wide. Leaves bear scattered stellate hairs and are usually cuneate at base. Native to China.

Diervilla P. Mill.   BUSH-HONEYSUCKLE   Family Caprifoliaceae

Three species are native to the SE, all shrubs with twig tips normally ending in clusters of capsular fruits or their remnants. Found in the Appalachian region and adjacent provinces with rocky habitats. Buds with several loosely imbricate scales; leaf scars opposite, with 3 bundle scars. The bark is shreddy on older stems. These shrubs form clumps and rarely exceed 1.5m in height.

1. Twigs distinctly pubescent over entire surface; leaves pubescent on lower surface
                                                                         1. *D. rivularis*
1. Twigs glabrous or with hairs only in a line connecting nodes; leaves glabrous on lower surface
   2. Twigs often angular or slightly squared in cross section, near nodes; leaf petiole less than 5mm   2. *D. sessilifolia*
   2. Twigs rounded, not as above; leaf petiole over 5mm   3. *D. lonicera*

1. *Diervilla rivularis* Gattinger   HAIRY BUSH-HONEYSUCKLE

A short shrub of rocky soils in the Appalachian region from NC, TN to n. GA, AL. Uncommon, mostly seen on bluffs and rocky outcrops near streams.

2. *Diervilla sessilifolia* Buckl.   SOUTHERN BUSH-HONEYSUCKLE   *Figure 158*

The most common species of *Diervilla* in the s. section of the Appalachians, commonly found in high elevation balds, roadsides, and rocky outcrops of the Blue Ridge in NC, TN to n. SC, GA, AL. Occasional in lower elevations nearly to the Piedmont, along streams on cool, rocky shorelines.

3. *Diervilla lonicera* P. Mill.   NORTHERN BUSH-HONEYSUCKLE

A shrub similar to above species, distributed from n. regions southward into the Appalachians and other uplands of the SE, from VA, WV, south to NC, TN, and west to KY, IN. Mostly in higher elevations in the Blue Ridge and in drier, rocky woodlands in areas to the north and west.

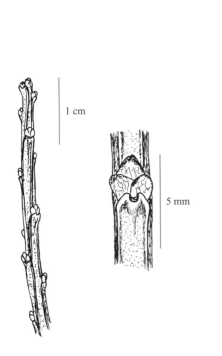

Figure 153. *Cytissus scoparius*

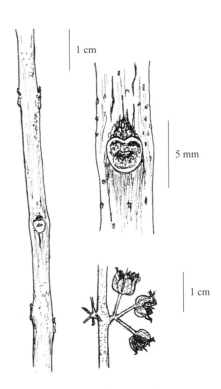

Figure 154. *Decodon verticillatus*

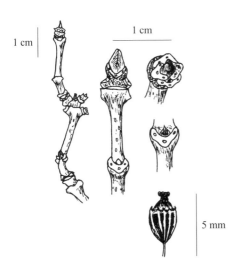

Figure 155. *Decumaria barbara*

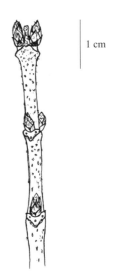

Figure 156. *Deutzia scabra*

Diospyros virginiana L. PERSIMMON   Family Ebenaceae   *Figure 159*

This tree is widespread across the entire SE, though rarely in high elevations of the Blue Ridge. Can be expected in various habitats, from sand ridges in the coastal plain to bottomlands. The twigs have no terminal bud, and the appressed lateral buds have 2 outer imbricate scales; 1 bundle scar in each leaf scar. The pith is diaphragmed. Bark fissured, becoming blockier and thicker with age. The fruit is a large berry, 2–4cm wide, astringent with tannins until fully ripened on the branches. Flesh of soft, fallen fruit orange and sweet. Seed flattened, brown.

Dirca palustris L. LEATHERWOOD   Family Thymeleaceae   *Figure 160*

A native shrub with flexible brown branches. The tough, fibrous bark accounts for the name. Distributed across the SE, but rather sporadic. Found in rich woodlands, mostly in circumneutral or calcareous soils. Twigs have a jointed appearance due to swollen nodes, with naked buds partially sunken and surrounded by the leaf scars. Flowers before leaf emergence in early spring; fruit appears in late spring to early summer, a greenish drupe about 1cm long.

Elaeagnus L. SILVERBERRY   Family Elaeagnaceae

Four species of these exotic shrubs/small trees have naturalized in the SE. All have silvery or brownish peltate scales on twigs, fruit, and leaves. Dotlike stellate hair clusters often are borne on the upper leaf surface. Terminal buds appear naked; laterals often collateral. Spiny spur shoots present on many species. One bundle scar visible in each leaf scar. Use of twigs alone to separate the species is difficult; remnant leaves are more useful.

1. Leaves evergreen, glossy above; flowers in autumn and fruits by spring; habit sprawling or vinelike   1. *E. pungens*
1. Leaves deciduous, dull or heavily punctate above; flowers in spring; habit more shrubby or treelike
    2. Leaves oblong or very elongate, 16mm wide or less, tip rounded, silvery below and grayish above; a tree with dark bark   2. *E. angustifolia*
    2. Leaves more ovate, commonly 19mm or wider; tip acute, punctate and green above; large shrubs
        3. Twigs brown; brown scales numerous on petioles and leaf midribs beneath
             3. *E. multiflora*
        3. Twigs silvery; leaves mostly with silvery scales only   4. *E. umbellata*

1. *Elaeagnus pungens* Thunb. THORNY ELAEAGNUS   *Figure 161*

A sprawling, sometimes vinelike shrub with elongate, arching branches covered with brownish peltate scales. Spinose spur shoots numerous. Leaves noticeably brown-dotted beneath. Most often seen near urban areas and waste places or near old homesites. Fruits red-dotted, about 1–1.5cm long, often with the persistent calyx tube adherent to the tip.

2. *Elaeagnus angustifolia* L. RUSSIAN-OLIVE

A large shrub or small tree, with thickly fissured bark on mature trunks. Older bark may have grayish plates with intervening areas blackish. Branches silvery-gray, conspicuous from a distance. Rarely naturalized east of KY and AR; with a preference for alkaline soils and drier climate, it is more commonly naturalized in the plains and w. US. Fruits yellow.

3. *Elaeagnus multiflora* Thunb. GOUMI-BERRY

A bushy shrub, to 3m, with numerous brown peltate scales covering the twigs, leaf undersides, midrib, and petiole. A rare escape, being less commonly cultivated in the SE than other species. Spinose spur shoots sparse or lacking. Fruit red-dotted, 15–20mm long, with a stalk 15–25mm long.

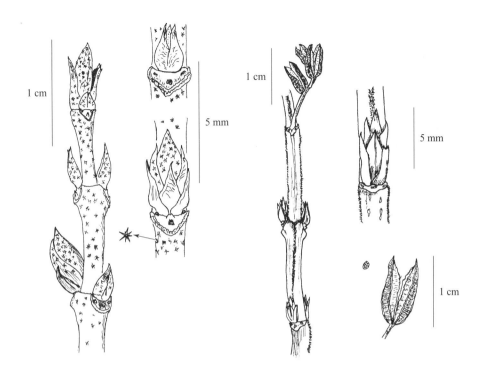

Figure 157. *Deutzia gracilis*

Figure 158. *Diervilla sessilifolia*

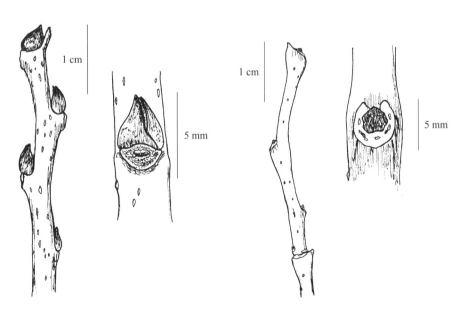

Figure 159. *Diospyros virginiana*

Figure 160. *Dirca palustris*

134 · Elaeagnus

4. *Elaeagnus umbellata* Thunb. AUTUMN-OLIVE *Figure 162*

A large shrub with silvery twigs and branchlets; commonly naturalized across the SE. The leaves bear an abundance of silvery peltate scales on the underside, and stellate hairs on top surface. Spiny spur shoots commonly present. Fruits red, to 1cm long, with a stalk to 12mm long. As with the other species, the fruits with seed are consumed and spread by birds.

***Elliottia racemosa* Muhl. ex Ell.** ELLIOTTIA Family Ericaceae *Figure 163*

A native shrub or small tree, rare and limited in range to e. GA and adjacent SC. Known only from a few populations in sandy woodlands and upland river bluffs in the coastal plain of these 2 states. The terminal buds are much larger than laterals, with outer scales running the length of the bud; bundle scars single. The fruit is a 5-lobed capsule. Bark grayish, shallowly furrowed.

***Epigaea repens* L.** TRAILING ARBUTUS Family Ericaceae *Figure 164*

This native diminutive shrub sprawls over the ground surface. Common in sunny, acidic upland forests over most of the SE to MS and AR. The evergreen leaves are cordate or rounded at the base, and conspicuously hairy beneath. The petioles and twigs are also hairy. Fruit is a small, rounded, hairy capsule.

***Erica* L.** HEATH Family Ericaceae

Low, bushy exotic shrubs that may naturalize or linger near areas of cultivation. Many species are used in cultivation, but the 2 species presented here are most likely naturalized in the region. The small, needlelike leaves are evergreen and whorled in groups of 4. Flowers small, elongate, borne in late summer or late winter.

1. Leaves glandular-ciliate on margin, 3–4mm long; twigs hairy 1. *E. tetralix*
1. Leaves not as above, mostly glabrous, 4–8mm long; twigs glabrous 2. *E. carnea*

1. *Erica tetralix* L. CROSS-LEAVED HEATH *Figure 165a*

A shrub rarely over 5dm high; known to naturalize in acidic, boggy or peaty sites in WV. The white to pink flowers are borne in terminal clusters in late summer.

2. *Erica carnea* L. SPRING HEATH *Figure 165b*

A shrub rarely over 3dm high, commonly cultivated. Its pink to reddish flowers are borne in axils of leaves in late winter.

***Erythrina herbacea* L.** EASTERN CORAL-BEAN Family Fabaceae *Figure 166*

A native shrub that attains more woody character and larger dimensions with more s. distribution, since stems freeze easily in cold winters. Found chiefly in the coastal plain from NC to FL and west to TX. Favors sandy soils and maritime woodlands. The green twigs bear prickles that are below the leaf scars and occasionally internodal. Leaf scars raised. Terminal bud present on twigs that have not flowered or died back. Scarlet flowers are showy; fruit is a plump legume with scarlet, beanlike seed, poisonous if eaten.

***Escobaria missouriensis* var. *similis* (Engelm.) N. P. Taylor** PRAIRIE CORY CACTUS
Family Cactaceae *Figure 167*

This small, native cactus occurs in the SE only in extreme w. sections of the region, in prairies of MO, OK, AR, TX and areas west and north. Clumplike and rarely over 6cm tall, it grows

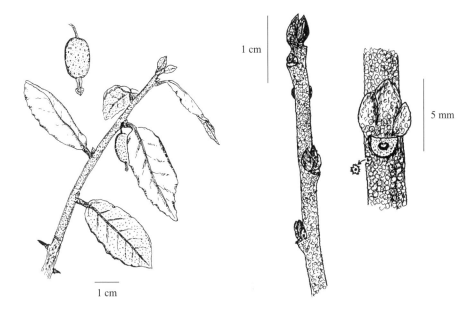

Figure 161. *Elaeagnus pungens*

Figure 162. *Elaeagnus umbellata*

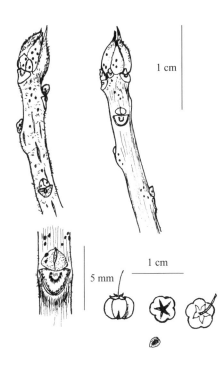

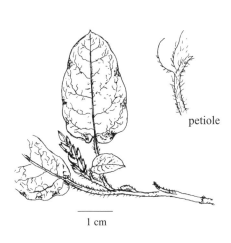

Figure 163. *Elliottia racemosa*

Figure 164. *Epigaea repens*

almost hidden in thin grasslands. Attains its full height when swollen with water, but shrinks close to the ground when dry; more evident when the yellow flowers open. The round red fruit is borne in the axils of the moundlike aereoles. Other varieties of the species occur in the Great Plains and Rocky Mountain regions. [*Coryphantha missouriensis* var. *caespitosa* (Engelm.) L. Benson].

Euonymus L.   EUONYMUS   Family Celastraceae

Three native and 5 naturalized species occur in the SE. All have opposite leaves, 1 bundle scar, and terminal buds. Twigs are green, lined or 4-ridged in some, winged in 1; pith is spongy or homogenous. Fruit is a capsule containing seed surrounded by a fleshy orange to red aril.

1. Leaves thick, evergreen
  2. Habit vinelike or a groundcover; twigs usually bear aerial rootlets   1. *E. fortunei*
  2. Habit shrubby; no aerial rootlets   2. *E. japonicus*
1. Leaves thin, deciduous
  3. Twigs or branches winged   3. *E. alatus*
  3. Twigs not winged
    4. Twigs distinctly 4-lined or squared, slender, and deep green; slender and spindly shrubs
      5. Shrub with erect stems over 5dm tall; leaves widest near middle of blade
           4. *E. americanus*
      5. Shrub sprawling over the ground, under 5dm; leaves widest beyond middle
           5. *E. obovatus*
    4. Twigs not as above, moderately stout, or light green; stiffly upright shrubs or small trees
      6. Bark thin, rarely fissured; arils scarlet; seed brown; leaves pubescent beneath
           6. *E. atropurpureus*
      6. Bark thicker, usually fissured; arils orange; seed white; leaves glabrous
        7. Leaves short-pointed; twigs mostly horizontal or erect   7. *E. europaeus*
        7. Leaves long-pointed; twigs drooping on mature plants   8. *E. bungeanus*

1. *Euonymus fortunei* (Turcz.) Hand.-Maz.   WINTER-CREEPER   Figure 168

A native of China; naturalized in various habitats of the SE but particularly conspicuous in areas west of the Appalachians. The vinelike habit enables access up tree trunks by aerial rootlets. Twigs weakly angled to rounded in cross section. Capsule pale or light pink; aril orange.

2. *Euonymus japonicus* Thunb.   JAPANESE EUONYMUS   Figure 169

A native of Japan; occasionally naturalized near urban areas and sources of its cultivation. An upright shrub with glossy foliage. Twigs 4-angled. Fruit similar to above species.

3. *Euonymus alatus* (Thunb.) Sieb.   WINGED EUONYMUS   Figure 170

A commonly planted Asian shrub, naturalized sparingly. Capsules often only 1- to 3-lobed; aril orange-red; seed light brown.

4. *Euonymus americanus* L.   STRAWBERRY-BUSH   Figure 171a

Native to the SE; distributed nearly throughout. A spindly shrub with green bark, common in woodlands from bottomlands to rocky uplands. Capsules pinkish, with tubercles over surface; aril scarlet; seed white.

5. *Euonymus obovatus* Nutt.   RUNNING STRAWBERRY-BUSH   Figure 171b

Native to the SE in the Appalachians from WV, VA south to GA, and west in uplands to MO. Occurs in moist rocky woods and along streams, often in dense shade from overhead tree

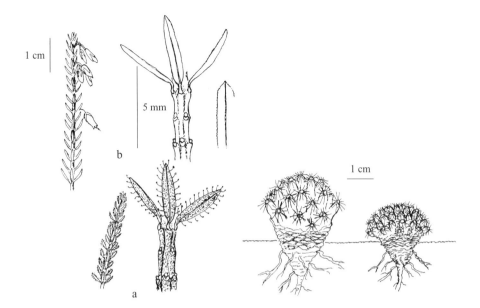

Figure 165. a: *Erica tetralix;* b: *E. carnea*   Figure 167. *Escobaria missouriensis* var. *similis*

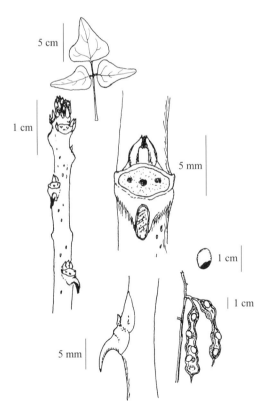

Figure 166. *Erythrina herbacea*

canopies. Sprawls over the ground, where it may often be overlooked in woodlands. Capsule often 3-lobed, tuberculate-surfaced; aril scarlet; seed white.

6. *Euonymus atropurpureus* Jacq.   WAHOO   *Figure 172*

Native to the SE and widely dispersed in woodlands nearly throughout. Most common in circumneutral or calcareous soils and in bottomlands. Rare in the Blue Ridge and similar areas of acidic soils. Capsules 4-lobed, smooth surfaced, pink to purplish, barely opening enough to see much of the scarlet aril. The flowers have 4 purplish petals.

7. *Euonymus europaeus* L.   EUROPEAN SPINDLE-TREE

A tall shrub or small tree from Europe and Asia. Most similar to the native species *E. atropurpureus* in winter, but twigs are generally darker green, bark thicker and slightly fissured. Capsules usually split wider and release the orange arils into view; seed are white.

8. *Euonymus bungeanus* Maxim.   CHINESE EUONYMUS

A native of China, rarely naturalized near areas of its cultivation. The gray bark becomes thick, with scaly ridges and furrows similar to bark of many willows. Twigs slender, pendulous. The pinkish capsules often persist on twigs into winter.

Exochorda racemosa (Lindl.) Rehd.   PEARLBUSH   Family Rosaceae   *Figure 173*

An exotic shrub, to 4m tall, naturalized sporadically across the SE, primarily in calcareous soils. Native to China; planted for its showy white spring flowers. The terminal buds are larger than laterals, with sharply tipped scales. Leaf scars are shallow, with 3 raised bundle scars that are sometimes partially hidden behind a ledge on the edge of the leaf scar. Fruit is a capsule, deeply 4-lobed and standing erect near the twig base.

Fagus grandifolia Ehrh.   AMERICAN BEECH   Family Fagaceae   *Figure 174*

A native tree, common across much of the SE except for se. GA. Grows in moist soils with loam or clay content, or in sandier soils if on cool, moist slopes or in ravines. A common component of Appalachian forests; ascends to the high elevations in the Blue Ridge. Buds are elongate, with several imbricate brown scales; stipule scars nearly encircle the twig at each node. Several bundle scars occur in the leaf scar. The fruit is a triangular nut borne in pairs within a 4-parted husk that is covered by nonrigid spines. Bark thin, gray, smooth.

Ficus L.   FIG   Family Moraceae

A large genus of tropical and subtropical species, none native to the se. region covered by this book. Of the many types grown for ornamentals, house plants, and fruit, only 2 are cold-hardy enough to be encountered to any extent in the SE. These 2 species may linger or spread from original sites of outdoor cultivation.

1. Leaves deciduous; a shrub 1. *F. carica*
1. Leaves evergreen; a vine 2. *F. pumila*

1. *Ficus carica* L.   COMMON FIG   *Figure 175*

A large, robust shrub widely grown across the Piedmont and coastal plain of the SE for its edible fruit. Usually the stout twigs die back following freezes but may remain intact in warmer zones. The milky sap may not be evident in winter. Terminal buds conical; laterals rounded. Stipule scars long and conspicuous. Fruit is the common edible fig of commerce.

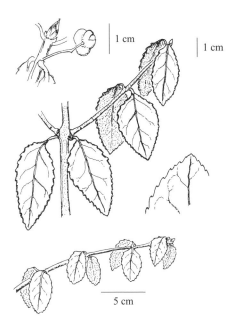

Figure 168. *Euonymus fortunei*

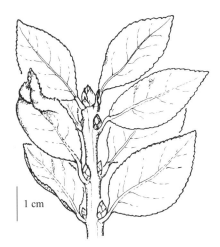

Figure 169. *Euonymus japonicus*

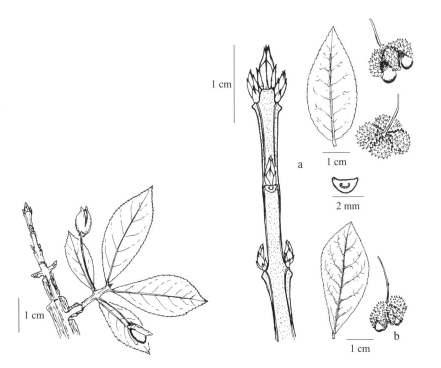

Figure 170. *Euonymus alatus*

Figure 171. a: *Euonymus americanus;* b: *E. obovatus*

2. Ficus pumila L. CLIMBING FIG *Figure 176*

A vine climbing by aerial rootlets. Popular as a covering for outdoor walls since it is evergreen and hardy over most of the outer coastal plain, from SC to TX. Leaves on juvenile plants are widely ovate, with asymmetrical bases, mostly under 3cm long. Mature plants, reached after many years of growth, have leaves oblong-elliptic, 6–10cm long. The fruit is 3–6cm long. Twigs often retain the conspicuous stipules, flanking the petiole bases.

Firmiana simplex (L.) W. F. Wight PARASOL-TREE Family Sterculiaceae *Figure 177*

An exotic tree with green bark and stout green twigs. Known to naturalize in the outer coastal plain, mainly in vicinity of maritime forest, marshes, barrier islands. May also naturalize near cultivated specimens inland to the Piedmont region. Terminal buds rusty-brown, hairy; leaf scars and stipule scars large. Fruit is a multiarmed bract bearing hard seed around its perimeter. The green bark becomes gray near old trunk bases but remains thin and relatively smooth.

Forestiera Poir. FORESTIERA Family Oleaceae

Four native species occur in the se. region covered by this book, 3 of which are primarily confined to the coastal plain. All have opposite leaves, twigs with superposed buds, 1 bundle scar, and short spurlike side shoots. Fruit is a drupe. Twigs alone can rarely separate these plants.

1. Twigs glabrous or with very fine or appressed hairs; leaves glabrous
    2. Twigs sometimes finely hairy; leaves with distinct petioles, thin, not punctate, long-tipped
        1. *F. acuminata*
    2. Twigs glabrous; leaves nearly sessile, thick, subevergreen, punctate below, tip blunt
        2. *F. segregata*
1. Twigs conspicuously pubescent; leaves hairy at least on veins below
    3. Twigs overall pubescent; leaf undersides hairy throughout; leaves mostly over 5cm long; rare
        3. *F. godfreyi*
    3. Twigs pubescent mostly on 2 sides; leaves hairy on veins, mostly under 5cm; widespread
        4. *F. ligustrina*

1. *Forestiera acuminata* (Michx.) Poir. SWAMP-PRIVET *Figure 178*

A large shrub or small tree widely distributed and common in the Mississippi River Valley. Less common but ranging eastward to FL and SC in the coastal plain. Typical of moist to wet lowland soils, especially in swamps and along swampy watercourses. Dioecious, the pistillate plants producing the 2–3cm, pinkish drupes.

2. *Forestiera segregata* (Jacq.) Krug & Urban FLORIDA-PRIVET *Figure 179*

A rare shrub of maritime forests and shores of marshes, typically near calcareous deposits or shell middens. Found from coastal SC to FL and along coasts of the FL peninsula where it also occurs in scrub habitats. Fruits blue-black, to 8mm.

3. *Forestiera godfreyi* L. C. Anderson GODFREY'S FORESTIERA *Figure 180*

A rare shrub known only from a few locations in n. FL, se. GA, and se. SC. Found in rich, moist woodlands, mainly in calcareous soils. Flowers in early spring; fruits blue-black, about 1cm long.

4. *Forestiera ligustrina* (Michx.) Poir. GLADE-PRIVET

A widespread shrub in the SE, found in the Piedmont and coastal plain from SC to FL, west to TX, and north to uplands of TN, KY. It occurs in upland rocky woods, along streambanks, and

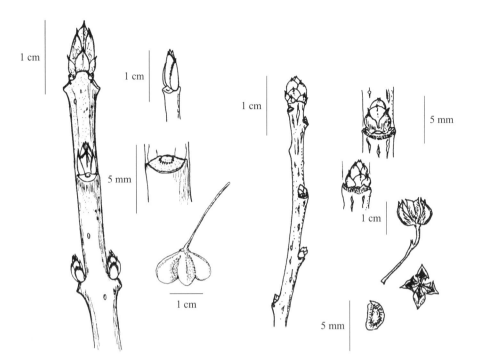

Figure 172. *Euonymus atropurpureus*

Figure 173. *Exochorda racemosa*

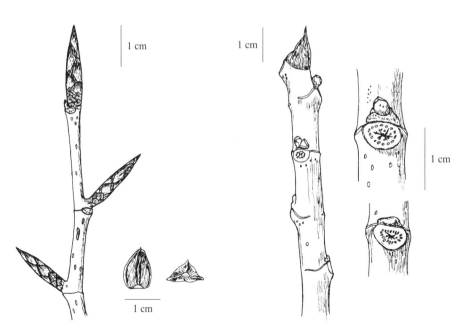

Figure 174. *Fagus grandifolia*

Figure 175. *Ficus carica*

Forsythia Vahl. FORSYTHIA Family Oleaceae *Figure 181*

Treated here are 3 species of these exotic shrubs, which are extensively cultivated in the SE. The following species may be encountered as occasional escapes from cultivation or persistent after original plantings. All have glabrous twigs with opposite leaf scars and prominent, pale lenticels. The pith of vigorously elongating twigs is useful in identification but mainly in internodes; pith in the nodes is often solid. Fruit is a small, elongate capsule.

1. Pith excavated (except at nodes); leaf bases broadly cuneate to rounded; leaves sometimes 3-parted 1. *F. suspensa*
1. Pith chambered (except at nodes) or occasionally excavated in basal internodes; leaves tapered at base
   2. Branches erect; leaves rarely 3-parted 2. *F. viridissima*
   2. Branches arching; leaves sometimes 3-parted (this is a hybrid of above 2 species) 3. *F.* × *intermedia*

1. *Forsythia suspensa* (Thunb.) Vahl FORSYTHIA

A native of China, with spreading or arching branches. Leaves serrate. Calyx lobes nearly as long as corolla tube.

2. *Forsythia viridissima* Lindl. FORSYTHIA

A native of China, with upright branches. Leaves serrate above middle of blade or nearly entire. Calyx lobes about ½ as long as corolla tube. Pith may be chambered even through some nodes.

3. *Forsythia* × *intermedia* Zab. FORSYTHIA

A hybrid of the above 2 species, and intermediate in characters. Calyx shorter than corolla tube.

Fothergilla L. FOTHERGILLA Family Hamamelidaceae

Two species are native to the SE. Both are shrubs with naked, scurfy terminal buds. The leaf scars are often darkened; bundle scars within may be obscure. Fruit is a beaked capsule that "shoots" out 2 brown, tapered, hardened seed. Twigs alone will not reliably separate the 2 species. Twigs are also similar in appearance to those of *Hamamelis;* if no other traits can be found to rule out this relative, leaf remnants can be used. Lowermost lateral veins are included within the blade margin in *Hamamelis* but extend for a short distance as part of the blade margin in *Fothergilla*.

1. Capsules 7–10mm long; robust twigs 1–2.5mm diam. at midpoint of 2d internode from twig apex 1. *F. gardenii*
1. Capsules 10–14mm long; robust twigs 3–4mm diam. at midpoint of 2d internode from twig apex 2. *F. major*

1. *Fothergilla gardenii* L. DWARF WITCH-ALDER *Figure 182*

A slender shrub, often multistemmed, usually under 1m high. Native to the coastal plain from NC to FL, west to AL. Uncommon; mostly seen in moist sands and peats in pine savannas, pocosins, bays, and bogs. Stoloniferous, sprouting from the spreading root system to form

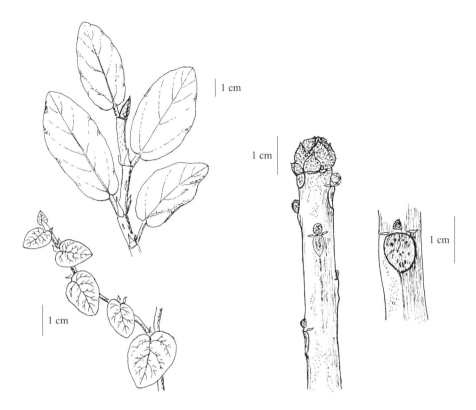

Figure 176. *Ficus pumila*

Figure 177. *Firmiana simplex*

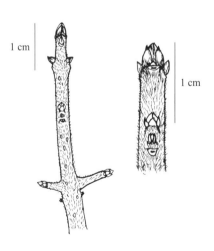

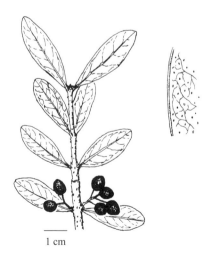

Figure 178. *Forestiera acuminata*

Figure 179. *Forestiera segregata*

colonies. When cultivated and removed from the impact of fire that favors the short, multi-stemmed habit so common in the wild, this species may exceed 1m and have a more clumplike habit. Leaves stellate-pubescent on both surfaces, though more so on underside. Largest leaves mostly under 5cm wide and to 6cm long in the wild, but to 8cm long in cultivation. Capsules mostly under 1cm. Seed usually under 6mm long.

2. *Fothergilla major* (Sims) Lodd.   WITCH-ALDER   *Figure 183*

A shrub, to 2.5m tall in competitive forest understories, though more frequently 1–2m. Distributed from the NC Piedmont west across the Appalachians to TN, AL, GA, and appearing again in AR. Uncommon; typically seen in dry or rocky woods and bluffs. Habit clumplike, not stoloniferous to the degree of previous species. Seed usually over 6mm long. Capsules mostly over 1cm long. Leaves nearly glabrous on top surface; stellate-pubescent mainly on veins and undersurface. Largest leaves mostly over 5cm wide, to 12cm long. Stipule scars and buds larger than in above species.

Frangula P. Mill.   BUCKTHORN   Family Rhamnaceae

Two species are treated here for the SE. One is native; the other, naturalized. Both are large shrubs or small trees. These species are transferred from the genus *Rhamnus* L. due to several important differences. Most notable in winter are the presence of naked terminal buds and no spines, but other floral differences, including 5-petaled flowers, support the division. Fruit is a berrylike drupe about 6–8mm diam., turning as it matures from green to red, then black; flesh soft, sweetish, not safe for consumption in quantity.

1. Twigs commonly lined from leaf scars; fruit 3-seeded; leaves minutely toothed above middle of blade   1. *F. caroliniana*
1. Twigs rarely lined; fruit 2-seeded; leaves entire   2. *F. alnus*

1. *Frangula caroliniana* (Walt.) Gray   CAROLINA BUCKTHORN   *Figure 184*

Usually a small tree; native to a wide area of the SE. Distributed from the Piedmont of NC to FL, west to TX and MO. Rare or lacking from most of the Blue Ridge and the coastal plain from VA to GA. Prefers calcareous soils or circumneutral soils, mostly in uplands. Common in TN, KY. Bark gray, thin, mostly smooth or shallowly fissured; inner bark yellowish. (*Rhamnus caroliniana* Walt.)

2. *Frangula alnus* P. Mill.   GLOSSY BUCKTHORN   *Figure 185*

A European species naturalizing extensively in the NE, where it has been cultivated for many years as a hedge plant. It should be expected from TN and VA northward as an escape in the SE. Birds eat the fruits, thus seed are widely dispersed. (*Rhamnus frangula* L.)

Franklinia alatamaha Bartr. ex Marsh.   FRANKLINIA   Family Theaceae   *Figure 186*

A native tree that is assumed to be extinct in the wild. Known historically from a single site on the Altamaha River in the coastal plain of GA but not seen there since 1790. Brought into cultivation at the time of its discovery; today widely planted but not naturalizing. The moderately stout twigs bear terminal buds with a few hairy scales and large leaf scars with a conspicuous bundle scar in each. Bark thin, smooth to lightly fissured. The fruit is a globular, woody capsule that splits from both ends.

Fraxinus L.   ASH   Family Oleaceae

Seven ashes are treated here, all native in the SE, and all trees with opposite leaf scars, terminal buds, and many bundle scars arranged in a line. The fruit, a samara, is as useful in determining

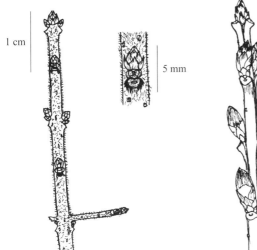

Figure 180. *Forestiera godfreyi*

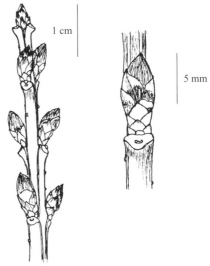

Figure 181. *Forsythia*

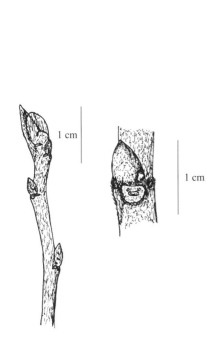

Figure 182. *Fothergilla gardenii*

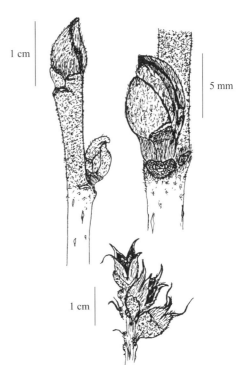

Figure 183. *Fothergilla major*

species as the twigs. In using twigs for winter identification, it is important to remember that on juvenile trees the buds, leaf scars, and certain other features may appear different from those on twigs of mature wood. The key is for twigs from mature branches.

1. Terminal bud scales truncated (flattened) at tip, on basal or outermost pair of scales (and often on next pair as well); bottomland or swamp trees, often with trunks fluted at base
   2. A large or tall tree; fruit winged above middle of seed   1. *F. profunda*
   2. A small tree; fruit winged to near base of seed   2. *F. caroliniana*
1. Terminal bud scales not truncated as above; upland trees, or trunks not fluted
   3. Outermost bud scales extending beyond others, often with elongate tips; fruit winged to below middle of seed
      4. Buds dark, almost black; lenticels large, elongate; n. species rare in SE   3. *F. nigra*
      4. Buds grayish; lenticels not as above; juvenile twigs 4-ridged; a central forest region species   4. *F. quadrangulata*
   3. Outermost bud scales shorter or without elongate tips; fruit winged only above middle of seed
      5. Top margin of leaf scar deeply notched (⅓ to ½ of total depth)   5. *F. americana*
      5. Top margin of leaf scar shallowly notched to nearly straight (less than ⅓ of total depth)
         6. Last pair of lateral buds not appreciably smaller than other laterals; twig tip usually swollen; fruit wing tapers abruptly from last ⅓ of plump seed
                                                                           6. *F. americana* var. *biltmoreana*
         6. Last pair of lateral buds (adjacent to base of terminal bud) smaller; twig tip not swollen; fruit wing tapers narrowly from middle of the narrow seed   7. *F. pennsylvanica*

1. *Fraxinus profunda* (Bush) Bush   PUMPKIN ASH   *Figure 187*

A large tree, scattered in the SE in the coastal plain from MD to FL, west to LA and up the Mississippi River Valley to OH. Occasionally in the Piedmont of NC, SC. Most frequently seen in swamps and other wet lowlands subject to flooding. The bark is gray, thin, and fairly smooth on young trees, becoming furrowed. Fruit 5–8cm long.

2. *Fraxinus caroliniana* P. Mill.   WATER ASH   *Figure 188*

A small tree of swamps and flood-prone lowlands from VA to FL, and west to TX, AR. Most common in the coastal plain. Trunk rarely exceeding 20cm diam., with grayish, shallowly furrowed bark. Usually seen as an understory tree standing in water during wet periods of the year. Fruit 3–5cm long, sometimes with 3 wings along the seed.

3. *Fraxinus nigra* Marsh.   BLACK ASH   *Figure 189*

A n. species reaching the SE only in WV, n. VA, and s. IN, IL. Height to 20m in optimum habitats but often shorter in marginal sites. Occurs in boggy or swampy acidic soils of mountain wetlands and streamsides. Bark grayish, with scaly ridges. Fruit 3–4cm long. Leaves have sessile side leaflets.

4. *Fraxinus quadrangulata* Michx.   BLUE ASH   *Figure 190*

A tree distributed west of the Blue Ridge, usually over limestone or in calcareous soils from w. WV and nw. GA to KS, OK. May be found in bottomlands in parts of its range. The 4-sided nature of the twigs, caused by 4 thin lines of cork, is most pronounced in juvenile trees. Bark grayish, with scaly plates or sometimes blocky. Fruit usually 3–4cm long.

5. *Fraxinus americana* L.   WHITE ASH   *Figure 191*

A large, common tree across most of the SE, though scarce in parts of the outer coastal plain of SC, GA, LA. Typical of rich upland soils, though also seen in lowlands not subject to flooding.

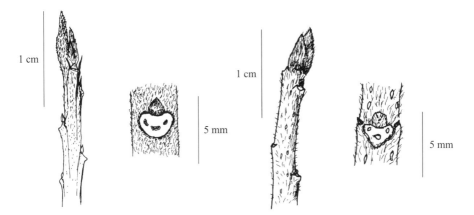

Figure 184. *Frangula caroliniana*   Figure 185. *Frangula alnus*

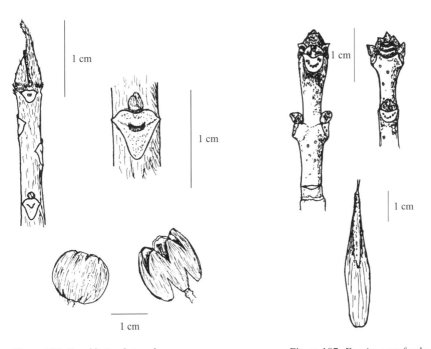

Figure 186. *Franklinia alatamaha*   Figure 187. *Fraxinus profunda*

Tolerates dry and thin soils over calcareous rock. Ascends to high elevations in mixed, mesic forests of the Blue Ridge. Bark gray, furrowed, with elongated ridges. Fruit usually 3–4.5cm long.

6. *Fraxinus americana* var. *biltmoreana* (Beadle) J. Wright   BILTMORE ASH   *Figure 192*

A large tree of sporadic but widespread range; found in the SE from VA to GA in the Piedmont and Appalachians, west to AR, MO, IL. Occurs in rich soils and bottomlands. Although often synonymized with white ash, the velvety twigs of Biltmore ash lack the characteristic notched leaf scar so often used to distinguish the former species. Biltmore ash also varies by its tendency to have brown bark with a softer or corkier texture and a denser, more rounded crown when open grown. Fruit 3–4.5cm long, sometimes with a proportionately wider wing than in white ash. (*F. biltmoreana* Beadle)

7. *Fraxinus pennsylvanica* Marsh.   GREEN ASH   *Figure 193*

Widespread and common across nearly all the SE, in habitats with moist soils. Most frequent in bottomlands, swamps, and along streams. Rarely seen in middle and higher elevations of the Blue Ridge. Bark is gray to brownish, fissured and with narrow, interlacing ridges. Fruit narrow, usually 3–5cm long, 5–7mm wide. The stalks of the lower leaflets are winged nearly to the base, a feature peculiar to this species. Twigs are typically pubescent in the originally described species of "red ash," which is var. *pennsylvanica,* though glabrous forms ("green ash") are just as common, if not more so, in many parts of the SE. These glabrous entities are var. *subintegerrima* (Vahl) Fern.

Garberia heterophylla (Bartr.) Merrill & F. Harper   GARBERIA   Family Asteraceae
Figure 194

A native shrub of central FL barely reaching the area of coverage of this book in its northernmost range into ne. parts of the FL peninsula. Found in sandy pinelands and upland scrub. Height rarely over 2m. The subevergreen leaves are cuneate at the base, rounded or notched at the tip, dull and grayish green. Twigs and leaves are rather scurfy and sticky-glandular. The pinkish-lavender flowers appear in late summer, and the heads of achenes mature in the winter months. Bark is shallowly fissured.

Gaultheria L.   TEABERRY   Family Ericaceae

Two native species in the SE, both of a creeping habit; upright parts less than 15cm in height. Both are evergreen and aromatic when crushed (wintergreen odor). Fruits are actually capsules that become surrounded by an enlarging, fleshy calyx, mimicking a berry.

1. Leaves over 1cm long, most 2–5cm; rhizomatous with upright, mostly unbranched stems; fruit red   1. *G. procumbens*
1. Leaves under 1cm; aboveground stems creeping and branched; fruit white   2. *G. hispidula*

1. *Gaultheria procumbens* L.   TEABERRY   *Figure 195*

Widespread across the SE, but most common in the Appalachian region and areas west and north. Occurs in the Piedmont and coastal plain from VA to SC, and rarely from GA to TX. Typically seen in acidic woodlands, from dry to mesic sites. Leaves are glossy above, remotely toothed. Fruit wintergreen flavored.

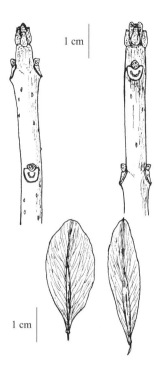

Figure 188. *Fraxinus caroliniana*

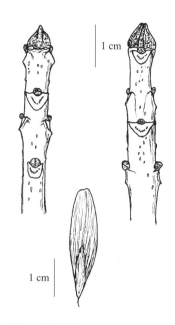

Figure 189. *Fraxinus nigra*

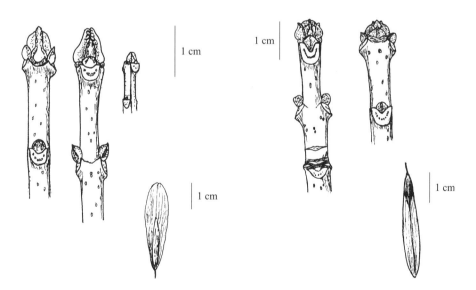

Figure 190. *Fraxinus quadrangulata*

Figure 191. *Fraxinus americana*

2. *Gaultheria hispidula* (L.) Muhl. ex Bigelow CREEPING SNOWBERRY *Figure 196*

A diminutive, matlike shrub barely entering the SE. More common in the North, but reaching the Appalachians of WV. Uncommon and limited to high elevations in WV, where it grows amid moss and near boggy soils, often under coniferous forests. Leaves 4–10mm long, with bristles beneath. The white fruit is larger than most of the leaves and is aromatic, edible.

Gaylussacia L. HUCKLEBERRY Family Ericaceae

Six species, with 1 of these having 2 additional varieties that are significant. All are shrubs with a liking for acidic soils. Twigs are slender, with small leaf scars, no terminal buds, and imbricate-scaled buds; determining all species using twigs alone is not possible, so leaves or their remnants should be sought for utilization of the key. The leaves have gland-dots on 1 or both surfaces in most species, best viewed with 5× or more magnification. The berrylike fruit contains 10 seed, turning from green to red and ultimately dark blue to blackish when fully ripe; edible and normally fallen by Oct.

1. Leaves evergreen, obscurely toothed from near base; twigs ridged below leaf scar
  1. *G. brachycera*
1. Leaves deciduous or late-deciduous, entire or minutely toothed only near tip; twigs not ridged
   2. Leaf tips often acuminate; largest leaves 6–10cm long; native only to Appalachians
     2. *G. ursina*
   2. Leaf tips acute, obtuse, or rounded; largest leaves mostly under 5cm long; not restricted to Appalachians
      3. Leaves gland-dotted on both surfaces; outer bud scales often extend full length of bud
        3. *G. baccata*
      3. Leaves gland-dotted below, rarely above; outer bud scales usually shorter than bud length
         4. Leaves dull green above; no stipitate glands on fruit or fruit stems
            5. Leaves glaucous; twigs glabrous or nearly so
               6. Habit tall, to 2m   4. *G. frondosa* var. *frondosa*
               6. Habit short, to 8dm   5. *G. frondosa* var. *nana*
            5. Leaves not glaucous; twigs hairy   6. *G. frondosa* var. *tomentosa*
         4. Leaves shiny green above; stipitate glands on fruit or fruit stems
            7. Twigs or leaves with short, curly hairs   7. *G. dumosa*
            7. Twigs or leaves with long, straight hairs   8. *G. mosieri*

1. *Gaylussacia brachycera* (Michx.) Gray BOX HUCKLEBERRY *Figure 197*

Height 1–4dm; stoloniferous, with short, erect stems from creeping underground parts. Occurs in the SE in WV, VA, KY, TN, and extends into states of the NE. Rare overall, but local colonies may be extensive. Typically on sandy or rocky slopes or ridges of dry woodlands. Leaves evergreen, glossy above, obscurely toothed, mostly 1–2cm long. Fruit glaucous.

2. *Gaylussacia ursina* (M. A. Curtis) T. & G. BUCKBERRY *Figure 198*

Height 5–15dm, in dry to moist forests of the s. Appalachians where it is endemic to sw. NC and adjacent TN, SC, GA. Often locally abundant, sometimes forming a continuous shrubby cover over the forest floor. Leaves thin, mostly glabrous, often acuminate at tip. Fruit not glaucous, usually shiny black when ripe, watery, without much flavor.

3. *Gaylussacia baccata* (Wang.) K. Koch BLACK HUCKLEBERRY *Figure 199*

Height 3–10dm, in dry to moist woodlands and openings. Most common in Appalachian region, but ranging in the Piedmont and parts of the coastal plain from MD to GA. Extends

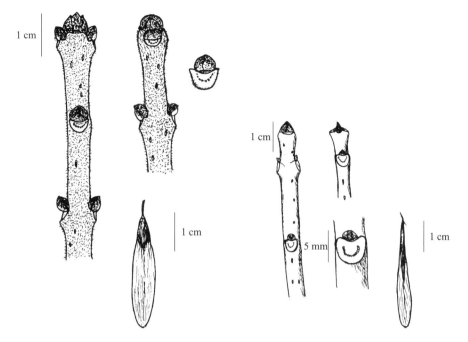

Figure 192. *Fraxinus americana* var. *biltmoreana*   Figure 193. *Fraxinus pennsylvanica*

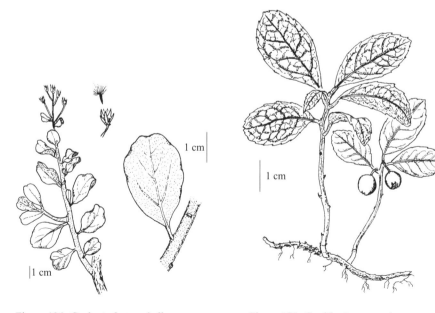

Figure 194. *Garberia heterophylla*   Figure 195. *Gaultheria procumbens*

westward to MO. Leaves usually dull and light green above, but with tiny, glistening resin dots. Twigs densely hairy to sparsely hairy. Fruit bluish to black; slightly glaucous or not.

4. *Gaylussacia frondosa* (L.) Torr. & Gray ex Torr.   DANGLEBERRY

Height usually 1–2m, in moist woodlands and pocosins in the SE, and drier sands or rocky habitats in the n. sections of its range. Ranges from the NE to FL, west to LA, and scattered uncommonly inland in the Piedmont and Appalachian regions. Leaves dull and usually glaucous. Twigs glabrous to hairy. Fruit glabrous, glaucous.

5. *Gaylussacia frondosa* var. *nana* Gray   DWARF DANGLEBERRY   Figure 200

Height usually 2–8dm, in moist to dry soils of pinelands and shrub-bays. Ranges from central FL north to s. GA and s. AL. Stoloniferous, with short, slender, and few-branched upright stems. Leaves dull above, usually glaucous. Fruit glaucous.

6. *Gaylussacia frondosa* var. *tomentosa* Gray   HAIRY DANGLEBERRY

Height usually 5–10dm, in dry to wet pinelands and shrub-bays. Ranges from SC to FL, west to s. AL. Leaves and twigs, especially when young, are tomentose; not glaucous. Fruit not glaucous.

7. *Gaylussacia dumosa* (Andr.) Torr. & Gray   DWARF HUCKLEBERRY   Figure 201

Height 1–5dm, stoloniferous, often forming large clones of slender upright stems. Widespread across most of the SE but most common in the coastal plain. Typically in dry, sandy uplands under pines or oaks, but tolerant of a variety of well-drained sites and even peaty soils in bogs. Leaves thick and shiny green. Young twigs and leaves bear short, curly hairs that may linger through maturity. Most parts also have glandular hairs (stipitate-glandular) intermixed, especially on flower, fruit stems. Fruit not glaucous. The variety *bigeloviana* Fernald occurs in peaty bogs of a few sites in NC, VA; otherwise this is a ne. variety with more glandular and hairy aspect to leaves and bracts, and more bushy-crowned habit than the species.

8. *Gaylussacia mosieri* Small   BOG HUCKLEBERRY   Figure 202

Height 4–15dm, in wet pine savannas and borders of bogs or bays. Scattered in the outer coastal plain from GA to LA, most common in the FL panhandle; recorded from 1 county in SC. Leaves late-deciduous, thick and glossy, dark green above, often with a few long white hairs around the margin or undersurface. The long hairs with red glands on their tips are conspicuous on young twigs, flower and fruit parts, and petioles in the growing season, though most slough by winter. Fruit not glaucous.

Gelsemium Juss.   CAROLINA JESSAMINE   Family Loganiaceae

Two native species in the SE, both twining vines of the coastal plain and Piedmont regions. Leaves evergreen, opposite, entire on margin. Twigs slender, glabrous, with leaf scars connected by a ridge or line and bearing a single bundle scar. The fruit is a small, flattened capsule with many tiny brown seed. Winter identification must utilize the fruit, as twigs and leaves are not diagnostic. Both species may climb high into the canopy of trees but are more familiarly seen among shorter trees, fence lines, and thickets. The tubular yellow flowers are showy, promoting extensive cultivation in the South.

1. Fruit 14–25mm long, 8–12mm wide, abruptly pointed; seed with a thin, distinct wing
   1. *G. sempervirens*
1. Fruit mostly 9–15mm long, 6–8mm wide, tapered to an elongate beak at apex; seed wingless
   2. *G. rankinii*

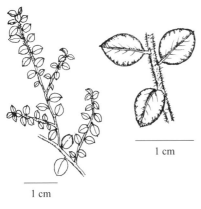

Figure 196. *Gaultheria hispidula*

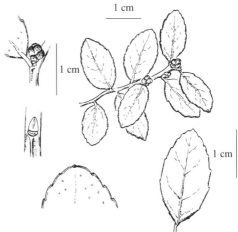

Figure 197. *Gaylussacia brachycera*

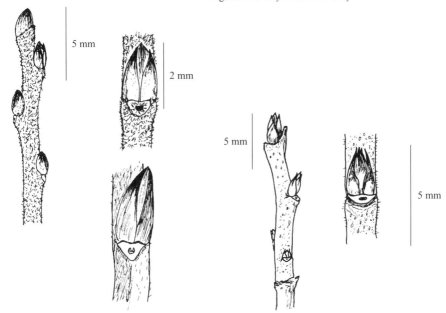

Figure 198. *Gaylussacia ursina*   Figure 199. *Gaylussacia baccata*

154 · Gelsemium

1. *Gelsemium sempervirens* St. Hilaire   CAROLINA JESSAMINE   *Figure 203a*

A widespread and common vine in the coastal plain and Piedmont, from VA to FL, west to e. TX, AR. Found in a variety of habitats from dry upland woods and fencerows to swamp forests. The calyx lobes normally drop early and are lacking on the mature fruit; this trait along with the fruit size and the small, 2mm or shorter point will usually distinguish this species from the next.

2. *Gelsemium rankinii* Small   SWAMP JESSAMINE   *Figure 203b*

An uncommon vine of wet lowlands in the coastal plain from NC to FL, west to LA. Typically in swamps and similar wet, forested habitats. The fruit is smaller than in the above species, has a beak about 3mm long, and calyx lobes usually persistent.

Ginkgo biloba L.   GINKGO   Family Ginkgoaceae   *Figure 204*

This is the sole living species of a primitive genus of extinct gymnosperms; today's plants derived from the only known source in e. China. Ultimately a large tree, fairly common in cultivation, naturalizing only on rare occasions, if at all (rodents sometimes "plant" the seed). The unique twigs bear 2 bundle scars, globular buds, and short, knobby spur shoots in mature specimens. Bark grayish to light brown, fissured or with elongated ridges. The fruit is oval, about 3cm long, yellowish in color, with a fleshy outer covering (aril) that wrinkles after maturity and is infamous for its foul odor. Seed whitish, about 2cm long; edible and often used for food in China. Plants dioecious, so only females bear fruit.

Gleditsia L.   HONEY-LOCUST   Family Fabaceae

Two native species in the SE, both thorny trees (although thornless varieties exist). Twigs of both species lack terminal buds, the laterals being sunken, and leaf scars have 3 groups of sometimes obscure bundle scars. Fruit is a flattened legume. The stiff, woody thorns may be branched on the trunk, often 1–2dm long. Young bark thin, dark, with prominent, raised lenticels; becomes scaly with age, into large, broad, loose plates. Separation of the 2 species is not possible with twigs alone.

1. Twigs mostly slender; legume less than 5cm long; usually a small tree, in understory
                                                                                      1. *G. aquatica*
1. Twigs moderately stout; legume over 10cm long; often a tall canopy tree   2. *G. triacanthos*

1. *Gleditsia aquatica* Marsh   WATER-LOCUST

Usually a small tree of the coastal plain from SC to FL, west to e. TX, north to s. MO, IL, IN. May reach 30m in height, 7dm diam. in the w. and n. portions of its range. Occupies wet lowland soils such as swamps, streamsides, and floodplains, though may appear in fairly well-drained deposits along major rivers in the w. parts of its range. The bark of young trunks is often quite lenticellate. Thorns usually more slender than in the next species, but similarly branched on the trunk. The legume is dry within, not filled with a pulp; seed flattened, nearly round, 1–3 per fruit.

2. *Gleditsia triacanthos* L.   HONEY-LOCUST   *Figure 205*

A medium to large tree (to 45m tall and over 1m diam.) of wide distribution in the SE, but much of its present range may represent naturalization after extensive cultivation. Original native range believed to be the Mississippi Valley region. Typically seen on rich soils of bottomlands or in uplands with deep soils. The elongate legumes are flat, usually 2–4dm long, straight, curved, or twisted, with a sweet pulp surrounding the oval, slightly flattened seed. The pulp in legumes not too dried out is edible and high in sugar content, and is 1 of the chief reasons for the early transport and cultivation of the tree as a food source for livestock and humans. Thorns on the trunk and lower limb bases may reach 4dm in length, but more commonly are 1–2dm

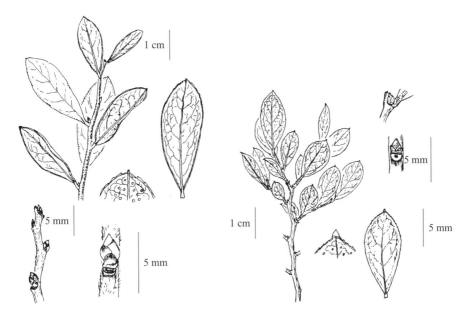

Figure 200. *Gaylussacia frondosa* var. *nana*   Figure 201. *Gaylussacia dumosa*

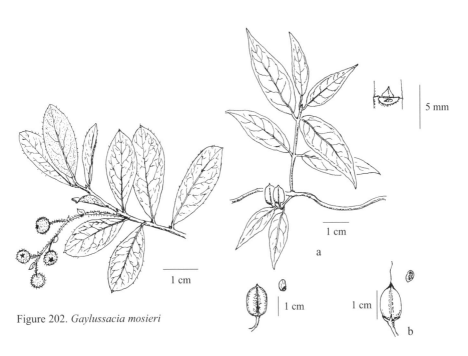

Figure 202. *Gaylussacia mosieri*

Figure 203. a: *Gelsemium sempervirens;* b: *G. rankinii*

long, branched, and may clothe the stem in masses. A widely planted thornless form (*var. inermis* Willd.) is often encountered near urban areas or habitations. Cultivars of this thornless form, though usually sterile, are commonplace as slender, round-headed, open-crowned lawn specimens.

Gordonia lasianthus (L.) Ellis   LOBLOLLY-BAY   Family Theacea   *Figure 206*

A slender-crowned evergreen tree, to 25m tall, native to the coastal plain from NC to FL and west to s. MS. Found in acidic and often peaty soil in and around bays, pocosins, depressions in pine savannas, and seepage slopes near streams. The thick, toothed leaves are usually 8–18cm long, glabrous, green beneath, not odorous when crushed, turning reddish before falling. Bark gray, thickly ridged. The fruit is a woody capsule about 18–20mm long, with a 4–7cm stalk. Known for its large (about 6cm wide) white flowers.

Gymnocladus dioicus (L.) K. Koch   KENTUCKY COFFEETREE   Family Fabaceae

Figure 207

A native tree that has naturalized in many parts of the SE due to seed spread from cultivated specimens. Original range thought to be west of the Blue Ridge, from w. WV to KS and nw. AL to LA, OK, and regions north. Typically in deep, rich soils. Twigs are stout, lenticellate, lack terminal buds, the lateral buds sunken in twigs. The pith is large, salmon or pinkish in color. The fruit is a plump legume, 8–20cm long, dark brown; seed rounded, hard, dark brown, surrounded by a green pulp that is bitter and inedible. Plants dioecious; fruits are borne only on female plants. Bark scaly, peculiarly thin-scaled when young, with whitish to pinkish colors showing between the curling plates. Mature and older trunks have dark bark with thick, scaly plates. Inner bark with odor reminiscent of green beans. Soaking the brown seed in hot water will render a coffee-colored liquid, made more bitter perhaps by cracking the seed coat; however, the seed have poisonous alkaloids within. The use of fruits or seed as a coffee substitute is not advisable, though reportedly practiced by early settlers.

Halesia Ellis   SILVERBELL   Family Styracaceae

Three species are native to the SE; 3 additional varieties are also described. Only 1 of the species has distinctive twigs for winter identification. All have 1 raised bundle scar in the leaf scars, few-scaled buds, and diaphragmed or chambered pith in the lower twig internodes or branchlets. Bark is furrowed into scaly plates on mature trunks. Fruit is an oblong seed covered by corky growth that is prolonged into 2 or 4 lateral wings. Flowers white, showy.

1. Terminal bud obviously lacking; laterals of similar size, with 3 or 4 imbricate, hairy scales; fruit 2-winged   1. *H. diptera*
1. Terminal bud often present or appearing so; laterals variously 2- to 4-scaled and of unequal sizes; fruit 4-winged
    2. Trees small or multistemmed; fruit widest near tip, gradually narrowing to stalk; wings narrow   2. *H. carolina*
    2. Trees medium to large; fruit tapering abruptly from about middle; wings broad
                                                                                                                                     3. *H. tetraptera*

1. *Halesia diptera* Ellis   TWO-WING SILVERBELL   *Figure 208*

A small tree of the coastal plain from s. SC to n. FL, and west to e. TX. Typically in moist to wet soils of swamps, floodplains, and rich slopes. The fruit is usually 4–5cm long, more greenish and a bit more succulent than is the case with other species, which tend to have brown and dry wings. Young bark not striped. The var. *magniflora* Godfrey is found in n. FL and adjacent GA; it bears larger flowers than the normal species.

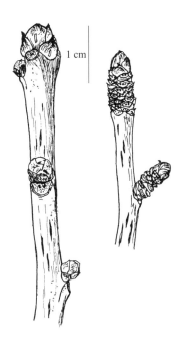

Figure 204. *Ginkgo biloba*

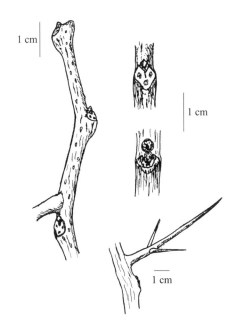

Figure 205. *Gleditsia triacanthos*

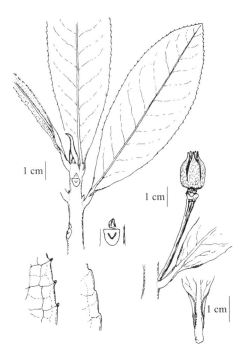

Figure 206. *Gordonia lasianthus*

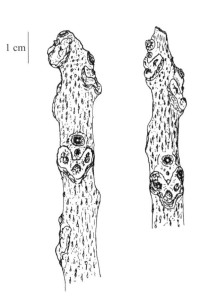

Figure 207. *Gymnocladus dioicus*

2. *Halesia carolina* L.   LITTLE SILVERBELL

A large shrub or small, multistemmed tree, to 9m, of the coastal plain and Piedmont from SC to FL and west to MS. Uncommon and occasional in sandy soils, rich woodlands, or moist lowlands. Fruit is less than 3cm long. An unusual twist of nomenclatural correction has changed the name for this species from *H. parviflora* Michx. The name *H. carolina* was applied apparently to the wrong entity for many years.

3. *Halesia tetraptera* Ellis   CAROLINA SILVERBELL   Figure 209

Scattered across much of the SE in a wide but discontinuous range; rare or missing mostly in lower coastal plain regions of NC to VA and s. LA. Most common in rich soils along major streams and in bottomlands of the Appalachians of NC to AL. Normally to about 15m tall in lower elevations, but transitions to a larger size in the Appalachians where it may reach a height of 30m in var. *monticola* (Rehd.) Reveal & Seldin. This mountain silverbell variety is also larger in overall dimensions of leaves, flowers, and fruit, and is chiefly found in moist, rich soils of coves and cool slopes above 1000m elevation, where it is a canopy species of the middle- and higher-elevation mixed hardwood forests. Fruit typically to 4cm long in the species, 4–5cm in var. *monticola*. Young bark thin, the beginning furrows imparting an orange-striped pattern. Older bark blue-gray to brown, with scaly plates; inner orange-red bark may still be visible in bottoms of some furrows. The name *H. carolina* was applied to this silverbell for many years, apparently in error.

Hamamelis L.   WITCH-HAZEL   Family Hamamelidaceae

Two native species in the SE, both large shrubs or small trees. Terminal buds are naked, scurfy; most buds stalked; 3 bundle scars. The unbeaked capsular fruit "shoots out" 2 black, lustrous seed. The widespread and common *H. virginiana* flowers in autumn–early winter, so winter twigs show stalked clusters of flowers already spent. The springtime witch-hazel, *H. vernalis*, blooms in late winter–spring, so it may have pendent, stalked flower buds present over the winter. The latter species is native mostly west of the Mississippi River, but is widely planted.

1. Buds sandy brown; flowers in late winter; stalked flower buds on twigs in winter
　　　　　　　　　　　　　　　　　　　　　　　　　　　　　　1. *H. vernalis*
1. Buds gray on at least one side; flowers in autumn; calyx remnant on twigs in winter
　　　　　　　　　　　　　　　　　　　　　　　　　　　　　　2. *H. virginiana*

1. *Hamamelis vernalis* Sarg.   SPRINGTIME WITCH-HAZEL   Figure 210

A large shrub, to 3m, clumplike and multistemmed. Native to rocky streamsides in AR, MO, OK, and sporadically east to LA, MS, AL, where it seems transitional to *H. virginiana*. Occasionally cultivated for its dense habit and sweetly aromatic, orange- or red-tinted flowers.

2. *Hamamelis virginiana* L.   WITCH-HAZEL   Figure 211

A large shrub or small tree, to 10m, with an open, coarse habit. Widespread across the SE, but uncommon or absent from areas near and west of the Mississippi River. Most common in moist woodlands and riparian zones, but also occurs in dry woodlands. Bark thin, gray, smooth. Inner bark used medicinally as an astringent.

Hedera helix L.   ENGLISH IVY   Family Vitaceae   Figure 212

An exotic, evergreen vine that climbs with aerial rootlets. The leaves are palmately lobed on juvenile wood, not so on mature plants. Terminal buds naked. Fruit berrylike, borne in umbellate

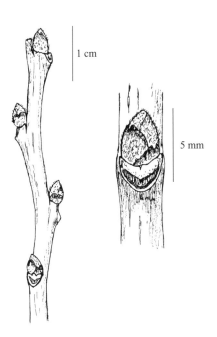

Figure 208. *Halesia diptera*

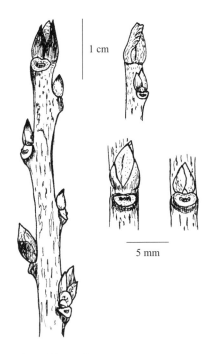

Figure 209. *Halesia tetraptera*

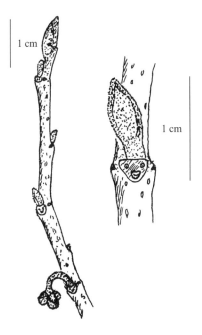

Figure 210. *Hamamelis vernalis*

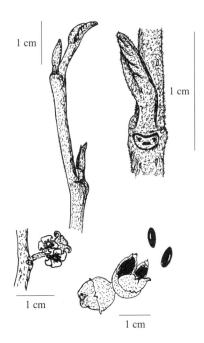

Figure 211. *Hamamelis virginiana*

clusters. Spreads from cultivation over the ground surface and up the trunks of trees. Occasionally spread by birds that transport fertile seed.

Hibiscus syriacus L.   SHRUB-ALTHEA   Family Malvaceae   *Figure 213*

An exotic shrub that commonly naturalizes near areas of its cultivation. The sunken buds are obscure, and leaf scars often are clustered at nodes. The largest leaf scar protrudes. Fruit is a capsule, with many hairy-fringed seed.

Hudsonia L.   GOLDEN-HEATHER   Family Cistaceae

Three species are native in the SE, all short, matlike shrubs rarely over 3dm high. The leaves are sufficiently persistent over the winter to allow identification by their features. Fruit is a small, 1-chambered capsule enclosed in the pubescent calyx, the calyx lobes extending past the end of the capsule. Small, showy yellow flowers appear in spring.

1. Leaves about 2mm long, scalelike, appressed to twig; fruit sessile or nearly so
   1. *H. tomentosa*
1. Leaves over 2mm, linear, not appressed; fruit with a stalk over 4mm long
   2. Leaves less than 5mm long; calyx lobes short-pointed; coastal species   2. *H. ericoides*
   2. Leaves mostly over 5mm long; calyx lobes long-pointed; only in s. Appalachians
   3. *H. montana*

1. *Hudsonia tomentosa* Nutt.   WOOLLY BEACH-HEATHER   *Figure 214a*

A densely hairy shrub generally of more n. distribution in deposits of sand near the coast and inland to the Great Lakes, reaching the SE in VA, NC. Also known from WV.

2. *Hudsonia ericoides* L.   GOLDEN-HEATHER   *Figure 214b*

A lightly hairy shrub of dry sandy soils, coastal or in pinelands inland. Occurs from rocky sites in the NE, near the coast, to DE; known in the SE only from a disjunct population in SC coastal plain.

3. *Hudsonia montana* Nutt.   MOUNTAIN GOLDEN-HEATHER   *Figure 214c*

A shrub endemic to a few isolated ledges and peaks in the Blue Ridge of NC. Habitat is sandy or thin, porous soils among rocky outcrops where naturally occurring fires reduce density of competing woody vegetation.

Hydrangea L.   HYDRANGEA   Family Hydrangeaceae

Four native species and 1 sparingly naturalized exotic are treated here for the SE. All are shrubs with opposite leaves, terminal buds, and clusters of small capsular fruits. There are normally 3 bundle scars, but 1 species may have more.

1. Terminal buds often over 10mm long, the outer scales brownish, hairy, elongated; bundle scars usually more than 3   1. *H. quercifolia*
1. Terminal buds 6mm or less, the outer scales not as above; bundle scars 3
   2. Bundle scars centrally situated in leaf scar, not raised; exotic species   2. *H. paniculata*
   2. Bundle scars raised or near raised edge of leaf scar; native species mainly identified by leaves
      3. Leaves green beneath and glabrous or nearly so   3. *H. arborescens*
      3. Leaves not as above; distinctly hairy beneath
         4. Leaves grayish beneath   4. *H. cinerea*
         4. Leaves white beneath   5. *H. radiata*

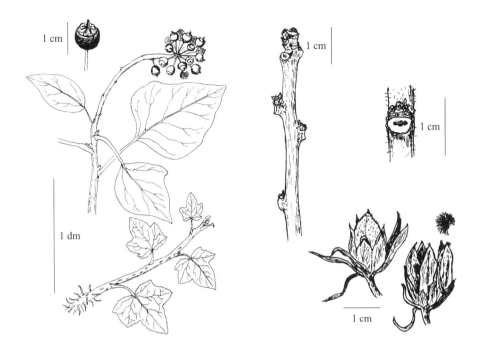

Figure 212. *Hedera helix* Figure 213. *Hibiscus syriacus*

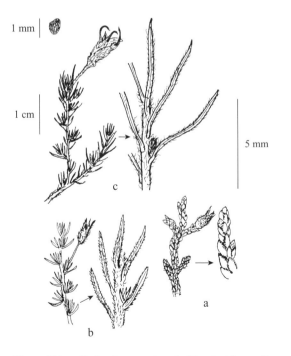

Figure 214. a: *Hudsonia tomentosa;* b: *H. ericoides;* c: *H. montana*

1. *Hydrangea quercifolia* Bartr. OAK-LEAF HYDRANGEA   *Figure 215*

A large shrub, widely planted and sometimes naturalizing out of its native range. Native to rich slopes in hardwood forests, from nw. SC across the Piedmont and coastal plain to FL, west to LA, and inland to TN. The red-brown, shreddy bark, showy panicles of white flowers, and large leaves all promote favored use as an ornamental across the SE.

2. *Hydrangea paniculata* Sieb. PEEGEE HYDRANGEA   *Figure 216*

A sprawling shrub native to Asia, occasionally persistent or spreading from cultivation. The large panicles of mostly sterile flowers are a common late summer sight in yards and urban areas across the South. Leaves sometimes whorled, and where the terminal flower clusters have been borne, twigs will not have terminal buds. Bark shreddy.

3. *Hydrangea arborescens* L. WILD HYDRANGEA   Figure 217

A widely distributed shrub across the SE, but found mainly in moist soils under hardwood forests. It often is common along woodland road banks and streams.

4. *Hydrangea cinerea* Small GRAY HYDRANGEA

Scattered in mostly upland sites and rocky outcrops in the interior regions of the SE from west of the Blue Ridge to MO, south to AR, AL, GA, and in the s. sections of the Blue Ridge from TN to SC. It is mostly found in circumneutral, basic, or calcareous soils.

5. *Hydrangea radiata* Walt. SNOWY HYDRANGEA

A conspicuous shrub with snowy white leaf undersides, endemic to the s. Blue Ridge from se. TN east to NC, nw. SC, and south to ne. GA. Typically near streams, rocky outcrops, and along roads, especially in the headwaters of the Chattooga River and adjacent watersheds to the north and northeast.

Hypericum L. HYPERICUM Family Clusiaceae (Hypericaceae)

Twenty-four native species and 2 additional varieties are treated here for the SE. These are considered the woody members of a large genus that contains additional herbaceous species. All have opposite leaves, and many in the se. coastal plain are either evergreen or the leaves are sufficiently persistent so that leaves and fruits are best used to distinguish species. Twigs alone are not diagnostic. On those more deciduous, the opposite leaf scars often have a ridge or line between them, 1 bundle scar, and partially sunken or concealed buds (leaf remnants may cover them). The clusters of capsular fruits terminate twigs where flowering has occurred, and remnant sepals at fruit bases, if present, will aid in identification. Short shoots with their accompanying small leaves, often in fascicles, are present in the axils of main stem leaves in many species; these may persist even when the larger leaves have fallen. Bark of most species is shreddy or exfoliating. Remnant styles at the fruit tip will hint at the number of separate pistils (carpels); the inner walls of the ovary in each carpel may not meet in the center and fruits will then have a single chamber (1-locular), or the walls may converge and divide the interior of fruit into 3 to 5 chambers (3- to 5-locular). Seed are tiny and numerous in each carpel.

1. Flower and fruit with 2 or 4 sepals
  2. Leaves broadly rounded to cordate or clasping at base
    3. Leaves clasping at base   1. *H. tetrapetalum*
    3. Leaves rounded to truncate at base   2. *H. crux-andreae*
  2. Leaves tapered at base, or basal part of leaf much narrower than rest of blade
    4. Outer sepals shorter than fruit; 3 styles on fruit   3. *H. microsepalum*

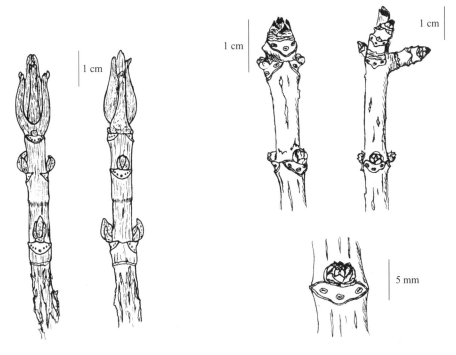

Figure 215. *Hydrangea quercifolia*

Figure 216. *Hydrangea paniculata*

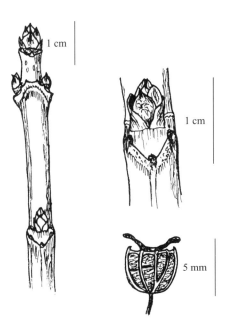

Figure 217. *Hydrangea arborescens*

4. Outer sepals as long or longer than fruit; 2 styles on fruit
  5. Flower stalks (and fruit stalks) 6–12mm long, reflexed, with tiny bractlets near stalk base   *4. H. suffruticosum*
  5. Flower stalks (and fruit stalks) 2–4mm long, erect, with tiny bractlets near base of sepals
    6. Habit erect or upright   *5. H. hypericoides*
    6. Habit matlike, spreading over the ground   *6. H. hypericoides* ssp. *muticaule*
1. Flower and fruit with 5 sepals
  7. Leaves needlelike (linear, mostly under 1.5mm wide)
    8. Leaves mostly 1cm long or less
      9. Twigs 2-winged; capsules 5mm or less; stems short, erect; of wet sandy soil   *7. H. brachyphyllum*
      9. Twigs 4- or 6-lined at nodes, not 2-winged; capsules mostly over 5mm; stems sprawling close to ground; of dry sandy soil   *8. H. reductum*
    8. Leaves mostly over 1cm long
      10. Shrubs rare in SE, endemic to FL panhandle only
        11. Bark thin, flaky, brown; habit 1m or less, slender-stemmed   *9. H. exile*
        11. Bark thick and shreddy or portions silvery or pinkish; over 1m tall
          12. Bark very thick, papery, inner parts pinkish   *10. H. chapmanii*
          12. Bark thin, shreddy, smooth portions metallic-silvery   *11. H. lissophloeus*
      10. Shrubs widespread in coastal plain, from Carolinas to AL
        13. Habit sprawling; of dry sands   *12. H. lloydii*
        13. Habit upright; of wet habitats
          14. Bark papery or shreddy   *13. H. fasciculatum*
          14. Bark thinly furrowed or narrowly flaky   *14. H. nitidum*
  7. Leaves not needlelike (not linear and over 2mm wide)
    15. Habit matlike, rarely over 15cm high; fruit 1-locular; rare mountain species   *15. H. buckleyi*
    15. Habit upright, often over 15cm; widespread, or a coastal plain species
      16. Leaves dull green to glaucous above; axillary fascicles of small leaves sparse or lacking
        17. Leaf bases cordate to clasping; sepals usually lacking on fruit   *16. H. myrtifolium*
        17. Leaf bases tapered; sepals persist on fruit
          18. Sepals leaflike, longer than fruit; bark papery, curling   *17. H. frondosum*
          18. Sepals not as above; bark flaky, with narrow shreds
            19. Capsule 4–6mm; sepals acute, widest near base   *18. H. nudiflorum*
            19. Capsule 8–10mm; sepals obtuse, widest near tip   *19. H. apocynifolium*
      16. Leaves lustrous or glossy green above; axillary fascicles of small leaves numerous
        20. Capsules 1-locular; habit slender and few-branched; woody near the base but often herbaceous above, rarely over 5dm high
          21. Capsule 4–5mm long, ovate to oblong; leaves point upward; coastal plain species   *20. H. cistifolium*
          21. Capsule 5–8mm, ovoid to subglobose; leaves spreading; rocky uplands
            22. Sepals 6–13mm long; stems rounded   *21. H. dolabriforme*
            22. Sepals 3–5mm long; stems angled   *22. H. sphaerocarpum*
        20. Capsules 3-locular or more; habit well-branched and woody; often over 5dm tall
          23. Seed less than 1mm long; largest leaves under 3cm long   *23. H. galioides*
          23. Seed mostly over 1mm; largest leaves exceed 3cm long
            24. Styles 4 or 5   *24. H. lobocarpum*
            24. Styles 3
              25. Leaves 2–7mm wide; capsule 3–6mm long   *25. H. densiflorum*
              25. Leaves 7–14mm wide; capsule 6–13mm long   *26. H. prolificum*

1. *Hypericum tetrapetalum* Lam.   LOW ST. PETER'S-WORT   *Figure 218*

A loosely branched shrub, to about 1m tall. Native to flatwoods and low, moist pinelands of FL and adjacent se. GA. Capsule about 6mm long; styles 3; outer 2 sepals large, cordate, inner 2 sepals narrow, as long as capsule or slightly longer. Twigs winged below each petiole.

2. *Hypericum crux-andreae* (L.) Crantz   ST. PETER'S-WORT   *Figure 219*

Similar in habit to above species but widely dispersed in the coastal plain from MD to FL, west to TX and north to OK, KY. Leaves are widest at base, narrowed toward tip. Capsule 7–10mm long, otherwise similar to preceding species; inner 2 sepals usually shorter than capsule. Twigs winged below each petiole.

3. *Hypericum microsepalum* (T. & G.) Gray ex S. Wats.   CROOKEA   *Figure 220a*

Rarely over 5dm tall, often only 1–3dm. Native to moist pinelands of the coastal plain of n. FL and sw. GA. Leaves rounded at apex, mostly 8–10mm long. Capsule about 4mm long; styles 3; the 4 narrow sepals nearly equal in size. [*Crookea microsepala* (T. &. G.) Small]

4. *Hypericum suffruticosum* P. Adams & Robson   PINELAND ST. ANDREW'S-CROSS
Figure 220b

Low, sprawling, under 15cm high, native to the coastal plain from NC to FL, west to LA. Found in sandy pinelands. Capsule about 4mm long; styles 2; outer sepals ovate, inner ones minute.

5. *Hypericum hypericoides* (L.) Crantz   ST. ANDREW'S-CROSS   *Figure 220c*

Bushy-branched, slender shrub usually under 1m high; distributed nearly throughout the SE. Typically in dry woods, but also occurs in many habitats from moist to dry. Capsules 4–9mm long; styles 2; inner sepals minute.

6. *Hypericum hypericoides* ssp. *multicaule* (Michx. ex Willd.) Robson   CREEPING ST. ANDREW'S-CROSS

A matlike shrub, otherwise generally similar to the above species. Leaves mostly obovate. Found in dry woods, woodland pathways and road banks, rocky or sandy soils, mostly within the range of *H. hypericoides,* but scarce to absent in the outer coastal plain. (*Hypericum stragalum* P. Adams & Robson.)

7. *Hypericum brachyphyllum* (Spach) Steud.   SHORTLEAF ST. JOHN'S-WORT   *Figure 221a*

A low, well-branched shrub with erect stems 5–12dm tall. Native to moist or wet sandy soils in the coastal plain from s. GA to s. MS, and nearly throughout FL. Largest leaves 6–10mm long; most under 8mm. Capsule about 4mm long, conical; styles 3.

8. *Hypericum reductum* (Svenson) P. Adams   LOW ST. JOHN'S-WORT   *Figure 221b*

A low, sprawling or matlike shrub with few erect stems above 2dm. Native to the coastal plain from NC to central FL, west to se. AL. Typically in dry, sandy pinelands, but also seen in lower and moister sands of pine flatwoods and savannas. Largest leaves 5–10mm, though most leaves are under 5mm. Capsule 4–8mm, conical; styles 3.

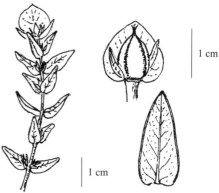

Figure 219. *Hypericum crux-andreae*

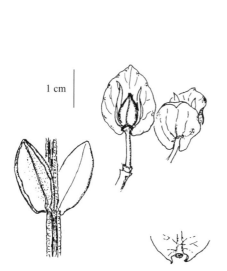

Figure 218. *Hypericum tetrapetalum*

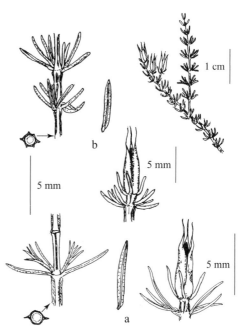

Figure 221. a: *Hypericum brachyphyllum;*
b: *H. reductum*

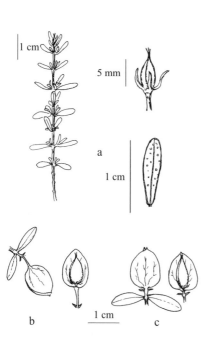

Figure 220. a: *Hypericum microsepalum;*
b: *H. suffruticosum;* c: *H. hypericoides*

9. *Hypericum exile* P. Adams   SLENDER ST. JOHN'S-WORT   *Figure 222a*

An upright shrub, to about 1m tall, well-branched only above its slender, limber main stem. Endemic to moist pinelands in FL panhandle. Leaves 12–25mm long. Capsules 5–7mm, conical; styles 3.

10. *Hypericum chapmanii* P. Adams   CHAPMAN ST. JOHN'S-WORT   *Figure 222b*

A stiffly erect shrub 2–3m tall, sometimes up to 4m and treelike in habit. Endemic to the FL panhandle, where it inhabits depressions in pine flatwoods or other areas where water may reside over the roots for part of the year. Bark is thick and spongy, with shredding layers that are light reddish-brown and striated between. Leaves mostly 12–25mm long. Capsules 5–6mm long, ovate; styles 3, but often break off due to their length (to 4mm).

11. *Hypericum lissophloeus* P. Adams   SINKHOLE ST. JOHN'S-WORT   *Figure 222c*

A slender shrub with a bushy crown, to 4m tall. Endemic to the FL panhandle, where it occurs along the edges of ponds and sinkholes; usually situated where water covers the lower parts of stems during wet seasons. Bark is mostly smooth between the peeling or ragged portions, with a brown to metallic-silvery luster. Slender, arching prop roots often develop on submerged stems. Leaves mostly 12–25mm long. Capsules 5–7mm long, conical; styles 3.

12. *Hypericum lloydii* (Svenson) P. Adams   CREEPING ST. JOHN'S-WORT   *Figure 223a*

A low, matlike shrub with few erect stems above 3dm tall. Native to the Piedmont and inner coastal plain, from NC to central GA. Typically in dry woods and sandy pinelands. Leaves 12–25mm long. Capsules 3–4mm long, ovoid; styles 3.

13. *Hypericum fasciculatum* Lam.   BUNCHLEAF ST. JOHN'S-WORT   *Figure 223b*

A stiffly erect shrub, to 2m tall, with bushy-branched crown. Native to the coastal plain from NC to s. FL, west to s. MS. Found along shores of ponds, lakes, wet depressions, and ditches, often where high water covers the lower stems. Bark shreddy and corky-thickened. Prop roots may develop on submerged parts. Leaves mostly 12–25mm long. Capsules about 4mm long, ovate to conic; styles 3, slender and spreading, but often break off due to brittle nature and length (to 4mm).

14. *Hypericum nitidum* Lam.   BLACKWATER ST. JOHN'S-WORT   *Figure 223c*

A stiffly erect shrub, to 3m tall, bushy-branched and often multistemmed. Native to the coastal plain from se. NC to n. FL, west to s. AL. Found in wet depressions in pinelands and blackwater wetlands. Bark thin, shallowly furrowed or shredding slightly in narrow flakes or strips. Leaves mostly 12–25mm long. Capsules about 4mm long, conical; styles 3, similar to those of the preceding species.

15. *Hypericum buckleyi* M. A. Curtis   BLUE RIDGE ST. JOHN'S-WORT   *Figure 224a*

A matlike shrub, rarely with stems over 3dm tall. Endemic to the Blue Ridge of sw. NC, nw. SC, ne. GA. Occurs near or on rock outcrops in balds, seepages, cliffs. Leaves mostly obovate, rounded at apex. Capsules 8–12mm long, ovoid, 1-locular; styles 3.

16. *Hypericum myrtifolium* Lam.   MYRTLELEAF ST. JOHN'S-WORT   *Figure 224b*

A slender, loosely branched shrub, to 1m tall. Native to the coastal plain from GA to se. MS, south to central FL. Occurs in moist to wet depressions in pinelands, flatwoods, and bogs.

168 · Hypericum

Leaves widest near base, ovate or triangular, narrowed to a blunt tip. Capsule about 8mm long, ovate to widely conical; styles 3.

17. *Hypericum frondosum* Michx.   GOLDEN ST. JOHN'S-WORT   *Figure 225*

A loosely to bushy-branched shrub, 1–2m tall. Widespread in calcareous regions from KY to AL, west to IN, TX, typically over limestone and in glades or barrens. A very glaucous form, with bluish foliage, is found in driest sites. Bark papery, shiny brown or metallic gray on smooth sections. Capsule 9–12mm long, ovoid conic; styles 3.

18. *Hypericum nudiflorum* Michx. ex Willd.   SMOOTH ST. JOHN'S-WORT   *Figure 226a*

Well-branched shrub with slender or multiple stems, to 1m tall. Native to the Piedmont and coastal plain from VA to FL, west to TX, n. to TN. Twigs usually winged below nodes. Capsule about 5 or 6mm long, ovoid; styles 3.

19. *Hypericum apocynifolium* Small   GULF ST. JOHN'S-WORT   *Figure 226b*

This species, which may be only a variety of *H. nudiflorum,* occurs in sw. GA, w. FL to TX and s. AR. It differs from *H. nudiflorum* by larger capsule size and sepal shape, as well as having seed to 2mm long (less than 1mm in *nudiflorum*).

20. *Hypericum cistifolium* Lam.   BOG ST. JOHN'S-WORT   *Figure 227*

A slender, semiwoody shrub with few-branched vertical stems from 2–10dm tall, but usually under 5dm. Native to the coastal plain from NC to FL, west to TX. Common in wet sites and near water, especially ditches, pineland depressions, bogs, savannas, wet flatwoods. Leaves sessile, oriented upward, light green and lustrous above. Capsule 4–5mm, ovoid or oblong; styles 3.

21. *Hypericum dolabriforme* Vent.   DRYLAND ST. JOHN'S-WORT   *Figure 228a*

Stems slender and few-branched, 2–5dm tall, often many originating from a trailing base or stolon. Native from KY to nw. GA, west to se. MO. Found in calcareous soils over limestone, in glades, or barrens. Leaves 2–4cm long. Capsule 4–8mm long, ovoid; styles 3, united into a stout beak; 1-locular.

22. *Hypericum sphaerocarpum* Michx.   ROUND-FRUITED ST. JOHN'S-WORT   *Figure 228b*

Similar to the above species but with stems finely angled or lined below nodes, sepals smaller, capsule subglobose. Native to uplands from KY to AL, west to MO, KS, AR, MS. In dry and rocky woods, glades, barrens, usually calcareous soils.

23. *Hypericum galioides* Lam.   SWAMP ST. JOHN'S-WORT   *Figure 229*

A slender, well-branched shrub, to 1–1.5m tall. Native to the coastal plain from NC to n. FL, west to e. TX. Occurs in moist or wet soils of hardwood and pine forests, along streams, ditches, and pond margins. Capsules 4–5mm long, conical; styles 3.

24. *Hypericum lobocarpum* Gattinger   INTERIOR ST. JOHN'S-WORT   *Figure 230a*

A bushy-branched shrub, to 2m. Native from KY to TN, west to se. MO, n. LA. Typical of acidic woodlands and barrens under dry to moist oak or oak-pine forests. Leaves narrow, obtuse at tip. Dense flower and fruit clusters similar to next species, but capsules 4- or 5-lobed, with 4 or 5 styles.

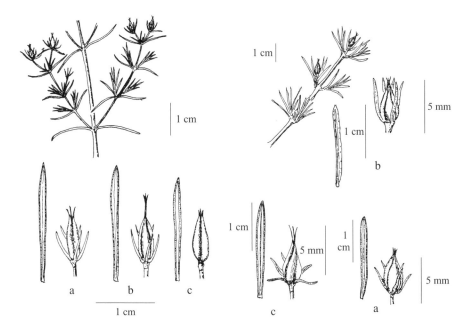

Figure 222. a: *Hypericum exile;*
b: *H. chapmanii;* c: *H. lissophloeus*

Figure 223. a: *Hypericum lloydii;*
b: *H. fasciculatum;* c: *H. nitidum*

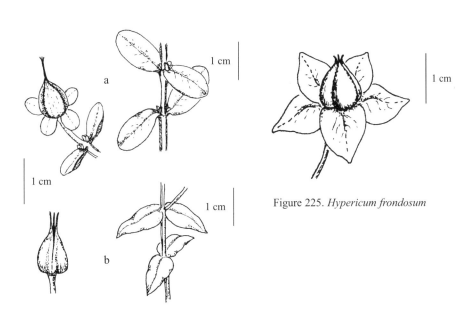

Figure 224. a: *Hypericum buckleyi;*
b: *H. myrtifolium*

Figure 225. *Hypericum frondosum*

170 · Hypericum

25. *Hypericum densiflorum* Pursh   DENSE ST. JOHN'S-WORT   *Figure 230b*

A bushy-branched shrub, 5–20dm tall. Native across much of the SE, from MD to SC, west to central AL, TX, MO. Common in the Appalachians, where it occurs on rocky outcrops, balds, bogs; scattered in other areas of the SE, mostly in wet, acidic soils. Capsules 4.5–6mm, conical; styles 3. Flower and fruit clusters crowded, mostly over 8 per inflorescence.

26. *Hypericum prolificum* L.   SHRUBBY ST. JOHN'S-WORT   *Figure 230c*

A bushy-branched shrub, 5–25dm tall. Distributed across most of the n. half of the SE, MD to central GA, west to n. LA. Occurs in a variety of habitats, including rocky woods, bogs, sandy acidic wetlands, meadows. Capsule 6–14mm long, conical; styles 3. Flowers usually up to 7 or 8 per inflorescence.

Ilex L.   HOLLY   Family Aquifoliaceae

Of the 17 species treated here as native or commonly naturalized in the SE, 8 are evergreen, with 2 of these exotic. Deciduous species cannot be identified by twigs alone, so fruits and remnant or fallen leaves are valuable to search for. All hollies generally have short terminal buds, tiny blackened stipule remnants, single bundle scars, and conspicuous lenticels. Spur shoots on mature wood are also commonly present, especially in the deciduous species. Bark is thin, gray, and mostly smooth. Fruit is a berrylike drupe with 4 to 7 seed, only produced on plants with functional pistillate flowers (plants either functionally male or female).

1. Leaves evergreen
  2. Leaf tip spinose, the spine rigid, over 2mm; if marginal teeth are present, they are similarly spinose
    3. Leaves mostly 3-spined at apex, main veins often impressed above; exotic shrub
          1. *I. cornuta*
    3. Leaves not as above, main veins not impressed; native tree   2. *I. opaca*
  2. Leaf tip not spinose, or small spinose point is 1mm or less; no long spines on leaf margins
    4. Leaves less than 8mm wide   3. *I. myrtifolia*
    4. Leaves over 8mm wide
      5. Leaves without dark dots (not punctate) on underside; fruit yellow or red
        6. Leaves entire or with a few sharp teeth   4. *I. cassine*
        6. Leaves crenate-toothed   5. *I. vomitoria*
      5. Leaves punctate on underside; fruit black
        7. Leaf serrations crenate, more than 3 per side; leaves under 3cm; calyx 4-lobed
            6. *I. crenata*
        7. Leaf serrations crenate and 3 or fewer per side, or serrations bristlelike; leaves mostly over 3cm long; calyx 5- to 9-lobed
          8. Leaf teeth blunt or rounded; most blades 2–5cm long   7. *I. glabra*
          8. Leaf teeth bristlelike; most blades 4–9cm long   8. *I. coriacea*
1. Leaves deciduous (beyond this point, twigs are not very useful for species identification)
  9. Fruit stem length (or flower stalks) mostly under 8mm
    10. Calyx lobes usually 5 to 7; seed smooth on rounded side, usually over 5 per fruit
      11. Calyx lobes obtuse, ciliate on margin; fruit opaque red; common species
            9. *I. verticillata*
      11. Calyx lobes acute, not ciliate; fruit translucent red; restricted range   10. *I. laevigata*
    10. Calyx lobes usually 4 to 5; seed ribbed or grooved on rounded side, 5 or fewer per fruit
      12. Leaf base cuneate; leaf teeth crenate, gland-tipped; calyx lobes usually 5
            11. *I. decidua*
      12. Leaf base acute to rounded; leaf teeth pointed or acute; calyx lobes usually 4
        13. Leaf blade length mostly between 2–7cm; petioles to 1cm   12. *I. ambigua*
        13. Leaf blade length mostly between 7–18cm; petioles to 2cm   13. *I. montana*

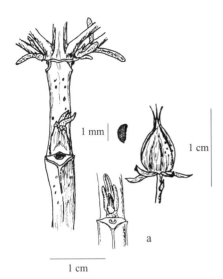
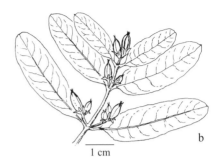

Figure 226. a: *Hypericum nudiflorum;* b: *H. apocynifolium*

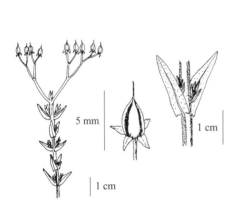
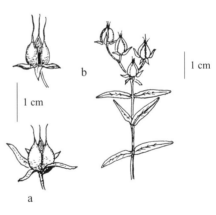

Figure 227. *Hypericum cistifolium*

Figure 228. a: *Hypericum dolabriforme;* b: *H. sphaerocarpum*

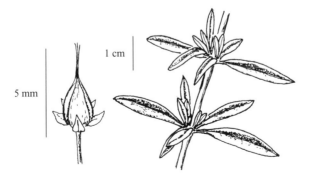

Figure 229. *Hypericum galioides*

9. Fruit stem length (or flower stalks) mostly over 8mm
   14. Leaf tips acuminate or mucronate; of high mountains or n. regions of SE
      15. Leaf tips acuminate; leaf teeth acute, prominent along most of margin   14. *I. collina*
      15. Leaf tips mucronate; leaf teeth obscure or margins mostly entire   15. *I. mucronata*
   14. Leaf tips acute to obtuse; native of lower mountain elevations to coastal plain in SE
      16. Leaf base cuneate; fruit shiny red; calyx lobes usually 5   16. *I. longipes*
      16. Leaf base acute to rounded; fruit dull red; calyx lobes 4   17. *I. amelanchier*

1. *Ilex cornuta* Lindley CHINESE HOLLY

A large evergreen shrub, to 3m, extensively cultivated and occasionally naturalizing near cultivated specimens and urban areas. Flowers borne on branchlets of the previous season, so the red fruits are not present on twigs of the current year.

2. *Ilex opaca* Ait. AMERICAN HOLLY   *Figure 231*

An evergreen tree, to 15m or more, native across most of the SE, except the Ozark region. Occurs in a variety of habitats, from dry soils to bottomlands. The var. *arenicola* (Ashe) Ashe occurs in peninsular FL, in deep scrub sands. Bark thin, gray, smoothish. The red fruits are borne on the twigs of the current year.

3. *Ilex myrtifolia* Walt. MYRTLE DAHOON   *Figure 232*

A small tree or large shrub of the coastal plain, from NC to FL, west to LA. Typically in wet soils of pine savannas, depressions, pond margins. Sufficiently similar to the next species that it is sometimes supposed only a variety of dahoon holly.

4. *Ilex cassine* L. DAHOON   *Figure 233*

A small tree, to 8m, native to the coastal plain from se. NC to FL and sparingly to e. TX. Typically in moist to wet, acidic soils. Leaves very narrow in parts of the range, approaching what is typical for the preceding species, but may be much wider and untoothed in s. portions of the range. Bark gray, sometimes with reddish blotches that are a type of crustose lichen.

5. *Ilex vomitoria* Ait. YAUPON HOLLY   *Figure 234*

A shrub or small tree, to 8m tall, common across the coastal plain from NC to FL and se. TX. Occasionally inland to n. AL, MS, and AR. Widely planted and sometimes naturalizing out of its natural range. Occurs in dry to moist, mostly sandy soils. Very common in maritime forests and other areas near the coast.

6. *Ilex crenata* Thunb. JAPANESE HOLLY   *Figure 235*

A shrub usually not over 5m tall, widely planted and occasionally naturalizing near sources of its cultivation. Many cultivars have been selected and are popular as ornamentals in the South.

7. *Ilex glabra* (L.) Gray LOW GALLBERRY   *Figure 236*

A shrub, to 3m, usually stoloniferous and forming colonies. Common in the coastal plain throughout the SE. Occurs in pinelands, bays, bogs, pocosins. The abundant fruit produced by this species is an important wildlife food in the lowlands.

8. *Ilex coriacea* (Pursh) Chapm. TALL GALLBERRY   *Figure 237*

A shrub, to 5m, native to the coastal plain from VA to FL and TX. Common; in similar habitats as the preceding species but more typical of wet sites and less common in drier pinelands.

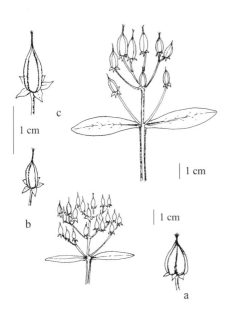

Figure 230. a: *Hypericum lobocarpum;* b: *H. densiflorum;* c: *H. prolificum*

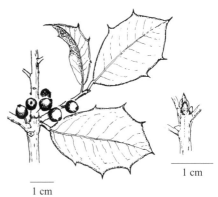

Figure 231. *Ilex opaca*

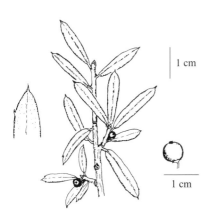

Figure 232. *Ilex myrtifolia*

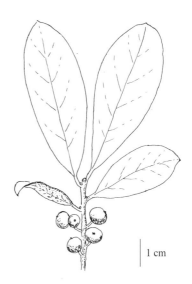

Figure 233. *Ilex cassine*

9. *Ilex verticillata* (L.) Gray   COMMON WINTERBERRY   *Figure 238*

A deciduous shrub, to 5m, occasionally taller. Widespread across the SE, though mostly in the e. half, from VA to n. FL, west to IN, MS. Typically in wet soils.

10. *Ilex laevigata* (Pursh) Gray   SMOOTH WINTERBERRY   *Figure 239*

A shrub, to 4m, native to the coastal plain and found in the SE from VA to SC. Uncommon and scattered along blackwater stream thickets and in acidic, peaty soils of pocosins. Similar to the preceding species in general appearance.

11. *Ilex decidua* Walt.   POSSUM-HAW   *Figure 240*

A shrub or small tree, to 10m, widespread across much of the SE, east and west of the Appalachian region. Typical of lowlands and moist soils; common in swamps, bottomlands, and wet pinelands with clay or high organic content in the soil. The var. *curtisii* Fern. of the n. FL Suwannee River area has leaves narrower than is typical of the species.

12. *Ilex ambigua* (Michx.) Torr.   CAROLINA HOLLY   *Figure 241*

A shrub or small tree, to 6m though more commonly under 3m. Distributed from n. NC to n. AR, south to central FL and to e. TX. Sporadic in dry woodlands or rich slopes. A similar described species, *I. buswellii* Small, has leaves mostly under 3cm long but is generally not recognized as distinct from *I. ambigua*. Another similar species that is arguably distinct but more often synonymized is *I. beadlei* Ashe, which has larger leaves than "normal" *I. ambigua,* approaching those of *I. montana*.

13. *Ilex montana* Torr. & Gray ex Gray   MOUNTAIN WINTERBERRY   *Figure 242*

A large shrub or small tree, to 12m tall, usually multistemmed or with some sprouts at base. Mostly occurs in the Appalachian region but extends sporadically through the Piedmont of VA and NC. Reported from mesic slopes of the Chattahoochee River as far south as n. FL. Common in the middle and higher elevations of the Blue Ridge, where it is found in moist hardwood forests, in balds or near rock outcrops, and boggy sites. [*I. ambigua* var. *montana* (T. & G.) Ahles.]

14. *Ilex collina* Alexander   LONGSTALK MOUNTAIN HOLLY   *Figure 243*

A large shrub, endemic to the high elevations of the s. Appalachians from WV, VA, to NC. Occurs in moist streamside thickets and bogs.

15. *Ilex mucronata* (L.) Powell, Savolainen, & Andrews   MOUNTAIN-HOLLY   *Figure 244*

A native shrub, to 3m tall, mostly a plant of the NE. Ranges into the SE in a few wet sites in the Appalachians of WV, n. VA. Twigs reddish to very dark, glabrous; spur shoots commonly on branchlets. Terminal buds with 3 to 4 reddish, rather triangular scales slightly overlapping and dark at base; lateral buds plump, reddish, with 3 to 5 lustrous scales, rarely superposed. Fruit dull red, about 8mm diam., with 4 or 5 seed-borne on a 1–3cm stalk. Known for many years as a representative of a separate genus from *Ilex,* now combined. [*Nemopanthus mucronatus* (L.) Loes.]

16. *Ilex longipes* Chapm. ex Trel.   GEORGIA HOLLY

A shrub, to 5m, distributed from the Piedmont region of NC and areas west of the Appalachians from TN to AR, south to the coastal plain of n. FL and se. TX. Occurs sporadically in dry to

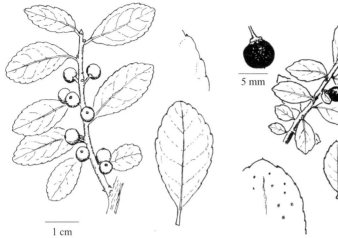

Figure 234. *Ilex vomitoria*

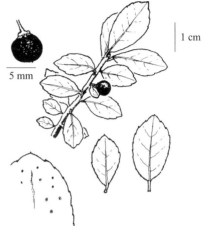

Figure 235. *Ilex crenata*

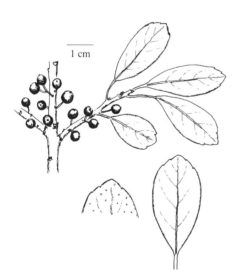

Figure 236. *Ilex glabra*

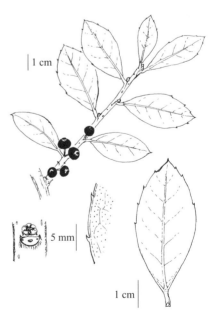

Figure 237. *Ilex coriacea*

moist woodlands, under pine or hardwood forests. Similar in general appearance to *I. decidua,* except more uniformly pubescent on leaf undersides, and fruit stems longer. A similar holly that has been described as *I. cuthbertii* Small has leaves pubescent on top surface and more obscure marginal teeth.

17. *Ilex amelanchier* M. A. Curtis ex Chapm.   SERVICEBERRY HOLLY   Figure 245

A shrub, to 5m, rare and local in distribution. Found in the coastal plain from NC to n. FL and west to e. LA. Occurs along sandhill streams, blackwater streams, and thickets near bays or cypress swamps. Leaves with prominent, reticulate venation.

Illicium L.   ANISE-TREE   Family Illiciaceae

Two species occur in the SE. Both are evergreen shrubs or small trees native to the Gulf states, but are widely planted. The spicy aroma of bruised leaves and twigs is distinctive. Fruits of both are green, circular, flat-topped clusters of follicles.

1. Leaf tips acute; branchlets and twigs mostly brown; terminal buds present   1. *I. floridanum*
1. Leaf tips obtuse or rounded; branchlets and twigs mostly green; terminal buds usually lacking   2. *I. parviflorum*

1. *Illicium floridanum* Ellis   FLORIDA ANISE-TREE   Figure 246

Ranging from w. FL to e. LA, and north into s. MS and central AL. Occurs in rich wooded ravines and along streams. Terminal bud to 15mm long, with 7 or 8 visible imbricate scales. Stalked flower buds may protrude from between scales of the terminal bud. The red flowers have numerous straplike petals, borne in leaf axils in spring.

2. *Illicium parviflorum* Michx. ex Vent.   YELLOW ANISE-TREE   Figure 247

Native to central FL, barely extending north to the area of coverage of this book. Occurs near streams, bays, hammocks. Twig tips bear several small, rounded buds, or a flat scar. Flowers are small, inconspicuous, with yellow petals; borne after twig growth, 1 often on terminal point of twig and thus leaving a branch or fruit scar on the twig rather than a terminal bud.

Indigofera suffruticosa P. Mill.   INDIGO   Family Fabaceae   Figure 248

An exotic shrub, to 2m, occasionally naturalized near areas of its cultivation mostly in the outer coastal plain. Other species of this genus are cultivated, but nearly all are predominately herbaceous and not native or naturalizing in the SE. Twigs lack terminal buds, the tip often dying back in winter; stipules are persistent but brittle and easily broken; leaf scar raised, with a thickened edge or hairs that partially hide the small bundle scars. Legumes 1–2cm long, plump, recurved, borne in clusters.

Itea virginica L.   VIRGINIA SWEETSPIRE   Family Saxifragaceae   Figure 249

A native shrub of moist to wet soils, throughout the SE. Absent from middle and higher elevations of the Blue Ridge. The terminal bud may appear naked or obscurely scaled. Tiny blackened stipules may persist, and there are 3 bundle scars. The small pith is chambered. Fruit is a 2-parted capsule borne in long racemes.

Iva L.   MARSH-ELDER   Family Asteraceae

Two woody species occur in the SE. Both are native shrubs of maritime habitats. Leaves often partially or wholly persisting in winter, opposite or alternate on the gray twigs. Twig tips may die back in winter freezes.

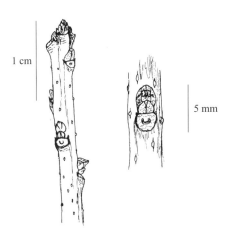

Figure 238. *Ilex verticillata*

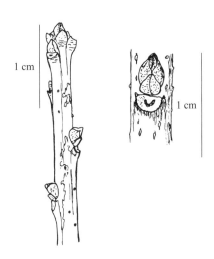

Figure 239. *Ilex laevigata*

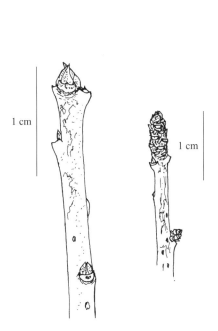

Figure 240. *Ilex decidua*

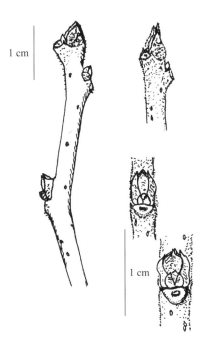

Figure 241. *Ilex ambigua*

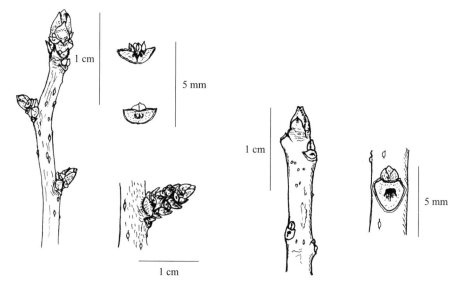

Figure 242. *Ilex montana*

Figure 243. *Ilex collina*

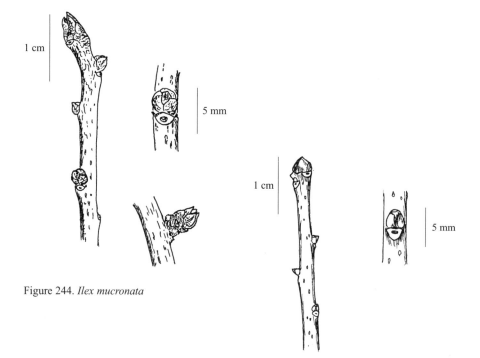

Figure 244. *Ilex mucronata*

Figure 245. *Ilex amelanchier*

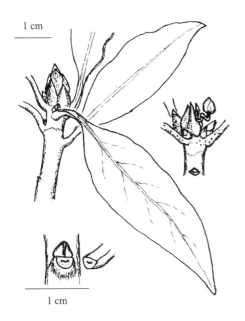

Figure 246. *Illicium floridanum*

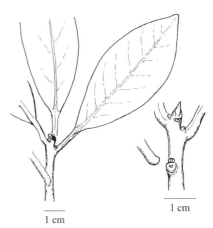

Figure 247. *Illicium parviflorum*

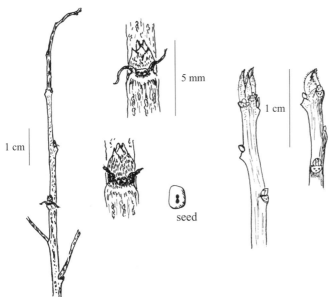

Figure 248. *Indigofera suffruticosa*

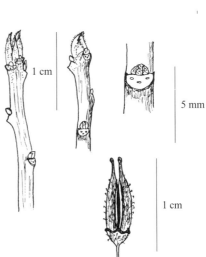

Figure 249. *Itea virginica*

1. Leaves firm but not succulent, often toothed, opposite on all but upper portions of twigs
   1. *I. frutescens*
1. Leaves thick, succulent, mostly untoothed, alternate on all but lower portions of twigs
   2. *I. imbricata*

1. *Iva frutescens* L.   MARSH-ELDER   *Figure 250*

A shrub, usually 1–2m high, of coastal zones throughout the SE. Occurs along or partially in brackish marshes and seacoast waterways, often in colonies or extensive groves. Leaves dull green, opposite on majority of the twig, though may be alternate in the inflorescence or near the twig tip. Leaf scars somewhat 3-lobed, with 3 bundle scars usually evident. Larger leaves mostly about 5cm long, though in var. *oraria* (Bartlett) Fernald & Gleason, the leaves are larger and more distinctly toothed; this form in NC and northward.

2. *Iva imbricata* Walt.   BEACH-ELDER   *Figure 251*

A weakly ascending shrub, rarely over 1m tall, the twigs semiwoody to herbaceous. Native to the coastal zone from VA to LA, mainly very near the beach. Typically in dunes and drier borders of beaches, above high tide level. Frequently seen partially buried by shifting sand. Leaf scars crescent-shaped, with obscure bundle scars, mostly alternate but may be opposite on lowermost parts of twig. Leaves glossy, light green, succulent.

Jasminum nudiflorum Lindl.   FLOWERING JASMINE   Family Oleaceae   *Figure 252*

A low, sprawling, exotic shrub that may persist after cultivation or spread near home sites on rare occasions. Leaves are opposite, trifoliate, sometimes persisting over part of the winter. Twigs are dark green and heavily lined, with no terminal bud. The dried petiole base persists over the leaf scar, but when removed the single bundle scar is seen. Flowers in late winter.

Juglans L.   WALNUT   Family Juglandaceae

Two species are native trees in the SE. Both are trees with stout twigs; the terminal buds partially valvate-scaled, scurfy; lateral buds often superposed. Pith is chambered. The fruit is a corrugated nut enclosed in a green, juicy husk that does not split along any sutures. Sap is aromatic, with an odor peculiar to this genus.

1. Leaf scars bear a hairy fringe over top margin; terminal buds tawny or light brown; nut oblong
   1. *J. cinerea*
1. Leaf scars not as above; terminal buds grayish; nut globular
   2. *J. nigra*

1. *Juglans cinerea* L.   BUTTERNUT   *Figure 253*

A tree, to 30m, scattered across the n. half of the SE. Typical of rich, deep soils. Bark is thin, gray, smooth on young trunks, becoming furrowed, with flat-topped gray ridges with age. The butternut wilt disease, a fungal infection that causes cankers and death of branches or trunks, has killed large numbers of trees and is particularly detrimental to seedling and sapling regeneration of this species in the SE.

2. *Juglans nigra* L.   BLACK WALNUT   *Figure 254*

A tree, to 40m, distributed across most of the SE except for parts of the coastal plain and poorly drained habitats. So extensively transported and planted over many human generations that it may have naturalized or expanded beyond an uncertain original range. Bark brownish and lightly scaly when young, becoming thick, dark, fissured and ridged with age.

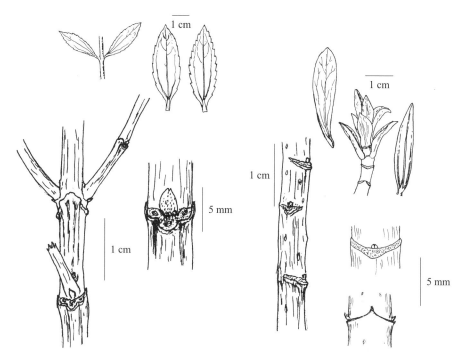

Figure 250. *Iva frutescens*

Figure 251. *Iva imbricata*

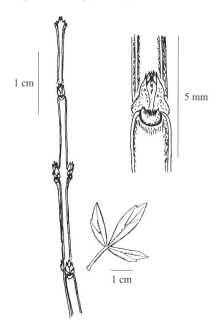

Figure 252. *Jasminum nudiflorum*

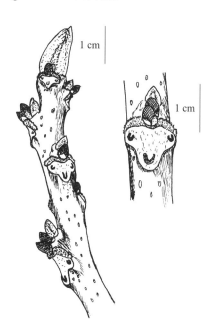

Figure 253. *Juglans cinerea*

182 · Juniperus

Juniperus L. JUNIPER Family Cupressaceae

Three species are native to the SE, with 1 additional variety of importance. All are evergreen conifers with fleshy cones (berrylike in appearance) that often appear blue due to a whitish, waxy layer. Leaves are scalelike (acicular) or flattish and needlelike (linear or subulate). Bark is shreddy, fibrous. Plants dioecious, fruit produced only on pistillate trees.

1. Leaves all needlelike or subulate, spreading from twig; a sprawling shrub
                                                                     1. *J. communis* var. *depressa*
1. Leaves subulate only on juvenile twigs, appressed and scalelike on mature twigs; a tree
    2. Leaf margins finely toothed (use 20× magnification); cones 6–9mm long; seed 4–6mm long; of w. portions of the SE   2. *J. ashei*
    2. Leaf margins not toothed; cones 3–6mm long; seed 1.5–4mm long; e. and widespread species
        3. Cones 4–6mm long; common tree   3. *J. virginiana*
        3. Cones 3–4mm long; confined to vicinity of coast   4. *J. virginiana* var. *salicicola*

1. *Juniperus communis* var. *depressa* Pursh. COMMON JUNIPER   *Figure 255*

A sprawling shrub in the SE, though upright forms from 3–10m are known in other regions. The species *J. communis* is widely distributed across N. Amer. and n. Europe, with 5 varieties described. Ranging into the SE is var. *depressa,* found only in a few places in VA, NC, SC, and GA. Habitats vary from dry, sandy soils in the coastal plain to rock outcrops in the Piedmont and Appalachians. Leaves opposite or whorled, mostly 8–12mm long, with 1 broad central white band on 1 side, tips sharply pointed. Twigs with jointed appearance, lined below each leaf. Cones 6–9mm long.

2. *Juniperus ashei* Buchh. ASHE'S JUNIPER   *Figure 256a*

A small tree, to 15m, native to a few w. states in the SE. Occurs on calcareous, rocky soils of MO, AR, OK, and TX. Highest concentrations are found in central TX. Bark brown, very shreddy.

3. *Juniperus virginiana* L. EASTERN RED-CEDAR   *Figure 256b*

A common and widespread tree, to 30m tall, across most of the SE. Seemingly replaced in the outer coastal plain, near the coast, by the next variety. Grows in a wide variety of habitats but most frequent on circumneutral to calcareous soils, old fields, and rock outcrops in upland woods. Leaves on juvenile shoots sharply tipped and subulate; those of mature shoots scalelike. Crown usually dense and conical, widening with age but rarely flattening out.

4. *Juniperus virginiana* var. *silicicola* (Small) E. Murray SOUTHERN RED-CEDAR
                                                                                                                      Figure 256c

A tree of the outer coastal plain from NC to FL, AL. Mainly restricted to the vicinity of the coast in hammocks and shell middens near brackish marshes, maritime forests, stable sands near dunes, and along rivers. Crown tends to be broadly conical when young, widening with age and often ultimately with a flattened and open aspect on old trees.

Kalmia L. LAUREL Family Ericaceae

Five native species are treated here for the SE. All are predominately shrubby in habit and occur in acidic soils. The fruit is a 5-parted, subglobose capsule that is usually persistent in winter.

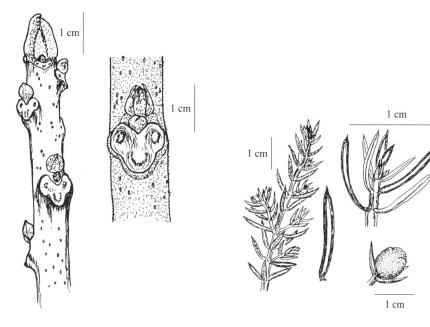

Figure 254. *Juglans nigra*

Figure 255. *Juniperus communis* var. *depressa*

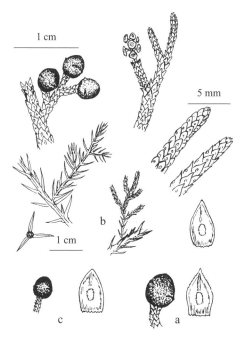

Figure 256. a: *Juniperus ashei;* b: *J. virginiana;* c: *J. virginiana* var. *silicicola*

1. Leaves deciduous; terminal bud obviously present, conical, scales reddish; a rare shrub of coastal plain .... 1. *K. cuneata*
1. Leaves evergreen; no terminal bud, or if present, rounded and greenish-scaled; widespread species
   2. Leaves less than 2cm long, all parts hirsute (with long, white, stiff hairs) ...... 2. *K. hirsuta*
   2. Leaves and twigs not as above
      3. Leaves alternate, often 3cm wide or more .... 3. *K. latifolia*
      3. Leaves opposite or whorled, rarely over 25mm wide
         4. Calyx and bracts with stalked glands; of VA coastal plain and north ...... 4. *K. angustifolia*
         4. Calyx and bracts not glandular; of areas mostly south of VA ...... 5. *K. carolina*

1. *Kalmia cuneata* Michx. WHITE-WICKY *Figure 257*

A low shrub, to 1.5m tall, stoloniferous and forming small colonies. Endemic to the coastal plain of NC, SC, where it is known only from a few populations affiliated with pocosin habitats. Capsules nodding on the slender pedicels. Most parts, including the twigs, are stipitate-glandular.

2. *Kalmia hirsuta* Walt. HAIRY-WICKY *Figure 258*

A low, stoloniferous shrub, rarely over 3dm tall, usually hidden among other low vegetation in pine flatwoods and savannas. Ranges in the coastal plain from SC to FL, west at least as far as AL. The tiny leaves are mostly 6–10mm long. Stiff white hairs occur on most parts. Capsule about 3mm wide, with 5 longer and persistent calyx lobes.

3. *Kalmia latifolia* L. MOUNTAIN-LAUREL *Figure 259*

A large shrub, occasionally with treelike habit, to 10m tall. Widespread in uplands of the SE, east of the Mississippi River. Rare or lacking in the outer coastal plain from NC to FL. Abundant in the Blue Ridge, seen at nearly all elevations. The elongate clusters of flower buds are stalked and visible over winter, but other buds are sunken and hidden underneath the petiole bases. Often planted as an ornamental for its showy flowers; many cultivars have been developed.

4. *Kalmia angustifolia* L. SHEEP-LAUREL *Figure 260*

A slender shrub, mostly 1m or less in height. Ranges from the NE, where it is most common, into n. and se. VA. Occurs in dry, sandy soils as well as wet, acidic sites.

5. *Kalmia carolina* Small CAROLINA SHEEP-LAUREL

Similar to the above species, but mostly south of its range. Occurs in the coastal plain from se. VA to SC, and in the Blue Ridge from sw. VA to ne. GA. Habitat is primarily acidic wetlands, but tolerates drier sandy soils in pinelands of the coastal plain.

Kerria japonica (L.) DC. EASTER-ROSE Family Rosaceae *Figure 261*

An exotic shrub that is occasionally persistent or spreading near old sources of its cultivation. Once a favorite ornamental due to its yellow spring flowers and green stems. Twigs slender, green, with persistent, dried stipules. Terminal bud lacking; collateral buds often present. Fruit rarely seen; spreads primarily by root sprouts.

Lagerstroemia indica L. CRAPE-MYRTLE Family Lythraceae *Figure 262*

This exotic shrub or small tree is extensively planted, and may linger or naturalize through sprouting long after cultivation, especially in the coastal plain. It is not known to spread by

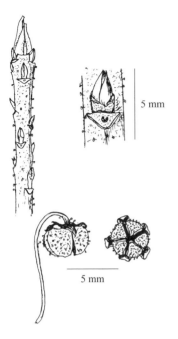

Figure 257. *Kalmia cuneata*

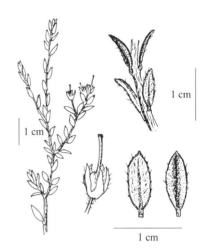

Figure 258. *Kalmia hirsuta*

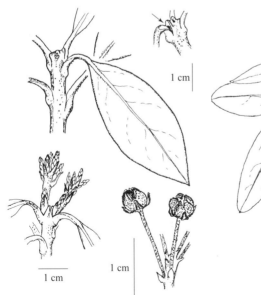

Figure 259. *Kalmia latifolia*

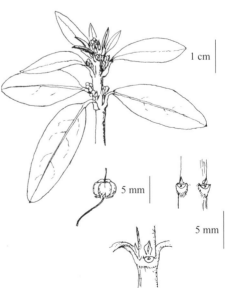

Figure 260. *Kalmia angustifolia*

seed. Twigs lack a terminal bud, and 4 ridges of brittle tissue often persist on the twigs. Bark is mottled; the scaling, darker outer bark reveals smooth, gray to pinkish inner bark. Fruit is a 5-parted capsule about 8mm diam.

Lantana L.   LANTANA   Family Verbenaceae

Four species are here considered woody and naturalized in the SE, mainly in the coastal plain from NC to TX. There are several more naturalizing and native species in subtropical FL. The opposite, toothed leaves, showy flowers, and heads of bluish berries are distinct from any other native shrub. Crushed parts are rankly aromatic.

1. Prickles often present on lower parts of stems
   2. Calyx lobes shorter than the calyx tube ..... 1. *L. horrida*
   2. Calyx lobes as long as or longer than tube ..... 2. *L. camara*
1. Prickles absent from all stems
   3. Habit upright or partially clambering ..... 3. *L. ovatifolia*
   3. Habit sprawling or prostrate and vinelike ..... 4. *L. montevidensis*

1. *Lantana horrida* H.B.K.   TEXAS LANTANA

An erect shrub, about 2m tall in the SE. Native to tropical Amer., possibly as far north as TX. Sparingly naturalized near plantings; most commonly seen in calcareous and rocky soils in TX and the sw. US. Twigs 4-sided and unarmed, or armed with stout curved prickles. Leaves mostly cordate at base, strongly and rankly odorous when crushed.

2. *Lantana camara* L.   LANTANA   *Figures 263 and 264*

An erect shrub, about 1–2m tall depending on severity of winter freezes and dieback. Native to tropical Amer.; naturalized mostly in vicinity of its cultivation. Usually seen in urban and waste areas; more common in FL. Twigs somewhat 4-angled; curved prickles may be present along the twigs. Buds small, hairy; leaf scars raised; bundle scar single, sometimes obscure. Scraped green twigs have strong, tomato-like odor. Pith white, striated or minutely spongy, interrupted through the nodes.

3. *Lantana ovatifolia* Britt.   LANTANA

An erect shrub, to 2m tall, sometimes clambering on adjoining vegetation. Native to tropical Amer.; naturalized sporadically in coastal areas of the SE, mainly near urban environments. Leaves mostly elliptic, base cuneate to rounded.

4. *Lantana montevidensis* (Spreng.) Briq.   WEEPING LANTANA

A sprawling or vinelike shrub that suffers dieback from winter freezes. Native to S. Amer.; naturalized near sources of its cultivation in urban areas and waste places. Twigs slightly hispid on tips, less so toward base; prickles lacking.

Larix laricina (DuRoi) K. Koch   LARCH   Family Pinaceae   *Figure 265*

A boreal native tree species, only reaching the SE by an occurrence in n. WV. This tree is widespread across n. N. Amer., in bogs and acidic swamps. The deciduous needles of this conifer leave peglike scars flanked by ridges, and the many-scaled globose buds are conspicuous. Spur shoots are stubby and often numerous on branchlets. Cones 2cm or shorter. Bark grayish to brown, scaly. Although no exotic larches are known to naturalize in the SE, they are more often planted and seen than the native species. Two of the most used are *L. decidua* Mill. (European

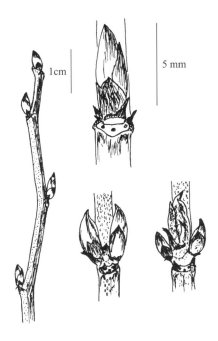

Figure 261. *Kerria japonica*

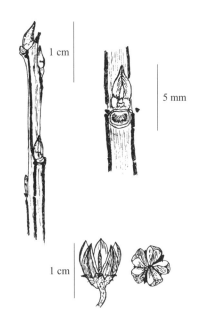

Figure 262. *Lagerstroemia indica*

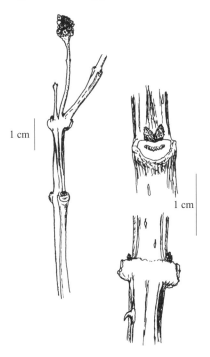

Figure 263. *Lantana camara*

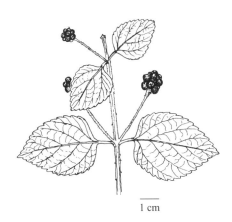

Figure 264. *Lantana camara*

larch), with 25–30mm cones that have unreflexed scale tips, and *L. leptolepis* (Sieb. & Zucc) Gord. (Japanese larch), with similar size cones but reflexed scale tips.

Leiophyllum buxifolium (Bergius) Ell.   SAND-MYRTLE   Family Ericaceae   *Figure 266*

An evergreen shrub distributed in the SE in the Appalachians of NC, KY, TN, SC, GA; also sporadic in the Piedmont and coastal plain of NC, SC. Occurs amid rock outcrops, dry sandy uplands, and in wet, peaty soils in pinelands. Habit may be prostrate to upright, 2m or less. Twigs lined, and leaf scars face upward since petioles lie against twigs for up to ⅔ of their length; 1 tiny bundle scar in leaf scar. Buds on twig tip resemble tiny clustered leaves. Leaves alternate, subopposite, or opposite, rarely over 2cm long (more commonly 8–15mm). The fruit is a small capsule, about 3mm long.

Leitneria floridana Chapm.   CORKWOOD   Family Leitneriaceae   *Figure 267*

An uncommon native shrub distributed from n. FL and s. GA to e. TX, and to e. AR, se. MO. Usually a coarsely branched upright shrub with numerous stems originating as a colony from a suckering root system, but sometimes treelike to 7m. Occurs in wet soils in swampy or marshy coastal plain habitats. Twigs moderately stout, with hairy catkin buds near the twig apex; terminal bud usually present; white hairs often obscure all but protruding tips of bud scales. Bark smooth and conspicuously lenticellate, with rank odor when scraped. Dioecious; the oblong, 2cm drupes borne on pistillate plants.

Lespedeza Michx.   SHRUB LESPEDEZA   Family Fabaceae

Of the many species in this genus, only 2 are woody enough to be included here, and both are Asian exotics. These are often mass-planted, through seed, for wildlife habitat improvement. They mature into clumplike, multistemmed shrubs, and spread from plantings into disturbed soils along fields, roads, and waste places. The stems usually die back in winter to varying degrees depending on the extent of freezing, but basal portions, at least, are woody. Collateral buds often present; bud scales 5 to 9, reddish, finely lined; leaf scars raised, bundle scars obscure. The fruit is a small, 1-seeded legume. Twigs alone cannot be used to separate the 2 species; seed, fruit with the remnant calyx lobes, or flowers are needed.

1. Calyx lobes mostly as long as calyx tube, or shorter; seed olive-green or mottled   1. *L. bicolor*
1. Calyx lobes mostly longer than tube; seed purple-brown   2. *L. thunbergii*

1. *Lespedeza bicolor* Turcz.   SHRUB LESPEDEZA   *Figure 268*

A commonly planted shrub, to 3m tall. Flowers in summer; the racemes of the panicle erect, each to 40cm long; flowers about 12mm long or less, petals dark pink to purplish. Color of winter twigs brown or purplish-brown on one side.

2. *Lespedeza thunbergii* (DC.) Nakai   JAPANESE LESPEDEZA

Less commonly planted intentionally than *L. bicolor*; to 2m tall. Flowers at same time or later than *L. bicolor*; the racemes of the panicle not as erect, each to 20cm long; flowers 12mm or a little longer, petals purplish. Twig color more purplish than above species.

Leucothoe D. Don   LEUCOTHOE   Family Ericaceae

Four native species occur in the SE, 2 of which are evergreen. All are shrubs with alternate leaves, white urceolate flowers, and 5-parted capsules persisting into winter. The common name of "dog-hobble" is often applied to the evergreen species, and "sweetbells" to the deciduous ones. Flower buds are formed by autumn, as terminal racemes or as axillary, cylindric buds; these are present all winter.

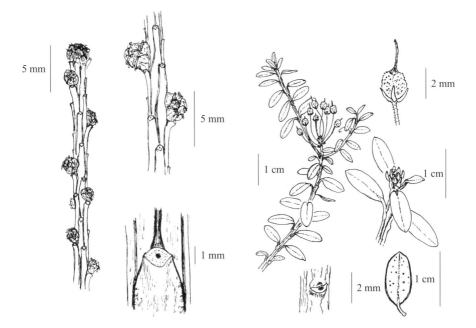

Figure 265. *Larix laricina*

Figure 266. *Leiophyllum buxifolium*

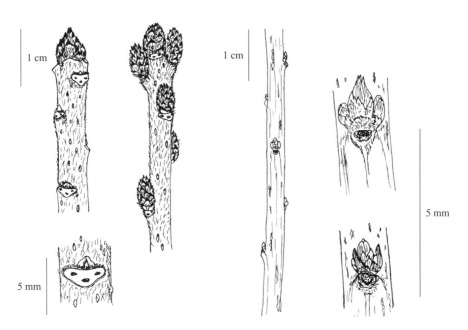

Figure 267. *Leitneria floridana*

Figure 268. *Lespedeza bicolor*

1. Leaves evergreen
    2. Leaf tips mostly acute; petioles 3–8mm; mostly of coastal plain 1. *L. axillaris*
    2. Leaf tips acuminate; petioles 8–15mm; Appalachians and adjacent Piedmont
 2. *L. fontanesiana*
1. Leaves deciduous
    3. Bracts on racemes of flower buds leaflike, numerous; capsules nearly round in cross section; seed not winged; mostly of Piedmont and coastal plain 3. *L. racemosa*
    3. Bracts linear, few or lacking; capsules angled or indented between sutures in cross section; seed winged; mostly of Appalachian mountains 4. *L. recurva*

1. *Leucothoe axillaris* (Lam.) D. Don   COASTAL DOG-HOBBLE   Figure 269a

A low, stoloniferous shrub, to 1.5m tall, with numerous, few-branched stems. Native to the coastal plain from VA to FL, west to LA. Typically in wet, acidic soils near streams, thickets, swampy woods, pocosins, bays.

2. *Leucothoe fontanesiana* (Steud.) Sleumer   MOUNTAIN DOG-HOBBLE   Figure 269b

Similar to above species in habit, but to 2m. Native to the Appalachian region from sw. VA to nw. GA and sparingly in the adjacent upper Piedmont of NC, SC. Commonly along streams, shady woodland roads, cool slopes, moist, acidic woodlands, and ericaceous thickets of nearly all mountain elevations. Often forms extensive colonies of arching and intertwined stems.

3. *Leucothoe racemosa* (L.) Gray   SWAMP SWEETBELLS   Figure 270

A slender shrub, to about 4m tall. Distributed primarily in the coastal plain and lower Piedmont from MD to FL, west to LA. Rare in the Appalachians. Occurs in wet, acidic soils, especially of bogs, bays, pocosins, swamps, and streamside thickets. The capsules are more rounded in cross section, with minimal indentions or lobing; the persistent calyx lobes are triangular, extending 1mm or more beyond capsule body outline.

4. *Leucothoe recurva* (Buckl.) Gray   MOUNTAIN SWEETBELLS   Figure 271

A slender shrub, to 4m tall. Distributed primarily in the Appalachian region from s. WV, e. KY to n. GA, and sw. VA to nw. SC. Rarely in the adjacent upper Piedmont. Typically in acidic woodlands, rocky slopes, openings in high elevations, ericaceous thickets, boggy sites, and occasionally in dry woods under oaks. The capsules are more angular and lobed than the above species, with ovate calyx lobes that barely extend beyond the capsule body outline.

Licania michauxii Prance   GOPHER-APPLE   Family Chrysobalanaceae   Figure 272

A low, stoloniferous shrub with numerous, relatively unbranched stems arising to a height of 8–20cm across most of its range, reaching 4dm or more in some s. sections (especially peninsular FL). Distributed in the coastal plain from se. SC to FL, west to LA. Occurs in deep, dry sands in pinelands or scrub, and along sandy roadsides. The extensive, running root system produces what appears to be a groundcover or mass of slender twigs and evergreen leaves. Leaves alternate, light green, glossy, with reticulate venation. Twigs slender, with small, persistent, blackened stipules and 3 bundle scars. The fruit is an oblong drupe to 3cm, reportedly eaten by the gopher tortoise.

Ligustrum L.   PRIVET   Family Oleaceae

No privet is native to N. Amer., but at least 12 species of these exotic shrubs are cultivated. Eight of these are currently known as having naturalized to some degree in the e. US and are treated here. If not extensively naturalized across the se. region, they are indicated as "rarely naturalized," though in some areas these may nevertheless be locally abundant and invasive.

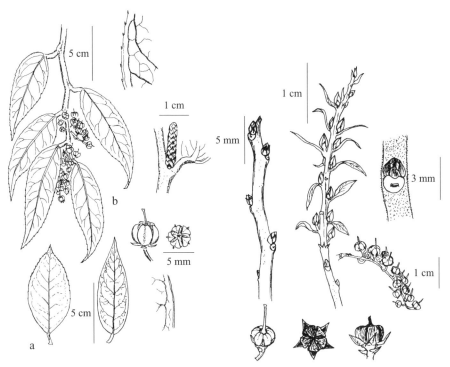

Figure 269. a: *Leucothoe axillaris*;
b: *L. fontanesiana*

Figure 270. *Leucothoe racemosa*

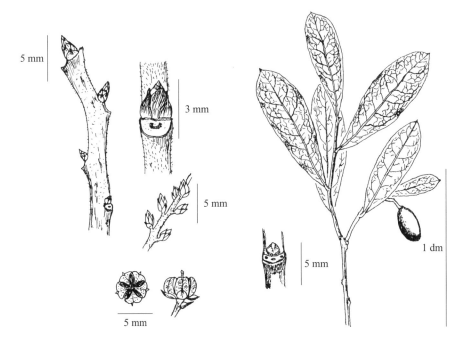

Figure 271. *Leucothoe recurva*

Figure 272. *Licania michauxii*

All bear opposite (rarely whorled), untoothed leaves; thin, gray bark that is smooth except for prominent, raised lenticels; and blue-black drupes. Fruit and seed spread by birds is the primary dispersal method in their naturalization. The white flowers are borne in late spring or summer, and the fruits persist into winter. Two species, *japonicum* and *lucidum,* are true evergreens, but many of the others may hold their leaves well into the winter, especially in milder climates of the SE. The twigs alone cannot be relied on to separate the species.

1. Branchlets or leaves glabrous
    2. Corolla tube over twice as long as the corolla lobes   1. *L. ovalifolium*
    2. Corolla not as above
        3. Leaves thick, glossy, evergreen
            4. Leaf tips bluntly pointed; corolla tube slightly longer than lobes; shrub rarely over 3m
                    2. *L. japonicum*
            4. Leaf tips acute or acuminate; corolla tube about equal to lobes; treelike, often over 4m
                    3. *L. lucidum*
        3. Leaves thin to firm, dull green, deciduous or late-deciduous
            5. Flowers sessile, open Aug–Sept; leaf petiole 1–3mm   4. *L. quihoui*
            5. Flowers stalked, open June–July; leaf petiole 3–10mm   5. *L. vulgare*
1. Branchlets pubescent; leaves pubescent, at least along midvein on leaf underside
    6. Corolla tube about as long as corolla lobes   6. *L. sinense*
    6. Corolla tube much longer than lobes
        7. Calyx glabrous   7. *L. amurense*
        7. Calyx pubescent   8. *L. obtusifolium*

1. *Ligustrum ovalifolium* Hassk.   CALIFORNIA PRIVET

A stiffly upright shrub, to 5m tall, with glossy leaves late-deciduous or subevergreen. Leaves 3–6cm; petioles 3–5mm long. Fruit 5–7mm diam. Native to Japan, not California. Most similar in general appearance to *L. japonicum*. Rarely naturalized.

2. *Ligustrum japonicum* Thunb.   JAPANESE PRIVET

A stiffly branched shrub usually to 3m, occasionally to 5m. Leaves evergreen, glossy, 4–8cm, veins indistinct; petiole 6–12mm. Fruit 4–7mm diam. Native to Japan, Korea. Occasionally naturalized, mostly near urban areas.

3. *Ligustrum lucidum* Ait. f.   GLOSSY PRIVET   *Figure 273*

Tall and treelike, to 10m in height. Leaves 8–12cm, thick, evergreen, dark green above but not as glossy as *L. japonicum,* despite the common name; veins distinct; petiole 10–20mm. Fruit 5–8mm diam. Native to e. Asia. An invasive species of bottomlands, lowland streambanks, and lands near urban areas in the Piedmont and coastal plain.

4. *Ligustrum quihoui* Carr.   WAXLEAF PRIVET

A spreading shrub, to 2m tall. Leaves 2–5cm long, mostly under 15mm wide, obtuse or notched at apex; petioles 1–3mm. Native to China. Rarely naturalized.

5. *Ligustrum vulgare* L.   EUROPEAN PRIVET   *Figure 274*

A spreading shrub, to 5m tall. Leaves 3–6cm long; petiole 3–10mm. Fruit 4–7mm diam., more lustrous than other species. Native to Europe, n. Africa. Rarely naturalized.

6. *Ligustrum sinense* Lour.   CHINESE PRIVET   *Figures 275 and 276*

A spreading shrub, to 4m tall. Leaves 2–6cm; petiole 2–6mm. Fruit 3–4mm diam. Twigs and petioles densely pubescent. Native to China. This is a commonly and widely naturalized exotic,

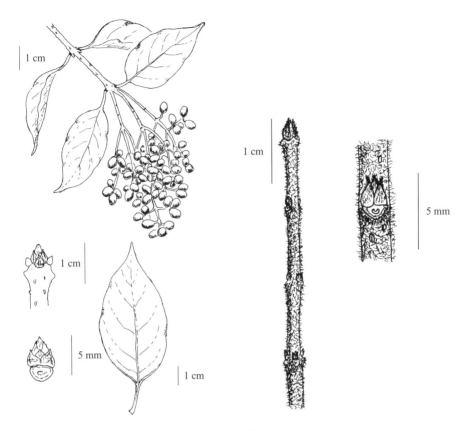

Figure 273. *Ligustrum lucidum*

Figure 275. *Ligustrum sinense*

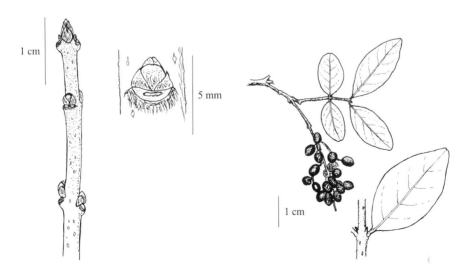

Figure 274. *Ligustrum vulgare*

Figure 276. *Ligustrum sinense*

7. Ligustrum amurense Carr. AMUR PRIVET

A dense shrub, to 5m tall. Leaves 3–6cm, sometimes lustrous; petiole 2–4mm, pubescent. Fruit 5–7mm diam., often glaucous, or with a slight bloom. Native to n. China. Rarely naturalized.

8. Ligustrum obtusifolium Sieb. & Zucc. BORDER PRIVET

A spreading shrub, to 4m tall. Leaves 2–6cm; petiole 1–5mm, pubescent. Fruit 3–5mm diam., sometimes glaucous. Twigs densely pubescent. Native to Japan. Most similar in general appearance to *L. sinense* and sharing with that species a tendency to invade bottomland forests, though not as extensively naturalized across the region.

Lindera Thunb.   SPICEBUSH   Family Lauraceae

Three species are native in the SE. All bear alternate leaves, the leaf scars usually with a single bundle scar (typical of most members of the laurel family), though occasionally this bundle trace may appear broken into segments of 2 or 3. Scraped parts are spicy-aromatic. Terminal buds are usually present, larger than laterals; superposed buds are common, and the spherical, stalked flower buds may be collateral in 2 species. The fruit is a red, lustrous drupe.

1. Terminal buds naked, brown-hairy; habit under 2m, colonial, with slender, multiple stems
   1. *L. melissifolia*
1. Terminal buds scaled; habit usually over 2m, clumplike and well-branched
   2. Fruit stalks 4mm long; sap with lemony scent; leaves thick, glaucous below, tip rounded; rare species   2. *L. subcoriaceae*
   2. Fruit stalks 2–4mm; sap spicy-scented; leaves thin, not glaucous below, tips mostly pointed; common and widespread   3. *L. benzoin*

1. Lindera melissifolia (Walt.) Blume   PONDBERRY   Figure 277

A low, slender shrub forming colonies near wet and swampy depressions and bogs in the outer coastal plain, usually 1m or less in height. Rare and local in distribution, known only from a limited number of sites from se. NC to sw. GA, and in s. AL, se. MO; historically from a higher number of sites, including FL, MS. The elongate leaves have prominent veins, hairy surfaces, and tend to droop from twigs. Twigs usually olive-green, with an odor when scraped reminiscent of sassafras. Fruit 9–12mm long.

2. Lindera subcoriacea B. E. Wofford   BOG SPICEBUSH

An upright shrub, to 3m tall, with ascending branches. Rare and local in known distribution, mostly in the coastal plain from se. VA to FL, west to LA. It resembles the next species most closely and may easily be overlooked. Occurs in acidic, boggy soils. Leaves much thicker in texture and paler underneath than *L. benzoin*, and usually rounded at the tip; the sap odor differs as well.

3. Lindera benzoin (L.) Blume   SPICEBUSH   Figure 278

A spreading shrub, to 4m tall, with multiple stems and a wide crown. Widely distributed across the SE; most common in deep, moist soils of basic or circumneutral nature. Frequent in bottomlands, on rich slopes, in coves, and by streamsides. Bark thin, dark, smooth but with raised, conspicuous lenticels. A hairy form, var. *pubescens* (Palmer & Steyermark) Rehd., is more common in the coastal plain.

Linnaea borealis var. *americana* (Forbes) Rehd.   TWINFLOWER   Family Caprifoliaceae
Figure 279

A diminutive shrub, trailing over the ground, with stems rarely ascending to 1dm. Primarily a boreal species, circumpolar in distribution, in n. Europe, n. Asia, and n. N. Amer. The variety described here occurs over most of N. Amer. and barely reaches the se. region by its occurrence in WV. Habitat is cool, moist woodlands, often associated with peat or moss. Leaves evergreen, opposite, rounded, with crenate teeth. Vertical flower stems (peduncles) typically fork into 2 pedicels, each with a nodding flower that later may form a fuzzy, 1-seeded capsular fruit.

Liquidambar styraciflua L.   SWEETGUM   Family Hamamelidaceae   *Figure 280*

A large native tree distributed throughout the SE except for middle and higher elevations in the Blue Ridge. Common in a wide range of habitats, but most typical of moist lowlands and deep, moist soils. Twigs have glossy, imbricate bud scales on the terminal and lateral buds, and 3 bundle scars that are white-bordered. Corky growth may occur on branchlets and branches of some trees. The compound fruit is spherical, about 3–4cm diam., derived from a head of flowers that develop into the multiple of capsules, each capsule with a beaked tip. Seed are winged, accompanied in the capsules by a mass of abortive seed resembling sawdust. Bark brown to grayish, roughly fissured and with elongated ridges.

Liriodendron tulipifera L.   TULIP-TREE   Family Magnoliaceae   *Figure 281*

A large native tree distributed across most of the SE, but uncommon to absent very near the coast and west of the Mississippi River. Habitats vary, but moist, aerated soils are preferred nearly throughout the range. All scraped parts spicy-aromatic. Twigs bear large terminal buds with 2 glabrous, valvate scales; stipule scars are ringlike, encircling twigs; bundle scars many; pith diaphragmed. The fruit is an oblong, somewhat conical aggregate of samaras, basal remnants of which persist on the twigs of mature trees over the winter. Bark of young trunks smooth, thin, gray, often with white striping as furrows develop; mature bark gray to brown, with narrow ridges; old bark thickly ridged and furrowed.

Litsea aestivalis (L.) Fern.   PONDSPICE   Family Lauraceae   *Figure 282*

A rare native shrub, to 3m, of swamp depressions and margins of ponds and sinks in the coastal plain, from se. VA to n. FL, west to LA. Known from relatively few locations within this range. The twigs are slender, glabrous, often zigzag, usually reddish, with terminal buds. Stalked, enlarged flower buds may be present. Bark gray, smooth. Leaves 1–3cm long. The fruit is a red drupe, 4–6mm long. Plants dioecious, so fruit produced only on pistillate specimens. There is little aroma to scraped twigs, compared to other members of the laurel family.

Lonicera L.   HONEYSUCKLE   Family Caprifoliaceae

Twelve species are treated here, as native or naturalized in the SE; 7 are shrubs (only 1 of these native), and 5 are vines (4 are native). There are 3 bundle scars, but often these are obscured by callousing of the leaf scar or by petiole remnants. Fruit is a reddish berry (black, in 1 species). Twigs alone cannot be used to identify many of the species; the use of leaves and flowers is a requirement and thus an impediment in winter identification. Flowers of several species from Asia turn from white to yellow as the flower ages past anthesis. The exotic bush-honeysuckles used in ornamental horticulture are numerous, and more species than the 6 mentioned here may become naturalized in the SE. Spread of these shrubs is most conspicuous near urban areas, closest to the concentrations of their cultivation. As birds continue to feed on the fruits and disperse the seed, the range of spread widens. Already, several of these Asian shrubs have become serious invasive weeds of woodland understories. They seem especially adapted to calcareous soils.

Figure 277. *Lindera melissifolia*

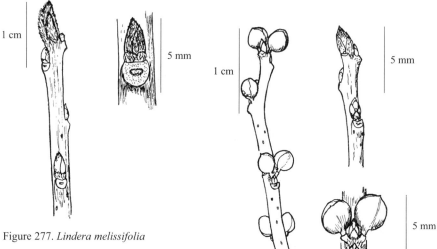

Figure 278. *Lindera benzoin*

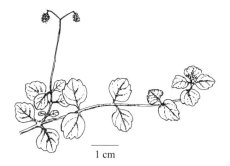

Figure 279. *Linnaea borealis* var. *americana*

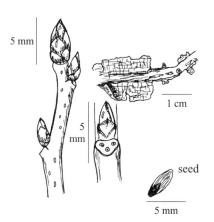

Figure 280. *Liquidambar styraciflua*

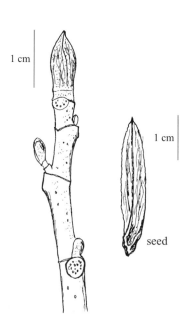

Figure 281. *Liriodendron tulipifera*

1. Habit vinelike
    2. Twigs densely pubescent; leaves usually persistent on some stems 1. *L. japonicum*
    2. Twigs only slightly pubescent or glabrous; leaves deciduous (twigs not diagnostic after this point)
        3. Flowers over 3cm long, slender and trumpet-shaped, corolla lips all nearly equal in size; a common, widespread species 2. *L. sempervirens*
        3. Flowers under 3cm, not as above, upper corolla lip broad, 4-lobed; not common species
            4. Connate leaves at twig tips oval or round, glaucous above, blunt or rounded at tips; flowers light yellow 3. *L. reticulata*
            4. Connate leaves oval or elliptic, green or only slightly glaucous above, blunt or pointed at tips; flowers yellow to reddish
                5. Flowers reddish, tube enlarged on 1 basal side; leaves white beneath 4. *L. dioica*
                5. Flowers yellow to orange, no tube enlargement; leaves grayish beneath 5. *L. flava*
1. Habit shrubby
    6. Branchlet pith solid (homogenous)
        7. Terminal buds usually present; leaves thin, deciduous, tip blunt; rare native species 6. *L. canadensis*
        7. Terminal buds usually lacking; leaves stiff, semievergreen, bristle-tipped; exotic 7. *L. fragrantissima*
    6. Branchlet pith hollow (excavated)
        8. Stalks bearing fruit (and flowers) shorter than leaf petioles; leaves long-pointed at apex 8. *L. maackii*
        8. Stalks of fruit longer than petioles; leaf tips rounded to bluntly pointed
            9. Leaves and peduncles of flowers and fruit glabrous or only minutely hairy; flowers pinkish
                10. Flower stalks 15–40mm long, flowers usually pink 9. *L. tatarica*
                10. Flower stalks 3–15mm, flowers usually pale pink, turning yellow 10. *L.* × *bella*
            9. Leaves and peduncles of flowers and fruit pubescent; flowers usually white, turning yellow
                11. Buds rounded, blunt; length of flower bractlets about equal to ovary 11. *L. morrowii*
                11. Buds elongated, pointed; bractlets of flowers shorter than ovary 12. *L. xylosteum*

1. *Lonicera japonica* Thunb. JAPANESE HONEYSUCKLE *Figures 283 and 284*

An exotic, subevergreen vine that sprawls over the ground or twines into shrubs and small trees. Distributed throughout the SE, and 1 of the most common and tenacious of our exotic "woody weeds." Berries black, lustrous; consumed and spread by birds. Terminal buds usually present. Bark grayish brown, shreddy. Native to Asia.

2. *Lonicera sempervirens* L. CORAL HONEYSUCKLE *Figure 285*

A widespread native vine in the SE but scarce in the Blue Ridge. Leaves white beneath. Flowers usually red. Berries orange-red to red, borne in whorled groups on a spike, with sections of naked stem between.

3. *Lonicera reticulata* Raf. GRAPE HONEYSUCKLE

An uncommon native vine, distributed mostly west of the Appalachians, scattered from KY, TN west to MO, AR. Typically in rocky woodlands in circumneutral to calcareous soils. Leaves are white beneath and strongly glaucous above. Flowers pale yellow. Fruit orange to red. (*L. prolifera* [Kirch.] Rehd.)

4. *Lonicera dioica* L. MOUNTAIN HONEYSUCKLE *Figure 286*

A native vine sparingly distributed in the Appalachians from MD to n. GA, and west to AR, OK, MO. Occurs in dry to moist woods, most commonly in circumneutral soils or rich, rocky

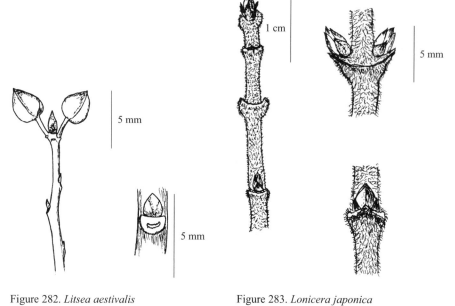

Figure 282. *Litsea aestivalis*

Figure 283. *Lonicera japonica*

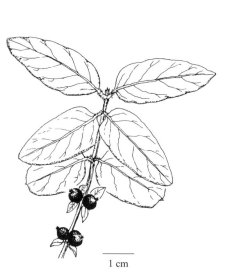

Figure 284. *Lonicera japonica*

Figure 285. *Lonicera sempervirens*

hardwood forests. Leaves white beneath, dark green above. Flowers purplish to reddish. Fruit orange-red to red. The var. *glaucescens* Rydb. has hairy leaves.

5. *Lonicera flava* Sims.　YELLOW HONEYSUCKLE

A native vine sporadically distributed in the southernmost part of the Appalachians and upper Piedmont of NC, SC, GA, AL, TN, and west in KY, AR, OK, MO. Mostly seen in circumneutral or calcareous soils, in rocky openings or bluffs. Leaves lightly glaucous to very pale green beneath. Flowers light yellow to orange-yellow. Fruit orange-red, often borne in 1 crowded whorl on the fruiting spike. (*L. flavida* Cockerell)

6. *Lonicera canadensis* Bartr. ex Marsh.　AMERICAN FLY-HONEYSUCKLE　*Figure 287*

A suckering shrub, to 2m tall, mainly more n. in its distribution but ranging southward in high Appalachian elevations from WV to n. GA. Uncommon, occurring on rocky, moist slopes and ridges or in high-elevation openings. Twigs glabrous, yellow-green to grayish, ridged below leaf scars, usually with a terminal bud; most buds elongate and sharply pointed. Leaves ciliate on margin. Flowers yellow-green to white. Berries red, borne paired at the end of the peduncle.

7. *Lonicera fragrantissima* Lindl. & Pax.　SWEET-BREATH-OF-SPRING *Figures 288 and 289*

A densely branched shrub, to 3m tall. Leaves subevergreen, fairly stiff, tip curved downward or apiculate. Twigs yellow-brown, glabrous; terminal bud present or absent; leaf scars raised, sometimes with petiole remnants attached. Yellowish-white flowers, sweetly fragrant, appear in late winter. Berry orange to red. Native to e. China.

8. *Lonicera maackii* (Rupr.) Maxim.　AMUR HONEYSUCKLE

A densely branched shrub, to 5m tall. A particularly aggressive and invasive weed in calcareous areas west of the Blue Ridge, where it may dominate forest understories to the exclusion of nearly all other plants. Leaves thin, acuminate at tip. Terminal buds present or not; buds rounded, with light brown scales. Flowers white, turning yellow. Berries dull, dark red until fully ripe, then glossy, bright red; appear close to twig due to short stalks. Bark of mature stems brown or gray, with flat ridges and fissures between. Native to Korea, Manchuria.

9. *Lonicera tatarica* L.　TATARIAN HONEYSUCKLE　*Figure 290*

A densely branched shrub, to 3m. Leaves thin, acute at tip, rounded to subcordate at base. Twigs as in previous species. Flowers pink, or if white, not turning yellow. Berries orange to red on long stalks. Native to s. Russia, Turkestan.

10. *Lonicera* × *bella* Zabel　BUSH-HONEYSUCKLE

A dense shrub, to 3m. A hybrid of *L. tatarica* and *L. morrowii,* with intermediate characters. Introduced from Asia.

11. *Lonicera morrowii* Gray　MORROW HONEYSUCKLE

A widely branching shrub, to 2.5m. Leaves thin, tip obtuse, base widely rounded. Terminal buds present or not; buds rounded or bluntly pointed. Flowers white, turning yellow. Berries red, sometimes yellow. Native to Japan.

12. *Lonicera xylosteum* L.   EUROPEAN FLY-HONEYSUCKLE

A dense shrub, to 3m tall. Leaves thin, acute at tip, rounded at base; gray-green above, paler and hairy beneath. Buds elongate and pointed; terminal bud present or not. Flowers yellowish-white, similar to *L. fragrantissima* but not as fragrant. Fruit dark red. Native to Eurasia.

Lupinus L.   LUPINE   Family Fabaceae

Of several species in this genus native to the SE, 3 are mentioned here. The consideration of these being woody enough to include in this treatment is warranted only by their tendency in some s. sites to have leafy stems over the course of the winter season, including buds that provide additional primary growth in spring. At the risk of setting a precedent for validating inclusion of all perennials that may be green in winter, the last 2 clumplike species of lupine are herein included as cespitose subshrubs. The first lupine is suffruticose, with more shrublike erect stems. The leaves of all are mostly elliptic, alternate, distinctly pubescent, and with a distinct petiole. Fruit a hairy legume, 2–5mm long,

1. Habit shrublike, with upright branching stems to 1m tall 1. *L. westianus*
1. Habit herblike, with cespitose, mostly herbaceous stems (but may be woody at base)
  2. Leaves finely short-pubescent; flowers blue 2. *L. diffusus*
  2. Leaves shaggy or woolly; flowers brownish-pink 3. *L. villosus*

1. *Lupinus westianus* Small   SMALL'S LUPINE   Figure 291

A loosely branched low shrub, to 1m, with some dieback in winter. Endemic to the FL panhandle, where it is found in dry sands of dunes, pinelands, scrub, or disturbed areas. Leaves gray-green, softly and densely pubescent on all surfaces. Flowers blue, with a reddish-purple interior spot. Legume about 2cm long, covered with silky hairs. A similar species of peninsular FL, out of range of this book, is the scrub lupine *Lupinus cumulicola* Small with legumes to 3cm.

2. *Lupinus diffusus* Nutt.   SPREADING LUPINE   Figure 292

Herblike and clumped, but with decumbent stems reaching 4dm in length. Scattered in the coastal plain and adjacent lower Piedmont from s. NC to FL, west to MS. Habitat is usually dry, sandy uplands in pinelands or sparse oak woodlands. Leaves are densely short-pubescent. Flowers blue, with a cream-colored interior spot. Legume about 4cm long, with short hairs.

3. *Lupinus villosus* Willd.   LADY LUPINE

Herblike and clumped, with decumbent stems to 3dm long. Scattered in the coastal plain from s. NC to FL, west to MS. Typically in dry, sandy pinelands and scrub. Leaves bear long, shaggy hairs. Flowers flesh-tinted to pinkish, with reddish-purple interior spot. Legume about 4cm long, shaggy.

Lycium L.   MATRIMONY-VINE   Family Solanaceae

Three species are treated here for the SE, 1 being native; the other 2, naturalized exotics. All are clambering, vinelike shrubs with red berries in autumn. The leaf scars are often clustered at the nodes, each with 1 bundle scar. Depending on species, spines or sunken buds on the twigs may aid identification.

1. Buds sunken, rarely visible; twigs not lined; leaves less than 1cm wide; native of coastal areas 1. *L. carolinianum*
1. Buds visible; twigs lined; leaves over 1cm wide; exotics of various habitats

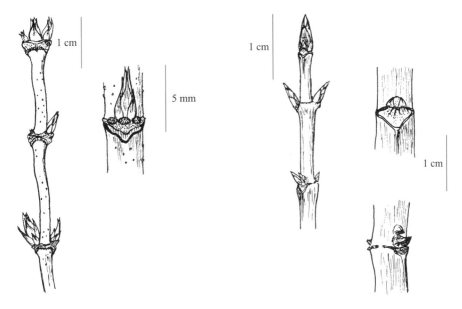

Figure 286. *Lonicera dioica*

Figure 287. *Lonicera canadensis*

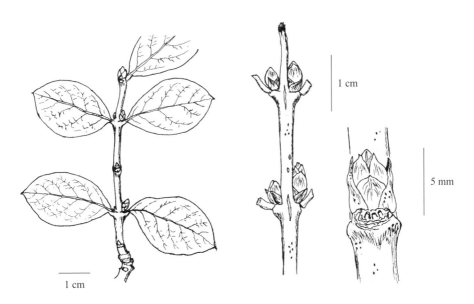

Figure 288. *Lonicera fragrantissima*

Figure 289. *Lonicera fragrantissima*

2. Spines occasionally terminate twigs; leaves mostly oblong-lanceolate, gray-green
                                   2. *L. barbarum*
  2. Spines lacking; leaves rhombic-ovate, bright green   3. *L. chinense*

1. *Lycium carolinianum* Walt. CHRISTMAS-BERRY   *Figure 293*

A clambering shrub, to 3m, usually shorter. Bears many short side branches from the few elongate, flexible main branches. Occurs in vicinity of the coast, along brackish marshes, shell mounds, even to salty soils of high tide shores. Somewhat sporadic in its range from GA and FL to TX, possibly east to SC; most common in FL and the West Indies. Twigs gray, with swollen nodes (these swellings actually spur shoots), and often spine-tipped. Buds sunken, with several leaf scars visible in the nodal swellings. Leaves narrowly cuneate to the sessile base, somewhat succulent, to 2.5cm long and 5mm wide, late-deciduous. Flowers mostly blue, 4-petaled, borne in late summer. Fruit shiny red, to 15mm long, ripe in early autumn.

2. *Lycium barbarum* L. MATRIMONY-VINE   *Figure 294*

An upright, clambering, or vinelike shrub. Known to naturalize sparingly in several regions of the SE, mostly in waste areas, road or railway margins, old home sites, or urban areas close to its main centers of cultivation. Twigs gray, lined, sometimes spinose. The small buds are visible above the leaf scars, which may crowd on the nodes and short spur branches. Leaves 2–7cm long, 1–3cm wide. Flowers purplish; fruit to 2cm long, borne on stalks 8–20mm long. Native to Europe, w. Asia; used ornamentally or for hedges.

3. *Lycium chinense* P. Mill. CHINESE MATRIMONY-VINE

A clambering shrub similar in most respects to the above species in habit and habitat preference. Twigs yellow-gray, usually lacking spines. Leaves 3–8cm long, 15–50mm wide, bright green and late-deciduous. Fruit, to 2.5cm, on 3–12mm stalks. Native to e. Asia.

Lyonia Nutt. STAGGERBUSH Family Ericaceae

Five species are native to the SE, of which 2 are deciduous. True deciduous species will bear no leaves on the twigs of the previous season (branchlets). Twigs lack terminal buds. The fruit is a 5-parted capsule, with a thickened suture or rib along the suture.

1. Leaves deciduous
 2. Buds 2-scaled; capsule globular; leaves toothed   1. *L. ligustrina*
 2. Buds with 3 or more scales; capsule somewhat pyramidal, or elongated at end; leaves entire   2. *L. mariana*
1. Leaves evergreen
 3. Lower leaf surface green, glabrous, but may be dotted; a vein runs along edge of leaf margin   3. *L. lucida*
 3. Lower leaf surface with many rusty or grayish peltate scales; not as above
  4. Capsules borne on branchlets; lower leaf surface often with large rusty scales and smaller gray scales; leaf margin strongly revolute   4. *L. ferruginea*
  4. Capsules borne on twigs; lower leaf surface mostly with only rusty scales; leaf margin revolute but strongly so only on sprouts   5. *L. fruticosa*

1. *Lyonia ligustrina* (L.) DC. MALEBERRY   *Figure 295*

A slender shrub, to 4m tall. Distributed throughout the SE. The typical variety occurs mainly in the upland and interior parts of the SE, from the Piedmont of SC and AL northward and east of the Mississippi River; habitats vary from moist, acidic soils to dry ridges, and to high elevations in the Blue Ridge. The var. *foliosiflora* (Michx.) Fern. occurs in the coastal plain throughout the

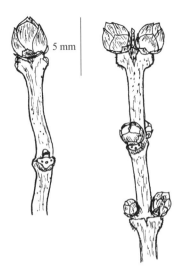

Figure 290. *Lonicera tatarica*

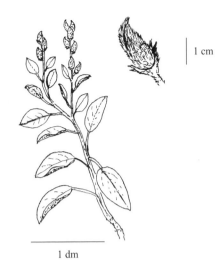

Figure 291. *Lupinus westianus*

Figure 292. *Lupinus diffusus*

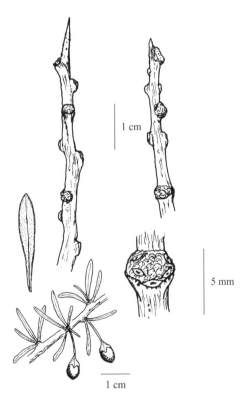

Figure 293. *Lycium carolinianum*

SE, and to OK, AR, MO; mostly in wet sites. Flowers of the latter have leaflike bracts partially obscuring them. All types have twigs varying from glabrous to hairy, often with a row of hairs down 1 or 2 sides. Capsule about 3mm.

2. *Lyonia mariana* (L.) D. Don   LOW STAGGERBUSH   Figure 296

A low shrub, rarely over 1m tall (but may reach 2m). Native to the coastal plain and lower Piedmont from MD to FL. Also appears from nw. LA to MO, west to adjacent TX, OK. Habitat is typically pinelands, in both dry and wet acidic conditions. Twigs ascending, often angular. Capsule about 5mm long.

3. *Lyonia lucida* (Lam.) K. Koch   FETTERBUSH   Figure 297

A multistemmed shrub, to 2m tall. Native to the coastal plain from VA to FL, west to LA. Typically in wet, acidic soils of pine flatwoods, savannas, bays, bogs, swampy streams, and thickets; also occasional in drier uplands. Twigs green, distinctly ridged. Leaves thick, evergreen, with a fine marginal vein. Capsule about 4–5mm long.

4. *Lyonia ferruginea* (Walt.) Nutt.   TREE LYONIA   Figure 298

A shrub or small tree, usually 2 or 3m in n. parts of its range, to 6m in the south. Native to the coastal plain from se. SC to FL, where it ranges far southward in peninsular FL. Occurs in moist sands or peaty soils, and on dry sands of pinelands and scrub. Rusty-colored peltate scales on leaf undersides may mostly slough on oldest leaves, but a mix of the 2 scale types is usually present. Twigs tend to be slender and not rigidly ascending, with no fruit borne on them (these being on branchlets). Capsule about 3 or 4mm, rounded or ovoid. Bark brown, thinly scaly.

5. *Lyonia fruticosa* (Michx.) G. S. Torr.   STAGGERBUSH   Figure 299

A multistemmed shrub, usually 2m tall or less. Native to the coastal plain from GA to FL, and south in peninsular FL. Habitat similar to above species, though more common in the moister situations. The leaf undersides of older leaves, in winter, may show only scattered rusty peltate scales over a glaucous or blue-green surface. The revolute margins are most noticeable from the blade base, less pronounced toward the tip. Twigs tend to be rather stiff and ascending, often with abundant capsules near the twig tip; the leaves also tend to be dramatically reduced in size from twig base toward its apex. Capsule about 5mm long, egg-shaped.

Macfadyena unguis-cati (L.) A. H. Gentry   CAT'S-CLAW-VINE   Family Bignoniaceae
Figure 300

An exotic vine from S. Amer., sporadically naturalized in the coastal plain from SC to FL, and possibly westward. Usually seen in or near urban areas. The compound leaves have 2 leaflets with a 3-clawed tendril between them. Some leaves may persist in winter, or at least the tendrils are present. Climbing is achieved by the tendrils, with aerial rootlets anchoring the larger stems to a vertical substrate. Flowers yellow, borne in summer; fruit a slim capsule to 4dm long, rarely seen except on high-climbing mature vines.

Maclura pomifera (Raf.) Schneid.   OSAGE-ORANGE   Family Moraceae   Figure 301

A tree originally native to w. sections of the SE but now widely naturalized due to past cultivation. It makes best development in deep, rich soils such as bottomlands, which typify its habitat in its original range. It may be encountered throughout the SE on other sites, including calcareous and rocky soils. The branchlets often have knobby spur shoots. Twigs usually bear thorns at

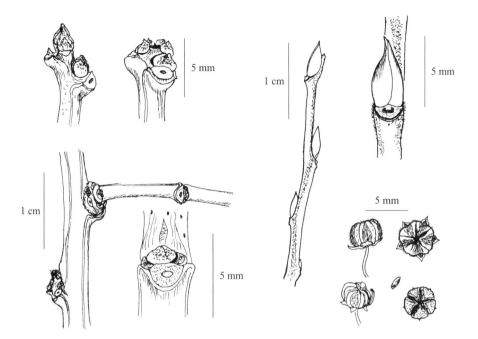

Figure 294. *Lycium barbarum*

Figure 295. *Lyonia ligustrina*

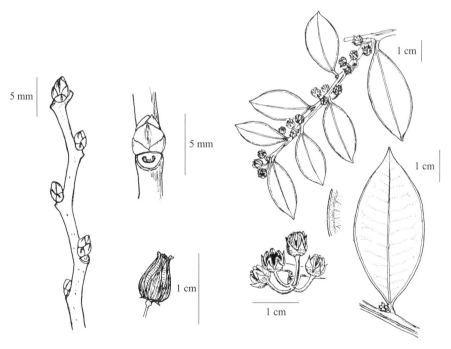

Figure 296. *Lyonia mariana*

Figure 297. *Lyonia lucida*

the nodes and obscure bundle scars; there is no terminal bud. Bark finely fissured when young, becoming thicker with age; old bark with scaly ridges. Inner bark yellow to light orange. The fruit is large and spherical, a multiple composed of drupelets encased by their fleshy calyx, 8–12cm diam.; not edible, and formed only by female trees. The milky sap that is so evident from cut parts in summer is not always evident in winter.

Magnolia L.   MAGNOLIA   Family Magnoliaceae

Eight magnolias are treated here as species native to the SE. Two of these (the pyramid and Ashe magnolias) are treated as varieties or subspecies by some taxonomists. One additional species may occasionally be encountered as a persistent specimen of cultivation, the saucer magnolia (*M.* × *soulangeana* Soul.), but no record of its naturalizing in the region is known. All magnolias have large, 1-scaled terminal buds, ringlike stipule scars, numerous bundle scars, and spicy-scented inner bark. Terminal buds that contain flowers are larger than terminal vegetative ones and are more ovoid or swollen near the base or middle. The conelike fruit is actually an aggregate of follicles derived from a compound pistil. Each follicle splits to release the seed, which is enclosed in a colorful and oily aril (fat rich and attractive to birds).

1. Leaves persistent over winter, or both twigs and buds greenish in color
   2. Twigs or buds have brown hairs; leaves 15cm or longer, green or brownish-velvety beneath   1. *M. grandiflora*
   2. Twigs or buds with pale hairs, to glabrous; leaves rarely over 15cm, whitened beneath   2. *M. virginiana*
1. Leaves deciduous; twigs brownish, or if green, then buds white-silky and over 25mm long
   3. Terminal buds covered with white- to gray-silky hairs
      4. Twigs stout, green; terminal buds often over 25mm, with pointed tip
         5. Habit tall, treelike; fruit 5cm diam. or more; widespread species   3. *M. macrophylla*
         5. Habit short or multistemmed; fruit 4cm diam. or less; rare species of FL   4. *M. ashei*
      4. Twigs only moderately stout, brown; terminal buds under 25mm, obtuse at tip
         6. Twigs glabrous   5. *M. acuminata* (cultivated *M.* × *soulangeana* will key here; however its bark is gray and smooth as opposed to the furrowed bark of *M. acuminata*)
         6. Twigs pubescent or roughened, at least near tip   5. *M. acuminata* var. *subcordata*
   3. Terminal buds glabrous
      7. Twigs stout, light brown, slightly fluted at areas of clustered leaf scars   6. *M. tripetala*
      7. Twigs slender to moderately stout, dark reddish-brown, very fluted at clustered leaf scar areas
         8. Fruit usually over 6cm long; common tree of Appalachian region   7. *M. fraseri*
         8. Fruit usually under 6cm; uncommon tree of coastal plain   8. *M. pyramidata*

1. *Magnolia grandiflora* L.   SOUTHERN MAGNOLIA   *Figure 302*

A large, evergreen tree with a dense crown of shiny, dark green foliage. Native to the coastal plain from NC to FL, west to TX, AR. Because of extensive cultivation, this tree has naturalized widely in regions to the interior of its original range. Typical natural habitat is maritime forests, hammocks, ravines, and slopes. Bark gray, smooth at first, becoming fissured into large, slightly scaly plates.

2. *Magnolia virginiana* L.   SWEETBAY   *Figure 303*

A large shrub or tree, deciduous to evergreen. Ranges from MA to FL, west to TX, and inland to TN, AR. Primarily a coastal plain species, though also seen in parts of the Piedmont region. Commonly in acidic, moist soils such as along streams, in bays, pocosins, savannas, and peaty, swampy sites. Habit more shrubby and deciduous in the n. parts of the range; evergreen and treelike, to 20m or more, in the southernmost sections. The glaucous or white undersides of leaves are a distinctive trait. Twigs and buds greenish, glabrous to silky, usually with white

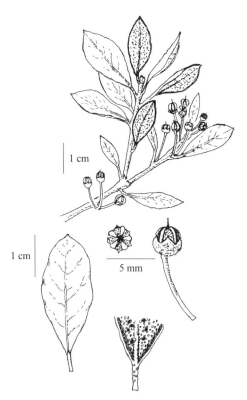

Figure 298. *Lyonia ferruginea*

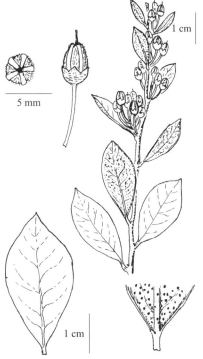

Figure 299. *Lyonia fruticosa*

hairs. Bark gray, smooth or with a few fissures on old, large trunks. Fruit mostly 2–4cm long, but to 5.5cm in some populations (such as var. *ludoviciana* of TX).

3. *Magnolia macrophylla* Michx. BIGLEAF MAGNOLIA *Figure 304*

A tree, to 20m or more, usually with an open, coarse crown. It is scattered across the SE, primarily west and south of the Blue Ridge, from e. KY, sw. VA to n. AR, and w. GA to LA. A few isolated populations occur east of the Blue Ridge. Typically a tree of moist, fertile soils of sheltered coves and ravines in uplands, such as the Cumberland Plateau, and on slopes and along small streams in lowlands, such as in the coastal plain. The large leaves, up to about 1m long, are whitened beneath and are conspicuous even when drying on the ground in winter. Bark gray, thin, mostly smooth. Fruit rounded, to 8cm long and wide.

4. *Magnolia ashei* Weatherby ASHE MAGNOLIA

A small tree, to 10m, often leaning or with multiple trunks. It is known only from the FL panhandle, in rich wooded forests near bluffs and in ravines. By nature an understory species. Leaves to 50cm. Fruit 3–6cm long, 2–4cm wide. Similar to *M. macrophylla* in general traits of twigs, bark, and leaves.

5. *Magnolia acuminata* (L.) L. CUCUMBER-TREE *Figure 305*

A tree, to 30m, usually with a long, straight trunk. A canopy species of mesic forests, most common in the Appalachian region and rich uplands westward to MO, AR. Also ranges into the Piedmont and coastal plain regions from VA to GA, west to LA, but merging in these more lowland parts of the range into the variety known as "yellow cucumber-tree." Habitat is normally rich, deep soils of good organic content. Twigs lustrous, dark brown, glabrous; terminal bud grayish, with silky hairs. Bark brown, furrowed and with elongated, narrow ridges; inner bark layers reddish-brown. Fruit 3–7cm long, usually knobby or bent; seed dark brown, with orange aril. The var. *subcordata* (Spach) Dandy, or var. *cordata* Michx., is known as yellow cucumber-tree due to its yellower flower color; its leaves typically are smaller, more pubescent, and twigs more pubescent than the species; it is also of smaller stature. These characters may only reflect influences of habitat and geographic range in *M. acuminata,* as intermediates are numerous and often inseparable.

6. *Magnolia tripetala* (L.) L. UMBRELLA MAGNOLIA *Figure 306*

A tree, to 15m, with a coarse, broad crown. An understory or subcanopy species of wide distribution in the SE. It is found in lower elevations of the Blue Ridge, westward to w. KY, TN, eastward and southward to the Piedmont of NC and nw. FL, LA; also seen in the Ozarks of AR. Occurs in moist, rich woodlands, especially near streams. Twigs stout, light brown, with a purplish-brown terminal bud. Bark gray, smooth. Fruit to 10cm long; seed corrugated, grooved, with an orange aril.

7. *Magnolia fraseri* Walt. FRASER MAGNOLIA *Figure 307*

A tree, to 30m in optimum sites, rarely over 15m in some regions. Endemic to the s. Appalachians, ranging from WV and w. VA south through w. NC, nw. SC to n. GA, and e. KY, e. TN. A canopy species typically in moist, rich soils of mesic forests, in coves, ravines, sheltered slopes. Occurring to high elevations in the Blue Ridge. Twigs glabrous, red-brown, moderately stout, with purplish terminal buds. Bark gray, thin, smooth. Fruit 6–11cm long, with elongated, curved tips on the follicles; seed black, smooth, in a pinkish to orange aril.

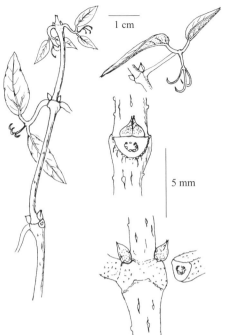

Figure 300. *Macfadyena unguis-cati*

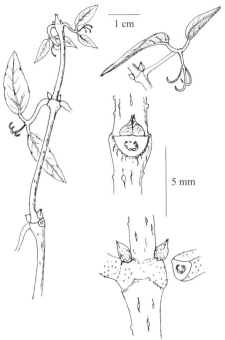

Figure 301. *Maclura pomifera*

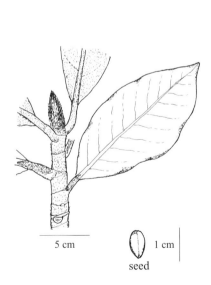

Figure 302. *Magnolia grandiflora*

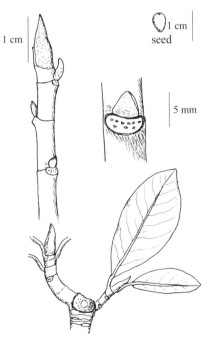

Figure 303. *Magnolia virginiana*

8. *Magnolia pyramidata* Bartr.   PYRAMID MAGNOLIA

A tree, to 20m in optimum sites, rarely over 12m in majority of its range. Overall uncommon and local in distribution, this subcanopy species is sporadic across the coastal plain, from SC to FL, west to e. TX. Also occurs in central SC, AL. Habitat is typically mesic woodlands, in ravines and on rich slopes, though also found on much drier and sandier upland sites in e. TX. Bark gray, thin, smooth. Twigs red-brown, glabrous, with purplish terminal buds. Fruit 4–9cm long, though usually under 6cm. This species is similar to *M. fraseri* and has been considered as a variety or southern form of it. The stamens are shorter in *M. pyramidata,* as well as a generally smaller aspect in other characters from *M. fraseri*. Leaves of *M. pyramidata* tend to have a wide-shouldered or pandurate shape in most of its range (not so pronounced in TX), and it may begin flowering only when 1 or 2m tall.

Mahonia bealei (Fortune) Carr.   OREGON GRAPE-HOLLY   Family Berberidaceae   *Figure 308*

An evergreen shrub, usually 1–2m tall (can reach 4m). Occasionally naturalizes, especially near sources of its cultivation. The stem is often naked and leafy only at the top. The pinnately compound leaves are rigid and thick, with stiff, sharp spines on the tip and margin. Leaflets sessile on the rachis, which appears jointed. Terminal bud present, large, or lacking if replaced by clusters of flower buds. Fruit is a blue-black berry, with a glaucous bloom, borne in racemes, ripe in summer. Inner bark yellowish. Though the popular name implies a connection to Oregon, this plant is a native of China. Native species from Oregon and other states of the w. US are sometimes cultivated in the SE [*M. aquifolium* (Pursh) Nutt. and *M. repens* (Lind.) G. Don.] but not known to naturalize; their leaves are thinner, spines not rigid.

Malus P. Mill.   APPLE   Family Rosaceae

Three species are treated here for the SE. The native species are known as crabapples, and the only exotic that is commonly naturalized is the common apple, long cultivated and extensively developed as to variations in fruit. Seedlings from cultivated trees are reversions toward the normal "unimproved" species. Due to the wide naturalization of the common apple, hybridization with the native *M. coronaria* may be encountered, this being *M.* × *platycarpa* Rehd. This hybrid bears varying degrees of intermediate characteristics of the parents. All apples have terminal buds, spur shoots, and large pomes as fruits. Twigs are not diagnostic in identifying the crabapples. A great number of cultivated crabapples are derived from oriental species and their hybrids, but no naturalization of any of these is evident.

1. Twigs moderately stout; spiny spur shoots rarely present; fruit over 5cm diam., pedicel shorter than fruit   1. *M. sylvestris*
1. Twigs more slender; spinose spur shoots present; leaves lobed on sprouts; fruit under 5cm diam., pedicel longer than fruit
   2. Calyx pubescent; leaves tomentose beneath; uncommon species of w. parts of SE
   2. *M. ioensis*
   2. Calyx glabrous or pubescent only on inner surface; leaves glabrous or nearly so; widespread across SE
      3. Leaves of spur shoots bluntly pointed to rounded at apex; margin crenate-serrate
      3. *M. angustifolia*
      3. Leaves of spur shoots acute or conspicuously pointed; margin coarsely, sharply serrate
         4. Leaves toothed but rarely lobed; fruit usually 4–5cm   4. *M.* × *platycarpa*
         4. Leaves often lobed; fruit 3.5cm or less   5. *M. coronaria*

1. *Malus sylvestris* (L.) P. Mill.   COMMON APPLE   *Figure 309*

A tree naturalized across the SE but most commonly seen in regions where apple orchards are commercially developed. Seedlings from cultivated trees may develop spinose spur shoots. Fruit size and condition of flesh varies, but first-generation seedlings usually bear fruit 5–7cm diam. Bark gray, in scaly plates, often lichen-covered.

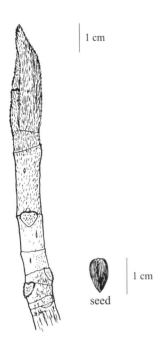

Figure 304. *Magnolia macrophylla*

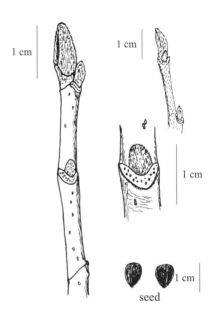

Figure 305. *Magnolia acuminata*

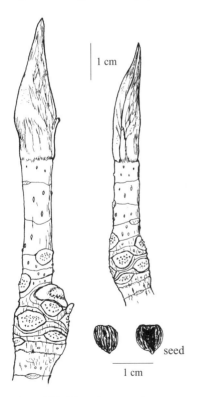

Figure 306. *Magnolia tripetala*

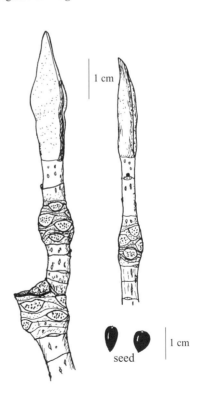

Figure 307. *Magnolia fraseri*

2. *Malus ioensis* (Wood) Britt. PRAIRIE CRABAPPLE

A small tree, to 6m tall, with an open crown. A midwestern species that occurs in the SE in states west of the Mississippi River, most commonly in MO. It is scattered in woodlands and woods edges in prairies, pastures, along streams, and bottomlands. Fruit to 3cm diam.

3. *Malus angustifolia* (Ait.) Michx. SOUTHERN CRABAPPLE

A small tree, to 8m tall, developing a dense crown or sometimes a colony of sprouts from the roots. Ranges over most of the SE but most common in the coastal plain and Piedmont. Typical of pastures, field edges, thin woodlands, and thickets. Fruit to 3cm diam.

4. *Malus* × *platycarpa* Rehd. HYBRID APPLE

Hybridizations between *M. sylvestris* and *M. coronaria* are encountered in apple-growing regions of the SE, where the 2 species grow in proximity. Leaves may appear more toothed and less often lobed than in normal *M. coronaria,* and fruits are larger, to 5cm diam., though appearing mostly greenish, with sour flesh and width equal to or exceeding length. Second-generation hybrids can inherit more traits from either parent that lessen the intermediate characters.

5. *Malus coronaria* (L.) P. Mill. SWEET CRABAPPLE *Figure 310*

A small tree, to 9m tall, with an open or irregular crown. Occurs mainly in the Appalachian region and areas west to MO. Habitat is moist forests with thin canopies, upland pastures, fence lines, thickets near streams. Fruit to 3.5cm diam. Flowers often lightly but sweetly fragrant, which accounts for the name. Bark scaly to shallowly furrowed. Several variations or types of this crabapple have been described. Two such entities bear leaves that are often distinctly lobed, widely rounded to cordate at the base, and with the lowermost main veins arising near the margin of the blade. One of these, called *M. glaucescens* Rehd., has glaucous leaf undersides; the other, green beneath, with angular fruit, is called *M. glabrata* Rehd. Additional types add to the complex, which is best simplified by considering all under the wide variation of 1 species, *M. coronaria.* What appear to be hybrids of this species with other *Malus* are sometimes encountered.

Manihot grahamii Hook. MANIHOT Family Euphorbiaceae *Figure 311*

An exotic shrub, to 6m or more; woodier and more treelike only in the s. and warmer parts of the coastal plain. Naturalized mainly near urban areas. The leaves remain on the large twigs until freezing weather. Leaves over 3dm long, minus the petiole, deeply palmately divided into 5 to 11 lobes. Petioles 2dm; sap milky. Stems green, with large pith. Leaf scars rounded, with obscure bundle scars. Fruit a round capsule. This is a related species to the manioc (*M. esculenta* Crantz), grown in the tropics for its starchy roots.

Melia azedarach L. CHINABERRY Family Meliaceae *Figure 312*

An exotic tree that has naturalized widely in the se. coastal plain. The stout twigs have no terminal bud; lateral buds rounded, hairy; leaf scars large, with 3 U-shaped bundle scars. Bark grayish, with scaly plates and ridges. The fruit is a whitish drupe borne in terminal clusters; poisonous if eaten. Native to Asia; long cultivated in the South as a shade and ornamental tree.

Menispermum canadense L. MOONSEED Family Menispermaceae *Figure 313*

A native twining vine; stems greenish to deep reddish-brown. Widespread across the SE, mostly in moist and rich soils such as bottomlands and streamsides. Twigs slender; leaf scars

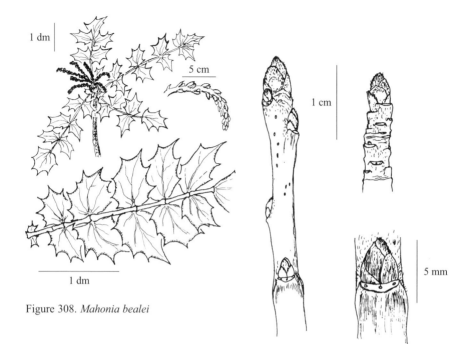

Figure 308. *Mahonia bealei*

Figure 309. *Malus sylvestris*

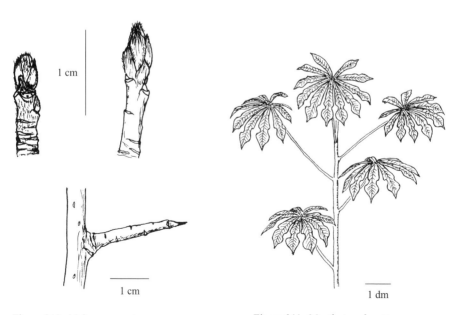

Figure 310. *Malus coronaria*

Figure 311. *Manihot grahamii*

large; buds hairy, sunken. The blue-black drupes resemble grapes, but the single seed is flattened, rough on the edge, and semicircular. Plants dioecious and staminate; plants seem to be more common across most of the SE.

Menziesia pilosa (Michx. ex Lam.) Juss. ex Pers.   MINNIE-BUSH   Family Ericaceae
Figure 314

A native shrub of 1–3m height, mostly of the high elevations of the Appalachians. Ranging from WV and VA along the mountains to NC, TN, SC, GA, in rocky woods, on cool slopes and balds. Twigs resemble those of azaleas, but bud scales are pale-margined, sometimes blunt at their apex, and fruit stems are longer than the 4-parted capsules. Bark is grayish to light brown, usually papery and exfoliating.

Mespilus canescens Phipps   STEARN'S MEDLAR   Family Rosaceae   *Figure 315*

A rare native shrub, only known from a single wet area in the AR coastal plain. Habit clumplike and multistemmed. Twigs red-brown; spiny spur shoots may be present, with 2 buds at the base on each side; sometimes a bud occurs along the length of larger spurs. Terminal buds are usually present; bud scales mostly 6 or more, reddish, abruptly pointed with a darkened tip. Pith light brown, homogenous. The fruit is a small pome with hard or dense flesh and hairy stalk, 6–9mm long.

Mitchella repens L.   PARTRIDGEBERRY   Family Rubiaceae   *Figure 316*

A diminutive shrub, creeping over the ground or rarely with ascending stems. Ranges throughout the SE in moist forests. Leaves evergreen, opposite, usually less than 2cm wide. Stems slender, vinelike. The red fleshy fruits are usually present in winter; edible but not highly flavored.

Morus L.   MULBERRY   Family Moraceae

One native species and 2 naturalized exotics occur in the SE. Twigs have no terminal bud; lateral buds have several imbricate bud scales. Cut twigs exude a milky sap, but mainly when sap is in flow during the growing season. Bark is scaly to ridged. The fruit is a multiple, derived from several pistils, the seed enclosed by a fleshy calyx. Maturing in summer, the sweet fruit is a popular food item with many birds.

1. Buds rounded or with a blunt tip; twigs gray to yellowish-gray; leaves glabrous   1. *M. alba*
1. Buds with short but sharp tip; twigs olive-green to reddish-brown; leaves hairy on both surfaces
   2. Leaves truncate to slightly cordate at base; a widespread, common native tree   2. *M. rubra*
   2. Leaves distinctly cordate at base; uncommon exotic tree   3. *M. nigra*

1. *Morus alba* L.   WHITE MULBERRY   *Figure 317*

A broad-crowned tree, to 15m tall, native to China but cultivated in many parts of the world as food for silkworms. Widely naturalized across the SE, mainly close to urban areas. Fruit white, pink, purplish, or nearly black, succulent but not as sweet as next 2 species. Bark grayish and slightly smooth on young stems, becoming fissured and scaly, ultimately ridged.

2. *Morus rubra* L.   RED MULBERRY   *Figure 318*

A broad-crowned tree, to 20m, native and fairly common throughout the SE. Most typical in deep, moist soils in the understory of hardwood forests. Bark gray to brown, scaly when

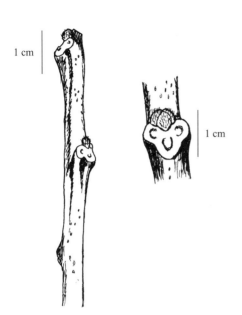

Figure 312. *Melia azedarach*

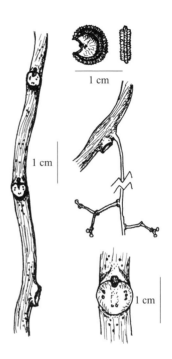

Figure 313. *Menispermum canadense*

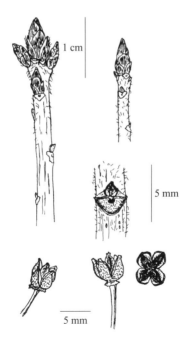

Figure 314. *Menziesia pilosa*

young, developing thick, scaly ridges with age. Inner bark and wood yellow, turning dark brown with prolonged exposure to air. Fruit 2–3cm long, reddish, purplish, or nearly black; sweet.

3. *Morus nigra* L.   BLACK MULBERRY

A broad-crowned tree, to 10m tall, sometimes planted for fruit; rarely seen as an escape in most of the SE. The fruit is 2–3cm long, dark red to nearly black, sweet and edible. Native to w. Asia.

Myrica L.   WAX-MYRTLE   Family Myricaceae

Five species are treated here as native in the SE. It is pertinent to point out that the taxonomy of this group is debatable. The species included here as members of 1 genus may also be divided into 2 or 3 genera, partly to reflect differences in flower and fruit structure. Wilbur (1994) and others have proposed that only *M. gale* be retained under the genus *Myrica*. To avoid confusion in cross-reference to other works, all species are here continued under *Myrica*. Twigs of the entire group have imbricate-scaled buds and 3 bundle scars in the alternate leaf scars. Twigs alone cannot be relied on to separate all the species, though leaves are usually late-deciduous to nearly evergreen on most taxa and can be used to aid identification. Fruit is a wax-covered, bumpy-surfaced drupe in all but 1 species. Bark is smooth, grayish or red-brown, and lenticellate.

1. Twigs of mature plants with catkin buds, or buds pointed; bark red-brown; fruit burrlike; of mountains 1. *M. gale*
1. Twigs lack catkin buds, all rounded or blunt; bark gray; fruit a drupe; of Piedmont or coastal plain
    2. Leaves distinctly punctate on both surfaces
        3. Twigs glabrous; leaf margins entire; no spicy aroma in crushed leaves   2. *M. inodora*
        3. Twigs usually hairy; leaf margins with a few teeth; aromatic when crushed.   3. *M. cerifera*
    2. Leaves chiefly punctate below
        4. Twigs mostly glabrous; leaves late-deciduous; fruits 5mm or wider; uncommon in SE
        4. *M. pensylvanica*
        4. Twigs hairy; leaves subevergreen; fruits usually under 5mm; common in SE
        5. *M. heterophylla*

1. *Myrica gale* L.   SWEETGALE   *Figure 319*

A shrub common in the North, where it ranges across N. Amer. Known to reach the SE in only 1 site, a bog in the NC Blue Ridge. Typically found in wet, boggy soils throughout its range.

2. *Myrica inodora* Bartr.   ODORLESS BAYBERRY   *Figure 320*

A tall shrub or small tree of the FL panhandle, west to s. MS. Leaves subevergreen, odorless when crushed, margins entire, glossy green above. Fruit 6–7mm diam., a little longer than broad.

3. *Myrica cerifera* L.   WAX-MYRTLE   *Figure 321*

A shrub or small tree, widespread and common across the coastal plain of the SE, from MD to FL, west to TX and inland to se. OK, AR. Often cultivated and seen in the Piedmont in plantings. Habitat varies from brackish coastal soils, wet depressions, and bogs to fairly dry upland sands under pine or pine-oak forests. Leaves late-deciduous to subevergreen in coastal areas. Fruit 2–3mm diam. A dwarf, stoloniferous form, var. *pumila* Michx., occurs in sandy soils of drier uplands or occasionally in moist sites but is 1m or less in height, with leaves mostly under 4cm long; it is considered a separate species by some botanists.

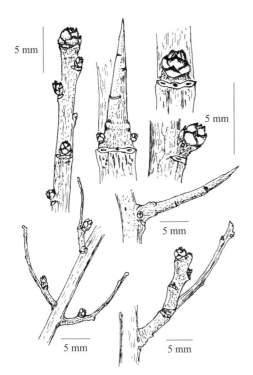

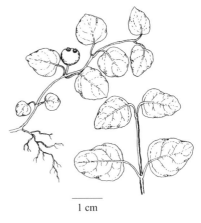

Figure 316. *Mitchella repens*

Figure 315. *Mespilus canescens*

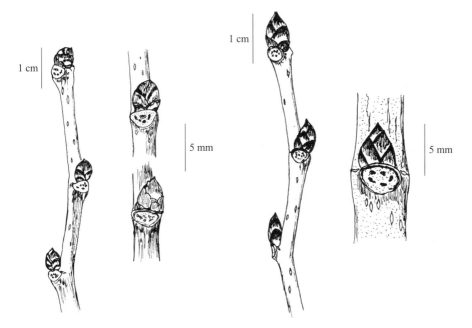

Figure 317. *Morus alba*

Figure 318. *Morus rubra*

4. *Myrica pensylvanica* Loisel.   NORTHERN BAYBERRY   *Figure 322*

A large shrub, with a wide crown and stoloniferous habit. Native mostly to the NE, but ranges as far south as NC along the Atlantic coast. Occurs in sandy soils in or near dunes, rarely more inland. Fruit 4–7mm diam.

5. *Myrica heterophylla* Raf.   POCOSIN BAYBERRY   *Figure 323*

A shrub widely dispersed across the SE, mostly in the coastal plain from se. VA to FL and LA, but ranging inland and north to the Piedmont of NC, n. MS, n. AR. Typically in wet pinelands, shrubby depressions, and acidic, boggy soils. The branchlets often are blackish, and leaves glossy, dark green. Fruits 2–5mm diam., often dark-colored.

Nandina domestica Thunb.   HEAVENLY-BAMBOO   Family Berberidaceae   *Figure 324*

An exotic shrub, to 2m tall, with slender, stiffly erect stems that are few-branched. Native to Asia; naturalized in woodlands and near urban areas, mainly from the fruits being dispersed by birds. Leaves are bipinnate, evergreen, borne on tips of the stems. The fruit is a red berry borne in terminal clusters.

Nestronia umbellula Raf.   INDIAN-OLIVE   Family Santalaceae   *Figure 325*

An uncommon native shrub that is parasitic on the roots of trees. It is rarely over 1m tall and usually forms colonies. Ranges from VA to AL, TN, KY; widely scattered within this range in coastal plain, Piedmont, and Appalachian regions. Twigs lenticellate, with opposite leaf scars, single bundle scars, no terminal bud, and keeled, reddish bud scales. The fruit is a greenish, rounded drupe, about 1cm long.

Neviusia alabamensis Gray   SNOW-WREATH   Family Rosaceae   *Figure 326*

A rare native shrub, to 2m tall, stoloniferous and usually forming a spreading colony. Found in calcareous soils in a few sites in the Appalachian region of nw. GA, AL, usually near streams. Twigs are slender, lined, with dark, persistent stipules, and tips may die back slightly in winter; terminal buds lacking. The white flowers lack petals; appear before leaves in early spring. Fruit is a small, drupelike achene.

Nyssa L.   TUPELO   Family Cornaceae

Four native species are treated here for the SE, with 1 additional (*N. ursina*) tupelo mentioned that may variously be considered a species or variety. All have terminal buds, 3 bundle scars in each of the alternate leaf scars, imbricate-scaled buds, and diaphragmed or occasionally chambered pith. The fruit is a fleshy drupe with sour flesh, produced on pistillate trees (plants dioecious).

1. Terminal buds conspicuously hairy, often leaning toward 1 side, with only 2 or 3 scales visible   1. *N. ogeche*
1. Terminal buds glabrous or sparsely hairy, usually vertical, with 3 or more scales visible
    2. Buds bluntly pointed or rounded, scales reddish; lateral buds much smaller than terminal
        2. *N. aquatica*
    2. Buds sharply pointed, with brownish or gray scales; some lateral buds at least ½ as large as terminal
        3. Trunk not buttressed at base; leaves usually acute to acuminate at tip; widely distributed
            3. *N. sylvatica*
        3. Trunk usually buttressed; leaves usually blunt at tip; of lowland swamps and wet sites
            4. *N. biflora*

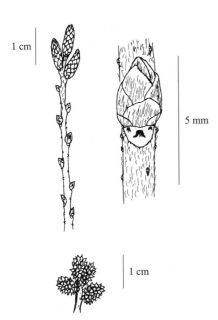

Figure 319. *Myrica gale*

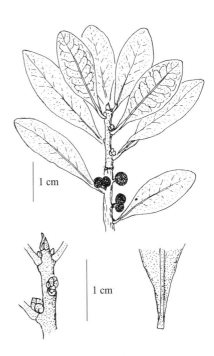

Figure 320. *Myrica inodora*

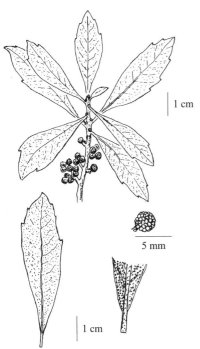

Figure 321. *Myrica cerifera*

Figure 322. *Myrica pensylvanica*

220 · *Nyssa*

1. *Nyssa ogeche* Bartr. ex Marsh. OGEECHEE TUPELO   *Figure 327*

A tree, to 20m tall, often shorter and with crooked, short, multiple or mangled trunks. Occurs in the coastal plain from se. SC to FL, west to s. AL. Typically seen lining blackwater streams in river swamplands. Twigs greenish or red-tinted on 1 side in juvenile plants, to red-brown on mature trees. Bark gray, smooth at first, becoming furrowed, often obscured by mud or silt adhering from floods. Fruit to 4cm long, red, on short stalks.

2. *Nyssa aquatica* L. WATER TUPELO   *Figure 328*

A tall tree, to 30m, with a long, clear trunk with a buttressed base. Usually seen in frequently flooded swamps and wet soils along lowland rivers. Occurs in the coastal plain from VA to FL, west to TX, north in the Mississippi embayment to s. IL. Twigs moderately stout, with leaf scars more crowded near the large terminal buds. Bark grayish, furrowed or in plates, often obscured by layers of silt or moss. Fruit to 3cm long, dark bluish to purple-black, prominently dotted, hanging on stalks longer than the fruit. Leaves mostly over 15cm long, glaucous beneath.

3. *Nyssa sylvatica* Marsh. BLACKGUM   *Figure 329*

A tree, to 40m, common and widespread across the SE. Occurs in a wide variety of habitats, from wet lowlands to dry upland forests. The trunk is not normally buttressed when growing in wet sites, and the bark is typically made up of blocky ridges. Twigs brown; terminal bud with 3 or 4 visible scales; pith diaphragmed. Fruit to 15mm long, black but with a glaucous bloom, borne 1 to 3 per stalk that is 2–3cm long.

4. *Nyssa biflora* Walt. SWAMP BLACKGUM   *Figure 330*

A tree similar in many respects to the above species, but mostly limited to the coastal plain and lowlands interior of it. Habitat is typically wet soils susceptible to frequent flooding, such as blackwater swamps, depressions in pine flatwoods, bays, bogs, pocosins, and headwater flats of blackwater streams. The trunk is usually buttressed, and bark is longitudinally furrowed but not blocky as in the above species. Fruit is usually borne in pairs, each slightly longer than broad. What may be a dwarfed form of *N. biflora* (or a separate species) occurs in the FL panhandle, *Nyssa ursina* Small; it grows in wet sites over a clay hardpan, rarely exceeding 4m in height; its fruit is typically more globose than in *N. biflora*.

Opuntia L. PRICKLY-PEAR   Family Cactaceae

Seven species of these cacti are native or naturalized in the SE, with 3 additional varieties mentioned here. Identification must use habit, spines, pads, and sometimes flowers or fruits. The flattened pads (joints) are modified stems, green with chlorophyll for photosynthesis. Pads are studded with areoles, areas corresponding to buds, and where true leaves, spines, and hairlike glochids are produced. True leaves are elongate, fleshy protuberances that may only occur in certain parts of the year. Spines vary from 1 to several, or may be lacking. Glochids are minutely barbed, hairlike structures easily dislodged into human flesh, where they can be most irritating. The fruit may similarly be decorated with glochids; it is derived from carpels within a floral tube, with many seed inside. The fruit's outer layer turns red and is edible, though glochids should be removed by flame or some other means before eating. Habitat of all *Opuntia* in the SE is dry, sandy, rocky, or gravelly soils, with ample sunlight and minimal competition.

1. Habit prostrate and spreading, rarely rising higher than length of 2 joints
  2. Joints elongated, slender, easily detached; spines barbed   1. *O. pusilla*
  2. Joints obovate to orbicular, not easily detached; spines not barbed

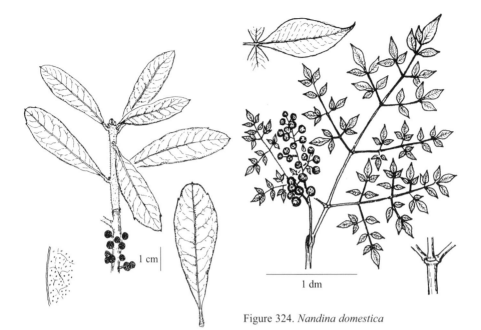

Figure 324. *Nandina domestica*

Figure 323. *Myrica heterophylla*

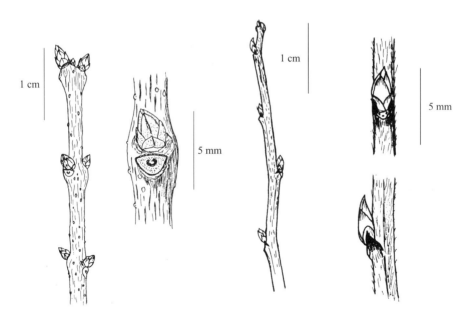

Figure 325. *Nestronia umbellula*

Figure 326. *Neviusia alabamensis*

3. Joints blue-green; spines mostly deflexed; main root tuberous; w. species
 2. *O. macrorhiza*
3. Joints light to dark green; spines mostly spreading; all roots fibrous; widespread, coastal species
 4. Spines 2–3cm long; joints 5–7.5cm long; widespread species
 3. *O. humifusa*
 4. Spines 3–8cm; joints 7.5–10cm long; coastal dunes, FL to TX
 4. *O. humifusa* var. *austrina*
1. Habit shrublike or trunk bearing, or if somewhat spreading, then often rising 3 or 4 joints high
 5. Spines rounded in cross section at base; of coastal dunes or interior pinelands of FL
 6. Joints light green; fruit less than 5cm long; plant rarely over 4 joints high
 5. *O. humifusa* var. *ammophila*
 6. Joints lustrous green, basally constricted; fruit 5cm or more long; plant to 3m high
 6. *O. monocantha*
 5. Spines elliptical to flattened in cross section at base, at least in larger spines; not restricted to FL
 7. Joints broadly obovate to orbicular; forms large clumps; of TX and LA grass and brushlands
 7. *O. engelmannii* var. *lindheimeri*
 7. Joints obovate to oblong; erect or somewhat sprawling; of FL and other parts of coastal plain
 8. Spines white to pale brown; fruit globular obovoid; often treelike, 3–7m with a trunk; exotic rarely naturalized, TX to FL
 8. *O. ficus-indica*
 8. Spines yellow; fruit base constricted into a short stalk; native to coastal plain, TX, FL, SC
 9. Spines lacking, or 1 per areole on margins of joints; spines 2cm long or less
 9. *O. stricta*
 9. Spines 1 to 11 per areole, to 6cm long
 10. *O. stricta* var. *dillenii*

1. *Opuntia pusilla* (Haw.) Nutt.   DUNE PRICKLY-PEAR   *Figure 331a*

A low, sprawling cactus, native to dunes and sandy soils very near the Atlantic coast. Distributed from NC to FL, west to TX. Joints elongate, 6cm or shorter. The spines are numerous, sharp, useful to the plant in spearing passing animals and thus spreading the easily detached joints to new locations. Spines are barbed and resist easy removal from flesh.

2. *Opuntia macrorhiza* Engelm.   PLAINS PRICKLY-PEAR   *Figure 331b*

A clump-forming cactus, under 15cm tall. Mostly a sw. species that ranges eastward and northward in the Great Plains, reaching the SE in OK, MO, AR, TX, LA. Joints usually glaucous (blue-green), broadly obovate, 10cm or less long, with 1 to 6 spines per areole, these usually deflexed. Fruit to 4cm long.

3. *Opuntia humifusa* (Raf.) Raf.   EASTERN PRICKLY-PEAR   *Figure 332a*

A low, spreading species that may form an extensive colony. The most common cactus in the SE, distributed throughout the region. Joints broadly obovate, mostly under 8cm long, spineless or with 1 spine per areole. Fruit to 4cm. Two additional varieties are recognized.

4. *Opuntia humifusa* var. *austrina* (Small) Dress   FLORIDA PRICKLY-PEAR   *Figure 332b*

Usually rather clumplike. Uncommon to the north and west of the predominately FL range, but reaching SC and TX. Most often seen near coastal zones and shell mounds. Joints obovate or elliptic, mostly 8–10cm long.

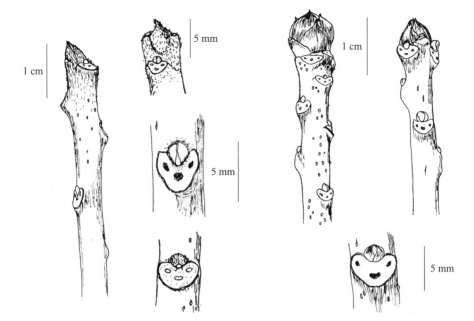

Figure 327. *Nyssa ogeche*

Figure 328. *Nyssa aquatica*

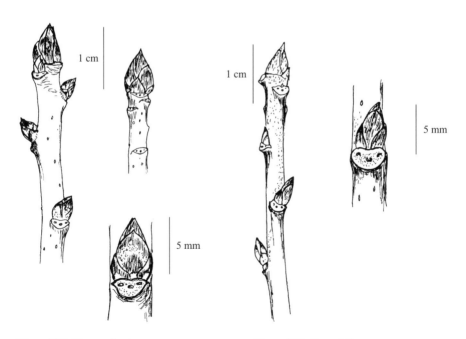

Figure 329. *Nyssa sylvatica*

Figure 330. *Nyssa biflora*

5. *Opuntia humifusa* var. *ammophila* (Small) L. Benson   SCRUB PRICKLY-PEAR
Figure 332c

An upright form, with 2 to 4 joints erect, occurring throughout FL. Joints elongate, to 17cm long; spines to 3cm long.

6. *Opuntia monocantha* (Willd.) Haw.   SOUTH AMERICAN PRICKLY-PEAR   *Figure 333a*

An upright species to 4m tall, often forming a trunk. Native to S. Amer.; occasionally seen naturalized in FL. Joints elongate to narrowly obovate, narrowed at the base, 10–30cm long, glossy green. Spines of joint not numerous, 1 or 2 per areole; more numerous and to 12 per cluster on trunk; larger spines to 7.5cm long and stout, to 1.5mm thick. Fruit 5–7.5cm long. (*O. vulgaris* P. Mill.)

7. *Opuntia engelmannii* var. *lindheimeri* (Engelm.) Parfitt & Pinkava   TEXAS PRICKLY-PEAR
Figure 333b

A spreading and somewhat erect species, forming large clumps. Native to the SW, but entering the SE in e. TX and w. LA. Joints broadly obovate, to 25cm long, Spines to 3cm, 1 to 3 per areole, yellow. Fruit to 5cm long.

8. *Opuntia ficus-indica* (L.) P. Mill.   INDIAN-FIG   *Figure 334a*

A large, treelike species, 3–6m tall, developing a trunk to 10dm diam. Presumed native to Mexico, long cultivated and transported amid regions with warmer climates as a food crop. Rarely naturalized in FL and TX. Joints variable from broadly obovate to oblong, 30–60cm long. Spines commonly absent, but sometimes 1 to 6 per areole, whitish or light brown. Fruit 5–10cm long, spineless or nearly so. Numerous horticultural selections are known, the more spineless forms most commonly planted, for the edible fruit.

9. *Opuntia stricta* (Haw.) Haw.   SPINELESS PRICKLY-PEAR   *Figure 334b*

A spreading or somewhat erect species, sometimes with a short trunk. Mostly limited to coastal zones in FL, ranging west to TX and northeast to SC. It was an important food source for several American Indian tribes, and it has likely been transported and cultivated for a long period. Habitat varies from rocky hardwood hammocks to sandy dunes and shell middens. Joints elongate, 10–25cm long. Spines lacking, or only 1 per a few areoles on margin of joint. Fruit 4–6cm long, with few areoles. One significant variety different in general appearance follows.

10. *Opuntia stricta* var. *dillenii* (Ker-Gaw.) L. Benson   COASTAL PRICKLY-PEAR
Figure 334c

Similar to the above species in habit and range. Joints obovate, 10–30cm long. Spines numerous, yellow, mostly 4cm or less, 1 to 11 per areole. Fruit as above species, but areoles more numerous.

Osmanthus americanus (L.) Benth. & Hook. f. ex Gray   DEVILWOOD   Family Oleaceae
Figure 335

A native evergreen shrub or small tree of the coastal plain. Distributed from se. VA to FL, west to LA, in maritime forests and variable habitats more inland, from sandy, well-drained soils to wet lowlands. Leaves are opposite, entire on the margin, 10–14cm long. Twigs brown; terminal buds with valvate scales. The clusters of dark drupes mature in winter, and at the same time the stalks of flower buds for next year can be seen on the twigs. Bark is thin, grayish, smooth.

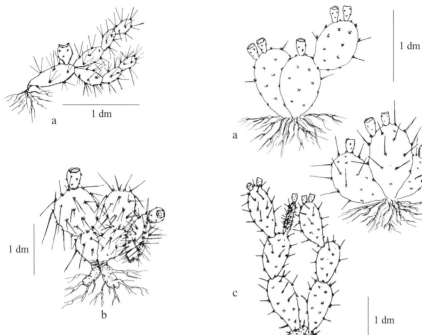

Figure 331.a: *Opuntia pusilla;*
b: *O. macrorhiza*

Figure 332. a: *Opuntia humifusa;*
b: *O. humifusa* var. *austrina;*
c: *O. humifusa* var. *ammophila*

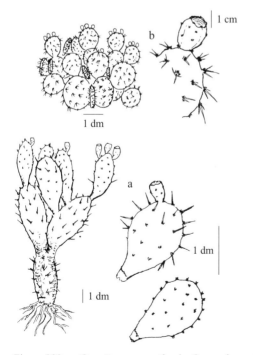

Figure 333. a: *Opuntia monocantha;* b: *O. engelmannii* var. *lindheimeri*

Ostrya virginiana (P. Mill.) K. Koch   HOPHORNBEAM   Family Betulaceae   *Figure 336*

A small understory tree of wide distribution in the SE, uncommon to absent only in parts of the coastal plain of NC to GA and LA. Native to a variety of habitats but most common in uplands in mesic to dry or rocky woods. Bark of small stems smooth, lenticellate, reddish-brown, quickly becoming browner and flaky or scaly. Twigs slender, may have catkin buds near apex in mature plants; buds pointed, the bud scales with dark margins; no terminal bud, though branch scar may be tiny or inconspicuous. The fruit is a hoplike cluster of membranaceous sacks each with 1 seed; bristling with stiff hairs that easily detach and may cause itching of tender skin if handled.

Oxydendrum arboreum (L.) DC.   SOURWOOD   Family Ericaceae   *Figure 337*

A small native tree of forest understories, ranging throughout the SE, mostly east of the Mississippi River. Common in well-drained, acidic woodlands of uplands. Twigs slender, moderately stout on sprouts, reddish or greenish; no terminal bud; laterals rounded or dome-shaped, with imbricate scales. A single bundle scar and fairly large, white, homogenous pith. Bark gray, deeply furrowed. The sour-tasting sap is noticeable in all parts when chewed, and the fragrance is distinctive, once familiar. Fruit is a small capsule, borne in drooping clusters, persistent all winter. The white, urn-shaped flowers hang from the twig tips in summer.

Pachysandra Michx.   PACHYSANDRA   Family Buxaceae

Two species treated here, 1 native and 1 a naturalized exotic. Both are subshrubs; low, rhizomatous, covering the ground surface but rising only 2dm or less in height. They tend to expand and form colonies or large clonal patches. Leaves are usually present into at least early winter.

1. Leaves subevergreen, pale dull green or mottled above, hairy beneath   1. *P. procumbens*
1. Leaves evergreen, dark glossy green above, glabrous   2. *P. terminalis*

1. *Pachysandra procumbens* Michx.   ALLEGHENY-SPURGE   *Figure 338*

A native species of rich woodlands, widely scattered across the interior regions of the SE. The range extends from KY to LA and from nw. SC to FL panhandle, though wide areas in this range are devoid of populations. The flower buds are formed on the stems well back from the leafy terminal section. Individual stems do not live beyond the second growing season.

2. *Pachysandra terminalis* Sieb. & Zucc.   ASIATIC PACHYSANDRA   *Figure 339*

A native of Asia; naturalized in areas close to its cultivation or spreading from cultivation into shady woodlands. The flowers are borne terminally on the erect stems.

Parkinsonia aculeata L.   JERUSALEM-THORN   Family Fabaceae   *Figure 340*

A small tree native mostly to arid regions of Mexico. It has been cultivated in warmer areas of the sw. and se. US, and has naturalized sparingly. In the SE it is usually seen near the coast, from TX to FL. It prefers dry, rocky or gravelly soils, especially when calcareous. The twigs have sunken buds and obscure leaf scars, and single or paired thorns at the nodes. The bark is green, smooth, though more grayish near the base of the trunk. The compound leaves have 2 to 4 pinnae, each with a winged, straplike rachis and tiny leaflets. The leaflets may be shed in dry seasons or by winter. The legume is 6–10cm long, constricted between seed. Flowers yellow.

Paronychia virginica K. Spreng   NAILWORT   Family Caryophyllaceae   *Figure 341*

A small, matlike subshrub of open, rocky sites, scattered from w. MD, VA, and WV down the Appalachians to AL and to AR, OK, TX. Uncommon to rare in the Appalachians, typically on

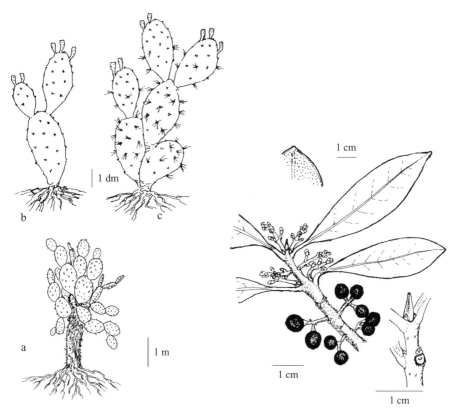

Figure 334. a: *Opuntia ficus-indica;* b: *O. stricta;* c: *O. stricta* var. *dillenii*

Figure 335. *Osmanthus americanus*

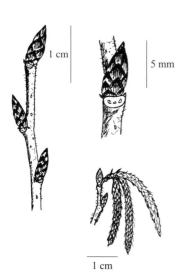

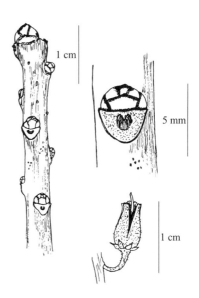

Figure 336. *Ostrya virginiana*

Figure 337. *Oxydendrum arboreum*

shale barrens, talus slopes, and dry, often calcareous rocky outcrops. This species is one of the most fruticose, or shrublike and woody, of a large number of herbaceous species. Twigs jointed, brittle, breaking easily at the joints. Silvery or whitish membranous bracts and stipules are conspicuous. The needlelike leaves are 15–30mm long, about 1mm wide, sessile, with thin, membranaceous stipules at their base. Flowers yellowish, borne in late summer. Fruit is elongate and bladderlike, termed a utricle.

Parthenocissus Planch.   VIRGINIA-CREEPER   Family Vitaceae

Two species are treated here for the SE, both native vines climbing by tendrils that attach to other surfaces. Aerial rootlets further anchor the vines as their stems enlarge. The twigs and bark of younger stems are lenticellate. Twigs lack terminal buds; branch scar large; bundle scars several; stipule scars slitlike, conspicuous. Fruit a blue-black berry, inedible, borne in clusters with red stems. Leaves palmately compound, usually with 5 leaflets.

1. Tendrils tipped with small disks; fruit 5–7mm diam.; a common, widespread species
 1. *P. quinquefolia*
1. Tendrils lack disks on tips; fruit 8–10mm diam.; a n. species, rare in SE   2. *P. inserta*

1. *Parthenocissus quinquefolia* (L.) Planch.   VIRGINIA-CREEPER   *Figure 342*

Distributed throughout the SE. A common vine of woodlands, fencerows, thickets, especially in moist soils. Climbs tree trunks, posts, and walls using disk-tipped tendrils. Often ascends high into the canopy. The flower clusters are paniculate-shaped, with a prolonged central stem.

2. *Parthenocissus inserta* (Kerner) K. Fritsch   NORTHERN-CREEPER

Primarily a n. species, reaching the SE perhaps in VA and WV. Usually spreading over the ground and in low shrubbery, rarely high-climbing, not clinging to walls or tree trunks. Tendrils tend to lack adhesive disks; instead they mostly twist around other vegetation. Flower clusters smaller and wider than in the above species, without the prolonged central axis. Leaves dark lustrous green above, lighter green and slightly lustrous below. Considered only a variation of the Virginia-creeper, by some botanists. [*Parthenocissus vitacea* (Knerr) A. Hitchcock]

Paulownia tomentosa (Thunb.) Sieb. & Zucc.   ROYAL PAULOWNIA   Family Scrophulariaceae   *Figure 343*

A short-trunked tree with a broad, open crown, native to China. Naturalized in many parts of the SE by wind-borne seed that are produced in multitudes. Often seen along rocky slopes and road cuts. Twigs stout, lenticellate; terminal bud lacking; lateral buds small, rounded; leaf scars large, concave; many bundle scars arranged in a circular pattern; pith excavated or chambered. Stalks of nodding, tomentose flower buds may be present in winter. The clusters of dried fruits (capsules) are distinct in winter; each capsule about 3–4cm long, filled with thousands of brown seed surrounded by a thin wing, each the size of a pinhead. Bark gray to brown, thin and lenticellate. Conspicuous in spring when the blue to purplish elongate flowers open.

Paxistima canbyi Gray   MOUNTAIN-LOVER   Family Celastraceae   *Figure 344*

An evergreen, stoloniferous shrub that sprawls over the ground and forms large colonies. Native to the Appalachian region from VA, WV, s. OH to KY and e. TN, mostly west of the Blue Ridge (though reported from NC). Typically found on calcareous soils overlying rocky bluffs. The opposite leaves bear sharp teeth, thickened margins, mostly under 25mm long and 6mm wide. Twigs slightly lined below nodes; terminal buds tiny, rounded, with reddish, imbri-

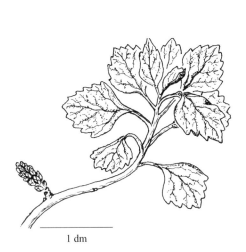

Figure 338. *Pachysandra procumbens*

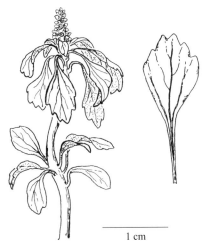

Figure 339. *Pachysandra terminalis*

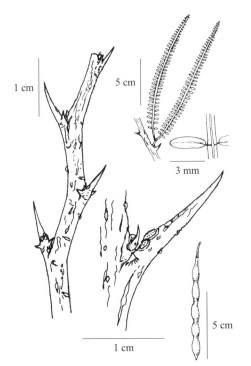

Figure 340. *Parkinsonia aculeata*

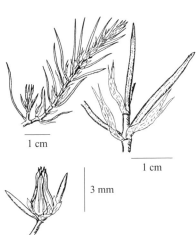

Figure 341. *Paronychia virginica*

cate scales. The low habit and sharp serrations account for the alternate name of "rat-stripper." Fruit is a small capsule.

Pentaphylloides floribunda (Pursh) A. Love SHRUB CINQUEFOIL Family Rosaceae
Figure 345

A low, bushy shrub, to 1m tall. A circumpolar species, found across n. sections of N. Amer. and Eurasia. Ranges into the SE in s. OH, IN, and perhaps MO. Occurs on rocky, often dry soils such as barrens, rocky ledges, and gravelly prairies. Terminal buds appear present; buds are very loose-scaled and white-hairy; base of the leaf petioles persist along with adnate stipules, reaching around the twig and hiding part of the bud; tiny lenticels apparent; 1 small, dark bundle trace. Bark shreddy, grayish. Leaves deeply cut to pinnately divided, about 3cm or less in length; sometimes adherent in dried condition to some winter twigs. (*Potentilla fruticosa* L.)

Persea P. Mill. BAY Family Lauraceae

Three species are treated here. All are small trees native to the coastal plain, with spicy-aromatic sap, grayish, furrowed bark, and alternate, evergreen leaves with entire margins. The terminal buds appear naked, hairy; a single bundle scar occurs in each leaf scar. The blue-black drupes are about 1cm long. Crushed leaves with fragrance similar to the related commercial bay leaf (*Laurus nobilis* L.).

1. Twigs green, glabrous; midrib and veins on leaf undersides not conspicuously hairy
 1. *P. borbonia*
1. Twigs brown or red-brown, due to abundance of hairs; midrib and veins of leaf undersides similarly covered by hairs
  2. Twigs and leaf veins bear reddish-brown hairs not appressed; common species
 2. *P. palustris*
  2. Twigs and entire leaf undersurface with brownish, silky, appressed hair; FL species
 3. *P. humilis*

1. *Persea borbonia* (L.) Spreng. REDBAY

A tree native to the coastal plain from NC to FL, west to TX. Typically in maritime forests, sandy soils of dunes, hammocks, and scrub near the coast. Leaves usually glossy green above, light green and with very fine, golden-colored, appressed hairs beneath, visible with a lens.

2. *Persea palustris* (Raf.) Sarg. SWAMPBAY *Figure 346*

Native to the coastal plain throughout the SE. The most common species of *Persea,* occurring in dry to wet maritime habitats and in low, wet sites inland. Typical of wet, swampy depressions in pinelands, peaty soils in swamps, bays, pocosins, and streamside thickets. The reddish-brown hairs are conspicuous, not appressed to the twigs or leaf undersides.

3. *Persea humilis* Nash SILKBAY

A small tree of scrub habitat in central peninsular FL, barely extending far enough north to reach this book's area of coverage. Leaf undersides of youngest leaves have a coppery sheen of silky, appressed hairs. These may be dull to a brown or gray on older leaves. Twigs appear dark, due to the same type of hairs.

Philadelphus L. MOCK-ORANGE Family Hydrangeaceae

Four major species in the SE, all shrubs. One is an exotic that spreads around urban areas and home sites (*P. coronarius*), apparently the only exotic 1 known to naturalize from many

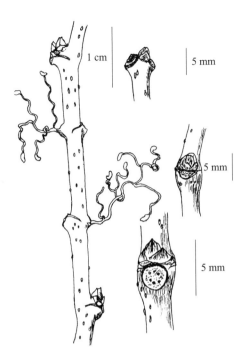

Figure 342. *Parthenocissus quinquefolia*

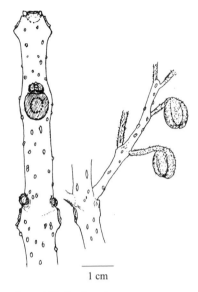

Figure 343. *Paulownia tomentosa*

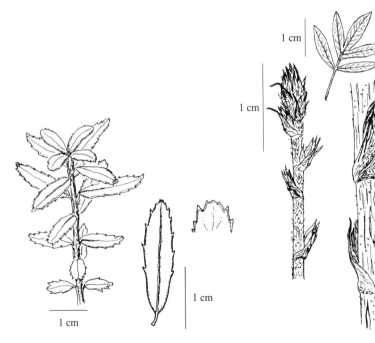

Figure 344. *Paxistima canbyi*

Figure 345. *Pentaphylloides floribunda*

cultivated exotic species. Leaf scars joined by a ridge. Fruit a rounded capsule. Cultivated for the showy white flowers in spring.

1. Bark of branchlets gray, tight and not shreddy or flaky   1. *P. pubescens*
1. Bark of branchlets reddish or occasionally grayish, shreddy or flaky
    2. Terminal buds usually present, or lateral buds not sunken; twigs rough-hairy (hirsute)
                                                                                                              2. *P. hirsutus*
    2. Terminal buds lacking; lateral buds sunken; twigs mostly glabrous
       3. Inflorescences with 1 to 3 flowers; flowers not aromatic; native shrub   3. *P. inodorus*
       3. Inflorescences with 5 to 7 flowers; flowers sweet-scented; cultivated exotic
                                                                                                                     4. *P. coronarius*

1. *Philadelphus pubescens* Loisel   DOWNY MOCK-ORANGE

Native to the interior of the SE, from KY to nw. GA, west to MO, OK, AR. Mainly seen on rocky bluffs and riverbanks with limestone outcroppings. The bark is gray, not shredding. Leaves pubescent beneath. Flowers borne in 5- to 9-flowered racemes; calyx pubescent. The var. *intectus* (Beadle) A. H. Moore has calyx and leaves essentially glabrous.

2. *Philadelphus hirsutus* Nutt.   HAIRY MOCK-ORANGE   *Figure 347*

A shrub of upland, usually rocky habitats from VA, KY to AR, AL, n. GA. Usually seen on or near bluffs, gorges, rock outcrops, and in rocky woods. Leaves pubescent beneath and often scabrous above; calyx pubescent; flowers usually 3 together, each about 3cm wide; calyx lobes in fruit are divergent or bent downward. Bark dark brown, shreddy on branchlets. A similar species likely only a form of *hirsutus* is *P. sharpianus* Hu; it is glabrous on calyx and less hairy on leaves; known from TN.

3. *Philadelphus inodorus* L.   ODORLESS MOCK-ORANGE   *Figure 348*

A native shrub of wide distribution, from the Appalachians to the coastal plain of VA to FL and to TN, AL, MS. Usually seen in rocky woodlands over circumneutral to calcareous soils. The calyx lobes in fruit tend to be erect and glabrous. Typically the flowers are 4–5cm wide; leaves are entire or remotely denticulate, glabrous or hairy only in vein axils beneath. Variations in these characters have been used to identify several other species in the past, but the true taxonomic relationships are not well understood. Some of the similar forms are: *P. floridus* Beadle, with calyx villous and leaves pubescent beneath; *P. grandiflorus* Willd., with flowers to 5.5cm wide, leaves larger, more dentate. Other entities include *P. gattingeri* Hu, *P. laxus* Schrad., *P. gloriosus* Beadle.

4. *Philadelphus coronarius* L.   COMMON MOCK-ORANGE

A tall shrub, to 3m, commonly seen around old home sites and urban areas. Native to s. Europe. Leaves usually denticulate, or with a few dentate teeth, glabrous or hairy in vein axils beneath. The aromatic flowers are borne 5 or more together; thus the fruits from pollination may also be in a group of 5 or more. Otherwise, difficult to separate in winter from the above species.

Phoradendron leucarpum (Raf.) Reveal & M. C. Johnston   MISTLETOE   Family Loranthaceae
                                                                                                                                      Figure 349

A parasitic evergreen that grows on branches in the crown of many species of hardwood trees. Seen throughout the SE. In the Appalachian region *Nyssa sylvatica* (blackgum) is 1 of the more common hosts. In the Piedmont and coastal plain, oaks are often used (particularly *Quercus nigra* in NC, SC). Similar preferences in host species may vary locally. The opposite leaves, jointed twigs, and white drupes in winter are characteristic, aside from the arboreal habit.

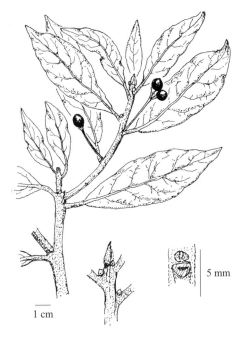

Figure 346. *Persea palustris*

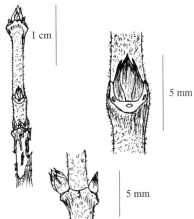

Figure 347. *Philadelphus hirsutus*

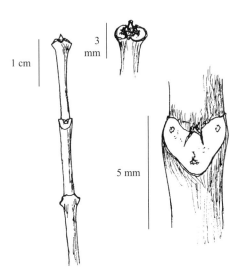

Figure 348. *Philadelphus inodorus*

Phyllostachys Sieb. & Zucc.   BAMBOO   Family Poaceae   *Figure 350*

A large genus, mostly native to China. Of many species introduced and used horticulturally in the SE, 3 are treated here. These are the most widely naturalized, though others may be locally encountered. Exotic bamboos usually naturalize through spreading of the underground rhizomes, away from cultivated specimens. With age, the rhizomes send up progressively larger and taller aboveground stems, called culms. The entire patch of culms forms a grove, and the density of culms and persistent foliage usually prevent competition of other plants from within. All species of the genus bear a vertical groove (sulcus) in the culm internodes, above each nodal joint. A nodal ridge also encircles the culm at the nodal joint, just above the sheath scar. Leaves vary from 5–15cm long; minutely serrate on one margin, entire or nearly so on the other. Groves typically die after flowering.

1. Culms with short, knobby internodes near the ground; culms swollen below joints throughout midstem   1. *P. aurea*
1. Culms not as above, without lumplike or swollen areas
   2. Culms may have kinks (geniculate) near the ground; sheaths white-lined   2. *P. aureosulcata*
   2. Culms rarely geniculate; sheaths mottled   3. *P. bambusoides*

1. *Phyllostachys aurea* Carr. ex A. & C. Rivière   FISH-POLE BAMBOO

A slender species, to 8m tall. Culms usually to 3, rarely to 4cm diam.; green, or yellowish in bright sun or with age. The lower portion of the culm has short, lumpy internodes; internodes above mostly 12–22cm long; small swellings also occur at the base of joints in culms throughout. Culm sheaths at first reddish, with purplish streaks, aging to yellow-green. Native to China.

2. *Phyllostachys aureosulcata* McClure   YELLOW-GROVE BAMBOO

A tall species, potentially to about 10m. Culms usually 2–4cm diam.; yellow-green, with more pronounced yellowish color in the groove above each node (the original colloquial name probably describing this trait as "yellow-groove" bamboo). The lower portion of culms may be slightly geniculate (zigzag) at the nodes. Culm sheaths light green, with pale stripes; leaf sheaths with 2 tufts of bristles on auricles (the small extension on each side of the sheath). Native to China.

3. *Phyllostachys bambusoides* Sieb. & Zucc.   JAPANESE BAMBOO

A tall species, usually between 8–20m. Culms 3–12cm diam., thick-walled; variable in color from green to yellowish, sometimes with a bloom; internodes straight throughout. Outer culms often lean toward light, on borders of groves. Culm sheaths mottled with purple; leaf sheaths with 2 tufts of bristles on auricles. Native to China, Japan. Used extensively in Japan for industrial purposes and construction, with many variations and selections known.

Physocarpus opulifolius (L.) Maxim.   NINEBARK   Family Rosaceae   *Figure 351*

A native shrub, to 3m tall. Distributed mostly in the interior sections of the SE, from Appalachians and Piedmont of VA to SC, west to AR, MO, though occasionally seen in the coastal plain to n. FL, s. AL. Usually seen in rocky woods, on rock outcrops with seepage and along streams. Twigs ridged below nodes; stipules often persistent; bundle scars 3; terminal buds usually present; bud scales imbricate, loose and rather brittle. Bark brownish, very shreddy or exfoliating. The fruit is a cluster of 3 or 4 inflated follicles.

Picea A. Dietrich   SPRUCE   Family Pinaceae

One native species occurs in the SE, and 1 sparingly naturalized exotic is also included here. Spruces have pointed needlelike leaves attached to the twigs on a peglike base; needles mostly

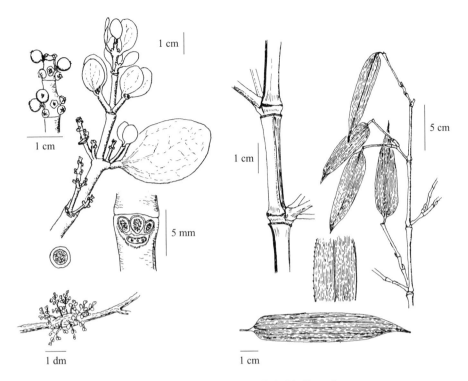

Figure 349. *Phoradendron leucarpum* Figure 350. *Phyllostachys*

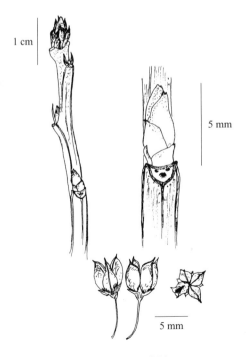

Figure 351. *Physocarpus opulifolius*

4-angled or ridged in cross section. Twigs are roughened by the raised needle attachment points; terminal buds brownish, with imbricate scales. Bark grayish, scaly. Cones are pendent, red-brown to gray.

1. Cones less than 5cm long; twigs horizontal throughout crown   1. *P. rubens*
1. Cones over 10cm long; twigs drooping on lower branches of mature trees   2. *P. abies*

1. *Picea rubens* Sarg.   RED SPRUCE   *Figure 352a*

Native to the NE and the Appalachians; ranging into the SE in the higher elevations of WV, VA, NC, TN. A major forest canopy component of the boreal zone of the Blue Ridge, mostly above 1200m (4000ft) elevation, but occasionally lower in boggy sites. Needles sharply pointed; twigs usually slightly hairy. Cones mostly 3–4cm long.

2. *Picea abies* (L.) Karsten   NORWAY SPRUCE   *Figure 352b*

A tall tree often planted in the n. or interior sections of the SE, rarely naturalized. In some areas of the Blue Ridge, planted stands are reproducing by seed and establishing seedlings. Juvenile plants are difficult to separate from the native species. Twigs are usually glabrous. Needles more bluntly pointed on mature trees, and cones are 10–15cm long. The drooping nature of twigs and branchlets on lower limbs is apparent only on mature specimens. Native to n. Europe.

Pieris D. Don   FETTERBUSH   Family Ericaceae

Two species are native to the SE, both evergreen shrubs. Racemes of flower buds are formed in late summer and visible on the twigs throughout winter. The fruit is a 5-parted, rounded capsule.

1. Leaf margins crenate, with black hairs; black hairs on petiole; leaves reticulate above; of Appalachians   1. *P. floribunda*
1. Leaf margins thickened, revolute, no hairs; leaves punctate above; of coastal plain
                                                                                             2. *P. phillyreifolia*

1. *Pieris floribunda* (Pursh.) Benth. & Hook. f.   MOUNTAIN FETTERBUSH   *Figure 353*

A bushy, erect shrub endemic to the Appalachian region, distributed from WV and VA to NC, TN, and ne. GA. Scattered throughout the range in local populations; overall rather uncommon. Occurs in forest openings and ericaceous thickets in high elevations of the Blue Ridge, and on dry ridges or rock outcrops in parts of the n. range and in the upper Piedmont. Leaves have stiff, black hairs (strigose trichomes) on leaf serrations, petioles, and twigs; these may slough to some degree with age. Leaf apex usually with a thickened, glandlike point, and venation is finely reticulate on the dark upper surface. Flower bud racemes are borne on the twig tips.

2. *Pieris phillyreifolia* (Hook.) DC.   VINE-WICKY   *Figure 354*

A rhizomatous or vinelike shrub, usually climbing the trunks of pond-cypress (*Taxodium ascendens*) trees, or with short, ascending stems to about 2dm tall from a rhizomatous habit on the ground near rotted logs or boles of trees. The climbing habit utilizes the rough plates of the host tree to shield the ascending stems, with short, horizontal leafy branches appearing out of the bark along the length of the host tree's trunk. Found in the coastal plain from SC to FL, west to s. AL. Associated with cypress ponds and similar wetlands in pinelands. Racemes of flower buds are in axils of the leaves.

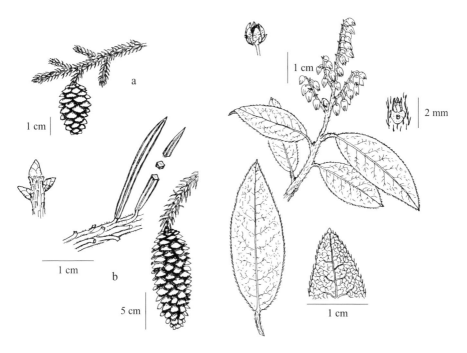

Figure 352. a: *Picea rubens;* b: *P. abies*   Figure 353. *Pieris floribunda*

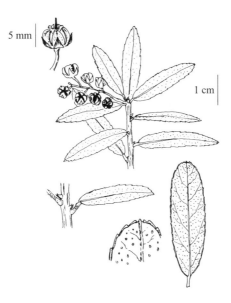

Figure 354. *Pieris phillyreifolia*

Pinckneya bracteata (Bartr.) Raf.   PINCKNEYA   Family Rubiaceae   *Figure 355*

A shrub or small tree with an open crown, to 8m tall. Native to the coastal plain from SC to n. FL, and west perhaps to s. AL. Rare and sporadic in its distribution, mostly in damp to wet, acidic soils such as bays, bogs, blackwater seepage wetlands, and swampy thickets. Twigs hairy, with elongated, pale lenticels; terminal bud reddish, 2-scaled; lateral buds much smaller and nearly hidden; leaf scars opposite, connected by stipule scars; bundle scars single. Bark brownish, smooth to shallowly fissured. The fruit is a capsule about 15mm long, with small bumps on the surface and with winged seed.

Pinus L.   PINE   Family Pinaceae

Twelve pines are native to the SE. Additionally, the Scots pine (*P. sylvestris*), an exotic that may occasionally be seen as an escape from nearby cultivation, is included in the key. Other exotic pines, only locally naturalized or persisting from cultivation, may be encountered in the SE on rare occasions. All pines have needlelike leaves held in bundles, and the number per bundle, length, and relative straightness or amount of twist per needle are factors used in identification. Range, habitat, cones, and bark features are also important. The pines are often described as belonging to 2 general groups, soft (white) pines and hard (yellow) pines. Only 1 white pine (*Pinus strobus*) is native in the e. US.

1. Needles 5 per bundle   1. *P. strobus*
1. Needles 2 or 3 per bundle (rarely 4)
  2. Needles nearly always 2 per bundle
    3. Needles straight, mostly over 10cm; upper trunk branches whorled; n. species
        2. *P. resinosa*
    3. Needles usually twisted, mostly under 10cm; no distinctly whorled branches; s. species
      4. Cones mostly 7cm or more long, with stout spines on scales   3. *P. pungens*
      4. Cones less than 7cm long, with scales bearing slender prickles, or unarmed
        5. Cone scale prickles weak or deciduous
          6. Bark furrowed, not scaly on young stems   4. *P. glabra*
          6. Bark reddish, scaly   5. *P. sylvestris*
        5. Cone scale prickles slender or stout, usually present; bark furrowed on mature trunks but scaly or flaky on young stems
          7. Prickles slender; cones open at maturity; upper trunk bark pinkish or reddish; common in Appalachians, Piedmont   6. *P. virginiana*
          7. Prickles small but stout; cones usually remain closed at maturity; upper trunk bark gray; native of FL, s. GA only   7. *P. clausa*
  2. Needles mostly 3 per bundle, but mixtures of 2- and 3-needled bundles may be present
    8. Needles mostly under 12cm long
      9. Cone scale prickles stout (if spurlike spines present, see *P. pungens*); base of opened cones flat; cones about as long as broad; most needles 3 per bundle   8. *P. rigida*
      9. Cones with weak prickles; opened cone base not flat, cones longer than broad; needles commonly 2 or 3 per bundle   9. *P. echinata*
    8. Needles mostly over 12cm long
      10. Needles commonly 2 or 3 per bundle; cones with a short but distinct stalk
        10. *P. elliottii*
      10. Needles mostly 3 per bundle; cones sessile
        11. Cones remain closed long after maturity; about as broad as long, under 6.5cm in length   11. *P. serotina*
        11. Cones open with maturity; longer than broad, often over 7cm long
          12. Needles often over 20cm long, basal sheath 1.5cm or more; cones 15cm or more long   12. *P. palustris*
          12. Needles under 20cm, sheaths usually under 1.5cm; cones under 13cm long
            13. *P. taeda*

1. *Pinus strobus* L.   EASTERN WHITE PINE   *Figure 356a*

A tall tree of uplands, mostly distributed in the SE along the Appalachians and upper Piedmont from WV and VA to n. GA. Occurs in a variety of habitats, from moist lowlands to dry, rocky ridges and gorge slopes. Usually intermixed with hardwoods, unless pure stands have been induced by disturbances to hardwood forests or through direct management of timberlands to favor its development or planting. An important timber species of the Appalachians and the NE. Crown typically pyramidal, with all branches in a whorled arrangement. Needles soft, not rigid, 7–12cm, 5 per bundle. Cones 10–20cm, cylindric; scales thin, lacking prickles but often with globules of resin. Bark dark greenish-brown and smooth when young, becoming fissured into scaly ridges.

2. *Pinus resinosa* Ait.   RED PINE   *Figure 356b*

A n. species barely ranging into the SE, reaching its s. limit in WV. Plantings elsewhere in the SE have apparently not naturalized. Needles break cleanly when nearly doubled between the fingers; 10–17mm long. Cones mostly 4–6cm, no prickles on scales. Bark scaly or with scaly plates.

3. *Pinus pungens* Lamb.   TABLE MOUNTAIN PINE   *Figure 356c*

A wide-crowned tree, rarely over 12m, native to the Appalachians. Found in the SE from WV and VA southward along the mountains to GA and SC. Typical of dry ridges, bluffs, and stony, open slopes. Adapted to release seed from serotinous cones after fire. Branches horizontally elongated, tough and flexible. Needles 3–6cm, strongly twisted, thick and dark green. Cones ovoid, 6–10cm; scales thick, and many with a strong, curved claw instead of a prickle. Bark gray, in scaly ridges.

4. *Pinus glabra* Walt.   SPRUCE PINE   *Figure 357a*

A coastal plain native, from SC to FL, west to LA. Scattered in distribution, mostly among hardwoods in bottomland forests or swamps. Needles dark green, 4–8cm. Cones mostly 4–7cm, with weak and easily detached prickles. Bark gray and fairly smooth on young trunks, becoming dark and longitudinally ridged, the ridges rough and slightly scaly.

5. *Pinus sylvestris* L.   SCOTS PINE   *Figure 357b*

An exotic tree, native to Eurasia. An occasional escape in the n. sections of the SE. Farther north, the trees have been planted more extensively and are perhaps more commonly seen naturalized. Needles 4–7cm, strongly twisted. Cones mostly 3–6cm, with no prickles. Bark reddish and papery-scaly when young, with scaly plates when older.

6. *Pinus virginiana* P. Mill.   VIRGINIA PINE   *Figure 357c*

Native to areas of the SE mainly inland of the coastal plain, especially from Piedmont and Appalachian regions north and west to the Ohio and Mississippi Rivers. Typically seen in well-drained upland soils, where it may form dense successional stands after disturbances to hardwood forests. Crown wide when open grown, but trunk long and frequently crooked and lined with dead branches when in competitive growth. Needles mostly 3–7cm, strongly twisted. Cones 3–7cm, with sharp prickles. Bark red-brown, thin, fairly smooth or with thin scales on upper sections of trunk; thicker and with grayish, scaly ridges on lower trunk.

7. *Pinus clausa* (Chapm. ex Engelm.) Vasey ex Sarg.   SAND PINE   *Figure 357d*

A tree primarily of peninsular FL, ranging to n. FL and west in the panhandle to sw. AL. Natural habitat is dry, sandy uplands and scrub. Sometimes experimentally planted outside its

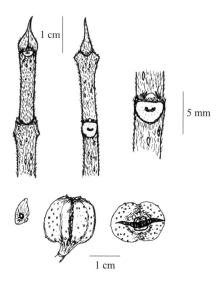

Figure 355. *Pinckneya bracteata*

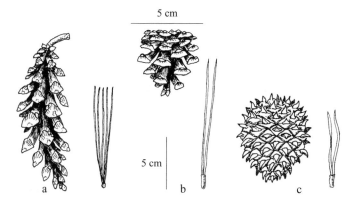

Figure 356. a: *Pinus strobus;* b: *P. resinosa;* c: *P. pungens*

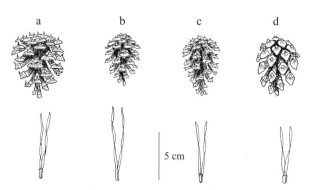

Figure 357. a: *Pinus glabra;* b: *P. sylvestris;* c: *P. virginiana;* d: *P. clausa*

native range for wood pulp production on poor, dry sands of s. GA to s. SC. Well adapted to fire, regenerating after the serotinous cones are opened by heat. Needles 5–9cm. Cones mostly 4–7cm, with a weak prickle. Bark gray, somewhat smooth on upper portions of trunk, lower trunk with scaly ridges.

8. *Pinus rigida* P. Mill. PITCH PINE   *Figure 358a*

A wide-crowned tree in the open, or narrower and with several dead lower limbs on the trunk when under competition. Distributed from the coastal plain of VA to the Piedmont of SC and GA, and areas more inland. Common in the Appalachian region, typically on well-drained sites but sometimes on acidic, boggy habitats or streamside flats. Needles 5–10cm, usually 3 per bundle. Cones 3–8cm long, egg-shaped when closed, with a wide, flat base when open; prickle sharp, downcurved. Bark dark gray to blackish, with scaly ridges or irregular plates; no resin pockets within the bark layers. Tufts of needles can sometimes be seen growing on the trunk (epicormic buds).

9. *Pinus echinata* P. Mill. SHORTLEAF PINE   *Figure 358b*

Distributed across most of the SE, in upland or well-drained soils. Needles 6–12cm long. Cones 4–6cm, with a sharp but rather weak prickle. Bark red-brown, with broad plates, resin pockets between the bark layers composing the plates. An important timber species in the interior uplands of the SE. Less common in the coastal plain.

10. *Pinus elliottii* Engelm. SLASH PINE   *Figure 358c*

Native to the coastal plain from se. SC to FL, west to e. LA. Planted widely for timber production both in and out of its natural range and natural habitat. By nature, occupying sites near and in maritime forests and in wet flatwoods, where it may be a primary canopy component. Needles 15–20cm. Cones mostly 10–18cm, with a stalk 2 or 3cm long; scales glossy brown, with a short prickle. Bark red-brown, in broad, scaly plates.

11. *Pinus serotina* Michx. POND PINE   *Figure 359a*

Native mostly to the coastal plain from MD to FL, west to s. AL, though also occurs less commonly in the Piedmont. Habitat is typically damp to wet lowlands, in pocosins, flatwoods, boggy depressions, savannas. A tall tree in fertile soils; dwarfed in acidic, wet, nutrient-poor sites such as pocosins. Well adapted to sprout from adventitious buds after fire and to release seed from the serotinous cones. Needles 15–20cm. Cones 5–8cm, egg-shaped, may remain closed for several years. Bark grayish to red-brown, with scaly ridges or irregular plates.

12. *Pinus palustris* P. Mill. LONGLEAF PINE   *Figure 359b*

Native to the coastal plain and parts of the Piedmont from se. VA to FL, west to se. TX. Historically dominant on a variety of lowland sites from moist to dry, where it was the primary canopy component of many savannas and flatwoods. This dominance has been severely reduced after several human generations, due to fire suppression and other site management practices that have been detrimental to the species' regeneration. Also occurs in dry, sandy uplands of the Piedmont and on the foothills of the Blue Ridge in east-central AL. Needles 20–45cm long. Cones 15–25cm, with short, stout prickles. Bark scaly or with scaly plates. The seedlings undergo a "grass stage," whereby stem growth is slow but root growth extensive; the juvenile plants resemble a grass clump during this phase.

13. *Pinus taeda* L. LOBLOLLY PINE   *Figure 360*

Widespread across most of the SE, east of the Blue Ridge and south of TN. Some of this range is due to naturalization after extensive planting. Historically, most common and typical of

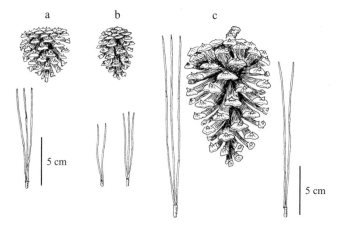

Figure 358. a: *Pinus rigida*; b: *P. echinata*; c: *P. elliottii*

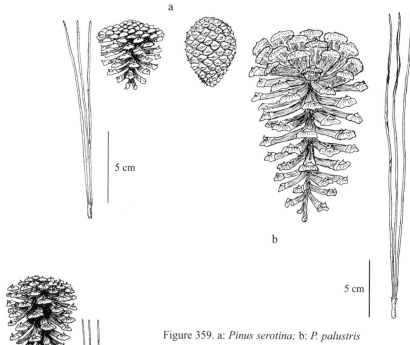

Figure 359. a: *Pinus serotina;* b: *P. palustris*

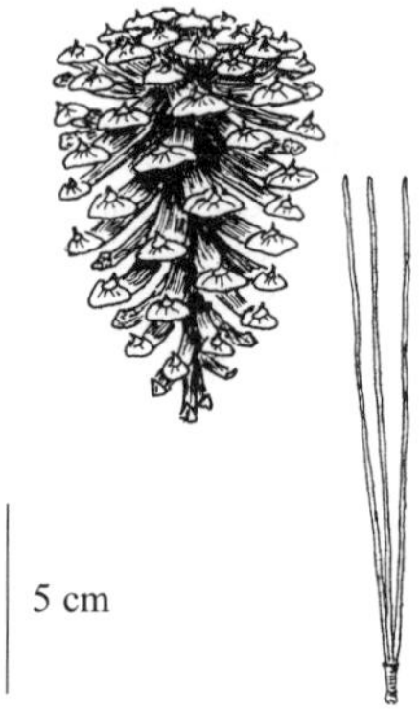

Figure 360. *Pinus taeda*

coastal plain and Piedmont lowlands. Needles mostly 12–20cm long. Cones 6–12cm, scales dull gray or brown, with stout, sharp prickles. Bark on mature trunks with wide, scaly plates. Probably the most extensively planted pine in the SE, for timber production.

14. *Pistacia chinensis* Bunge   CHINESE PISTACHIO   Family Anacardiaceae   *Figure 361*

A tree, to 25m, with a broad crown. Planted as a shade tree and occasionally naturalized in the SE, mainly in the coastal plain. Twigs stout, dense and rigid, roughened with numerous raised lenticels. Terminal buds large, conical, with several reddish, rigid, imbricate bud scales; outer scales usually with truncate tips or keeled. Leaf scars bear several bundle scars, sometimes arranged in 3 groups. Inner bark aromatic, resembling odor of sumac. Bark gray, scaly or furrowed into scaly ridges. Large panicles of small, hard fruits are attractive to birds, and hence the tree is spread.

Planera aquatica J. F. Gmel.   PLANER-TREE   Family Ulmaceae   *Figure 362*

A small tree, of 15m or less, with a short or crooked trunk. Native to the coastal plain from NC to FL, west to TX and north in the Mississippi River lowlands to s. IL. An understory tree of swamps and frequently flooded bottomlands; often seen leaning over swampland sloughs. Twigs are slender, zigzag, lenticellate; terminal bud lacking; lateral buds often with a swollen basal bud scale, or collateral buds present. The leaf scars are small, dark; bundle scars obscure. Bark scaly, with outer gray scales revealing reddish-brown inner bark when they curl or fall. The fruit is a brown nutlet covered with fingerlike growths, maturing in late spring.

Platanus occidentalis L.   AMERICAN SYCAMORE   Family Platanaceae   *Figure 363*

A large native tree distributed throughout the SE, but rare or absent in se. GA and e. sections of n. FL. Common in bottomlands and near streams. Twigs moderately stout, zigzag; terminal bud lacking; lateral buds conical, 1-scaled, nearly encircled by leaf scars; a ringlike stipule scar encircles twig. Bark smooth, whitish or mottled on upper stems, more brown and scaly below, thicker and furrowed on trunks of old trees. The fruit a globular head of achenes, disintegrating when mature in late autumn; each seed with a brownish fringe of hairlike fibers.

Polygonella Michx.   JOINTWEED   Family Polygonaceae

Three species are treated here as woody members of a group of mostly herbaceous annual or perennial plants of the coastal plain. Even these 3 species are not entirely woody, some of the twigs usually dying back in winter or after flowering and setting fruit. At least in the s. parts of their range, these 3 species tend to have a woody base, and some green leaves may be found into the winter. Twigs have a jointed appearance due to sheathing stipules (ocreae) at each node. White flowers appear in summer to autumn, and the triangular nutlet (achene) grows mostly hidden within the enlarging sepals into late autumn. The whitish sepals may change to pinkish with fruit maturation.

1. Leaves mostly over 1cm wide, obovate; often to 1m tall; native of FL   1. *P. macrophylla*
1. Leaves all under 1cm wide, not obovate; nonflowering stems rarely over 3dm; widespread species
   2. Leaves mostly linear, 1mm wide, nearly terete; fruit sepals 3–5mm; seed 3mm; flowers perfect   2. *P. americana*
   2. Leaves more spatulate, 1–5mm wide, flat; fruit sepals 2–3mm; seed 1.5mm; plants dioecious   3. *P. polygama*

1. *Polygonella macrophylla* Small   GULF JOINTWEED

A stiffly erect but few-branched shrub, the main woody portion to 1m tall, with flowering stems rising to 1.5m. Native to deep sands of scrub and dunes near the Gulf Coast in n. FL west to se.

AL. Leaves reaching 5cm long, 2cm wide. Older stems distinctly roughened or with ringed scars from sloughing ocreae. Seed about 3.5mm long.

2. *Polygonella americana* (Fisch. & C. A. Mey.) Small   JOINTWEED   Figure 364

A semiwoody shrub, to 3dm; flowering stems to 8dm. Distributed from NC to s. GA, west to TX, north to se. MO; in dry, sandy soils. Most common in sandy scrub and pinelands of the coastal plain. Leaves 4–15mm long, about 1mm wide, and nearly as thick as they are wide. The erect flowering stems rise well above the branched leafy portion of the plant. Flowers June–Sept.

3. *Polygonella polygama* (Vent.) Englm. & Gray   OCTOBER-FLOWER

A low, bushy, semiwoody shrub, to 2dm tall, but with flowering stems to 6dm. Distributed from se. VA to FL, west to se. TX, mostly in deep sands. Leaves are normally narrowly spatulate, 5–20mm long and 2–5mm wide, though in the sandhills the var. *croomii* (Chapm.) Fernald bears leaves that are mostly 5–12mm long, 0.5–2mm wide.

Poncirus trifoliata (L.) Raf.   TRIFOLIATE ORANGE   Family Rutaceae   Figure 365

An exotic shrub or small tree with green bark and spiny, contorted branches. Twigs dark green; leaf scars small with a single bundle trace; no terminal bud; lateral buds small, rounded. Large, flattened thorns occur at each node, between bud and leaf scar. The twigs are also angular, a ridge extending below each of the hefty thorns, likely a structural reinforcement. The fruit is a typical citrus; a yellow berry to 5cm diam., with firm, sour, acidic flesh; not palatable.

Populus L.   POPLAR   Family Salicaceae

Ten species are included here, for the SE. Five of these are native species. The exotics that have naturalized do so through root sprouts. Since poplars are dioecious, only female trees can produce seed. All have terminal buds and lateral buds with a single large basal scale spanning the entire width of the bud. Lateral flower buds are larger and much plumper than ordinary vegetative buds; they contain the undeveloped catkins that emerge in early spring. Leaf scars show 3 bundle scars or 3 groups of bundle scars. The pith is angled in cross section. These are fast-growing but relatively short-lived trees. The fruit is a capsule filled with downy material between seed, accounting for the popular name of "cottonwood." The species with whitish bark will appear green when wet.

1. Terminal bud less than 12mm long
    2. Buds not resinous when crushed; bark smooth, whitened or greenish-gray on all but lower part of trunk
        3. Twig apex or bud scales with distinct grayish hairs, bud scales mostly dark-margined; lowermost bud scale of lateral bud usually covers ¼ or less of bud length, with 4-lobed margin
            4. Bark whitened; twigs often downy; leaves white beneath   1. *P. alba*
            4. Bark greenish-white; twigs lightly hairy; leaves green or gray-hairy beneath
                                                                         2. *P.* × *canescens*
        3. Twig apex and bud scales not as above; bud scales light-margined; lowermost bud scale of lateral bud ⅓ or more of bud length, with 2- or 3-lobed margin
            5. Bark whitened; bud scales shiny; leaf teeth small, more than 20 per side
                                                                         3. *P. tremuloides*
            5. Bark greenish-white; bud scales gray; leaf teeth large, 5 to 18 per side
                                                                         4. *P. grandidentata*
    2. Buds slightly resinous when crushed; bark furrowed or roughened over most of trunk

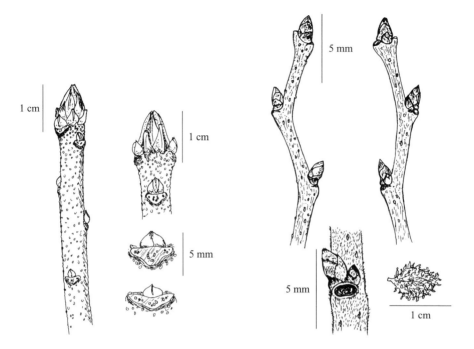

Figure 361. *Pistacia chinensis*

Figure 362. *Planera aquatica*

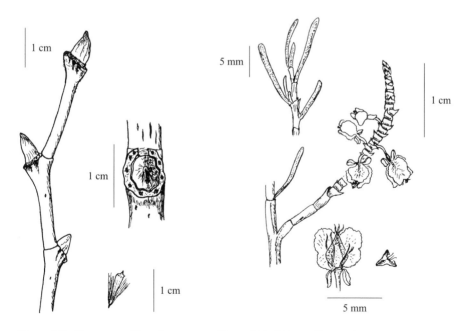

Figure 363. *Platanus occidentalis*

Figure 364. *Polygonella americana*

246 · *Populus*

    6. Terminal buds about as long as wide or wider; lateral buds much smaller (except flower buds), with 1 or 2 visible scales; native tree of wet sites   5. *P. heterophylla*
    6. Terminal buds much longer than wide; most lateral buds nearly as long as terminal, with 3 or 4 scales; exotic tree with narrow (fastigiate) growth habit   6. *P. nigra* var. *italica*
1. Terminal bud over 12mm
    7. Buds dark red-brown, very resinous and aromatic when crushed; twigs brown
        8. Twigs glabrous; a native n. tree, very rare in SE   7. *P. balsamifera*
        8. Twigs with a few hairs; occasionally seen in Appalachians of SE   8. *P.* × *jackii*
    7. Buds greenish to light brown, only slightly resinous and barely aromatic; twigs yellow-brown
        9. Larger lateral buds have margin of largest bud scale hidden behind bud, its smooth outer surface positioned directly above basal scale; stems of robust growth heavily lined below nodes; wide-crowned native tree   9. *P. deltoides*
        9. Larger lateral buds have margin of largest bud scale (the one above basal scale) visible above middle of basal scale; stems lightly lined or unlined on robust growth; narrow-crowned hybrid tree spreading by root sprouts   10. *P.* × *canadensis*

1. *Populus alba* L. WHITE POPLAR

An exotic tree, to 30m, normally with an open and coarse crown, though there is a narrow, fastigiated form. The bark is whitish on upper parts of trunk, greenish-white farther down, the trunk base roughly furrowed and black. Twigs slender, whitish pubescent to white-woolly near the tip. Leaves white and velvety beneath, usually 3- to 5-lobed, suggesting maple leaves in general shape. Native to Eurasia. Tends to sprout extensively from the root system, forming colonies or groves of slender saplings. Both pistillate and staminate trees may be encountered, but rarely together; viable seed production therefore uncommon (and fortunately so). Occasionally planted in the SE, mostly seen as an escape near sources of cultivation.

2. *Populus* × *canescens* (Ait.) Small GRAY POPLAR   *Figure 366*

A presumed hybrid of *P. alba* and *P. tremula* L. (European aspen), originating in Europe or w. Asia and widely planted in the interior uplands of the SE, at least in years past. Naturalized through prolific root sprouting, forming groves. This species is frequently mistaken for the native *P. grandidentata* due to a few aspenlike traits. Crown slender, open, with greenish to greenish-white smooth bark on upper parts; trunk furrowed and blackened near base. Twigs slender, grayish-hairy only near the tip. Leaves less lobed than in *P. alba;* those on twig tip mostly triangular and gray-hairy beneath; those of lower nodes to nearly orbicular, with large, irregular, blunt to rounded teeth, green and glabrous beneath.

3. *Populus tremuloides* Michx. QUAKING ASPEN   *Figure 367*

A native tree widely known and characteristic in n. and w. forest regions, but barely reaching the SE. Found as a few local populations in WV, VA, TN, and MO. Sometimes planted, and in cooler climates of the SE, may persist or spread from the roots. Crown slender, open; bark whitish, smooth, but blackened and rough near trunk base. Twigs slender, glabrous, glossy; buds sharply pointed, with glossy scales. Leaves with many small teeth.

4. *Populus grandidentata* Michx. BIGTOOTH ASPEN   *Figure 368*

A native tree, to 25m, with an irregular, oblong crown. Distributed from e. VA and central NC west to IL, TN; scattered and fairly uncommon across this range, more common in the NE. Occurs on rocky ridges or disturbed slopes in moist to dry soils. Twigs moderately slender, grayish to light brown; buds sharply pointed, with pale brown, dull scales that may have appressed gray hairs near their base. Bark greenish-gray or gray, smooth on upper trunks,

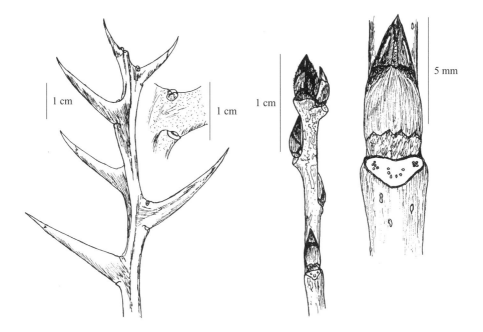

Figure 365. *Poncirus trifoliata*

Figure 366. *Populus* × *canescens*

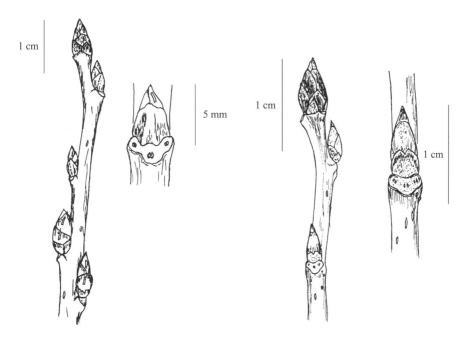

Figure 367. *Populus tremuloides*

Figure 368. *Populus grandidentata*

sometimes with white lenticels, becoming furrowed on basal portions of the trunk. Leaves with several large, pointed teeth.

5. *Populus heterophylla* L.   SWAMP COTTONWOOD   Figure 369

A tree, to 30m, the crown irregular and coarse when forest grown. Native to coastal plain lowlands across the SE, though scattered and local in distribution; absent from wide areas of seemingly suitable habitat. Typically in swamps and poorly drained soils in bottomlands. Twigs fairly stout, brown, with pale lenticels; terminal bud wide-based; pith brownish. Bark grayish brown to red-brown, with thick, scaly ridges or plates, or deep furrows. Leaf blades mostly over 10cm long, thick, with crenate teeth, rounded base, blunt tip; petiole not flattened near blade.

6. *Populus nigra* var. *italica* Du Roi   LOMBARDY POPLAR   Figure 370

The species of *P. nigra* L., the European black poplar, is rarely seen in cultivation in the SE, due to poor vigor and health in the region. It is a species better adapted to a more n. climate, and where the incidence of fungal cankers is not so prevalent. One variety that is more often planted across the SE, especially in the Appalachian region, is the Lombardy poplar. The fastigiate habit (compact and pyramidal) of numerous, short, suckerlike branches that are all up-reaching is characteristic. All trees are staminate, and spread of this tree is purely by root sprouts. Rarely naturalized, but often persistent or resprouting after cultivation. Twigs moderately slender, yellow-brown; buds red-brown, slightly resinous and aromatic. Bark brown, irregularly furrowed.

7. *Populus balsamifera* L.   BALSAM POPLAR

A native tree widespread across the n. sections of N. Amer., but very rare in the SE, known only to reach as far south as WV. Twigs moderately stout, red-brown, glabrous; terminal and lateral buds dark red-brown, with sticky and highly aromatic resin, or balsam, within. Leaves rounded to very slightly cordate at base, glabrous, glaucous beneath. Bark brownish or gray-brown, furrowed, except smooth on young stems. Old trunk bark may become thick, with broad, scaly ridges and deep furrows.

8. *Populus* × *jackii* Sarg.   BALM-OF-GILEAD   Figure 371

A tree of uncertain origin, assumed to be a variant clone of the above species or a hybrid. Occasionally seen in the Appalachian region of the SE, where it was once more widely planted for its aromatic buds, used medicinally. It persists after cultivation through root sprouts. All trees are pistillate, and though fruit may be produced, viable seed are not present; even with nearby potential male pollinators of other related species, it appears sterile. Twigs and buds similar to above species, but scattered hairs are present. Leaves mostly cordate at base, with hairs beneath, along the margin, and on petiole. Bark brown, irregularly furrowed, with rough or scaly plates and ridges. (*P. candicans* Ait.; *P.* × *gileadensis* Rouleau)

9. *Populus deltoides* Bartr. ex Marsh.   EASTERN COTTONWOOD   Figure 372

A large native tree with a wide crown, to 45m tall in some sites. Widespread across the SE, except absent from many parts of the Appalachians, se. GA, and ne. FL. Typical of lowlands with moist soils and plenty of light after disturbances, where the trees can make rapid early growth without competition from other hardwoods. Old stands may often be seen in silty river floodplains. Twigs greenish to light brown, 3-ridged or lined below the leaf scars; buds conical, pointed, with greenish to light brown scales; slightly resinous within. Leaves widely triangular, light green on both sides, with incurved teeth; petiole flat near blade. Bark grayish brown,

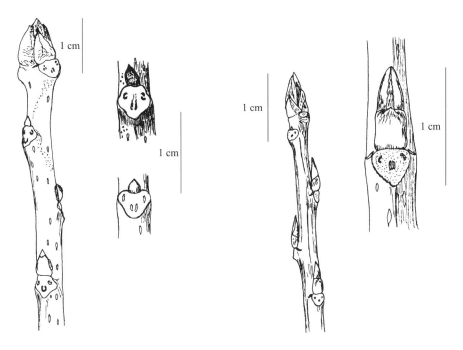

Figure 369. *Populus heterophylla*

Figure 370. *Populus nigra* var. *italica*

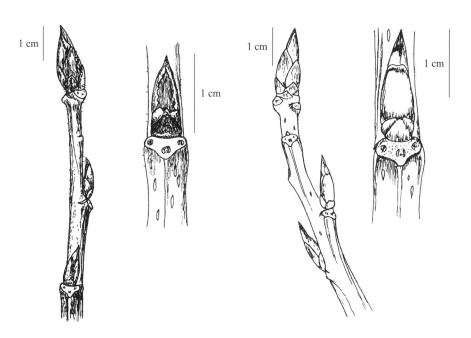

Figure 371. *Populus* × *jackii*

Figure 372. *Populus deltoides*

deeply furrowed. Along the stream courses into the prairies and plains of the Midwest, this tree merges into the plains cottonwood (*P. sargentii* Dode.); that entity now considered *P. deltoides* ssp. *monilifera* (Ait.) Eckenwalder; differing in proportionately wider leaves with larger teeth, twigs more yellow-brown.

10. *Populus* × *canadensis* Muenchh.   NORWAY POPLAR   *Figure 373*

A hybrid of *P. deltoides* and *P. nigra;* the first type originated in France, but numerous other hybridizations between the same parent species are known. These hybrids are often promoted as superior fast-growing trees and have been planted across the SE, particularly in the northernmost sections of the region. Rarely naturalizing but sometimes resprouting from roots to appear naturalized. Crown pyramidal or oblong, not as wide-spreading as in *P. deltoides*. Twigs similar to that species, but rarely lined, more yellow-brown, usually with largest bud scale of lateral bud facing opposite direction. Leaves dark green above, with rounded, slightly incurved teeth, base of blade sloping at an angle from petiole that is greater than the near 90-degree aspect of *P. deltoides*. Bark brown, shallowly furrowed into irregular ridges. Not long-lived in the SE. (*P. euramericana* Guinier)

Prosopis glandulosa Torr.   HONEY MESQUITE   Family Fabaceae   *Figure 374*

A shrub or small tree, to 6m tall, with a broad or rounded crown with drooping outer branchlets and short, usually forked or multiple trunks. Native to areas mostly west of the region of coverage of this book, but ranging into e. TX on dry plains and sandy uplands. Twigs light brown or grayish, moderately stout, often with spines at nodes. Terminal bud absent; laterals small, with many long-pointed brown scales and stipular points on edges of some leaf scars. Spur shoots commonly present, with several raised leaf scars each with stipular projections (appearing as if many small buds are clustered). Bark grayish, roughened by scaly plates or ridges; inner bark yellowish, aromatic. Leaves compound, with 2 pinnae; leaflets gray-green to light green, 5cm or less wide, 12 to 20 per pinnae. Legume narrow, mostly 10–20cm long, constricted between seed.

Prunus L.   CHERRY, PLUM, PEACH   Family Rosaceae

Of the 25 species treated here for the SE, 17 are native; the others naturalized to varying extents in various parts of the region. The fruits of all are fleshy drupes, with a single large seed commonly referred to as a "pit" or "stone." The twigs are not diagnostic for all species, so the following key must take into account some leaf or fruit characteristics to help separate out the members. Fallen or remnant leaves are important items to search for in field identification over winter. Scraped twigs, if fresh, may reveal the bitter-almond odor that is conspicuous in many of the species in this genus. Bark of young and fast-growing trunks is generally smooth, but with conspicuous horizontally elongated lenticels. This is a large genus, composed of at least 150 species dispersed mostly in temperate regions of the world. The species are separated into several subgenera and sections, based on flower and fruit characters. Of the species in the SE, there are 5 subgenera, 2 composed of the plums and 3 of cherries. Plums are generally defined by the lengthwise-grooved fruit that is often glaucous (with a bloom) and with compressed seed. Cherries generally have smaller, glossy fruit without a groove, with plump seed. In the SE, plums generally lack true terminal buds and cherries generally have them present (there are 2 or 3 exceptions). Also, plums tend to have spinous spur shoots. Subgenus *Amygdalus* (L.) Focke includes plumlike members with nearly sessile flowers and pubescent fruit (the peach); subg. *Prunophora* Focke includes all our other plums; subg. *Cerasus* Pers. includes the cherries with flowers solitary or in few-branched clusters; subg. *Padus* (Moench) Koehne includes the cherries with flowers in elongate racemes; subg. *Laurocerasus* Koehne includes the evergreen cherries.

1. Twigs greenish, glabrous; buds with large, grayish, hairy scales; fruit 2.5cm diam. or more and surface velvety-pubescent; seed surface deeply pitted   1. *P. persica* (peach)

Prunus · 251

1. Twigs and buds not as above; fruit smaller or glabrous; seed not pitted
    2. Twigs slender (internodes 2.5mm diam. or less), no terminal bud (except in *P. domestica,* where twigs moderately stout and terminal bud usually present); spinous spur shoots commonly present; fruit longitudinally grooved on one side, surface usually with a powdery bloom; plums
        3. Leaf serrations sharp, acute or acuminate
            4. Petioles glabrous; leaf tips usually acuminate; leaves mostly obovate or widest beyond middle of blade; small trees
                5. Petiole glands usually present; leaves rugose, usually pubescent below, base often subcordate; native to areas west of Blue Ridge 2. *P. mexicana*
                5. Petiole glands usually lacking; leaves not rugose; glabrous or pubescent only along veins below, base narrowed to rounded; widespread range 3. *P. americana*
            4. Petioles pubescent; leaf tips gradually acuminate, acute, or blunt; leaves mostly elliptic or widest at middle; shrubs or occasionally small trees
                6. Twigs velvety; a shrub of coastal areas, VA to ME 4. *P. maritima*
                6. Twigs glabrous or only slightly hairy; shrub or small tree of inland areas
                    7. Petioles rarely glandular; of Appalachians north of NC 5. *P. alleghaniensis*
                    7. Petioles commonly glandular; of Piedmont, coastal plain 6. *P. umbellata*
        3. Leaf serrations glandular, crenate, or obtuse
            8. Serrations glandular; leaves usually folded lengthwise or trough-shaped (if glandular and leaves flat, see *P. hortulana*)
                9. Leaves 6–11cm long
                    10. Glands at tips of leaf serrations; veins prominent below; fruit matures in autumn
  7. *P. hortulana*
                    10. Glands near base of leaf serrations; veins not prominent; fruit matures in summer
  8. *P. munsoniana*
                9. Leaves mostly 2–6cm long
                    11. Leaf tips blunt to acute, leaves glabrous below; calyx lobes eglandular; widespread across SE 9. *P. angustifolia*
                    11. Leaf tips acuminate, leaves slightly pubescent below; calyx lobes glandular; only in TX, OK 10. *P. rivularis*
            8. Serrations not glandular; leaves flat, not folded lengthwise
                12. Leaves reticulate or rugose above or below
                    13. Twigs moderately stout, glabrous, usually with terminal bud 11. *P. domestica*
                    13. Twigs slender, pubescent; terminal bud lacking
                        14. Leaves 4–8cm long; fruit bluish black; exotic tree 12. *P. insititia*
                        14. Leaves 2–5cm long; fruit red; rare native shrub of TX, OK, AR 13. *P. gracilis*
                12. Leaves not reticulate or rugose
                    15. Leaves 6–10cm; n. tree barely reaching SE 14. *P. nigra*
                    15. Leaves 3–6cm long, often reddish-tinted; exotic 15. *P. cerasifera*
    2. Twigs slender to fairly stout (internodes 3mm or more diam.), terminal bud present; no spinose spur shoots; fruit not with a groove, surface glossy, without a bloom; cherries (except *P. domestica,* a plum repeated here)
        16. Leaves evergreen 16. *P. caroliniana*
        16. Leaves deciduous
            17. Leaves lanceolate, long-tipped 17. *P. pensylvanica*
            17. Leaves not as above
                18. Petioles mostly under 1cm long
                    19. Twigs, petioles glabrous 18. *P. pumila* var. *susquehanae*
                    19. Twigs, petioles hairy 19. *P. tomentosa*
                18. Petioles over 1cm long
                    20. Twig tips and nodes swollen; leaves rugose 11. *P. domestica*
                    20. Twigs not as above; leaves not rugose
                        21. Leaves broadly ovate, about as wide as long 20. *P. mahaleb*
                        21. Leaves ovate or elliptical, longer than wide
                            22. Leaf teeth small, sharply pointed; trunk slender, habit shrubby 21. *P. virginiana*

22. Leaf teeth large or blunt; habit treelike
   23. Bark of mature trunk scaly; twigs slender; buds sharply pointed; native species
      24. Leaf tips blunt; leaves whitish below   22. *P. alabamensis*
      24. Leaf tips pointed; leaves green to pale but not whitish below   23. *P. serotina*
   23. Bark smoothish; twigs moderately stout; buds bluntly pointed; exotic species
      25. Leaves with 10 to 14 pairs of main veins; a tall, straight tree   24. *P. avium*
      25. Leaves with 6 to 8 pairs of main veins; a short, spreading tree 25. *P. cerasus*

1. *Prunus persica* (L.) Batsch   PEACH   Figure 375

A small tree, to 8m tall, widely cultivated and sporadically naturalized across the SE. Twigs are greenish or with reddish tinges; terminal bud present; bud scales loose, ragged, with gray hairs; leaf scars dark; bundle scars obscure. Bark gray, thin, mostly smooth. Leaves thin, lanceolate, usually folded or trough-shaped; marginal teeth crenate.

2. *Prunus mexicana* S. Wats.   MEXICAN PLUM

A small tree, to 10m, usually with a single trunk and tending not to produce peripheral root sprouts. Occurs west of the Appalachians, from s. OH to AL, west to KS, TX. Found in varied habitats, both rocky uplands and moist lowlands. Twigs slender, often with spiny spur shoots; no true terminal bud; lateral buds 3–4mm long. Leaves often with glands on petiole, near blade; veins rugose on top surface; marginal teeth sharp, doubly toothed; petiole and underside usually hairy when young, glabrous or nearly so on old leaves. Fruit to 3cm, reddish, sour.

3. *Prunus americana* Marsh.   AMERICAN PLUM   Figure 376

A shrub or small tree, to 10m, sometimes sprouting from the root system and forming small groups or groves. Distributed in the SE from the Piedmont of VA to n. FL, west to sw. AR, MO. Most common in the Appalachian areas and westward in the interior uplands of the region. Habitat varied, from thinly forested lowlands to rocky uplands. Similar to the above species in general appearance; leaves less often with petiolar glands, less often rugose. The var. *lanata* Sudw. is more pubescent than the species.

4. *Prunus maritima* Marsh.   BEACH PLUM

A low, dense shrub, rarely over 2m tall, often with branches spreading onto the ground. Occurs in sandy soils near the Atlantic coast, entering the SE in coastal VA (from there northward to ME). Leaves acute at tip; petiole usually glandular, pubescent, to 6mm long. Fruit globose, about 2cm diam.; usually purplish and with a bloom.

5. *Prunus alleghaniensis* Porter   ALLEGHENY PLUM

An uncommon shrub or small tree, to 5m tall, often sprouting from the roots. Native to the Appalachian region, mainly the Allegheny province in PA to w. MD, e. WV; also occurs in w. VA. Habitat is varied, from moist streamsides to rocky uplands and thickets. Leaves long-pointed (or acuminate); marginal teeth small, sharp; petiole rarely glandular, to 15mm long. Fruit subglobose, 1–2cm diam.; dark reddish-purple to purple, with a glaucous bloom.

6. *Prunus umbellata* Ell.   FLATWOODS PLUM

A shrub or small tree, to 6m tall, usually single-stemmed and rarely spreading by root sprouts. Fairly common across the coastal plain and Piedmont of the SE, from s. NC to FL, west to AR, TX. Occurs in sandy uplands under pine or scrub, in mesic hammocks, or in moist lowlands where there is sufficient light in the understory. Leaves acute at the tip; margins with fine, acute teeth; petiole often glandular, 5–10mm long. Fruit subglobose, to 2cm long, dark purplish, rarely red or yellow, with a glaucous bloom. The var. *injucunda* (Small) Sarg. is more pubes-

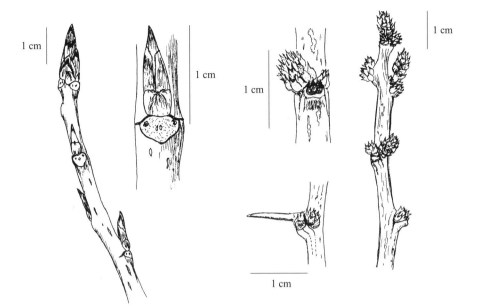

Figure 373. *Populus × canadensis*

Figure 374. *Prosopis glandulosa*

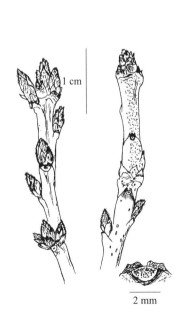

Figure 375. *Prunus persica*

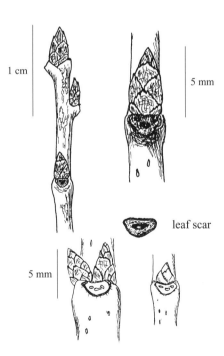

leaf scar

Figure 376. *Prunus americana*

cent on twigs and leaves; flowers slightly earlier than the species; typically in sandhills or scrub from NC to MS; sometimes considered a separate species.

7. *Prunus hortulana* Bailey   HORTULAN PLUM   Figure 377

A small tree, to 10m tall, usually single-stemmed but may form groves or thickets in ample sunlight. Not common in the SE, being mostly a midwestern species. Ranges east to s. OH, w. KY and south to TN, AR, ne. OK. Occurs in moist lowlands to calcareous uplands. Leaves folded lengthwise in bud (conduplicate), flat and glossy when mature, 7–12cm long, with sunken or distinct veins on top surface; margins crenate-serrate; petioles glandular, 15–25mm long. Fruit globose, to 3cm diam., red and fairly lustrous (with little bloom); seed pointed on both ends, slightly reticulate.

8. *Prunus munsoniana* W. Wight & Hedrick   WILDGOOSE PLUM

A small tree, to 8m tall, often sprouting from the roots to form clumps or groves. Scattered across the SE, west of the Blue Ridge and AL, to KS and TX. Typically occurs in prairies and openings near streams and rich uplands. Leaves emerge rolled inward lengthwise (convolute), staying somewhat folded or troughlike when mature, 6–10cm long, with veins not distinct as in above species; margins crenate-serrate; petioles glandular, to 2cm long. Fruit usually longer than wide, to 25mm, red, lustrous, only with a slight bloom; seed often blunt to flat on one end.

9. *Prunus angustifolia* Marsh.   CHICKASAW PLUM

A shrub, to 4m tall, with multiple stems from the spreading root system, forming thicketlike groves or colonies. Widespread across most of the SE but rare in the Blue Ridge. Typically in sandy soils of uplands, roadsides, and old fields. Leaves similar to above species, but smaller; petioles glandular or not. Fruit oval, to 25mm long, red, with a bloom.

10. *Prunus rivularis* Scheele   CREEK PLUM

A shrub, to 3m tall, thicket forming as in the above species. Occurs in TX, OK over calcareous soils or along streams in limestone areas. Similar to above species, and perhaps only a variety of it.

11. *Prunus domestica* L.   GARDEN PLUM   Figure 378

An exotic small tree, to 10m tall, sometimes sprouting from the roots and forming groves where there is sufficient light. Native to w. Asia, with various selections and hybrids used for fruit production. Also used as rootstock for budding cultivars of orchard plums. Rarely naturalized in the SE, but occasionally seen in areas where plums are cultivated. Leaves coarsely crenate-serrate, acuminate; veins distinct and impressed above, reticulate below. Fruit oval, to 3cm long, blue-black, with a bloom.

12. *Prunus insititia* L.   DAMSON PLUM

An exotic shrub or small tree, to 6m tall, sometimes thicket forming. Native to Europe; cultivated for the fruits and occasionally naturalizing near areas of its cultivation. Similar to above species; sometimes considered only a variety of it. Twigs tend to be more pubescent and spinose. Also used as rootstock for grafting various cultivar plums.

13. *Prunus gracilis* Engelm. & Graebn.   OKLAHOMA PLUM

A shrub, to 2m tall, usually sprouting from the roots to form thickets or groves. Native to w. AR, e. OK, TX. Uncommon and sporadic, in sandy to rocky upland soils. Leaves acute at tip;

margin finely serrate; underside reticulate-veined and hairy; petiole pubescent, usually not glandular. Fruit subglobose, to about 15mm diam., red, with a bloom.

14. *Prunus nigra* Ait.   CANADA PLUM

A small tree, to 6m tall, usually single-stemmed. Mostly found in the Great Lakes region, but may reach the SE in n. WV or s. IN. Leaves acuminate at tip; margin with crenate or obtuse teeth; petiole usually glandular, 15–25mm long. Fruit oval, to 3cm long, red, with a slight bloom.

15. *Prunus cerasifera* Ehrh.   PURPLELEAF PLUM

A small tree, to 8m tall, usually single-stemmed. Native to w. Asia; cultivated for fruit and as a flowering specimen. Rarely naturalized near sources of its cultivation. One of the most common forms planted in the SE is the var. *atropurpurea* Jaeg.; it has dark red or purplish foliage, dark red twigs, and wine red fruits. Leaves acute at tip; margin with obtuse teeth. Fruit subglobose, to 3cm diam., reddish, with a bloom; seed almost round.

16. *Prunus caroliniana* (P. Mill.) Ait.   CAROLINA LAURELCHERRY   *Figure 379*

A small tree, to 12m tall, usually single-stemmed. Native to the coastal plain from NC to FL, west to e. TX. The original range and habitat is obscured, as the species has spread widely after disturbance and cultivation across the SE. It is common in vacant lots and other fringes of urban cultivation. It also occurs from maritime forests and sandy hammocks to moist hardwood or pine-hardwood understories, along streams, or scrub. Leaves evergreen, lustrous; margins entire to remotely toothed. Twigs odorous when scraped; stipules persist on endmost nodes. Bark gray, smooth, often with sapsucker holes. Fruit dull black, about 1cm long, subglobose but pointed; persist over winter.

17. *Prunus pensylvanica* L. f.   FIRE CHERRY   *Figure 380*

A tree, to 15m, usually shorter, with an open, oblong crown. Extends from its primarily n. range into the SE along the higher elevations of the Appalachians, from WV and w. MD to ne. GA. Occurs in moist mountain forests, where it may form dense stands after disturbances; it is relatively short-lived. Twigs reddish or red-brown, glabrous; buds usually clustered at twig apex, with red bud scales. Leaves lanceolate, 8–15cm long. Bark thin, smooth and sometimes shiny red or red-brown on young trunks, aging dark gray or blackish, with a few horizontally curling bark strips. Fruit globose, 5–8mm diam., shiny red.

18. *Prunus pumila* var. *susquehanae* (hort. ex Willd.) Jaeg.   SAND CHERRY

A stoloniferous shrub, to 2m, with many stiffly erect stems and branches. This variety and the species are n. plants, barely reaching the SE in VA, NC, and perhaps WV. Typically in sandy or rocky, acidic soils of uplands. Leaves elliptic to oblong, blunt at the tip, crenate-toothed, glaucous beneath. Fruit globose, purple-black, 1–2cm long.

19. *Prunus tomentosa* Thunb.   NANKING CHERRY

An exotic shrub, rarely over 3m tall. Native to China and Japan; planted occasionally and rarely naturalizing in the SE. Twigs slender, hairy. Leaves mostly widely elliptic, 5–7cm long, tip acuminate, margin coarsely serrate; rugose above, hairy below. Petiole hairy, 2–4mm long. Fruit red, shiny, about 1cm wide.

20. *Prunus mahaleb* L.   PERFUMED CHERRY

An exotic tree, to 10m tall. Native to Eurasia; planted occasionally and used as rootstock for grafted ornamental cherries. Naturalized sporadically across the SE, mainly near urban areas,

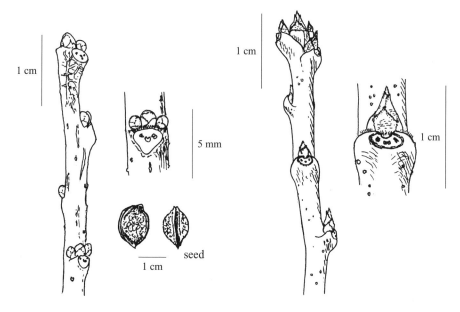

Figure 377. *Prunus hortulana*

Figure 378. *Prunus domestica*

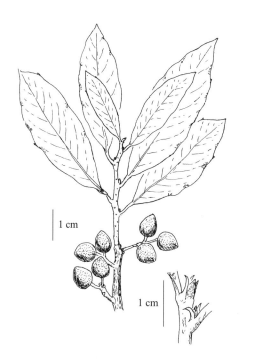

Figure 379. *Prunus caroliniana*

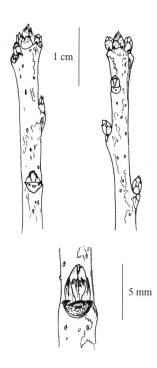

Figure 380. *Prunus pensylvanica*

roadsides, forest edges. Twigs hairy; scraped bark very aromatic. Bark dark, mostly smooth. Leaves nearly round, mostly 3–5cm long; tip acuminate, margin crenate-serrate, petiole glandular. Fruit globose, shiny black when ripe, about 1cm wide.

21. *Prunus virginiana* L.   CHOKECHERRY   Figure 381

Mostly a shrub, to 5m tall, stoloniferous and forming thickets. Widespread across the n. and w. parts of N. Amer., but limited in the SE to the Appalachians from w. MD to sw. NC, and scattered in areas west, in TN, KY, s. IL, MO. Twigs grayish brown, with pointed buds; bud scales light brown, rounded at tips. Bark smooth, thin, nearly black. Leaves with fine, sharp teeth on margin; underside pale; petiole glandular. Fruit globose, about 1cm long, shiny red, becoming purplish-red after ripening; hang in racemes.

22. *Prunus alabamensis* C. Mohr   ALABAMA BLACK CHERRY

A small tree, to 10m tall, but usually 4–6m. Native to the coastal plain from se. NC to FL, west to AL. Typically in dry or sandy soils of uplands. Twigs hairy when young, sometimes so in winter; strongly odorous when scraped. Leaves usually lustrous above, with blunt tips and pale to whitened undersides; midrib with brownish hairs; margins crenate-serrate; petioles usually hairy, 6–12mm long. Fruit turns from shiny red to black when ripe, each about 8–10mm diam.; in racemes. Bark light grayish, scaly, darkening with age to nearly black on lower trunk. (*P. cuthbertii* Small)

23. *Prunus serotina* Ehrh.   BLACK CHERRY   Figure 382

A tree varying in height and form as to site quality, from 20–30m. Widespread across the SE. Most common and of largest sizes in the moist forests of the Appalachians. Twigs glabrous, red-brown; buds pointed; scales red-brown; strongly odorous when scraped. Leaves smooth above, paler green below, with a few hairs along the midrib; margin crenate-serrate; petioles glabrous, glandular, 1–2cm long. Fruit similar to above species. Bark dark gray to black, smooth only on small stems, scaly on mature trunks.

24. *Prunus avium* (L.) L.   MAZZARD CHERRY   Figure 383

A tall tree, to 25m tall. Native to Eurasia; extensively planted and naturalized sporadically across the SE. Most often seen near urban areas. Trunk usually long and straight. Twigs moderately stout, red-brown; buds blunt, with dark brown scales. Bark gray, smooth but with conspicuous horizontally elongated lenticels; sometimes with a few horizontally curling strips or scales on old, darker trunks. Leaves 6–15cm long, dull green above, slightly hairy below; margin with large or dentate teeth; petiole usually glandular, to 4cm long. Fruit subglobose, slightly cordate at base, to 25mm diam., shiny red to nearly black. The juicy, sweet fruit ripens in summer and is relished by many birds.

25. *Prunus cerasus* L.   SOUR CHERRY

A broad-crowned tree, to 10m tall. Native to Eurasia; extensively planted for fruit production in various areas across the SE. Naturalized sporadically, especially in urban areas, with the aid of birds that eagerly consume the summer fruits. The trunk is usually short and divided within 3m of the ground. Twigs and bark similar to above species. Leaves 5–10cm long, lustrous green above, glabrous below, margin finely serrate, petiole glandular, to 2cm long. Fruit similar to above species, but rarely over 15mm; shiny red, flesh sour.

Pseudosasa japonica (Sieb. & Zucc. ex Steud.) Makino ex Nakai   ARROW BAMBOO
Family Poaceae   Figure 384

An exotic bamboo that is sparingly naturalized in parts of the SE. Forms a thick grove of very slender and closely spaced culms to 5m tall and 2cm diam., though more commonly about 3m tall. Culms are rounded in cross section, often with a waxy bloom. The sheaths are persistent, sparingly hispid, about as long as culm internodes; a leaflike tip extending outward also persistent. Nodes of culms not swollen, usually with only 1 branch produced. Leaves mostly 8–20cm long, 2–4cm wide, lustrous above; minutely serrulate on one margin, mostly entire on the other. The slender but strong culms, with their relatively smooth and unswollen nodes, are "straight as an arrow," hence the name.

Ptelia trifoliata L.   HOPTREE   Family Rutaceae   *Figure 385*

A native shrub or small tree, widespread across most of the SE. May be locally common, while absent over wide intervening areas. Typically over limestone or granitic outcrops, glades, and in thin woods over circumneutral to calcareous soils. Twigs brown, finely pubescent; terminal bud absent; lateral buds sunken, almost surrounded by the leaf scars. The inner bark has a pungent or rank odor reminiscent of citrus; skunklike to some noses. Bark brownish, thin, mostly smooth but slightly fissured with age. The fruit is a rounded samara about 25mm wide.

Pueraria montana var. *lobata* (Willd.) Maesen & S. Almeida   KUDZU   Family Fabaceae
Figure 386

An exotic, twining vine that has naturalized widely across the South. Young stems with coarse hairs become brown and very finely fissured with age. Some winter dieback of twigs often occurs. No terminal buds; lateral buds paired, on each side of a peglike projection and with a large outer scale covering most of bud. The dried stipules or their scars are large and conspicuous. The fruit is a hispid legume, to about 8cm long.

Pyracantha coccinea M. Roemer   FIRETHORN   Family Rosaceae   *Figures 387 and 388*

Of the many species of *Pyracantha* used as ornamentals, this species is known to naturalize frequently. Other species may become naturalized locally, usually near sources of cultivation, but are not included here. All are European or Asian species used in ornamental horticulture. Spinous spur shoots and orange-red berries are usually conspicuous winter characters. Habit is shrublike or sprawling, sometimes partially reclining on other plants. Twigs hairy to nearly glabrous, spine-tipped or with a hairy-scaled bud. Some leaves may persist into winter; apex rounded to notched; 1–5cm long; undersides woolly, hairy, or nearly glabrous. Fruit about 8mm wide; seed black, shiny.

Pyrularia pubera Michx.   BUFFALO-NUT   Family Santalaceae   *Figure 389*

A coarse shrub, to 3m tall, parasitic on the roots of several species of deciduous trees, especially oaks. Native to the Appalachian region from WV and VA to n. GA, mostly in moist deciduous forests. Twigs grayish to light brown, glabrous; buds 5–12mm long, with several greenish, imbricate scales; terminal buds often appear to be present; leaf scars pale, with 3 bundle scars. Bark gray, smooth, with prominent lenticels. The fruit is a greenish drupe, about 2–2.5cm long, flat on the base; seed spherical, oily and sweetish-smelling when crushed, poisonous if eaten. Inner bark also has a sweet odor.

Pyrus L.   PEAR   Family Rosaceae

Three species are encountered in the SE as naturalized exotics. All are spread by discarded cores, or the seed are dispersed by wildlife feeding on the fruit. The most common habitat where these plants become established are roadsides, field edges, and in open disturbed areas in the vicinity of cultivated orchards or lawn specimens. The fruits of all are pomes, varying in size and shape but with the characteristic "grit cells" in the flesh. Leaves are lustrous or glossy, unlike the dull green leaves of apple trees, for which these trees may be mistaken at a distance.

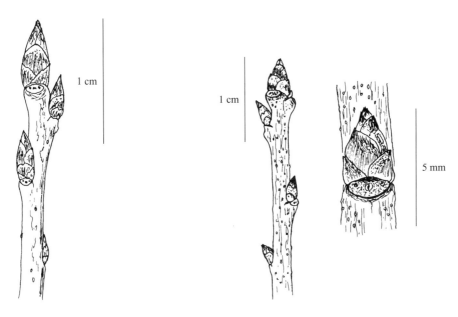

Figure 381. *Prunus virginiana*

Figure 382. *Prunus serotina*

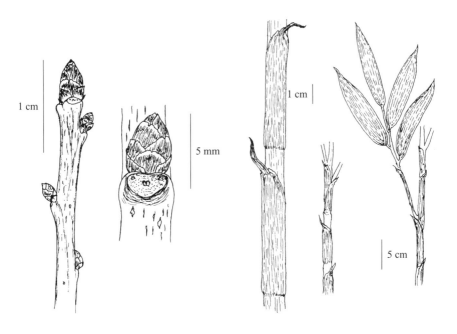

Figure 383. *Prunus avium*

Figure 384. *Pseudosasa japonica*

260 · Pyrus

Branches tend also to be more rigidly upright than in apples, with a brown or yellow-brown rather than reddish color. Bark grayish, scaly on mature trunks.

1. Twigs moderately stout, glabrous; bud scales tightly imbricate, slightly keeled
   1. *P. communis*
1. Twigs more slender, slightly hairy; bud scales not as above
   2. Buds over 8mm long on mature branches; bud scales loose; fruits less than 15mm wide
      2. *P. calleryana*
   2. Buds on mature branches smaller; bud scales tight; fruit over 20mm wide   3. *P. pyrifolia*

1. *Pyrus communis* L.   COMMON PEAR   *Figure 390*

A tree, to 15m tall, with ascending branches and a short, thick trunk. Twigs fairly stout, with stubby spur shoots that usually end in conical buds, but sometimes spines. Terminal buds mostly 5–8mm long, conical, with tightly imbricate, glabrous, brownish scales. Leaves crenate-serrate, glabrous. Fruit 1.5–8cm wide; usually over 3cm wide in first-generation seedlings from cultivated pears planted for fruit (some of these orchard trees of hybrid origin). Native to Eurasia.

2. *Pyrus calleryana* Decne.   CALLERY PEAR   *Figure 391*

A small tree, to 10m tall, with slender, upswept branches. This is the parent species of the Bradford pear, a widely planted clonal selection used as an ornamental tree, with spineless twigs. Seed from the Bradford pear is spread by birds that feed on the small fruits, and seedlings of these "reversions" to the parent species have formed thickets on rich soils in various areas of the SE, particularly the French Broad River Valley in w. NC. Twigs moderately slender, with spinose spur shoots. On mature branches, the buds are 10–15mm long, with loose, hairy, grayish scales; leaf scars small, darkened, lined beneath. Fruit globose, about 1cm diam., brown with pale dots. Leaves crenate-serrate, glabrous. Native to China.

3. *Pyrus pyrifolia* (Burm. F.) Nakai   ORIENTAL PEAR

A tree, to 12m tall, with slender, upright branches. Twigs moderately slender, with spinose spur shoots. Buds 5–8mm long, with reddish-brown scales hairy on margins; leaf scars raised, not conspicuously lined onto internode. Fruit globose, 2–4cm wide, brown with pale dots. Leaves hairy when young, margins sharply serrate. Native to China.

Pyxidanthera barbulata Michx.   PYXIE-MOSS   Family Diapensiaceae   *Figure 392a*

A diminutive, creeping shrub of the coastal plain. In the SE, it occurs uncommonly in VA, NC, SC. Typical habitat is moist sandy or peaty soils in pinelands. Leaves persistent, alternate, sessile, mostly 4–9mm long, hairy near the base, crowded on the slender prostrate twigs. Fruit a capsule, to 2mm long. The var. *brevifolia* (B. W. Wells) Ahles is rarer; occurring in dry, upland sands of the Sandhills region in NC, SC. It varies in having leaves 1–4mm long, hairy over the whole upper surface (Figure 392b).

Quercus L.   OAK   Family Fagaceae

There are 44 taxa of native oak treated here for the SE, with additional significant varieties mentioned for some of the species. One additional exotic species is included. In the SE our species fall into 2 taxonomic categories within the genus; the white oaks and the red oaks. The white oaks are designated as section *Quercus* L. (or subgenus *Lepidobalanus* Endl.); the red oaks, as section *Lobatae* Loud. [or subgenus *Erythrobalanus* (Oerst.) O. Schwarz]. Regardless of the taxonomic nomenclature, the 2 groups are easily defined by several characters of foliage,

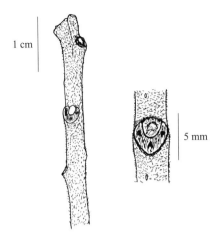

Figure 385. *Ptelia trifoliata*

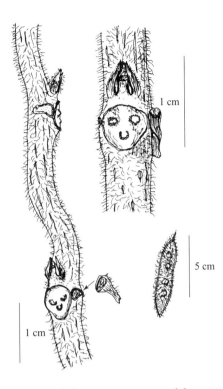

Figure 386. *Pueraria montana* var. *lobata*

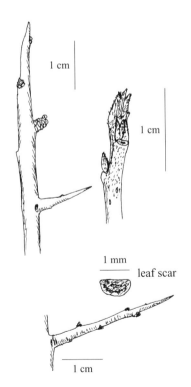

Figure 387. *Pyracantha coccinea*

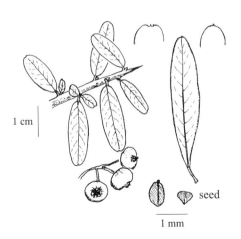

Figure 388. *Pyracantha coccinea*

fruit, and wood structure. The white oak group contains 20 of the species; the red oak group, 24. Identifying factors in the field include: (1) leaf tips or lobe points of red oaks have bristles (awns); white oaks do not; (2) white oaks ripen acorns in 1 growing season; red oaks in 2 (with 1 exception, *Q. pumila*); (3) inner lining of acorn velvety in red oaks; (4) bark scaly in many white oaks (some exceptions). Determination of oak species in winter, based on twigs alone, is usually sufficient in localized regions where the number of oak species may be limited, but use of a winter key to all species across the entire SE is difficult. Remnant leaves from the branches or under the tree are valuable features to observe, as are fallen acorns or acorn fragments. Due to the slow degeneration of oak leaves and acorns, these parts are usually available under mature trees. Twigs of all oaks show clustered end buds (1 is a true terminal bud), many imbricate bud scales, alternate leaf scars, many bundle scars, and an angular pith (normally 4- or 5-sided). On the red oaks, immature acorns persist over the winter months on the twigs, if flowers were pollinated. Threadlike stipules may persist at the base of some of the endmost buds in some oaks. Bark may be variable even within a species, depending on age and growth rate. The fruit of all oaks is a nut, held in a husk (involucre) made up of overlapping scales; commonly known as an acorn. For ease of reference in this largest group of se. woody plants, the list of species following the key is alphabetical.

1. Leaves evergreen or subevergreen, nearly all nodes with a persistent leaf
  2. Leaf margins strongly revolute; twigs distinctly hairy
    3. Buds acute; leaves lustrous green beneath   29. *Q. myrtifolia*
    3. Buds obtuse or rounded; leaves pale, finely pubescent beneath   13. *Q. geminata*
  2. Leaf margins slightly or not revolute; twigs glabrous or nearly so
    4. Leaves hairy or pale beneath, or with distinct, reticulate veins; no awns on leaf tips
      5. Buds acute; a stoloniferous shrub usually under 1m tall   26. *Q. minima*
      5. Buds obtuse; tall shrubs or trees, or not stoloniferous
        6. Acorn rounded on apex; widespread in coastal plain   45. *Q. virginiana*
        6. Acorn tapered to apex; of limestone soils in TX, OK   12. *Q. fusiformis*
    4. Leaves green, smooth and glabrous beneath; may have awn on leaf apex
      7. Leaf apex blunt or rounded; surface dull green; in moist soil   20. *Q. laurifolia*
      7. Leaf apex pointed; surface lustrous; in sandy uplands   15. *Q. hemisphaerica*
1. Leaves deciduous, or twigs with only a few leaves persisting into winter
  8. Twigs conspicuously hairy, visible without use of lens (buds may be hairy or not)
    9. Buds long-tapered, to 6mm or more, length over twice the width
      10. Twigs slender, 1–2mm diam.; buds brownish, to 6mm   18. *Q. incana*
      10. Twigs stouter, 3mm or more; buds grayish, angled, over 6mm   24. *Q. marilandica*
    9. Buds not as above, 5mm or less in length
      11. Twigs moderately stout, over 2.5mm diam.; largest buds 3–5mm; tall trees
        12. Buds acute, angle away from twig; upper trunk bark furrowed; leaf lobes awned
          13. Lobes of leaf near apex are longer than others; buds rarely angled   11. *Q. falcata*
          13. Lobes of leaf nearly all the same length; buds usually angled   32. *Q. pagoda*
        12. Buds obtuse, not as above; upper trunk bark scaly; leaf lobes rounded, no awns
          14. Leaves with 2 large lobes near middle; in uplands   42. *Q. stellata*
          14. Leaves mostly 3-lobed near tip; tree of wet lowlands   39. *Q. similis*
      11. Twigs slender, 2mm or less diam.; largest buds 2–3mm; shrubs or short trees
        15. Buds obtuse, leaf lobes not awned; leaves green or yellowish below; bark scaly
          16. Petiole 5mm or more; of inner coastal plain, Piedmont   7. *Q. boyntonii*
          16. Petiole less than 5mm long; of outer coastal plain   9. *Q. chapmanii*
        15. Buds usually acute; leaf points awned; leaves gray below; bark smooth or furrowed
          17. Petiole under 5mm; leaves unlobed; of coastal plain   36. *Q. pumila*
          17. Petiole over 5mm; leaves lobed; of mountains, Piedmont   16. *Q. ilicifolia*
  8. Twigs glabrous, or essentially so, to naked eye
    18. Buds conspicuously hairy, visible without use of lens
      19. Buds elongate, mostly over 5mm long, length at least twice the width

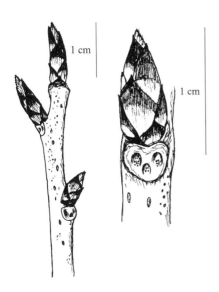

Figure 389. *Pyrularia pubera*

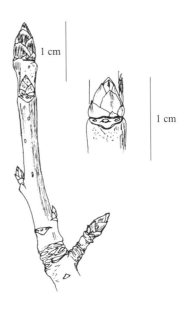

Figure 390. *Pyrus communis*

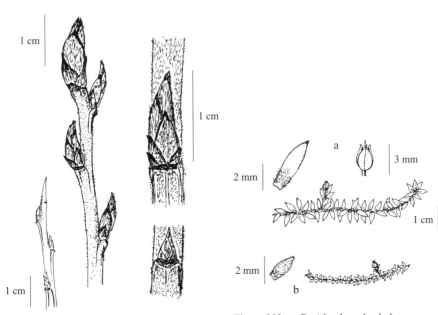

Figure 391. *Pyrus calleryana*

Figure 392. a: *Pyxidanthera barbulata;* b: *P. barbulata* var. *brevifolia*

20. Buds 5–7mm, brown; leaves toothed, not lobed — 2. *Q. acutissima*
20. Buds mostly 6–13mm, gray or rusty-brown; leaves lobed
  21. Buds strongly angled, gray-hairy — 44. *Q. velutina*
  21. Buds not angled, rusty-hairy — 19. *Q. laevis*
19. Buds ovoid, not as above, mostly 5mm or less
  22. Stipules often persist at end buds; no awns on leaves; upper trunk bark scaly
    23. Twigs grayish or light brown, lenticels not conspicuous; leaves lobed, pale below — 21. *Q. lyrata*
    23. Twigs red-brown, lenticellate; leaves unlobed, green-stellate below — 31. *Q. oglethorpensis*
  22. Stipules usually lacking; awns on leaf points; upper trunk smooth or furrowed
    24. Buds obtuse, grayish or hairy mostly beyond middle of bud
      25. Twigs slender, 2mm or less diam.; acorn cup less than 5mm deep; leaves spatulate — 30. *Q. nigra*
      25. Twigs over 2mm diam.; acorn cup over 5mm deep; leaves deeply lobed — 10. *Q. coccinea*
    24. Buds acute or uniformly red-brown and slightly hairy throughout
      26. Twigs moderately stout, 2.5mm diam. or more
        27. Buds angled; nut under 15mm long; leaves softly hairy below — 32. *Q. pagoda*
        27. Buds not angled; nut larger; leaves glabrous below — 37. *Q. rubra*
      26. Twigs slender, mostly 1–2.5mm diam.
        28. Acorn cup 4–6mm deep; leaves lobed — 14. *Q. georgiana*
        28. Acorn cup 5–9mm deep; leaves unlobed
          29. Buds angled; leaves widest near middle; of interior region — 17. *Q. imbricaria*
          29. Buds rarely angled; leaves widest beyond middle; of coastal plain — 4. *Q. arkansana*
18. Buds essentially glabrous to naked eye
  30. Leaves bear awns on apex or lobe tips; twigs may bear juvenile acorns (red oaks)
    31. Buds mostly 4–7mm long (largest buds); nuts often 15mm or longer
      32. Buds dark red-brown; twigs lustrous red-brown; leaves lobed ½ to midrib — 37. *Q. rubra*
      32. Buds gray or gray-brown; twigs gray-brown or dull red-brown; leaves lobed nearly to midrib
        33. Acorn cup margin thin, without pronounced shoulder — 43. *Q. texana*
        33. Acorn cup margin thick, with a shoulder
          34. Leaves as wide as long, or wider; rare tree of AR — 1. *Q. acerifolia*
          34. Leaves much longer than wide; not rare tree
            35. Buds slightly angled, grayish; hair tufts in leaf vein axils below; widespread tree — 38. *Q. shumardii*
            35. Buds not angled, sometimes red-brown; hair tufts tiny or lacking; tree of TX, OK — 8. *Q. buckleyi*
    31. Buds mostly 2–5mm long; nuts mostly under 15mm long
      36. Leaves deeply lobed
        37. Bud scales ciliate (use lens); acorn cup covers ⅓ of nut; rare s. tree of uplands — 14. *Q. georgiana*
        37. Bud scales glabrous; acorn cup covers ¼ of nut; common n. tree of wet soils — 33. *Q. palustris*
      36. Leaves unlobed
        38. Twigs slender, 2mm or less diam.; leaves 2cm wide or less; petiole under 5mm long — 34. *Q. phellos*
        38. Twigs usually 2–2.5mm diam.; leaves over 2cm wide; petiole over 5mm
          39. Buds slightly angled; leaves widest near middle; of interior region of SE — 17. *Q. imbricaria*

39. Buds rarely angled; leaves widest beyond middle; of coastal plain
4. *Q. arkansana*
30. Leaves lack awns; twigs never bear juvenile acorns over winter (white oaks)
40. Acorn cup usually over 1cm deep
41. Buds often 5–6mm long, acute, red-brown; leaves toothed, not lobed; stipules rarely persist at twig tip
42. Bark of upper trunk scaly; acorn cup scales pointed; lowland tree
25. *Q. michauxii*
42. Bark of entire trunk furrowed; acorn cup scales knobby; upland tree
27. *Q. montana*
41. Buds mostly 2–5mm long, obtuse, gray-brown; leaves lobed or with a few large teeth; stipules often persist at base of end buds
43. Buds glabrous; nut 2cm or less; cup 10–15mm deep 6. *Q. bicolor*
43. Buds finely hairy or ciliate on margins (use lens); nut over 2cm long; acorn cup over 15mm deep
44. Twigs or branchlets usually with corky growth; acorn cup margin fringed with threadlike scales 22. *Q. macrocarpa*
44. Twigs not as above; acorn cup scales knobby, none threadlike; acorn cup encloses most of nut 21. *Q. lyrata*
40. Acorn cup 1cm or less deep
45. Buds acute; acorn cup covers ⅓ to ½ of nut
46. Twigs yellow-brown; stipules rarely persistent; buds red-brown with pale margins; leaves with large, pointed teeth
47. Leaves with 10 or more teeth per side; a tree 28. *Q. muhlenbergii*
47. Leaves with 9 or fewer teeth per side; a shrub 35. *Q. prinoides*
46. Twigs brown or red-brown; stipules often persistent; buds brown, not pale-margined; leaves entire or with blunt or rounded lobes
48. Buds projecting at a low angle, close to twig; leaves glabrous and smooth beneath; uncommon tree 5. *Q. austrina*
48. Buds projecting at a greater angle outward from twig; leaves usually with some stellate hairs (use lens) or slightly raised veins beneath; common tree
23. *Q. margarettiae*
45. Buds obtuse or rounded; acorn cup usually covers ⅓ or less of nut
49. Twigs gray, 2mm diam. or less; nut often flat-based
50. Acorn cup thin, 2–5mm deep; an uncommon tree scattered in coastal plain
40. *Q. sinuata* var. *sinuata*
50. Acorn cup thick, 3–8mm deep; shrubby, of TX, OK uplands
41. *Q. sinuata* var. *breviloba*
49. Twigs brownish or red-brown, 2–3mm diam.; nut rounded at base
51. Twigs usually lack persistent stipules; acorn cup thick, with knobby scales, covers ¼ of nut; nut 15–21mm long; leaves lobed, glabrous 3. *Q. alba*
51. Twigs often with persistent stipules; acorn cup thin, with flat scales, covers ⅓ of nut; nut 9–12mm long; leaves unlobed, stellate-hairy below
31. *Q. oglethorpensis*

1. *Quercus acerifolia* (Palmer) Stoynoff & Hess   MAPLE-LEAVED OAK

A tree, to 15m tall, rare and localized in its distribution. Known only from the Ouachita Mountain region of AR, where it occurs on a few mountainous sites in rocky, dry, or calcareous soils. Similar to *Q. shumardii* in most respects; has been considered a variety of it by some botanists. The wide leaves, suggesting maple, are the most significant identifying trait. Buds are gray, larger ones 3–6mm long. Acorn cup 10–20mm wide, covering about ⅓ of nut; nut 10–20mm long. Bark grayish or darker, furrowed and roughly ridged. A red oak.

2. *Quercus acutissima* Carruthers SAWTOOTH OAK *Figure 393*

A tree, to 20m tall, introduced from Asia and planted extensively across the SE, both as a horticultural specimen and for wildlife food enhancement. Occasional naturalized seedlings may be encountered, and planted specimens may appear naturalized due to their scattered locations. Buds 4–7mm long, brown, with slightly hairy scales. Acorn cup about 3cm wide, with a fringed rim of elongated scales, covering ⅔ of nut; nut 20–25mm long. Bark brown, furrowed. Leaves resemble those of *Castanea,* with awn-terminated marginal teeth; petiole 15–25mm long. A red oak.

3. *Quercus alba* L. WHITE OAK *Figure 394*

A tree, to 30m, distributed throughout the SE and one of the most common species of oak. Occurs on a range of sites, from moist to dry loamy soils, but not swampy sites or xeric sands. Buds obtuse or rounded, dark brown; largest ones about 3mm long, with glabrous scales. Acorn cup 15–20mm wide, rather thick, scales knobby, covering ¼ of nut; nut 15–25mm long. Bark ashy gray, scaly, becoming thickly ridged on older, lower sections of trunk. Leaves variously lobed, with rounded lobes, glabrous beneath.

4. *Quercus arkansana* Sarg. ARKANSAS OAK *Figure 395*

A tree, to 18m, overall uncommon and scattered over the coastal plain from s. GA to e. TX. May be locally abundant in sandy, well-drained uplands. Buds red-brown, sometimes hairy, larger ones mostly 3–4mm long. Acorn cup 10–15mm wide, covering ¼ to ½ of nut; nut 10–15mm long. Bark grayish, smooth at first, becoming furrowed with age. Leaves widest toward tip, shallowly lobed to nearly entire; petiole 10–25mm. A red oak.

5. *Quercus austrina* Small BLUFF OAK *Figure 396*

A tree, to 25m, uncommon and sporadic in the coastal plain from NC to FL, west to MS. Occurs in rich and moist sandy soils with high organic content, typically on river bluffs, ravines, bottomlands. Buds brown, acute at tip, larger ones 3–5mm long; all tend to point upward or at a low angle from twig, not very divergent. Acorn cup 10–12mm wide, with thin, appressed, pointed scales, covering ⅓ to ½ of nut; nut 12–18mm long. Bark light gray, scaly, furrowed on older trunks. Leaves entire to shallowly lobed, the lobes blunt or rounded; underside smooth, green, glabrous. A white oak.

6. *Quercus bicolor* Willd. SWAMP WHITE OAK *Figure 397*

A tree, to 30m, mostly concentrated in states north of NC, TN, AR, but sporadic as far south as the coastal plain of NC, n. SC, n. AL. Typically occurs in poorly drained soils, often heavy in clay or with calcareous hardpan. Buds brown, obtuse or rounded, glabrous, larger ones 2–4mm; stipules often persist at bud bases near twig apex. Acorn cup 15–25mm wide, with scales appressed but slightly elongate-tipped near rim, the cup covering ½ to ¾ of nut; stalk 2–6cm long; nut 15–23mm long. Bark grayish, scaly or with thick, scaly ridges. Leaves variously lobed, sometimes appearing more toothed, the points of lobes or teeth obtuse to acute, not awned; lower surfaces whitish on mature trees.

7. *Quercus boyntonii* Beadle BOYNTON OAK *Figure 398*

A shrub or small tree, usually sprawling, stoloniferous, and under 2m tall, though may rise to 5m. Rare and local in distribution, known only from e. TX and n. AL. Occurs in shallow soils over sandstone and in acidic sands along rocky streams and understories of pine forests. Buds obtuse, red-brown, 2–3mm long. Acorn cup with appressed gray scales, about 1cm wide,

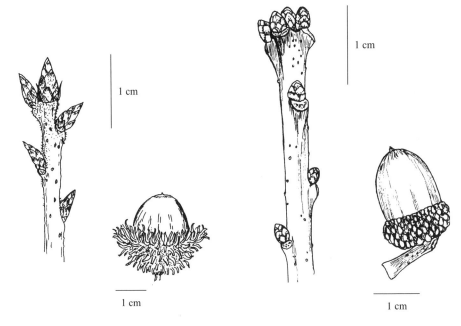

Figure 393. *Quercus acutissima*

Figure 394. *Quercus alba*

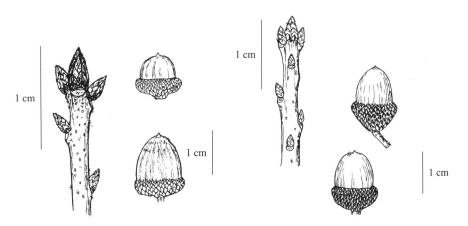

Figure 395. *Quercus arkansana*

Figure 396. *Quercus austrina*

covering ⅓ to ½ of nut; nut 10–15mm long. Bark gray, scaly. Leaves thick, weakly 3-lobed near apex, lobes rounded; veins distinct, raised below; underside yellowish and stellate-pubescent. A white oak. [*Q stellata* var. *boyntonii* (Beadle) Sarg.]

8. *Quercus buckleyi* Nixon & Dorr   TEXAS OAK   *Figure 399*

A tree, to 15m, native to calcareous uplands and along rocky streams in central TX to central OK, barely reaching the area of coverage of this book. Similar in general appearance to *Q. shumardii*. Buds gray to red-brown, 3–6mm long. Acorn cup 12–18mm wide, covering ¼ to nearly ½ of nut; nut 12–20mm long. Bark grayish, furrowed and roughly ridged. Leaves deeply lobed, the points awned; undersides glabrous or with tiny hair tufts in vein axils. A red oak.

9. *Quercus chapmanii* Sarg.   CHAPMAN OAK   *Figure 400*

A shrub or small tree, rarely over 5m tall. Native to the outer coastal plain from s. SC to central FL, and localized along the Gulf Coast in w. FL, s. AL. Occurs in deep, sandy upland soils of pinelands and scrub. Resprouts vigorously after fire. Buds obtuse or rounded, about 2mm long. Acorn cup 10–15mm wide, with thick but pointed scales, covering ⅓ to ⅔ of nut; nut 12–18mm long. Bark gray, scaly. Leaves late-deciduous; variously and shallowly lobed to nearly entire; petiole mostly 1–4mm long. A white oak.

10. *Quercus coccinea* Muenchh.   SCARLET OAK   *Figure 401*

A tree, to 30m, widely distributed in the SE, but most common in the Piedmont, Appalachian, and interior plateaus. Ranges sporadically to the coastal plain of e. NC, s. GA, s. AL, s. MS, and west to AR, MO. Occurs on well-drained uplands, mostly in drier soils. Buds red-brown near base, white-hairy near tip, mostly 4–6mm long. Acorn cup thick, 15–30mm wide, covering ⅓ to ½ of nut; nut 15–20mm long, usually with concentric rings at apex. Bark gray, rather thin and smooth at first, becoming furrowed and ridged with age. Leaves deeply lobed, the points awned. A red oak.

11. *Quercus falcata* Michx.   SOUTHERN RED OAK   *Figure 402*

A tree, to 30m, distributed across most of the SE, but uncommon or absent in the Appalachian region of VA, WV. Occurs mostly in well-drained uplands. Buds red-brown, usually not angled in cross section, mostly 4–5mm long; laterals tending to diverge or angle away from twig. Acorn cup 10–15mm wide, covering ⅓ to ½ of nut; nut 10–15mm long, the internal "meat" dark orange. Bark grayish to black, roughly ridged and furrowed. Leaves 3- to 7-lobed, awned, with terminal lobe and adjacent 2 lateral lobes longest; outline of leaf orbicular or obovate; underside grayish and densely pubescent. A native red oak also known in some locales as Spanish oak.

12. *Quercus fusiformis* Small   TEXAS LIVE OAK   *Figure 403*

A shrub or tree, sometimes to 25m tall, but more often shorter and forming clonal groups of shrubby sprouts around the older plants (locally called "copses" or "shinneries"). Occurs in TX and OK in open woodlands and grassland, usually over limestone or calcareous soils. Similar to *Q. virginiana* in many respects of twigs, foliage, and bark; sometimes treated taxonomically as only a variant of that species. Twigs grayish, finely hairy; buds dark brown, rounded, 2 or 3mm long. Acorn cup about 1cm wide, turbinate, covering about ⅓ of nut; nut 20–30mm long, typically narrowing to a point from about the middle. Bark brownish, scaly, becoming furrowed to blocky with age. Leaves subevergreen, entire or with a few sharp, short points or teeth; no awns. A white oak.

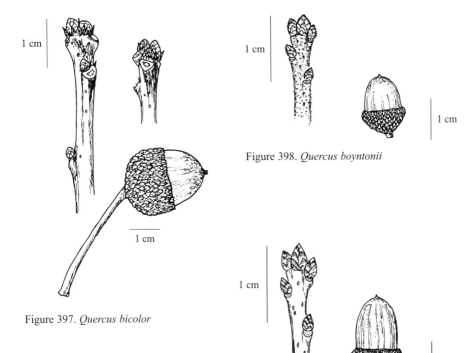

Figure 397. *Quercus bicolor*

Figure 398. *Quercus boyntonii*

Figure 399. *Quercus buckleyi*

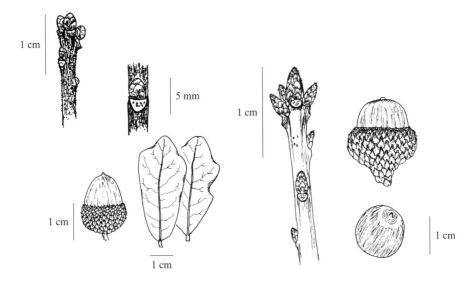

Figure 400. *Quercus chapmanii*

Figure 401. *Quercus coccinea*

13. *Quercus geminata* Small   SAND LIVE OAK   *Figure 404*

A shrub or tree, potentially to 25m tall but quite variable in habit. Distributed in the outer coastal plain from NC to FL, west to s. MS. Inhabits dry, sandy uplands in pinelands, sandhills, and scrub. In most xeric or fire-frequent habitats, a rhizomatous habit induces clonal patches of shrubby stems to develop. Twigs usually gray and hairy, with dark brown, rounded buds about 2mm long. Acorn cup 8–12mm wide, turbinate, covering about ⅓ of nut; nut 15–20mm long, rounded to short-pointed at apex, often borne in pairs on a stalk 2–5cm long. Bark gray-brown, scaly at first, becoming furrowed and ridged. Leaves subevergreen, usually unlobed, margin heavily revolute; veins distinct and reticulate both sides. A white oak. [*Q. virginiana* var. *geminata* (Small) Sarg.]

14. *Quercus georgiana* M. A. Curtis   GEORGIA OAK   *Figure 405*

A small tree, potentially to 15m tall but more often 3–6m. Rare and localized in only a few Piedmont sites in SC, GA, AL. Occurs in thin soils over granite outcrops and in dry oak-pine forests on ridges and slopes. Buds acute, red-brown; scales glabrous or finely ciliate, mostly 2–5mm long. Acorn cup thin, 8–12mm wide, covering about ⅓ of nut; nut 10–14mm long. Bark gray or darker, with rough or slightly scaly ridges and plates. Leaves deeply lobed, the lobes rarely with more than 1 or 2 points, these awned; undersurface glabrous, lustrous green. A red oak.

15. *Quercus hemisphaerica* Bartr. ex Willd.   DARLINGTON OAK   *Figure 406*

A tree, to 35m, native to the coastal plain from se. VA to FL, west to TX. Occupies sandy soils that are well drained. Buds red-brown, acute, 3–5mm long. Acorn cup thin, 10–15mm wide, covering ¼ to ⅓ of nut; nut 8–15mm long, about as wide as long. Bark gray, smoothish at first, becoming shallowly furrowed. Leaves subevergreen, lustrous on both surfaces, entire or with a few large teeth, points awned; leaf apex usually pointed and base rounded to obtuse. A red oak.

16. *Quercus ilicifolia* Wang.   BEAR OAK   *Figure 407*

A shrub or small tree, rarely over 3m but potentially to 6m tall. Distributed from the NE southward along the Appalachians of MD, VA, WV, with a few isolated populations on some NC Piedmont mountains. Typically a scrubby oak of dry, sandy or rocky barrens and slopes, where fire reduces competition and induces prolific resprouting and regeneration. Buds red-brown, 2–4mm long, acute or bluntly pointed, finely hairy. Acorn cup 10–15mm wide, covering ¼ to almost ½ of nut; nut 10–15mm long. Bark gray, mostly smooth, but shallowly furrowed on old trunks. Leaves lobed, the points awned, undersides grayish, pubescent. A red oak.

17. *Quercus imbricaria* Michx.   SHINGLE OAK   *Figure 408*

A tree, to 20m, primarily of regions in the interior and n. states of the SE. Ranges sporadically east in lower elevational parts of the Appalachians and in the Piedmont of VA, NC, and south to TX, LA. Occurs mostly on bottomlands in e. portions of its range, but also in dry uplands in the w. portions. Buds brown, 3–6mm long, acute, slightly angled. Acorn cup 10–15mm wide, covering ⅓ to ½ of the nut. Bark gray to blackish, roughly ridged and furrowed. Leaves unlobed, tip awned, underside densely pubescent. Often a favored host for twig gall insects. A red oak.

18. *Quercus incana* Bartr.   BLUEJACK OAK   *Figure 409*

A tree, to 10m, but often much shorter. Native to the coastal plain from se. VA to FL, west to TX. Occurs on well-drained sandy uplands, often in dry sandhills, pinelands, and scrub. Buds

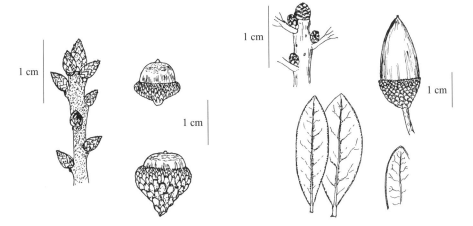

Figure 402. *Quercus falcata*

Figure 403. *Quercus fusiformis*

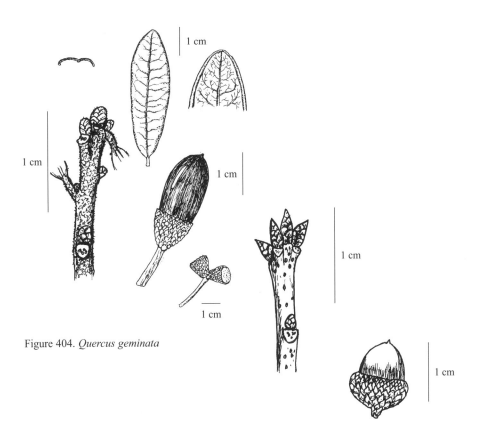

Figure 404. *Quercus geminata*

Figure 405. *Quercus georgiana*

brown, elongated and sharply pointed, 4–7mm long. Acorn cup thin, 10–15mm wide, covering ¼ to nearly ½ of the nut; nut 10–15mm long and as wide. Bark grayish to blackish, roughly ridged. Leaves blue-green above, gray and hairy below, unlobed, with awned tip; petiole 2–10mm. A red oak. (*Q. cinerea* Michx.)

19. *Quercus laevis* Walt. TURKEY OAK *Figure 410*

A small tree, rarely over 15m tall. Native to the coastal plain from se. VA to FL, west to e. LA. Typical of dry upland sites with deep sand, such as sandhills and scrub. Buds rusty red-brown, hairy, conical, mostly 6–10mm long; leaf scars often as wide as bud base. Acorn cup thick, with a wide shoulder and marginal scales rolled over onto interior side, 15–25mm wide, covering about ⅓ of nut; nut often 20–25mm long, blackish when mature and striped or ringed. Bark gray to blackish, roughly ridged or blocky. Leaves deeply lobed, awned, the middle or lower lobes longest, lustrous green and glabrous on both sides. A red oak.

20. *Quercus laurifolia* Michx. LAUREL OAK *Figure 411*

A tree, to 40m tall. Native to the coastal plain from se. VA to FL, west to e. TX. Occurs in moist to wet soils, such as swamps, bottomlands, and low areas in maritime forests. Buds red-brown, mostly 3–6mm long, slightly angled. Acorn cup thin, 12–16mm wide, covering ¼ to ½ of nut; nut mostly 10–15mm long and nearly as wide. Bark gray, smooth, with age becoming furrowed and with wide ridges. Leaves subevergreen, dark dull green above, veins not conspicuous, typically with an obtuse to rounded tip, cuneate to acute base; margin entire and rarely toothed even on juvenile trees. The awn at the leaf tip is usually lacking, falling away early. A red oak.

21. *Quercus lyrata* Walt. OVERCUP OAK *Figure 412*

A tree, to 25m, widespread across the SE in lowlands of the coastal plain and Piedmont, from se. VA to FL, west to e. TX, and north in the Mississippi Valley to s. IN. Occurs in poorly drained soils, swamps, and bottomlands that are frequently flooded. Buds rounded, about 3mm long. Acorn cup retains the nut, nearly or completely surrounding it; the whole acorn 2–5cm wide, a little less in length. Acorns are adapted to float on water, for dispersal amid floodwaters. Bark grayish to brown, scaly or with thick, scaly ridges or plates. Leaves variously lobed or with a few large, bluntly pointed teeth, no awns, lower side pale. A white oak.

22. *Quercus macrocarpa* Michx. BUR OAK *Figure 413*

A tree, to 40m, mostly of the Plains and Midwest, reaching the SE in the Ohio River Valley region and in outlier pockets in MD, WV, AL, LA. A prairie and bottomland species, seen in moist to wet lowlands or calcareous upland clays. Buds obtuse, 3–5mm long; endmost ones often flanked by persistent stipules; twigs and branchlets often developing corky excrescences. Acorn cups thick, with large, thickened scales that have elongated tips near the cup margin (giving a fringed collar). Cup covers ½ or more of the nut; entire fruit 25–60cm long and as wide. Bark brown, roughly ridged into irregular, scaly plates and ridges. Leaves variously lobed, widest above middle, variable in size from 8–30cm long. A white oak.

23. *Quercus margarettiae* Ashe ex Small SAND POST OAK *Figure 414*

A tree, to 15m, native of the coastal plain from se. VA to FL, west to TX, north to n. AL, s. AR, s. OK. Typically on dry uplands, in sandy or gravelly soils in pinelands, scrub, or barrens. Buds brown, obtuse, mostly 3–4mm long, sometimes flanked by small, slender persistent stipules; twigs gray to brown, glabrous. Acorn cup 10–18mm wide, covering ½ to ⅔ of nut; nut 15–22mm long. Bark grayish, slightly scaly on upper trunk, furrowed on lower and older parts.

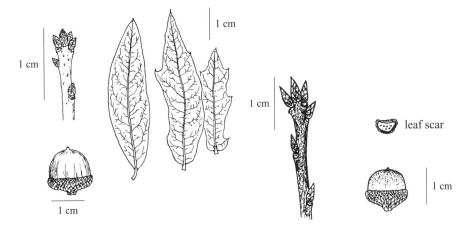

Figure 406. *Quercus hemisphaerica*

Figure 409. *Quercus incana*

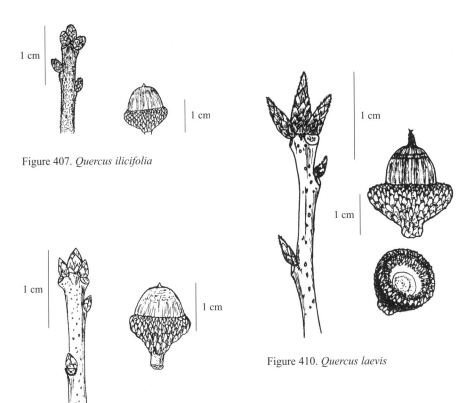

Figure 407. *Quercus ilicifolia*

Figure 410. *Quercus laevis*

Figure 408. *Quercus imbricaria*

Leaves rarely over 8cm long, with a pair of larger lateral lobes remotely suggesting a cruciform shape as in *Q. stellata,* but lobes smaller and more rounded. Leaves sparsely stellate-hairy when young; old leaves often nearly glabrous. A peculiar clonal form of this species is occasionally seen in frequently burned pinelands, where numerous, short, seemingly sterile stems 1–5dm tall are produced from a rhizomatous root system. A white oak. [*Q. stellata* var. *margarettiae* (Ashe ex Small) Sarg.]

24. *Quercus marilandica* Muenchh.   BLACKJACK OAK   *Figure 415*

A tree, to 15m, with a coarse crown of darkened, stiff, and somewhat crooked branches. Widely distributed across the SE; scarce or absent in parts of the higher Appalachians, WV, n. KY, peninsular FL. Typically on dry, gravelly or sandy uplands, or over clays prone to dry out in summer months. Buds elongated, 6–10mm long, angled, red-brown with grayish hairs. Acorn cup thick, finely hairy, 12–18mm wide, covering about ⅓ of nut; nut 12–20mm long, often with a pronounced apical point. Bark blackish, deeply furrowed into blocky ridges and plates. Leaves thick, broadest near the tip, unlobed to shallowly lobed along the wide apex, with awns, 7–20cm long, hairy and yellow-green beneath. The var. *ashei* Sudworth occurs in s. MO, AR, TX, and OK; leaves are smaller (5–7cm), more commonly 3-lobed at the tip, gray-hairy in the vein axils beneath; it is found on limestone uplands, most common in central TX. A red oak.

25. *Quercus michauxii* Nutt.   SWAMP CHESTNUT OAK   *Figure 416*

A tree, to 30m, widely distributed across lowlands in the SE mostly in the Piedmont and coastal plain; absent in most of the Appalachians and WV. Typically on moist to wet bottomlands, alluvial flats, and rich, moist soils not prone to prolonged periods of flooding. Buds red-brown, 5–8mm long. Acorn cup thick, with stout, pointed scales, 2–4cm wide, covering ¼ to ⅓ of nut; nut 25–40mm long. Bark gray to gray-brown, scaly on upper trunk; thicker with age, furrowed on old trunks. Leaves toothed, the teeth fairly pointed or occasionally rounded but not awned; veins reticulate, underside pale, pubescent. A white oak.

26. *Quercus minima* (Sarg.) Small   DWARF LIVE OAK   *Figure 417*

A rhizomatous shrub, rarely reaching 1m or a little more. Forms dense colonies of slender stems from a spreading root system, especially when fires reduce competition. Native to the outer coastal plain from e. NC to FL, west to s. AL. Occurs in pine flatwoods, savannas, and pinelands near the coast that generally have a high frequency of fire. Twigs gray; buds rounded, brown, 1–2mm long. Acorn cups 10–15mm wide, funnel-shaped, with a thickened rim and scales slightly keeled, covering about ⅓ of nut; nut 15–25mm long. Leaves evergreen, often with large points along the margin; yellow-green or grayish and closely hairy beneath. A white oak. (*Q. virginiana* var. *minima* Sarg.)

27. *Quercus montana* Willd.   CHESTNUT OAK   *Figure 418*

A tree, to 30m, distributed mostly in the Piedmont, Appalachian, and interior plateaus of the SE. Typically seen on steep, rocky, or dry mountain slopes and ridges, and similar woodlands where the soil is thin and fast draining. Buds 5–8mm long, scales brown with pale margin. Acorn cup 18–25mm wide, gray, inner surface funnel-like, tapered without much concavity; cup margin thin, scales bumplike or fused, cup covering about ⅓ of nut; nut 2–4cm long. Bark grayish or brown, deeply furrowed into elongate ridges that are wide at the base. Leaves toothed, the teeth rounded to obtuse, no awns; underside pale, glabrous, veins distinct. A white oak. (*Q. prinus* L.)

28. *Quercus muhlenbergii* Engelm.   CHINQUAPIN OAK   *Figure 419*

A tree varying in habit from as short as 3m on poor sites to 25m on fertile sites. Scattered in distribution across much of the SE, except the Piedmont of VA and NC and the coastal plain from

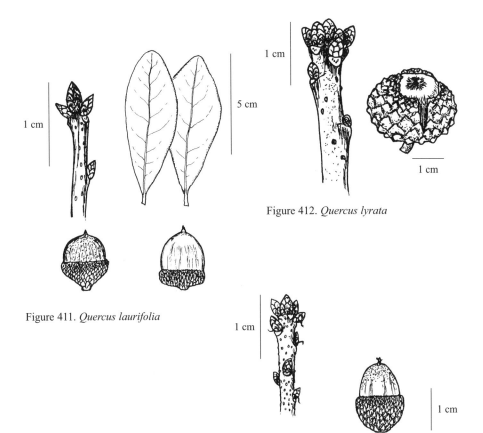

Figure 411. *Quercus laurifolia*

Figure 412. *Quercus lyrata*

Figure 414. *Quercus margarettiae*

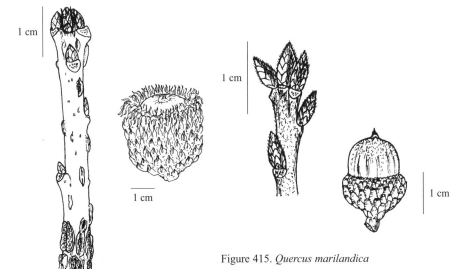

Figure 413. *Quercus macrocarpa*

Figure 415. *Quercus marilandica*

VA to FL, and w. FL to TX. Most often seen in circumneutral to calcareous soils, where in the extreme of thin, dry soils over limestone it may be little more than 4m in height. Buds bluntly pointed, 2–4mm long, scales dark brown with pale margins; twigs yellow-brown or grayish. Acorn cup thin, mostly 10–20mm wide, with slightly knobby scales, covering ⅓ to ½ of nut; nut mostly 15–20mm long. Bark gray, scaly. Leaves thin, toothed; teeth pointed, not awned. A white oak.

29. *Quercus myrtifolia* Willd.   MYRTLE OAK   *Figure 420*

A shrub or small tree, rarely over 10m tall. Native to the coastal plain from SC to peninsular FL and near the Gulf Coast from w. FL to s. MS. Occurs on dry, sandy uplands in pinelands, scrub, dunes near the coast, hammocks, and ridges. Buds mostly 2–4mm; scales brownish-red, nearly glabrous; twigs usually hairy. Acorn cup thin, 10–15mm wide, enclosing ⅓ or less of nut; nut 10–14mm long. Bark grayish, smooth, fissured or ridged on older trunk portions. Leaves evergreen, usually 4cm or less in length, margin entire and distinctly revolute, apex rounded and awn present or deciduous, undersurface green and mostly glabrous. A red oak.

30. *Quercus nigra* L.   WATER OAK   *Figure 421*

A tree, to 30m tall. Distributed throughout the coastal plain from se. VA to central FL, e. TX, and inland as far north as n. GA to sw. KY, se. MO. Perhaps the most common oak of the southernmost states of the region; found in a diverse range of habitats, including floodplains, moist slopes, mesic forests, and uplands with clay or sandy soils. Buds red-brown at base, paler and hairy near tip, 3–5mm long. Acorn cup thin, 10–16mm wide, covering about ¼ of nut; nut mostly 10–14mm long and as wide. Bark gray, fairly smooth, becoming ridged and shallowly furrowed with age. Leaves deciduous or subevergreen, mostly 5–10cm long, entire to 3-lobed near tip (with awns), variable in shape but typically widest near apex and cuneate at base (often spatulate). Juvenile trees may have narrow leaves with additional long, narrow lobes along the margin. A red oak.

31. *Quercus oglethorpensis* Duncan   OGLETHORPE OAK   *Figure 422*

A tree, to 25m tall, of fragmented distribution in the SE. Largest populations occur in the lower Piedmont of w. SC and e. GA, with other populations known in MS and LA. Overall a rare oak, though it may seem locally common. Occurs in rich, moist lowlands such as bottomlands and streamside flats; occasionally on higher adjacent grounds. Buds obtuse or rounded, brown, slightly hairy, often flanked by persistent stipules, 2–3mm long. Acorn cup thin, 8–10mm wide, enclosing about ⅓ of nut; nut 10–14mm long, often with silky, appressed hairs near apex. Bark gray, scaly. Leaves usually entire, lobed occasionally on juvenile growth, apex bluntly pointed, underside yellow-green and conspicuously pubescent, mostly 6–12cm long. A white oak.

32. *Quercus pagoda* Raf.   CHERRYBARK OAK   *Figure 423*

A large tree, reaching 30–40m in height. Primarily found in the coastal plain from se. VA to FL, west to e. TX, inland to s. IL, but also in the Piedmont regions of SC to AL. Typically in moist lowlands, including alluvial bottomlands, terraces of swamps, riverbanks, mesic slopes. Buds red-brown, usually angled in cross section, 4–6mm long, hairy. Acorn cup thick, 10–18mm wide, covering ⅓ to ½ of nut; nut 10–15mm long, about as wide. Trunks of young trees or upper portions of mature trees have narrow, flaky or scaly ridges, somewhat suggestive of *Prunus serotina* (black cherry) bark. Older trunk bark thicker, furrowed and ridged. Leaves pale and closely pubescent beneath; marginal lobes mostly straight, 4 or more per side, awned, the outline of leaf elliptical or ovate. A red oak.

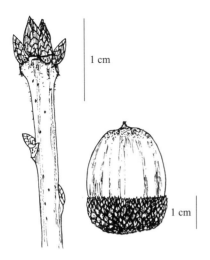

Figure 416. *Quercus michauxii*

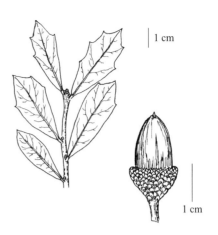

Figure 417. *Quercus minima*

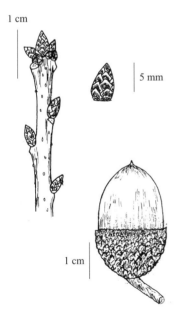

Figure 418. *Quercus montana*

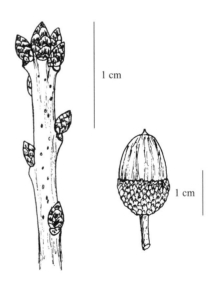

Figure 419. *Quercus muhlenbergii*

33. *Quercus palustris* Muenchh.   PIN OAK   *Figure 424*

A tree, to 25m, most common in the upper Mississippi River and Ohio River lowlands. Ranges into the SE to e. OK, central AR, TN, and the Piedmont of NC; not in the Blue Ridge of TN, NC. Occurs in poorly drained sites, often with acidic clay substrates, but also in alluvial bottomlands. Buds red-brown, glabrous, 3–5mm long. Acorn cups thin, 10–15mm wide, covering about ¼ of nut; nut 10–15mm long and about as wide. Bark gray, thin, fairly smooth on young trunks or fast-growing trees, becoming fissured and ridged with age. Drooping lower branches are a conspicuous trait of open-grown trees. Leaves deeply lobed, with awns on points. Widely planted as a lawn and street tree. A red oak.

34. *Quercus phellos* L.   WILLOW OAK   *Figure 425*

A tree, to 30m, with drooping lower branches when open grown (similar to above species). Mostly a tree of the coastal plain and Piedmont regions throughout the SE and inland to s. IL, s. KY. Peculiarly scarce or absent from se. GA through peninsular FL. Most common in moist lowlands such as bottomlands and swamp terraces, but also seen in wet flats in uplands and occasionally in well-drained soils. Often planted as a shade or specimen tree. Buds mostly 2–3mm long, sharply pointed. Acorn cups 8–12mm wide, covering ⅓ or less of nut; nut 8–12mm long. Bark grayish, thin and barely fissured on young or fast-growing trunks; deeply furrowed and ridged on older trunks. A red oak.

35. *Quercus prinoides* Willd.   DWARF CHINQUAPIN OAK   *Figure 426*

A shrub, 1–3m tall, sometimes taller but usually multistemmed. Of sporadic occurrence in a range that extends from n. VA to n. GA, west to s. OK and areas north to the Great Lakes and the NE. Mostly seen over acidic, sandy or shaley slopes and barrens. Differs from the similar but more treelike *Q. muhlenbergii* by the tendency to sprout from the roots and form colonies (aided perhaps by the incidence of fire), by the tendency to begin fruit production when only 3 or 4dm in height, and in the preference for acidic soils as opposed to calcareous substrates. Other differences from *Q. muhlenbergii* are 1–3mm long buds, 9 or fewer teeth per each side of the leaves, and a higher proportion of leaves less than 10cm long (10–18cm leaves are more typical in *Q. muhlenbergii*). A white oak.

36. *Quercus pumila* Walt.   RUNNER OAK   *Figure 427*

A rhizomatous shrub, 1m or less in height, forming colonies of bushy-branched stems. Native to the coastal plain from se. NC to s. FL, west to s. MS. Occurs in pine flatwoods and savannas in sandy soils with some organic content. Buds brown, 2–4mm long, endmost ones often flanked by persistent stipules. Acorn cup 10–15mm wide, rather thick, covering about ½ of nut, which is 10–15mm long and often flat about the apex. Unusual among the red oak group in ripening its acorns in a single season; sometimes flowering twice or some stems flowering at later stages in the same season; in such cases both mature and juvenile acorns may be present from late summer to autumn. Leaves usually unlobed, awned at tip, grayish and pubescent on the undersurface. A red oak.

37. *Quercus rubra* L.   NORTHERN RED OAK   *Figure 428*

A tree, to 30m tall, widely distributed in the e. US, most conspicuous in n. states. Ranging into the SE in the interior plateaus, Appalachians, and Piedmont from ne. NC to s. AL, west to s. AR and areas northward. Typically in mesic forests, occasionally on drier and rocky soils in n. parts of the range. Buds usually acute, red-brown, 5–7mm long. Acorn cup thick, 20–35mm wide, covering ¼ or less of nut, which is 20–30mm long. The var. *borealis* (Michx. f.) Farwell [*var. ambigua* (Gray) Fern.] produces acorns with cups 15–20mm wide, covering about ⅓ of

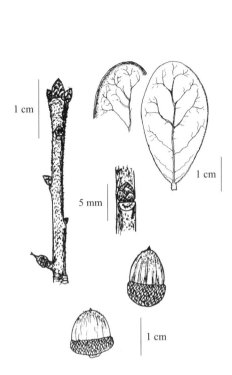

Figure 420. *Quercus myrtifolia*

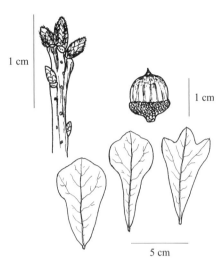

Figure 421. *Quercus nigra*

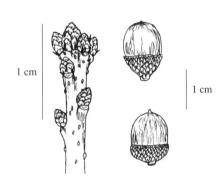

Figure 422. *Quercus oglethorpensis*

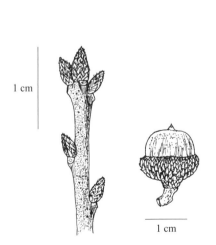

Figure 423. *Quercus pagoda*

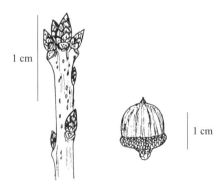

Figure 424. *Quercus palustris*

nut, which is 15–20mm long; buds may be rather blunt and slightly hairy. This latter variety mainly replaces the typical species in higher elevations of the Appalachians, though not always easily separated in intermediate habitats. Bark grayish or darker, with flat-topped, elongated ridges. Leaves rather evenly lobed, the sinuses extending about halfway to midrib, glabrous beneath, awned on points.

38. *Quercus shumardii* Buckl.   SHUMARD OAK   *Figure 429*

A tree, to 35m, distributed across all regions of the SE west of the Blue Ridge, and in the coastal plain and Piedmont from n. GA to central NC. Scarce or lacking in most of VA, WV, e. KY, as well as the mountains of NC, TN. Occurs in moist lowlands such as alluvial bottomlands and mesic slopes, as well as rocky, dry, calcareous uplands. Buds gray or light grayish-brown, pointed but fairly plump, slightly angled, 4–6mm long. Acorn cup thick, 15–30mm wide, covering ⅓ or less of nut, which is 15–30mm long, usually flat-based. The var. *schneckii* (Britt.) Sarg. has thinner acorn cups, covering ⅓ to nearly ½ of nut. Bark grayish or darker, with flat-topped ridges thickening with age. Leaves deeply lobed with many awned points. A red oak.

39. *Quercus similis* Ashe   DELTA POST OAK   *Figure 430*

A tree, to 25m, sporadic across the coastal plain from SC to GA, west to e. TX and AR, most common in LA. Occurs in moist to wet bottomlands and flatwoods, usually over poorly drained clay substrates. Buds about 3mm long, blunt, slightly hairy, endmost ones often flanked by persistent stipules; twigs usually hairy. Acorn cup thin, 10–12mm wide, covering ⅓ to ½ of nut, which is about 14mm long. Bark grayish or gray-brown, slightly scaly on upper portions, furrowed on most of main trunk, with slightly scaly ridges. Leaves mostly 3-lobed above the middle, but variable and not with a uniform shape as is more typical of *Q. stellata*. Young leaves stellate-pubescent below; older leaves sparsely so. Growth habit in e. portions of the range is coarse, trees smaller and more often confined to calcareous marl soils than in w. portions of the range. A white oak. (*Q. stellata* var. *paludosa* Sarg.; *Q. mississippiensis* Ashe)

40. *Quercus sinuata* Walt.   DURAND OAK   *Figure 431*

A tree, to 25m tall, though often shorter on harsh sites. Uncommon and widely scattered in a fragmented range from SC to n. FL, west to TX, mostly in the coastal plain but also known from the Piedmont of SC, GA, and s. Cumberland Plateau of AL. Occurs in a variety of habitats with circumneutral to calcareous soils. Seen in bottomlands, wet flats, streamsides, and among limestone outcrops and prairies. Buds blunt, brown, 2–3mm long, endmost ones often flanked by persistent stipules. Acorn cup thin, very shallow, 10–15mm wide, covering ¼ or less of nut, which is 10–15mm long and as wide, with a flat base. Occasional trees may have acorns slightly longer than wide; rarely the acorn cup covers ⅓ of nut. Bark gray, scaly. Leaves 6–12cm long; shape variable, from wavy-margined to entire, with lobes rounded to acute, the undersides green and essentially glabrous in juvenile and shaded plants to pale and closely appressed-pubescent in mature or open-grown plants; look for the minute stellate hairs on undersurface with 10× magnification. A white oak. (*Q. durandii* Buckl.)

41. *Quercus sinuata* var. *breviloba* (Torr.) C. H. Muller   BIGELOW OAK   *Figure 432*

A shrubby variety of the preceding species, usually forming clumps of stems that rise 2–5m tall. The clonal habit produces large expanses or colonies in parts of its range. Distributed from central TX to s. OK over limestone, plains, and scrublands. Differs from the above species (apart from habit) in acorn cups being thicker, nuts to 25mm long, and smaller leaves, under 6cm, more often wavy-margined or with rounded lobes. A white oak.

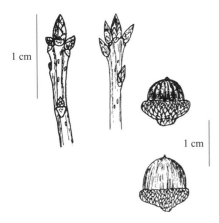

Figure 425. *Quercus phellos*

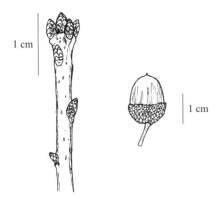

Figure 426. *Quercus prinoides*

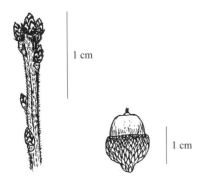

Figure 427. *Quercus pumila*

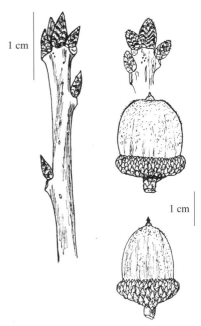

Figure 428. *Quercus rubra*

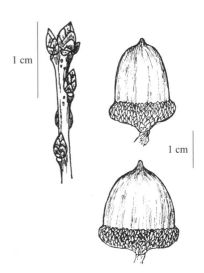

Figure 429. *Quercus shumardii*

42. *Quercus stellata* Wang. POST OAK *Figure 433*

A tree, to 25m, though often shorter on harsh sites. Distributed throughout the SE. Occurs on well-drained soils; most frequently seen in dry, stony, sandy, or infertile upland sites. Twigs rather stout, grayish, and pubescent. Buds brown, blunt or rounded, angled, 3–5mm long. Acorn cup fairly thin, 10–15mm wide, covering ⅓ to ½ of nut, which is 10–20mm long. Bark gray or light brown, scaly on upper parts, furrowed and ridged on lower or main trunk. Leaves thick, rough on both surfaces due to harsh stellate hairs, lobes rounded. A pair of large lobes at the middle or just above the middle of the leaf jut outward at a right angle and may be truncate or indented at the tip, imparting a cruciform pattern to the general leaf shape. This species hybridizes with many other oaks, such hybrids usually inheriting the conspicuous trait of rough stellate hairs. Intermediate forms of post oak with *Q. margarettiae* may occasionally be encountered, possibly due to hybridization. Much of the post oak common in central and e. TX has been called *Q. drummondii* Liebmann and has characters similar to both *Q. stellata* and *Q. margarettiae*. A white oak.

43. *Quercus texana* Buckley NUTTALL OAK *Figure 434*

A tree, to 25m, native to the lower Mississippi Valley from s. IL to e. TX, east to central AL. Occurs in bottomlands and alluvial floodplains. Buds gray or grayish-brown, mostly 3–5mm long. Acorn cup thin on the margin, 15–20mm wide, enclosing ¼ to nearly ½ of nut, which is 15–20mm long. Bark grayish, thin and fairly smooth on young or fast-growing trunks, thickening and fissured with age. Leaves deeply lobed. A red oak. (*Q. nuttallii* Palmer)

44. *Quercus velutina* Lam. BLACK OAK *Figure 435*

A tree, to 25m, distributed across nearly all the SE, though absent from portions of the coastal plain near the Gulf Coast and in se. GA. Occurs in a variety of habitats, but most common in well-drained uplands. Buds grayish to light brown, angled, hairy, 6–12mm long. Acorn cup thick, scales near the rim free at the tips, forming a small fringe, 15–25mm wide, covering about ½ of nut; nut is 15–22mm long, often with brownish pubescence on outer surface. Bark dark gray to black with thick, rough, irregular ridges or sometimes merely deeply furrowed and ridges flat-topped. Inner bark yellow, unusual since other red oaks have pink to reddish-brown inner bark. Leaves lobed and awned, lustrous on top surface, paler, dull, and hairy on or in forks of veins beneath. Juvenile trees and lower branches of mature trees have shallowly lobed leaves, deeply lobed ones occurring on upper branches. A red oak.

45. *Quercus virginiana* P. Mill. LIVE OAK *Figure 436*

Normally a tree, to 35m tall, but occasionally shrubby and dwarfed in deep sands near the coast. Native to the coastal plain from se. VA to FL, west to TX. Typical of maritime forests, hammocks, and more fertile areas of sandy scrublands. Forms nearly pure stands behind dunes on barrier islands, as it is more resistant to salt spray than other native hardwood trees. In dry, sterile sands, sands impacted by wind or salt spray, or in frequently burned habitats, growth may become shrublike or rhizomatous. Tubers often form on juvenile plants in such locations, probably a food storage response. Buds brown, blunt or rounded, about 2mm long. Acorn cups turbinate or tapered conically, 8–15mm wide, covering about ⅓ of nut, which is 15–25mm long. Bark grayish to brown, deeply furrowed into irregular ridges or blocky plates. Leaves subevergreen, thick, margin usually entire and barely revolute, underside pale and closely, minutely pubescent. A white oak.

Rhamnus L. BUCKTHORN Family Rhamnaceae

Four species are treated here for the SE, 2 of which are native. A taxonomic change has sent 2 other species formerly included in this genus (*R. caroliniana* Walt. and *R. frangula* L.) to the

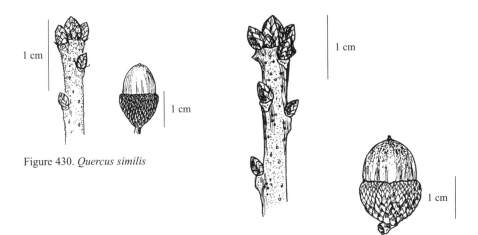

Figure 430. *Quercus similis*

Figure 433. *Quercus stellata*

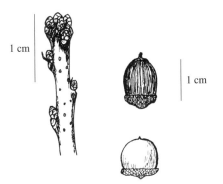

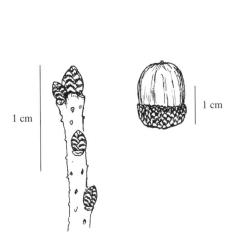

Figure 431. *Quercus sinuata*

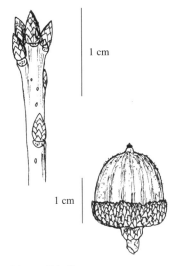

Figure 432. *Quercus sinuata* var. *breviloba*

Figure 434. *Quercus texana*

genus *Frangula*. All species in *Rhamnus* can now be characterized by having buds with imbricate scales, whereas *Frangula* bear naked buds. Spinose twig tips may be present on the 2 exotic species of *Rhamnus*. Bundle scars are in 3 groups or obscurely divided. The inner bark is yellowish and somewhat rankly odorous. Fruit is berrylike, black when ripe.

1. Leaf scars alternate; no spiny spur shoots; native shrubs
  2. Twigs reddish-brown; buds pointed, usually single per node; fruit 3-seeded   1. *R. alnifolia*
  2. Twigs grayish; buds blunt, collateral buds often present; fruit 2-seeded   2. *R. lanceolata*
1. Leaf scars often opposite or subopposite; spiny spur shoots commonly present; exotic shrubs or small trees
  3. Twigs slightly hairy; leaves dull green above   3. *R. cathartica*
  3. Twigs glabrous; leaves glossy above   4. *R. davurica*

1. *Rhamnus alnifolia* L'Her.   ALDERLEAF BUCKTHORN   *Figure 437*

A low native shrub, rarely over 1m tall, forming colonies. Ranges across the n. regions of N. Amer.; reaches the SE only in high mountain wetlands of WV, VA, and TN. Occurs in boggy seeps, wet meadows, and wetland borders, usually over circumneutral or calcareous soils. The twigs often retain small, blackened stipules at bud bases. Seed flattish, weakly grooved.

2. *Rhamnus lanceolata* Pursh.   NARROW-LEAF BUCKTHORN   *Figure 438*

A shrub, to 2m tall. Native to the Appalachians and regions west, to IN, MO, and AL to TX. Occurs sporadically in calcareous soils, from wet to dry. The seed are broadly grooved.

3. *Rhamnus cathartica* L.   EUROPEAN BUCKTHORN   *Figure 439*

A shrub or tree, to 5m tall. Uncommonly naturalized in the n. states of the SE. Short twigs may bear terminal buds; leaf scars vary from alternate to subopposite or opposite. Leaves thin, to 7cm long, length not over twice the width; dull green above, light green below. A native of Eurasia.

4. *Rhamnus davurica* Pallas   DAHURIAN BUCKTHORN   *Figure 440*

A large shrub or shrubby tree, to 10m tall, often sprouting from the roots. Naturalized in n. states of the SE. A native of e. Asia. Leaves thick, to 10cm long, over twice as long as wide, gray-green beneath, glossy above.

Rhapidophyllum histrix (Pursh) H. Wendl. & Drude ex Drude   FLORIDA NEEDLE PALM
Family Arecaceae   *Figure 441*

A native palm of the coastal plain, from se. SC to FL, west to s. AL. The stem is short, rarely over 1m high, and produces many palmately cleft leaves on long petioles, most over 1m long. The central mass of the plant is composed of a thick layer of old leaf stalk bases and sheaths, along with slender, black, needlelike spines protruding outward. There is rarely a visible vertical or horizontal trunk section. Leaves silvery-pubescent beneath, dark glossy green above.

Rhododendron L.   RHODODENDRON   Family Ericaceae

This genus is treated here as composed of 19 species in the SE, divided between 2 distinct groups, the evergreen rhododendrons (subgenus *Eurhododendron* Endl.) and the deciduous azaleas (subgenus *Pentanthera* G. Don.). All have terminal buds with several imbricate scales, and often several smaller lateral buds are clustered around the terminal bud at twig tips. Leaf scars bear a single bundle scar. The fruit is an elongated, 5-parted capsule. Bark is brown, smooth on young stems, shallowly fissured to scaly on older stems. The evergreen rhododen-

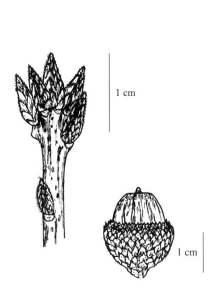

Figure 435. *Quercus velutina*

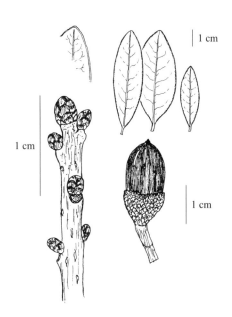

Figure 436. *Quercus virginiana*

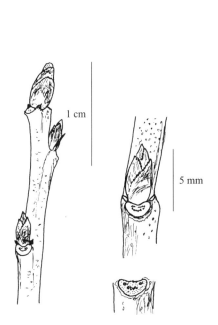

Figure 437. *Rhamnus alnifolia*

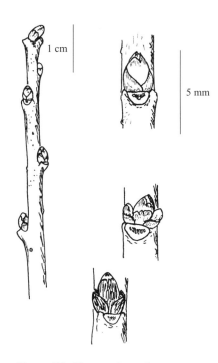

Figure 438. *Rhamnus lanceolata*

drons are composed of at least 3 species, the taxonomy being debatable whether 1 of these, *R. minus*, should be divided further into 3 separate species (*R. minus, carolinianum,* and *chapmanii*), as is done in the following treatment. Fourteen species of azaleas are native to the SE, best identified when in flower. In winter months, identification by twigs alone is difficult. Use of a hand lens (8× or 10× magnification) to inspect bud scale margins of the larger, terminal flower buds is useful, as is the condition and types of hairs on the fruits. Care in handling fruit may preserve important diagnostic characters such as the presence of glandular hairs (hairs with swollen tips); these are prone to become dry and brittle after fruit maturation in autumn.

1. Leaves evergreen, entire on margin
  2. Leaves not dotted or scaly beneath, most over 10cm long
    3. Leaves light green below; tips of lowermost bud scales of terminal bud elongate, leaflike
                                                                                                                1. *R. maximum*
    3. Leaves glaucous below; terminal bud scales all triangular, not as above 2. *R. catawbiense*
  2. Leaves with brown scales or dots below, mostly less than 10cm long
    4. Brown scales densely cover bud scales and twigs; leaves cupped; stiffly erect shrub of FL panhandle region   3. *R. chapmanii*
    4. Brown scales sparse on buds, twigs; leaves usually not cupped; spreading shrub of more inland areas
      5. Flowers usually white, corolla to 2cm long; of Appalachians only   4. *R. carolinianum*
      5. Flowers pink to lavender, corolla often over 2cm; mountains to coastal plain
                                                                                                          5. *R. minus*
1. Leaves deciduous, ciliate on margin (azaleas)
  6. Capsule bears glandular hairs (best viewed with lens on mature but not very old fruits)
    7. Flower buds widely ovate to nearly globular, base broadly rounded; bud scales rounded, about as wide as long; rare species of higher elevations in NC Blue Ridge 6. *R. vaseyi*
    7. Flower buds and bud scales longer than above; not restricted to high mountains of NC
      8. Habit stoloniferous, with many stems, rarely over 1m in height
        9. Capsules often glaucous, with spreading glandular hairs; sepals to 5mm
                                                                                                 7. *R. atlanticum*
        9. Capsules not glaucous, with appressed glandular hairs; sepals to 2mm 8. *R. viscosum*
      8. Habit not stoloniferous, usually taller than 1.5m
        10. Flower bud scale margins glandular, not ciliate; capsule 4 to 5 times as long as wide; native to Gulf vicinity   9. *R. austrinum*
        10. Flower bud scale margins ciliate, at least near the tip; capsule not as above
          11. Twigs glabrous, usually yellow-brown   10. *R. arborescens*
          11. Twigs with sparse to numerous hairs
            12. Capsules sparsely glandular-hairy to glabrous; flower bud scale margins lack glands; of mountains   11. *R. prinophyllum*
            12. Capsules very glandular-hairy; flower bud scale margins dark, with some glands; widespread   8. *R. viscosum*
  6. Capsule with nonglandular hairs only (use lens)
    13. Flower bud scale surfaces distinctly hairy; capsules densely pubescent 12. *R. canescens*
    13. Flower bud scale surfaces slightly hairy to glabrous
      14. Flower bud scale margins glandular, not ciliate, or ciliate only near tip
        15. Sepals mostly 2–4mm, margins glandular-hairy   13. *R. calendulaceum*
        15. Sepals mostly under 2mm long, hairs not glandular; uncommon species
          16. Sepals 1–2mm; pedicels to 7mm; near mountains   14. *R. cumberlandense*
          16. Sepals 0.5–1mm; pedicels to 12mm; of SC coastal plain   15. *R. eastmanii*
      14. Flower bud scale margins ciliate
        17. Capsules 5–7mm wide
          18. Twigs glabrous or nearly so; capsules sparsely hairy   16. *R. prunifolium*
          18. Twigs hairy; capsules closely or conspicuously hairy   17. *R. flammeum*
        17. Capsules less than 5mm wide

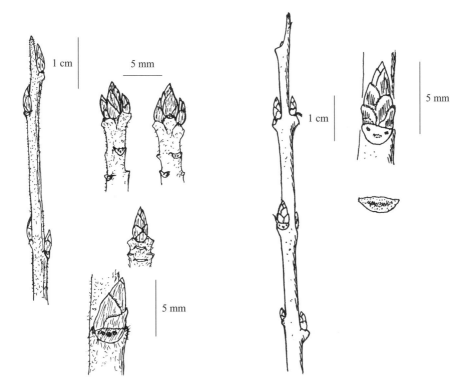

Figure 439. *Rhamnus cathartica*   Figure 440. *Rhamnus davurica*

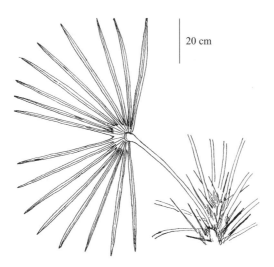

Figure 441. *Rhapidophyllum histrix*

19. Capsules cylindric, not swollen near base; sepals to 1mm; twigs usually hairy
18. *R. alabamense*
19. Capsules ovate-elongate, with swelling near base; sepals to 3mm; twigs sparsely hairy to glabrous
19. *R. periclymenoides*

1. *Rhododendron maximum* L.   ROSEBAY RHODODENDRON   *Figure 442*

A large shrub or small, bushy tree, to 8m tall. Distributed throughout the Appalachian region of the SE, with a few outlier populations in adjacent provinces such as the Piedmont and interior plateau. Common on moist and cool slopes, along streams, in swampy sites, and low to middle elevation heath balds. Often forms dense thickets. Flowers white to pale pink with a greenish internal blotch, appearing June–Aug after new leaves have matured. Sepal length 4mm or more.

2. *Rhododendron catawbiense* Michx.   CATAWBA RHODODENDRON   *Figure 443*

A large shrub or bushy tree, rarely over 5m tall. Distributed mostly along the Appalachians from WV and VA to AL, GA. Occasionally in outlier populations of the Piedmont of NC, SC. Common in the higher elevations of the Blue Ridge, though also seen in cooler, sheltered ravines and slopes of the Piedmont. A major component of high elevation heath balds. Flowers pink to purplish, with greenish internal blotch, appearing Apr–June (depending on elevation) as new leaves emerge. Sepals 1mm or less in length.

3. *Rhododendron chapmanii* Gray   CHAPMAN RHODODENDRON   *Figure 444*

A coarse shrub, to 3m tall, the branches stiffly upright. Rare and endemic; only known from n. FL. Occurs in moist pine flatwoods and near shrub bays or boggy thickets. Flowers pink, with greenish internal blotch, often over 2cm long, appearing Mar–Apr, mostly before new leaves emerge. Leaves usually strongly cupped or with wavy, revolute margins. The brownish-peltate scales are numerous on leaf undersides, petioles, twigs, buds.

4. *Rhododendron carolinianum* Rehd.   CAROLINA RHODODENDRON

A shrub, to about 2.5m tall. Endemic to the Blue Ridge of NC, TN, GA, SC. Scattered in the middle to higher elevations overall, but sometimes locally abundant down to the foothills of the Appalachians. Inhabits moist soils over rocky peaks, slopes, and shrub balds, or on drier woodland slopes. Flowers white to rosy-pink (usually white or pale pink), with greenish internal blotch, usually less than 2cm long, appearing Apr–May as new leaves emerge. (*R. punctatum* Small)

5. *Rhododendron minus* Michx.   PIEDMONT RHODODENDRON   *Figure 445*

A shrub, to 3m tall. Distributed along the easternmost portions of the Blue Ridge from nw. NC to AL and in the Piedmont of AL, GA, SC. Occasional in the coastal plain of SC, NC. Occurs in sandy or rocky uplands and on steep, rocky slopes or ledges in the mountains. Flowers rosy pink with greenish internal blotch, usually over 2cm long, appearing Apr–June (depending on elevation) after new leaves have emerged. (*R. punctatum* Andr.)

6. *Rhododendron vaseyi* Gray   PINK-SHELL AZALEA   *Figure 446a*

A shrub, to 3m tall. Rare and localized in the higher elevations of the Blue Ridge of NC. Occurs in moist woodlands, shrub balds, boggy sites, rocky peaks, road banks. Differs from all other azaleas native to the SE in its corolla tube being shorter than the corolla lobes. Flowers not fragrant, pink, with orange internal dots, appearing May–June before leaves emerge. Twigs glabrous; buds widely ovate, light brown. Capsule cylindrical, with glandular hairs.

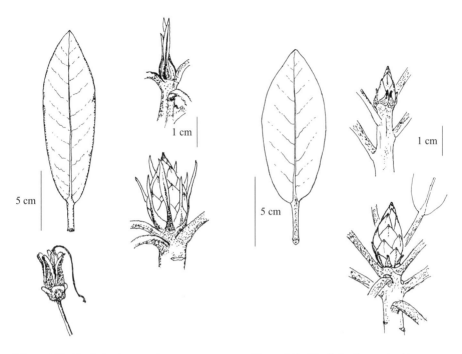

Figure 442. *Rhododendron maximum*

Figure 443. *Rhododendron catawbiense*

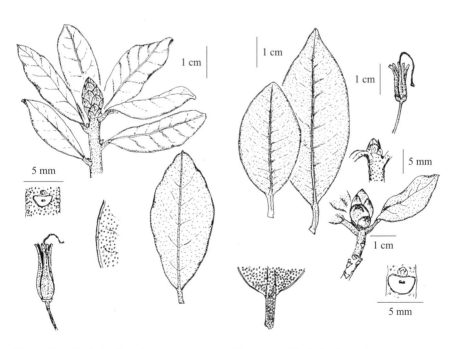

Figure 444. *Rhododendron chapmanii*

Figure 445. *Rhododendron minus*

7. *Rhododendron atlanticum* (Ashe) Rehd.   DWARF AZALEA

A low, rhizomatous shrub, rarely over 1m tall. Native to the coastal plain from VA to se. GA. Typically in moist pinelands and pocosins. Flowers fragrant, white, appearing in Apr as leaves emerge; filaments white to brown-tinged. Twigs hairy; buds greenish to light brown.

8. *Rhododendron viscosum* (L.) Torr.   SWAMP AZALEA

A shrub, to 6m tall, or occasionally rhizomatous and under 2m. Widespread across the coastal plain from MD to FL, west to e. TX, and in the interior to the Appalachians of VA to GA, n. AR. Typically in moist soils such as swampy thickets, streamsides, bogs, pocosins, shrub balds. Flowers fragrant, white, sticky due to glandular hairs, appearing May–July after leaves are mature. Twigs hairy; bud scales greenish or yellow-brown with distinct brown margins; flower buds usually with 8 to 12 bud scales. Other similar azaleas once described as separate species but now considered merely variants of this wide-ranging and variable species are *R. serrulatum* (Small) Millais of a more s. distribution, with 15 to 20 flower bud scales that have more pronounced brown margins; *R. oblongifolium* (Small) Millais of e. TX, with pubescent bud scales; *R. coryi* Shinners of e. TX, with a strongly rhizomatous habit and larger flowers.

9. *Rhododendron austrinum* (Small) Rehd.   FLORIDA AZALEA

A tall shrub, to 5m. Native to the coastal plain in the FL panhandle and adjacent sw. GA, s. AL, se. MS. Occurs in moist forests, often near river bluffs and on mesic slopes near bottomlands and swamps. Flowers fragrant, yellow to orange, appearing Mar–Apr before or as leaves emerge. Twigs usually pubescent; buds gray-brown.

10. *Rhododendron arborescens* (Pursh) Torr.   SWEET AZALEA   *Figure 446b*

A tall shrub, to 6m. Native to the Appalachian region from MD to central AL and the Piedmont of NC, SC, GA. Occurs in moist soils, especially along streams, seepage slopes, mountain shrub thickets, and balds. Flowers fragrant, white with a yellow internal blotch, appearing May–Aug (depending on elevation) after leaves are mature; filaments pinkish. Twigs glabrous, yellow-brown; buds greenish to yellow-green.

11. *Rhododendron prinophyllum* (Small) Millais   ELECTION-PINK

A shrub, to 3m tall. Mostly a ne. species that ranges into the SE in the Appalachians of MD, WV, VA, and NC. Also occurs in uplands of se. MO to central AR and adjacent OK, and in a few KY and s. OH sites. Occurs in upland woods and along streams, and in moist boggy sites and high elevations in the Appalachians. Flowers fragrant, pink, with pointed petals; filaments brownish, appearing Apr–May before or as leaves emerge. Twigs hairy; buds brownish, hairy. [*R. roseum* (Lois.) Rehd.]

12. *Rhododendron canescens* (Michx.) Sweet   PIEDMONT AZALEA

A tall shrub, to 6m. Widespread across most of the se. states between se. NC to e. TX, and to n. TN. Most common in coastal plain woodlands, near streams, savannas, pocosins, and moist slopes. Flowers slightly fragrant, white to pink, appearing Mar–Apr as leaves emerge. Twigs hairy; buds yellow-brown, hairy.

13. *Rhododendron calendulaceum* (Michx.) Torr.   FLAME AZALEA   *Figure 447*

A tall shrub, to 8m. Native to the Appalachian region from WV to ne. GA and occasionally the adjacent Piedmont of VA to GA. Occurs in moist to dry woodlands, streamsides. Flowers not fragrant, yellow to red (orange most common) with orange internal blotch, appearing May–June as leaves emerge. Twigs hairy; buds greenish with dark brown margins.

14. *Rhododendron cumberlandense* E. L. Braun   CUMBERLAND AZALEA

A shrub, to 3m tall, often rhizomatous. Native to the Cumberland province of se. KY, e. TN, ne. AL. Also sporadic in the Blue Ridge of e. TN and adjacent NC, n. GA. Occurs in balds, open or wooded peaks and ridges, and mesic forests. Flowers not fragrant, yellow to red (usually reddish), appearing June–July after leaves are nearly mature. Twigs slightly hairy; buds greenish to light brown, darker-margined. (*R. bakeri* Lemmon)

15. *Rhododendron eastmanii* Kron & Creel   EASTMAN AZALEA

A tall shrub, to 5m. Rare and localized in the coastal plain of central SC. Known only from a few wooded slopes with nearly circumneutral soils in 2 counties. Flowers slightly fragrant, white with yellow internal blotch, appearing in May after leaves are mature; filaments reddish. Twigs hairy; buds light brown.

16. *Rhododendron prunifolium* (Small) Millais   PLUMLEAF AZALEA

A shrub, to 5m tall. Rare and localized in the coastal plain and lower Piedmont of extreme w. GA and adjacent AL. Occurs in wooded ravines, slopes, and along streams. Flowers not fragrant, orange to red, appearing June–Sept after leaves are mature. Twigs glabrous; buds light brown.

17. *Rhododendron flammeum* (Michx.) Sarg.   OCONEE AZALEA

A shrub, rarely over 2m tall. Sporadic in the Piedmont and coastal plain of GA and adjacent SC. Occurs in dry woodlands to mesic sites near river bluffs or streams. Flowers not fragrant, yellow to red with orange internal blotch, appearing in Apr as leaves emerge; filaments pink. Twigs hairy; buds brownish, with sharp, dark scale tips. [*R. speciosum* (Willd.) Sweet]

18. *Rhododendron alabamense* Rehd.   ALABAMA AZALEA

A tall shrub, to about 4m. Native across most of AL and adjacent parts of MS, FL, GA, and across central TN. Occurs in upland woods, slopes, and along streams. Flowers fragrant, white with a yellow internal blotch, appearing Apr or May as leaves emerge; filaments usually white. Twigs usually hairy; buds brownish.

19. *Rhododendron periclymenoides* (Michx.) Shinners   PINXTER-FLOWER

A shrub, to 5m tall. Widespread across the e. states of the SE, mainly from WV to central TN and MD to central SC. Also scattered in GA, AL. Occurs in upland woods and along streams. Flowers fragrant, white to pink, appearing Apr–May as leaves emerge. Twigs slightly hairy to glabrous; buds light brown. [*R. nudiflorum* (L.) Torr.]

Rhodotypos scandens (Thunb.) Makino   JETBEAD   Family Rosaceae   *Figure 448*

An exotic shrub, to about 2m in height. Sparingly naturalized, spreading from cultivation through seed. Twigs grayish brown, often with tip dying back; terminal bud lacking; lateral buds with 8 or more dark-tipped scales, sometimes slightly stalked and collateral buds partially covered by a wide basal scale. Bundle scars 3; leaf scars connected by a ridge or line; dried stipules often persist. The shiny black seed are borne in a dry covering, 2 to 4 together in a terminal cluster. Native to Japan, China.

Rhus L. SUMAC Family Anacardiaceae

Five species are treated here as native to the SE. Terminal buds are lacking; laterals are naked, usually hairy, and sometimes partially sunken in the twigs. Leaf scars bear several bundle scars. The fruit is a red drupe with a thin, often hairy skin and single brown seed; borne in terminal clusters (or lateral in 1 species). Plants are dioecious, with fruits formed only on plants with functional pistillate flowers. In some texts, the related poisonous species may be included in this genus, but see *Toxicodendron* in this book for discussion of those plants. No species of *Rhus* treated here is poisonous or causes the same severe skin reactions as in *Toxicodendron*.

1. Catkin buds present, or lateral buds completely hidden between raised leaf scar and twig
   1. *R. aromatica*
1. Catkin buds lacking; lateral buds hairy and visible within or above leaf scar
   2. Twigs glabrous   2. *R. glabra*
   2. Twigs hairy
      3. Leaf scars surround ½ or less of bud circumference; twigs rarely over 1cm diam.
         3. *R. copallinum*
      3. Leaf scars surround ¾ or more of bud circumference; twigs often over 1cm diam.
         4. Twigs zigzag or knobby due to swollen nodes; rare shrub, under 1m tall   4. *R. michauxii*
         4. Twigs not as above; a widespread tall shrub or tree, over 2m tall   5. *R. typhina*

1. *Rhus aromatica* Ait.   FRAGRANT SUMAC   Figure 449

A rhizomatous or dense spreading shrub, to 2m tall, often forming colonies by root sprouts. Widespread in the SE from the Piedmont of VA to nw. FL and regions to the west and north. Occurs in dry woodlands, usually in sandy or rocky soils and frequently over calcareous rock. The catkin buds are borne near twig tips on mature plants; twigs lightly hairy or nearly glabrous. Bark gray to brown, thin, lenticellate, shallowly fissured or with thin scales at base, smooth beyond; inner bark aromatic. Fruit hairy, borne in small clusters along the twig or on the branchlets. Leaves trifoliate. Flowers appear before leaves emerge. This species has several named varieties, with 3 in the SE: the typical var. *aromatica* being most common; var. *illinoensis* (Greene) Rehd. of IL, MO has very hairy twigs and leaves; var. *serotina* (Greene) Rehd. of MO, AR, TX has smaller leaves, and flowering occurs with the emergence of leaves.

2. *Rhus glabra* L.   SMOOTH SUMAC   Figure 450

A large shrub, to 6m tall, usually forming colonies by root sprouts. Widespread across most of the SE, except not common in the outer coastal plain. Common in the Appalachians and interior uplands. Typical of field edges, roadsides, forest edges. Twigs light brown, stout, glabrous, usually somewhat angular in cross section, with a large, soft pith. Bark thin, lenticellate, mostly smooth. The pyramidal fruit clusters persist into winter, drupes with short, appressed hairs on the surface. Leaflets white beneath.

3. *Rhus copallinum* L.   WINGED SUMAC   Figure 451

A shrub or small tree, to 8m tall, sometimes forming colonies of root sprouts around older, established individuals. Widespread across the entire SE; found in dry woodlands, clearings, roadsides. Twigs hairy, smaller and more rounded in cross section than the 3 other pinnate-leaved sumacs. Bark thin, mostly smooth but conspicuously lenticellate. Fruit clusters often arched or crooked. Leaves pinnately compound, with a winged rachis. Several varieties have been named; 2 in the SE that seem distinct are var. *latifolia* Engler, with wider leaflets rounded on 1 side of the base, mostly found in the w. half of the SE, and var. *lanceolata* Gray, with narrow falcate leaflets, of e. TX west to NM.

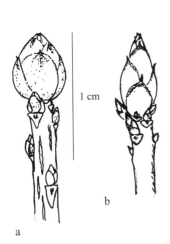

Figure 446. a: *Rhododendron vaseyi;* b: *R. arborescens*

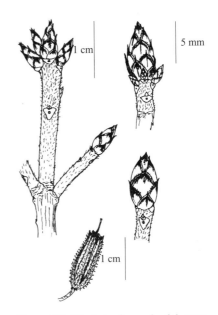

Figure 447. *Rhododendron calendulaceum*

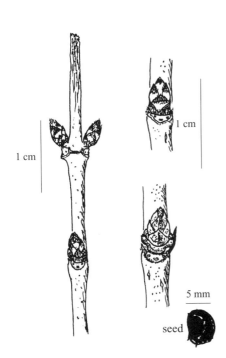

Figure 448. *Rhodotypos scandens*

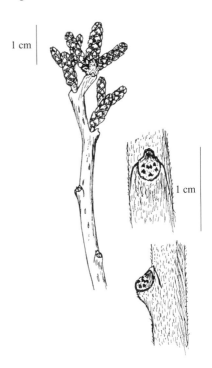

Figure 449. *Rhus aromatica*

4. Rhus michauxii Sarg.   MICHAUX SUMAC   *Figure 452*

A short, rhizomatous shrub, to 1m tall. Rare; known only from a few localities in the Piedmont and coastal plain of VA to GA. Occurs in open woodlands and areas historically susceptible to fire in organically enriched sandy or clayey soils of basic or circumneutral nature. Twigs with conspicuous lenticels; leaf scars very nearly encircling buds.

5. Rhus typhina L.   STAGHORN SUMAC   *Figure 453*

A tall shrub or small tree, with an open, coarse crown of ascending branches, to 10m tall. The most treelike of the sumacs, often with a long, clear trunk. Distributed mostly in the Blue Ridge from MD to GA, and interior regions westward and northward. Twigs distinctly hairy. Bark grayish, smooth except for conspicuous lenticels; inner bark yellow-brown. Fruits covered with long hairs. Leaves pinnately compound, whitened beneath. [*R. hirta* (L.) Sudw.]

Ribes L.   GOOSEBERRY   Family Grossulariaceae

Fourteen species are treated here as native or naturalized in the SE. The 3 exotic species included are sometimes cultivated for fruit production, particularly in more n. states, and seedlings may rarely become established in the vicinity of plantings. Eradication of many cultivated and naturally occurring *Ribes* plants was done in the 1930s in an effort to control the spread of white pine blister rust, a fungal disease for which *Ribes* serves as an alternate host. The genus is divided into at least 2 groups, the gooseberries and the currants. The gooseberries of subgenus *Grossularia* (Mill.) Richard usually have spur shoots and have nodal spines on the twigs; the fruit are borne singly or in groups of up to 4 per a short-branched raceme. The currants of subgenus *Ribesia* Berl. are spineless; spur shoots are less commonly produced; and fruits are often 4 or more in a more elongate raceme. The bristly currant (*R. lacustre*) is often considered to belong to a 3d subgenus, *Grossularioides* Jancz., with characters of both currants and gooseberries. The fruits of all are berries, rarely persistent into winter. Terminal buds are present, with thin, imbricate scales; bundle scars 3.

1. Nodal spines lacking; no internodal prickles (currants)
  2. Twigs or buds slightly hairy; branchlets and main stems ridged, or roughened by lenticels; fruit yellow or black
    3. Inner bark not distinctly odorous; no resin glands on bud scales; of w. states
                                                                                            1. *R. odoratum*
    3. Inner bark strongly aromatic; resin glands on bud scales (use lens); of ne. uplands
      4. Twigs finely lined; branchlets ridged; buds short-pointed; native of mountains
                                                                                       2. *R. americanum*
      4. Twigs not lined; branchlets very lenticellate; buds sharply pointed; exotic   3. *R. nigrum*
  2. Twigs and buds glabrous; branchlet bark smooth, shreddy, or peeling; fruit red or dark red
    5. Buds reddish; "skunky" odor in inner bark; bark smooth; fruit glandular-bristly
                                                                                          4. *R. glandulosum*
    5. Buds grayish to brown; inner bark not so aromatic; bark usually peeling or shreddy; fruit smooth
      6. Buds grayish; bark peeling; habit sprawling; native of high mountains   5. *R. triste*
      6. Buds brownish; bark finely shreddy; habit stiffly upright; exotic   6. *R. sativum*
1. Nodal spines present (all gooseberries except first 1)
  7. Internodes densely bristly; scraped bark with rank odor   7. *R. lacustre*
  7. Internodes sparsely bristly or not bristly; no rank odor
    8. Nodal spines stout, usually 1–2cm long; exotic, or west of Appalachians
      9. Spines on older stems stout, 3-parted; bark gray-striped; exotic   8. *R. grossularia*
      9. Spines and bark not as above; native west of Appalachians   9. *R. missouriense*
    8. Nodal spines slender or mostly less than 1cm; native to Appalachians or areas north or south

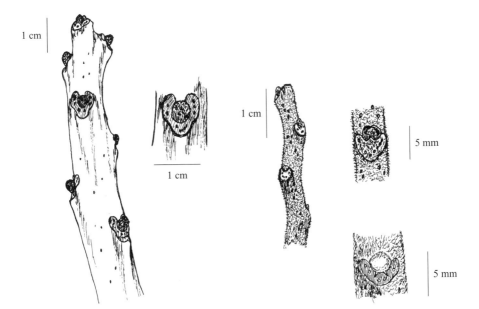

Figure 450. *Rhus glabra*

Figure 451. *Rhus copallinum*

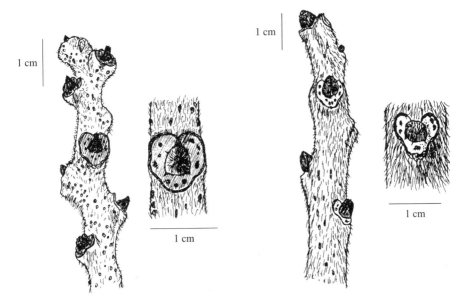

Figure 452. *Rhus michauxii*

Figure 453. *Rhus typhina*

10. Fruit spiny or bristly
   11. Twigs gray; fruit bristles gland-tipped; leaves unfold in autumn; rare, of coastal plain of SC, FL   10. *R. echinellum*
   11. Twigs yellow-brown; fruit bristles not gland-tipped; leaves deciduous in autumn; common in Appalachians and areas north and west   11. *R. cynosbati*
10. Fruit smooth
   12. Spines reddish; leaves punctate-glandular beneath; of mountains and Piedmont of GA, AL   12. *R. curvatum*
   12. Spines more grayish; leaves not as above; of Appalachians and areas north
      13. Prickles commonly present on internodes; rare n. species barely reaching SE   13. *R. hirtellum*
      13. Prickles absent; common Appalachian species   14. *R. rotundifolium*

1. *Ribes odoratum* Wendl.   BUFFALO CURRANT

A bushy, upright shrub, to 2m tall. Distributed mostly in the Plains region, reaching the SE in MO, AR, TX. Occurs in rocky or alkaline soils. The aromatic yellow flowers account for the name *odoratum*. Fruit about 1cm, usually black and smooth, occasionally yellow. (*Ribes aureum* Pursh var. *villosum* DC.)

2. *Ribes americanum* P. Mill.   EASTERN BLACK CURRANT

An upright shrub, to 1.5m tall, mostly n. in distribution. Ranges into the SE in the higher mountains of WV, VA. Occurs in moist to wet forests and marshy soils. Twigs usually finely lined, more distinctly ridged on branchlets or older stems; bud scales usually have a few resin dots and hairs; inner bark odorous. Leaves resin-dotted on both sides, though more distinctly beneath. Fruit black, smooth.

3. *Ribes nigrum* L.   EUROPEAN BLACK CURRANT

A stiffly upright shrub, to 2m tall. Native to Eurasia; planted occasionally and rarely naturalized. Twigs brown or red-brown; lateral buds only slightly smaller than terminal bud, all sharply pointed. Bark dark, roughened by many conspicuously raised lenticels, with strong odor when scraped. Fruit smooth, black.

4. *Ribes glandulosum* Grauer   SKUNK CURRANT   *Figure 454*

A sprawling shrub or with short erect stems, to 1m tall. Distributed mostly across n. N. Amer., reaching the SE in the high elevations of the Appalachians in WV, VA, NC, TN. Occurs in the boreal forest zone amid moist to wet rocky slopes and seepages. Twigs glabrous, with rank odor when scraped; terminal buds much larger than laterals, with reddish scales. Fruits glandular-bristly, purplish-black.

5. *Ribes triste* Pall.   SWAMP RED CURRANT

A sprawling shrub, rarely to 1m tall, often rooting from trailing stems and forming small colonies. Widely dispersed across n. regions of N. Amer. and Asia. Ranges into the SE in the high elevations of WV. Occurs in cool, moist to wet boreal and subalpine slopes, rocky outcrops, swamps, bogs. Buds grayish; bark dark red-brown, often exfoliating. Fruit smooth, red, somewhat firm to hard.

6. *Ribes sativum* Syme   RED GARDEN CURRANT

An upright shrub, to 1.5m tall. Native to Eurasia; planted occasionally and naturalized rarely. Twigs or bud scales sometimes hairy, especially early in the season. Leaves cordate or nearly so at base. Fruit smooth, red, with a 5-angled remnant of the flower at its base. Similar is *R. rubrum* L. (n. red currant), a native of Eurasia that is glabrous, leaves mostly truncate at base, and lacking the 5-angled structure to the flower receptacle; rarely seen.

7. *Ribes lacustre* (Pers.) Poir.   BRISTLY SWAMP CURRANT   *Figure 455*

A sprawling or spreading shrub, to 1m tall. Widely distributed across n. N. Amer., ranging into the SE in the higher elevations of WV, VA. Occurs in this region in moist to wet woodlands and swampy soils. Twigs slender, with clustered slender nodal spines to 1cm long and numerous internodal bristles. Fruits bristly, purplish to black.

8. *Ribes grossularia* L.   EUROPEAN GOOSEBERRY

An upright shrub, to 1m tall. A native of Europe; planted occasionally and naturalizing rarely. Twigs grayish; spines stiff and strong, 10–15mm, often 3-parted on robust stems. Bark often prickly, with longitudinal strips of bark rendering a gray-striped pattern. Fruits smooth to glandular-bristly, green to yellow or red. Many fruit selections of garden gooseberries have originated from this species. (*R. uva-crispa* L.)

9. *Ribes missouriense* Nutt. ex Torr. & Gray   MISSOURI GOOSEBERRY

An upright shrub, to 2m tall. Distributed mostly west of the Appalachians, from WV to MO, south to TN, AR. Occurs in dry or moist woodlands, usually in rocky or calcareous soils. Twigs grayish; spines stiff and strong, mostly 1–2cm long; bark smooth to slightly exfoliating. Fruits smooth, red to purplish after maturity.

10. *Ribes echinellum* (Coville) Rehd.   MICCOSUKEE GOOSEBERRY   *Figure 456*

A spreading shrub, to 1m tall, forming small colonies. Rare; known only from 2 coastal plain areas of SC and n. FL. Occurs in moist hardwood forests, where it will leaf out over winter months due to greater availability of sunlight at this time of year and the mild climate. Twigs slender; spines slender, 5–15mm long; bark grayish. Fruits very spiny and glandular, greenish at first, turning reddish to nearly black after maturity.

11. *Ribes cynosbati* L.   PRICKLY GOOSEBERRY

A low or weakly erect shrub, to 1.5m tall, often with arching branches, occasionally sprawling over rocks. A widely distributed species from the NE down the Appalachians to n. GA, west to OK, MO. Typically in moist soils amid rocks under mesic forests. Twigs slender; spines slender, 5–10mm long; bark grayish. Fruits green to dark wine red, with prickles on the surface.

12. *Ribes curvatum* Small   GRANITE GOOSEBERRY

A low shrub, to 1m tall, often sprawling over rocks and forming small colonies. Distributed from GA to TX in mountains, Piedmont uplands, and rocky slopes or woods. Twigs slender; spines slender, about 5mm long; bark reddish-brown. Fruits smooth, greenish.

13. *Ribes hirtellum* Michx.   SMOOTH GOOSEBERRY

A spreading, mostly upright shrub, to 1m tall. Mainly a n. species that barely reaches the SE in the mountains of WV. Occurs in moist, rocky woods or openings in high elevations. Twigs often lacking spines on upper nodes; spines slender, 3–8mm long. Bark commonly shreddy or exfoliating. Fruits smooth, greenish becoming purplish to black after maturity.

14. *Ribes rotundifolium* Michx.   ROUNDLEAF GOOSEBERRY   *Figure 457*

A spreading, mostly upright shrub, to 1m tall. Native to the Appalachians from WV and VA to TN and NC, northeast to MA. Occurs in higher elevations in moist soils of rocky outcrops, balds, rocky slopes. Twigs gray to brown, slender; spines slender or lacking on endmost nodes, 2–5mm long. Bark brown, slightly exfoliating. Fruit smooth, green to purplish.

Robinia L.   LOCUST   Family Fabaceae

Three species are native in the SE, but the 2 shrubby, pink-flowered species have several important varieties. Juvenile plants usually bear paired spines at the nodes (modified stipules), whereas twigs from mature plants, especially in *R. pseudoacacia,* may be spineless. Twigs of all lack terminal buds, and the lateral buds are sunken within the leaf scar or just above it. The fruit is a flattened legume, 10–15mm wide, 5–15mm long. Bark of the shrubby, pinkish-flowered species is thin, dark, and smooth except for small, raised lenticels. A majority of the shrubby species seem to have ranges centering in the s. Appalachians, where they are suspected to have hybrid origins. Here they continue to hybridize and occasionally develop local clones or colonies in disturbed areas where there is abundant sunlight.

1. Twigs smooth and glabrous on internodes; bark brown, thickly furrowed; a tree
                               1. *R. pseudoacacia*
1. Twigs with glands, hairs, bristles, or nearly smooth, and bark thin, grayish to blackened; shrubs or small trees
 2. Twigs viscid (covered with sticky glands)   *R. viscosa*
  3. Glands sessile or nearly so   2. var. *viscosa*
  3. Glands stalked, some to 1 or 2mm long   3. var. *hartwegii*
 2. Twigs with hispid (bristlelike) hairs   *R. hispida*
  4. Branchlets conspicuously hispid
   5. Plants sterile (no fruits)   4. var. *hispida*
   5. Plants fertile   5. var. *fertilis*
  4. Branchlets sparsely hispid or not hispid
   6. Plants low, rarely to 1m tall, of dry or sandy soil, mostly coastal plain   6. var. *nana*
   6. Plants reaching heights over 1m, of uplands in Appalachians
    7. Stipular spines short or lacking; plants sterile (no fruit)   7. var. *rosea*
    7. Stipular spines slender, usually present; plants fertile   8. var. *kelseyi*

1. *Robinia pseudoacacia* L.   BLACK LOCUST   *Figure 458*

A tree, to 25m tall on optimum sites, or mature at heights of 3m on harsh sites. Native to the Appalachian region from WV to AL, and west to MO, AR. Through extensive cultivation, now naturalized widely over most of the SE. Optimum habitat is mesic slopes and coves of middle elevations of the Appalachians. Occurs on a wide variety of sites from dry to moist, especially common along roads, field edges, and disturbed forests. Twigs brown, glabrous, with paired stipular spines on juvenile plants. Bark gray to light brown, deeply furrowed; inner bark with odor similar to fresh green beans. Legume to 10cm long, glabrous.

2. *Robinia viscosa* Vent. var. *viscosa*   CLAMMY LOCUST   *Figure 459a*

A shrub or small tree, to 10m tall, usually forming colonies from root sprouts. Native to the Appalachian region from WV, VA to GA, AL. Scattered in the Piedmont and other areas across the SE due to naturalization after cultivation. Twigs covered with sticky glands. Legume to 10cm, glandular-hispid.

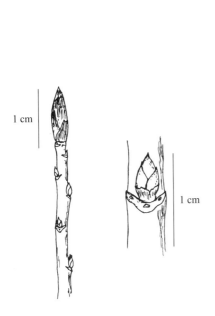

Figure 454. *Ribes glandulosum*

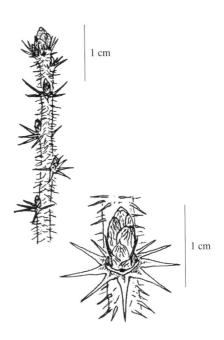

Figure 455. *Ribes lacustre*

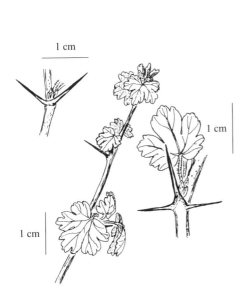

Figure 456. *Ribes echinellum*

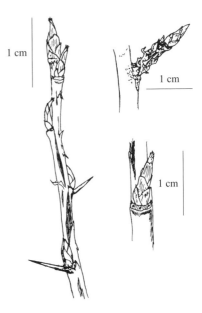

Figure 457. *Ribes rotundifolium*

3. *Robinia viscosa* var. *hartwegii* (Koehne) Ashe   HARTWEG LOCUST   *Figure 459b*

A large shrub or small tree, to about 5m tall, forming colonies from root sprouts. A rare native of the Blue Ridge of sw. NC and adjacent SC. Occurs on rock outcrops and forest edges in high elevations, known from a relatively small number of sites. Twigs covered by short-stalked sticky glands; stipular spines weak and slender. Legume to 14mm long, glandular-hispid.

4. *Robinia hispida* L. var. *hispida*   BRISTLY LOCUST   *Figure 459c*

A shrub, to 4m tall, but more often shorter and forming colonies from root sprouts. Original native range is uncertain, likely being the s. Appalachians. Encountered across other provinces of the SE occasionally, due to supposed naturalization. Twigs densely bristly, with many bristles remaining to the second year (branchlets bristly). Stipular spines short and weak. The typical var. *hispida* is apparently sterile, not known to produce fruit.

5. *Robinia hispida* var. *fertilis* (Ashe) Clausen   ARNOT BRISTLY LOCUST

Similar in nearly all respects to the above entity, except plants are fertile, producing fruit. Tends to produce colonies of stems 1–2m tall. Legume to 8cm long, densely bristly.

6. *Robinia hispida* var. *nana* (Elliott) DC.   DWARF BRISTLY LOCUST

A low, rhizomatous shrub, usually 3–6dm tall, though occasionally reaching 1m height. Native to the Piedmont and coastal plain from NC to AL. Typically in dry, sandy or rocky soils, especially in the Sandhills region. Twigs sparsely bristly to essentially glabrous; stipular spines slender, weak. Rarely fruiting, with hispid legume to 4cm. [*R. elliottii* (Chapm.) Ashe ex Small]

7. *Robinia hispida* var. *rosea* Pursh.   BOYNTON LOCUST

A shrub, to 3m tall, sometimes rhizomatous. Native to the s. Appalachian region of NC and TN to GA, AL. Uncommon and sporadic along roads, ridges, and sparsely forested slopes. Twigs glabrous or with a few sparse bristles; stipules short, barely rigid enough to be considered spinose. Apparently sterile, as no fruits are known to be produced. (*R. boyntonii* Ashe)

8. *Robinia hispida* var. *kelseyi* (Cowell ex Hutchinson) Isely   KELSEY LOCUST *Figure 459d*

A shrub, to 4m tall, usually forming colonies by root sprouts. Native to the s. Blue Ridge of sw. NC and adjacent SC. Rare and sporadic on mountain ridges, rock outcrops, and roadsides. Twigs glabrous or with a few bristles; stipular spines slender, weak. Legume to 6cm long, glandular-hispid.

Rosa L.   ROSE   Family Rosaceae

Of the 18 species included here as native or naturalized in the SE, 10 are exotics. Native roses are, for the most part, shrubby or slenderly erect plants that are often rhizomatous. Exotic species are more often clambering or vinelike. The twigs of all are greenish and prickly and/or bristly. Prickles are usually paired at the nodes and additionally scattered on internodes in some species. Bundle scars 3; terminal buds usually lacking. The fruit is a fleshy receptacle bearing the achenes (seed) inside; called a "hip." Leaves are alternate and compound. The following key takes into account characters of leaves and stipules where there are insufficient differences in twigs for winter identification.

1. Stems climbing, clambering, or long and arching (all exotics except for *R. setigera*)
    2. Habit mostly climbing or clambering over the ground or other plants.

3. Stems tomentose   1. *R. bracteata*
   3. Stems glabrous
      4. Leaflets mostly 7 to 11; stipules distinctly toothed   2. *R. wichuraiana*
      4. Leaflets mostly 3 or 5; stipules remotely toothed
         5. Leaves evergreen, glabrous; leaflets 3   3. *R. laevigata*
         5. Leaves deciduous; stipules glandular-hispid; leaflets 3 or 5   4. *R. setigera*
   2. Habit semierect, with long, arching stems
      6. Stipules fringed and ragged-toothed; fruit oval, less than 1cm long   5. *R. multiflora*
      6. Stipules finely toothed or entire; fruit elliptic, over 1cm long
         7. Leaves glabrous and nonglandular beneath; stipules not glandular-ciliate   6. *R. canina*
         7. Leaves pubescent or glandular beneath; stipules glandular-ciliate
            8. Leaflets obtuse, both sides glandular; sepals usually missing from fruit; styles glabrous   7. *R. micrantha*
            8. Leaflets acute, glandular only below; sepals often persist on fruit; styles pubescent
                   8. *R. eglanteria*
1. Stems stiffly erect or at least not long-arching
   9. Leaves rugose (deeply impressed vein network above)
      10. Twigs tomentose or densely pubescent and bristly   9. *R. rugosa*
      10. Twigs not tomentose, but with numerous unequal bristles   10. *R. gallica*
   9. Leaves not rugose
      11. Nodal prickles broad-based and stout, recurved
         12. Fruit elliptic, over 1.5cm long; an exotic   11. *R. damascena*
         12. Fruit rounded, under 1.5cm; native species
            13. Leaves coarsely toothed beyond middle (includes 5 to 15 teeth)   12. *R. virginiana*
            13. Leaves finely toothed beyond middle (includes 12 to 25 teeth)   13. *R. palustris*
      11. Nodal prickles slender, straight, or lacking
         14. Bristles present on twigs or stems
            15. Nodal prickles usually present; common low shrub of dry soils   14. *R. carolina*
            15. Nodal prickles often lacking; uncommon in SE
               16. Leaflets usually 9 or 11; fruit rounded, 10–15mm; in w. sections of SE
                      15. *R. arkansana*
               16. Leaflets usually 5 or 7; fruit pear-shaped, 15–20mm; a n. rose barely reaching SE
                      16. *R. acicularis*
         14. Bristles lacking; prickles weak; stem sometimes prickly only at stem base
            17. Prickles sometimes present at nodes; leaflets 7 to 11, narrow and glossy; of AR, OK, TX   17. *R. foliolosa*
            17. Prickles mostly lacking except at stem base; leaflets usually 5 or 7, dull; n. and midwestern species   18. *R. blanda*

1. *Rosa bracteata* J. C. Wendl.   MCCARTNEY ROSE   *Figure 460a*

A clambering shrub or semivine with evergreen or semievergreen leaves. A native of China; naturalized sporadically in the SE, mostly Piedmont and coastal plain. Twigs hairy, stipitate-glandular, with strong, curved prickles. Leaflets 5 to 9, thick, glossy above, remotely toothed. Stipules deeply cut (pectinate). Fruit to 3cm long, hairy, globose; sepals ovate, mostly entire, persistent. A cluster of toothed bracts usually persist at base of fruit.

2. *Rosa wichuraiana* Crepin.   MEMORIAL ROSE   *Figure 460b*

A trailing or vinelike shrub with semievergreen leaves. Native to Asia; naturalized and widely scattered across the SE. Typically seen along roads and railways and in waste places. Twigs glabrous, with strong, straight to curved prickles. Leaflets 5 to 9, lustrous. Stipules deeply toothed. Fruit ovoid, about 1cm long, smooth; sepals triangular.

3. *Rosa laevigata* Michx. CHEROKEE ROSE   *Figure 460c*

A vinelike shrub often clambering over other plants; with evergreen leaves. Native to China; naturalized widely across the SE, mostly coastal plain. Twigs glabrous, with stout, curved prickles. Leaflets 3, lustrous. Stipules remotely toothed to entire. Fruit pyriform, to 4cm long, bristly; sepals entire, mostly triangular but sometimes with an expanded tip.

4. *Rosa setigera* Michx. PRAIRIE ROSE

An arching or semivine-like shrub with deciduous leaves. Native to a broad area of the SE, from west of the Appalachians to FL and TX. Occurs in moist woodlands, streamsides, and thickets. Twigs glabrous, with sparse, curved prickles. Leaflets 3 or rarely 5, fairly lustrous above. Stipules nearly entire. Fruit subglobose, to 1cm, remotely stipitate-glandular; sepals long-pointed, mostly entire.

5. *Rosa multiflora* Thunb. ex Murr. MULTIFLORA ROSE   *Figure 460d*

A spreading shrub, or with climbing or clambering stems. Native to Japan and Korea; naturalized widely across much of the SE. Typically seen along roads, field edges, fencerows. Twigs glabrous, with stout, curved prickles. Leaflets 7 or 9, thin, not lustrous. Stipules deeply cut (pectinate). Fruit subglobose to ellipsoid, smooth, 6–9mm long; sepals lanceolate, long-tipped, sometimes lobed, often deciduous by fruit maturation.

6. *Rosa canina* L. DOG ROSE

A spreading or suckering shrub, with arching stems and deciduous leaves. Native to Europe; sporadically naturalized in some parts of the SE, mainly in the n. portions. Twigs glabrous, with stout, hooked prickles. Leaflets 5 or 7, glandless, smooth, and bright green above. Stipules glandular-serrate, narrow on lower nodes. Fruit ellipsoid, to 2cm long, smooth; sepals lobed, usually deciduous by fruit maturation.

7. *Rosa micrantha* Borrer ex Smith SMALL-FLOWERED ROSE

An upright, spreading or arching shrub with deciduous leaves. Native to Europe; naturalized over the n. and more interior portions of the SE. Typically seen in pastures, fencerows, road banks. Twigs glabrous, with curved prickles of uniform shape and length. Leaflets 5 or 7, mostly ovate, acute at tip; glandular-serrate on margin, glandular on lower surface. Stipules glandular-ciliate. Fruit subglobose to ovoid, to 2cm long, sparsely glandular-hispid near base; sepals entire to lobed, glandular-ciliate, usually deciduous at maturation of fruit.

8. *Rosa eglanteria* L. SWEETBRIAR ROSE

An upright, spreading shrub with deciduous leaves. Native to Europe; naturalized sporadically across the SE. Most similar to above species. Twigs glabrous, with stout, curved prickles and some intermixed slender prickles or bristles. Leaflets 5 or 7, oval or round, obtuse at tip; glandular-serrate on margin, glandular on both surfaces; with applelike odor when crushed. Stipules and fruit similar to above species, but sepals usually more persistent on mature fruit.

9. *Rosa rugosa* Thunb. RUGOSA ROSE   *Figure 461*

A stiffly erect, rhizomatous shrub, to 2m tall, more commonly 1m, typically forming dense clumps or colonies. Native to e. Asia; occasionally spreading from cultivation. Twigs and stems hairy, densely prickly and bristly, the prickles mostly straight or slightly curved. Leaflets 5 to 9, dark and lustrous, fairly thick, rugose above, margin curved downward. Stipules nearly entire, hairy. Fruit depressed-globose, to 25mm wide, smooth; sepals long-tipped, persistent.

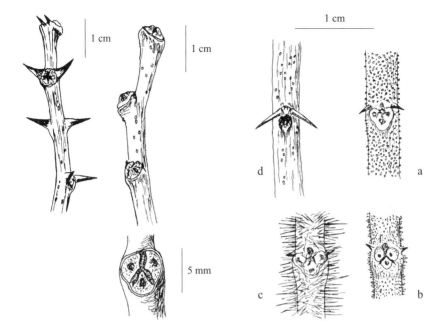

Figure 458. *Robinia pseudoacacia*

Figure 459. a: *Robinia viscosa* var. *viscosa;* b: *R. viscosa* var. *hartwegii;* c: *R. hispida* var. *hispida;* d: *R. hispida* var. *kelseyi*

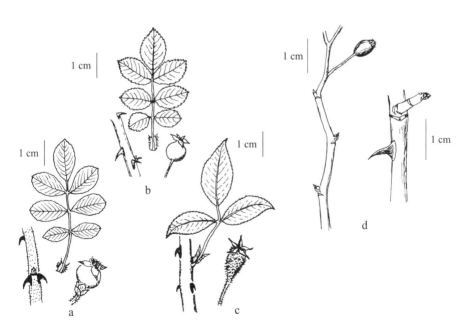

Figure 460. a: *Rosa bracteata;* b: *R. wichuraiana;* c: *R. laevigata;* d: *R. multiflora*

10. *Rosa gallica* L.   FRENCH ROSE

A stiffly erect, rhizomatous shrub rarely reaching 1m tall, forming colonies. Native to Europe; occasionally spreading from cultivation. Twigs stipitate-glandular, most of stems densely prickly and bristly, the prickles stout, curved. Leaflets usually 5, rarely 3 or 7, dark green, rugose above, margin curved downward. Stipules narrow, glandular-ciliate. Fruit subglobose, slightly glandular-hispid; sepals sometimes lobed, often deciduous from mature fruit.

11. *Rosa damascena* P. Mill.   DAMASK ROSE

A stiffly erect, rhizomatous shrub, to 2m tall. Native to Asia; occasionally spreading from cultivation. Twigs nearly glabrous, but older parts of stems have many stout, hooked prickles and are sparsely glandular-hispid. Leaflets 5 or 7, slightly rugose. Stipules ciliate or slightly glandular. Fruit obovoid, to 25mm long, bristly; sepals long-tipped, glandular.

12. *Rosa virginiana* P. Mill.   VIRGINIA ROSE

A slender, spreading shrub, mostly clumplike or weakly rhizomatous. Native from VA westward across uplands and the interior to MO; of sporadic occurrence in moist to dry sites. Twigs and stems with strong often curved prickles at nodes, and few to no prickles on internodes. Leaflets mostly 7 to 9, coarsely toothed. Stipules glandular, widened near tips on upper nodes. Fruit subglobose, to 15mm wide, smooth; sepals elongate-tipped.

13. *Rosa palustris* Marsh.   SWAMP ROSE *Figure 462a*

A slender, rhizomatous shrub, to 2m tall. Widespread across much of the SE in moist to wet soils. Twigs and stems with curved prickles at nodes, usually no prickles on internodes. Leaflets mostly 7 to 9, finely toothed. Stipules narrow, troughlike, margin nearly entire. Fruit depressed-globose to ovoid, to 12mm, smooth or sparsely stipitate-glandular; sepals with elongated tips.

14. *Rosa carolina* L.   CAROLINA ROSE *Figure 462b*

A slender, rhizomatous shrub, rarely to 1m tall, usually 2–5dm. Widespread throughout the SE, usually in dry upland soils. Twigs and stems with slender, fairly straight prickles on nodes and internodes. Leaflets mostly 5 to 7, coarsely toothed. Stipules narrow, entire to glandular-toothed. Fruit subglobose, to 1cm, smooth or sparsely stipitate-glandular; sepals with elongated tips.

15. *Rosa arkansana* T. C. Porter   DWARF PRAIRIE ROSE

A small, slender, rhizomatous shrub, rarely over 5dm tall. Native in the westernmost sections of the SE, where it occurs in prairies, rocky and dry soils, and scrubland from MO to TX. Twigs and stems with slender, straight prickles on nodes and internodes, intermixed with bristles. Leaflets mostly 9 to 11, coarsely but sharply serrate. Stipules entire to glandular-toothed near tip, hairy. Fruit subglobose, to 15mm, mostly smooth; sepals elongated, usually persistent on fruit.

16. *Rosa acicularis* Lindl.   BRISTLY ROSE

A slender, rhizomatous shrub, to 1m tall. Native across n. N. Amer. and Eurasia, ranging into the SE only in WV. Twigs and stems covered with numerous slender, weak prickles and bristles. Leaflets mostly 5 or 7, coarsely toothed; rachis glandular. Stipules glandular-ciliate on margin. Fruit pyriform, to 2cm long, slightly beaked below sepals, which are often erect and persistent.

17. *Rosa foliolosa* Nutt. ex Torr. & Gray   WHITE PRAIRIE ROSE

A low, rhizomatous shrub, rarely over 5dm tall. Native to AR, OK, TX, in sandy or gravelly soils of prairies and hills. Twigs and stems weakly prickly to nearly glabrous, the prickles slightly curved. Leaflets mostly 7 or 9, very narrow, finely toothed, lustrous above. Stipules narrow. Fruit subglobose, to 8mm, smooth to sparsely stipitate-glandular; sepals entire; deciduous before fruit matures.

18. *Rosa blanda* Ait.   SMOOTH ROSE

A slender, rhizomatous shrub, to 2m tall, usually about 1m. Forms clumps or close colonies on prairies, rocky summits or hills, and sandy shores or dunes. Native primarily to the NE; ranging into the southeast region covered by this book perhaps in MO. Twigs and stems have a few scattered slender prickles or bristles near the base (not at nodes) or are entirely glabrous. Leaflets 5 or 7, coarsely toothed. Stipules mostly entire, hairy. Fruit subglobose, smooth, to 15mm.

Rubus L.   BRAMBLE   Family Rosaceae

Twenty-two taxa are treated here for the SE: 4 native raspberries, 3 exotic raspberries, 4 native dewberries, 8 native blackberries, and 3 exotic blackberries. The majority of species produce prickles or bristles on the stems and leaves, and all have in common the aggregate type of fruit that is derived from several ovaries in 1 flower. Basal portions of petioles often persist at the node in winter, covering the leaf scar. The genus is divided into several subgenera, with these species included under the following: subgenus *Anaplobatus* Focke, with unarmed stems and simple leaves (*R. odoratus* only); subgenus *Cylactis* (Raf.) Focke, with unarmed, mostly herbaceous stems and compound leaves (*R. pubescens* only); subgenus *Idaeobatus* Focke, with the fruit hollow after separating freely from the receptacle (true raspberries); and subgenus *Eubatus* Focke, with fruit not separating from the receptacle (blackberries). The latter group is further separated into 2 general types based on growth habit; the dewberries, which creep and have few or no erect stems except for fruit-bearing side shoots, and the "true" blackberries, with erect or arching stems at least 5dm or more in height. The blackberries are an extremely variable and complex group, with hybridization and apomyxis contributing to taxonomic problems with species designation. Clonal populations are sometimes built up locally by plants of uncertain origin and intermediate characters of some of the main species treated here. All tend to form colonies through suckering, rhizomes, or rooting of stems that arch and touch the soil. Most are early successional in habitat preference, tending to establish themselves on disturbed lands such as roadsides, fields, forest openings and edges, and waste places. Fruits of all are edible, though palatability varies among the species and sometimes among local populations and habitats. Not all species can be identified in winter using twigs alone, so foliage features are used when necessary. Many of the species produce elongating stems called primocanes, which typically do not flower in the same growing season they appear. Flowering in such cases occurs with the onset of secondary shoots from these stems in the 2d year, these in entirety called floricanes. Individual floricanes tend to die after fruiting.

1. Stem bark exfoliating or shreddy 2
1. Stem bark smooth to bristly or prickly, not exfoliating 4
  2. Twigs glandular, no bristles or prickles; leaves simple, palmately lobed, green beneath
      1. *R. odoratus*
  2. Twigs bristly, or lower stems with bristles or prickles; leaves trifoliate or pinnate, white beneath 3
3. Twigs bristly but not glandular-hispid; escape from cultivation   2. *R. idaeus* var. *idaeus*
3. Twigs glandular-hispid; of high mountains   3. *R. idaeus* var. *strigosus*
  4. Stems with whitish powdery bloom   4. *R. occidentalis*
  4. Stems not with a bloom 5
5. Twigs glandular-hispid   5. *R. phoenicolasius*

5. Twigs not glandular-hispid ... 6
  6. Stems trailing and mostly herbaceous and unarmed; n. species ... 6. *R. pubescens*
  6. Stems trailing and prickly or bristly, or erect ... 7
7. Leaves pinnate, with 5 to 9 narrow leaflets ... 7. *R. illecebrosus*
7. Leaves trifoliate or palmate, with 3 to 5 leaflets (dewberries, blackberries) ... 8
  8. Stems trailing; erect stems none or under 2dm tall (dewberries) ... 9
  8. Stems erect or arching, over 2dm tall (blackberries) ... 12
9. Bristles present between prickles ... 10
9. Bristles lacking ... 11
  10. Bristles glandular ... 8. *R. trivialis*
  10. Bristles nonglandular ... 9. *R. hispidus*
11. Leaflet tips blunt or base tapered; prickles slender, rarely stout or curved ... 10. *R. enslenii*
11. Leaflet tips pointed or bases rounded; prickles stout, curved ... 11. *R. flagellaris*
  12. Stem prickles weak or sparse; stems glabrous ... 12. *R. canadensis*
  12. Stem prickles numerous or stems bristly ... 13
13. Prickles slender; stems thickly bristly ... 13. *R. setosus*
13. Prickles stout-based; stems lack bristles ... 14
  14. Stems within fruit clusters with few or weak prickles; canes arching but not horizontally scrambling ... 15
  14. Stems within fruit clusters with numerous, stout prickles; canes scrambling horizontally (exotics) ... 20
15. Leaflets gray to white-tomentose beneath; widest beyond middle ... 14. *R. cuneifolius*
15. Leaflets green and lightly hairy to glabrous beneath ... 16
  16. Hairs of twigs glandular ... 17
  16. Hairs not glandular ... 18
17. Leaflets mostly oblong-ovate; width ⅗ or less of length; common in Appalachians ... 15. *R. allegheniensis*
17. Leaflets widely ovate to nearly circular; width ¾ or more of length; rare in SE ... 16. *R. orarius*
  18. Terminal leaflet base cuneate, underside glabrous to lightly hairy; stem prickles usually straight ... 17. *R. argutus*
  18. Terminal leaflet base rounded or cordate, underside softly hairy; stem prickles usually curved or hooked ... 19
19. Leaflets broadly ovate, widest near base; bract at base of nearly every fruit stalk ... 18. *R. pensilvanicus*
19. Leaflets oblong-ovate, widest near middle; bracts only at lowermost fruit stalks ... 19. *R. ostryifolius*
  20. Leaves green beneath; deeply cut and divided ... 20. *R. laciniatus*
  20. Leaves whitened or grayish tomentose beneath ... 21
21. Prickles mostly straight; twigs nearly glabrous ... 21. *R. bifrons*
21. Prickles curved; twigs pubescent ... 22. *R. discolor*

Group 1: fruits separating from receptacle (raspberries or similar to raspberries)

1. *Rubus odoratus* L. PURPLE-FLOWERING RASPBERRY   *Figure 463*

A shrub, to 2m tall, found in the Appalachian region of the SE from NC, TN northward. Occurs in moist, shaded habitats. Stems lack prickles; bark brown, exfoliating. Twigs and fresh growth glandular-hairy. Fruit purplish-red, not very flavorful, about 2cm wide. Leaves simple, palmately lobed, thin and prominently veined.

2. *Rubus idaeus* var. *idaeus* L. RED RASPBERRY

A suckering shrub, with erect to leaning stems to 2m, covered by nonglandular hairs and bristles. Native to Eurasia; extensively cultivated for fruit production as the common red raspberry.

Occasionally spreading from cultivation into adjacent areas. Prickles straight, weak. Fruit red, sweet. Leaflets 3 or 5, white beneath.

3. *Rubus idaeus* var. *strigosus* (Michx.) Maxim.   RED RASPBERRY   Figure 464

A suckering shrub with erect stems covered with glandular hairs and bristles. A circumboreal species, distributed widely across n. sections of N. Amer. and Asia. Ranges into the SE in the high mountain elevations of WV, VA, TN, NC. Prickles straight, weak. Fruit red. Leaflets normally 3, rarely 5; white beneath.

4. *Rubus occidentalis* L.   BLACK RASPBERRY

A suckering shrub with stems glabrous, glaucous, prickly, purplish when bloom is rubbed away. Most common in the Appalachians in the SE, but ranging west in the interior to AR, MO, and eastward into the Piedmont. Prickles stout, wide at base, slightly curved. Leaflets 3, white beneath. Fruit black when ripe, sweet.

5. *Rubus phoenicolasius* Maxim.   WINEBERRY

A suckering shrub with long, arching stems covered by glandular hairs and bristles. Native to e. Asia; sometimes cultivated for its fruit. Naturalized sporadically across the SE, sometimes locally abundant. Prickles slender, weak. Fruit red, sweet, covered with sticky, glandular hairs. Leaflets 3, white beneath.

6. *Rubus pubescens* Raf.   DWARF RASPBERRY

A slender, creeping semiwoody species rarely with stems more than 5dm tall. A n. species reaching the SE only in damp, boggy soils in the high elevations of WV. Stems mostly herbaceous, without prickles, with broad, persistent stipules. Fruit red. Leaflets 3.

7. *Rubus illecebrosus* Focke   BALLOONBERRY

A spreading shrub, to 1m tall. Native to Japan; cultivated for the fruit and occasionally naturalized near plantings. Stems glabrous, without glandular hairs or bristles. Prickles few but curved, wide-based. Fruit oblong, red, to 3cm, not highly flavorful. Leaflets 5 to 9, narrow and lanceolate, mostly glabrous.

Group 2: fruits not separating from receptacle; habit creeping (dewberries)

8. *Rubus trivialis* Michx.   SAND DEWBERRY

Stems slender, creeping, hispid, covered with glandular bristles. Distributed from e. MD to FL, west to TX, and in the interior to MO; mostly a coastal plain species. Typically in dry or sandy soils. Prickles few, short, stout, curved. Leaflets mostly 5, subevergreen, glossy.

9. *Rubus hispidus* L.   SWAMP DEWBERRY   Figure 465a

Stems slender, creeping, hispid; bristles not glandular. Distributed widely in the SE from e. VA to n. SC, west to MO and areas north. Typically seen in moist soils and bogs. Prickles slender, weak, barely curved. Leaflets 3 or 5, subevergreen, glossy.

10. *Rubus enslenii* Tratt.   SOUTHERN DEWBERRY

Stems slender, creeping, sometimes only semiwoody. Distributed from the NE south to GA, west to LA, s. IN. Prickles slender, slightly curved, weak or rarely stouter. Leaflets usually 3, thin, dull, mostly obovate or oblong.

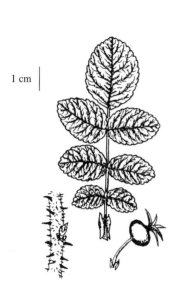

Figure 461. *Rosa rugosa*

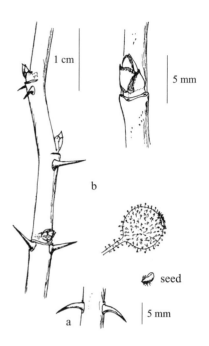

Figure 462. a: *Rosa palustris;* b. *R. carolina*

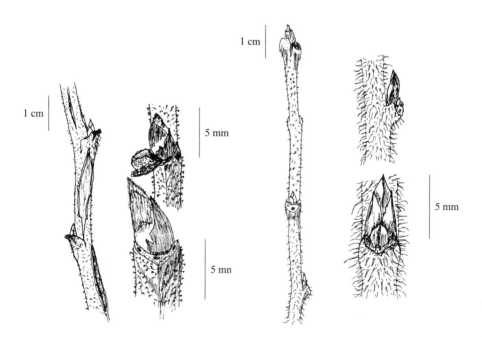

Figure 463. *Rubus odoratus*

Figure 464. *Rubus idaeus* var. *strigosus*

11. *Rubus flagellaris* Willd.   NORTHERN DEWBERRY

Stems creeping, with stout, curved prickles and no bristles. Widely distributed and common across the SE. Typically in old fields, along roads, and in woodland openings. Leaflets 3 to 5, ovate, often shiny but thin in texture.

Group 3: fruits not separating from receptacle; habit upright (native blackberries)

12. *Rubus canadensis* L.   SMOOTH BLACKBERRY

A suckering shrub with upright, weakly armed stems. Ranging from the NE along the Appalachians to n. GA. Typically in high elevation balds, forest openings. Prickles straight, weak, sparse to nearly entirely lacking on stems. Leaflets 5, glabrous; terminal 1 with rounded to cordate base.

13. *Rubus setosus* Bigelow   BRISTLY BLACKBERRY

A suckering shrub with upright, prickly and bristly stems to 1m tall. A n. species ranging south in the Appalachians to WV and VA. Prickles nearly straight, slender, weak; bristles numerous. Prefers moist or wet soils. Leaflets 3 to 5, acute to acuminate, dull, with bristly petioles.

14. *Rubus cuneifolius* Pursh   SAND BLACKBERRY   Figure 465b

A suckering shrub, to 1m tall, with prickly stems. Distributed from MD to FL, west to AL, mainly in the coastal plain and lower Piedmont. Typically in dry or sandy soil. Prickles numerous, stout, straight to curved. Leaflets 3 to 5, bluntly tipped and widest beyond middle; base cuneate; underside gray or whitish, densely hairy.

15. *Rubus allegheniensis* Porter   ALLEGHENY BLACKBERRY

A suckering shrub, to 2m tall. Distributed mostly in the Appalachians and adjacent Piedmont of VA to GA and in the interior westward to KY. Prickles nearly straight. Glandular hairs present on twigs and other parts. Leaflets 3 to 5.

16. *Rubus orarius* Blanch.   BLANCHARD'S BLACKBERRY

A suckering shrub similar to the above species. Distributed from the NE to MD, WV, and west in the interior to MO. Leaflets mostly 3, broadly ovate to nearly round.

17. *Rubus argutus* Link.   COMMON BLACKBERRY   Figure 466

A suckering shrub with erect stems to 2m tall. Widespread and common across much of the SE. Prickles straight on stems; twigs hairy to nearly glabrous. Leaflets 3 to 5, sharply toothed; terminal 1 of primocanes cuneate-based. Includes the species *R. betulifolius* Small, which has more glabrous leaves, mostly limited to the coastal plain and Piedmont from MD to FL, west to MS.

18. *Rubus pensilvanicus* Poir.   NORTHERN BLACKBERRY

A suckering shrub with erect to arching stems to 2m tall. Distributed from the NE south to VA, WV, west to MO, OK, also southward in the interior and west of the Blue Ridge from IN to AL. Prickles straight to curved. Leaflets 3 to 5, elliptic to nearly round, softly pubescent beneath, base rounded to cordate. Each fruit stalk may have a bract at base.

19. *Rubus ostryifolius* Rydb.   HIGHBUSH BLACKBERRY

A shrub similar to above species; distributed mostly from the NE to KS, ranging southward sporadically to WV, KY, MO, perhaps NC. Prickles numerous, stout, curved. Flower and fruit stalks bracted in middle or lower sections of inflorescence only. Leaflets 3 to 5.

Group 4: fruits not separating from receptacle; habit upright or sprawling (exotic blackberries)

20. *Rubus laciniatus* Willd.   CUT-LEAVED BLACKBERRY

A spreading shrub with scrambling or sprawling, coarse, very prickly stems. Native to Europe; naturalized sporadically in the SE, mainly in n. sections. Prickles strong, curved. Leaves subevergreen to evergreen, leaflets 3 to 5, deeply cut or cleft, green beneath, shiny green above.

21. *Rubus bifrons* Vest. ex Tratt.   HIMALAYA-BERRY

A scrambling, coarse shrub with very prickly stems. Native to Europe; naturalized sporadically across the SE, especially near urban areas. Prickles strong, wide-based, straight. Leaflets 3 to 5, white beneath.

22. *Rubus discolor* Weihe & Nees   HIMALAYAN BLACKBERRY

Similar to above species, also native to Europe. Stems with curved prickles. The twigs are usually hairy, especially toward the tip or when young. Leaflets 3 to 5, white beneath.

Sabal Adanason   PALMETTO   Family Arecaceae

Three native species of these evergreen palms occur in the region of the SE covered by this guide. All have large, palmately cleft leaves on long petioles. The fruit is fleshy, drupelike, borne in clusters.

1. Threadlike fibers extend from between leaf segments
  2. Habit treelike, with trunk present; leaf petioles over 5dm long   1. *S. palmetto*
  2. Habit not treelike, trunk lacking or very short; leaf petioles under 5dm long; of FL only
                                                                                         2. *S. etonia*
1. Threadlike fibers lacking   3. *S. minor*

1. *Sabal palmetto* (Walt.) Lodd ex Schult.   CABBAGE PALMETTO   *Figure 467*

A tree, to 25m tall, distributed near the coast from se. NC to FL. A conspicuous component of maritime forests and dunes near the beach; most common and widely dispersed in FL. Leaves to 1.5m wide, with a petiole to 15dm long. Young trunks covered with sheath fragments; these gradually weather and fall to reveal a brown, fibrous trunk appearing smooth from a distance. Drupelike fruits black, about 1cm wide, borne in large clusters in autumn.

2. *Sabal etonia* Swingle ex Nash   SCRUB PALMETTO   *Figure 468*

A shrubby palm with stems mostly underground. Native to FL scrublands; barely entering the range of this book. Leaves rarely over 6dm wide, with a petiole to 4dm long. Fruits to 15mm long.

3. *Sabal minor* (Jacq.) Pers.   DWARF PALMETTO   *Figure 469*

A shrubby palm over most of its range. Native from ne. NC to FL, west to TX. Occurs in coastal plain lowlands such as swamps and bottomlands, maritime forests, marsh borders, and

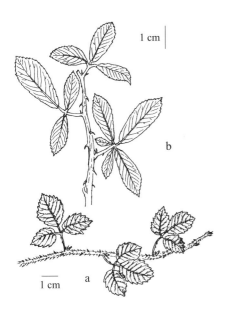

Figure 465. a: *Rubus hispidus;*
b: *R. cuneifolius*

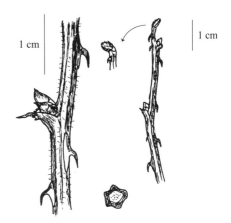

Figure 466. *Rubus argutus*

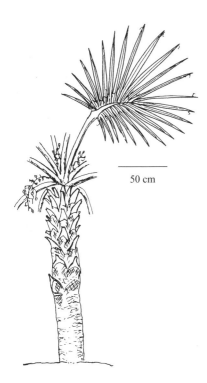

Figure 467. *Sabal palmetto*

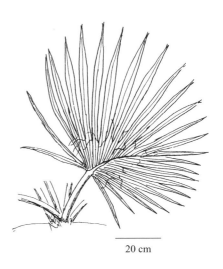

Figure 468. *Sabal etonia*

moist or mesic hammocks. Leaves about 1m wide, with a petiole to 1.5m. Fruits to 1cm long. Rarely develops any trunk aboveground, except in LA trunks may reach 3m, and these palms are sometimes considered a separate species, *S. louisiana* (Darby) Bomhard.

Sageretia minutiflora (Michx.) C. Mohr SHELLMOUND-BUCKTHORN Family Rhamnaceae
Figure 470

An uncommon native shrub with a sprawling, clambering, or vinelike habit. Distributed mostly in the vicinity of the coast from SC to FL, and west to s. MS. Typically on or near shell middens or calcareous hammocks near marshes. Twigs are slender, lined, gray, and sometimes with spinous spur shoots. Small, narrow, blackened stipules may persist; nodes may bear collateral buds, and leaf scars have 1 bundle scar. Fruit is a shiny black drupe to about 8mm wide. Leaves are small, glossy, and late-deciduous.

Salix L. WILLOW Family Salicaceae

Of the 22 willow taxa treated here for the SE, 13 are native. Many more exotic species, hybrids, and cultivars are used in horticultural plantings in the SE, but most of these seldom naturalize and are not included in this text. All willows have slender to moderately slender twigs and narrow leaf scars with 3 bundle scars. The buds are covered with a single caplike scale, with its margin fused together in most species but slightly overlapping and not fused in a few species. Stipule scars are present. Some willows have twigs that snap off branchlets easily from their base; these are described as "brittle at the joints." This adaptation enables the twigs felled by wind or other stresses to be swept by water to a patch of bare soil, where if kept moist the twigs root easily and quickly grow into another plant. Most willows have a tendency to grow in moist sites or near water; also being pioneer species in disturbed sites where they can grow freely without overhead competition. The genus is much more complex to the north.

1. Leaf scars variously alternate, subopposite, or opposite 1. *S. purpurea*
1. Leaf scars only alternate
  2. Twigs grayish green, closely and densely pubescent
    3. Habit a large shrub or tree; most buds wider than twigs
      4. Buds mostly shiny; a n. species rare in SE 2. *S. bebbiana*
      4. Buds dull or hairy; more common in SE
        5. Branchlets glabrous; buds nearly glabrous; young stem wood smooth beneath its bark
          3. *S. caprea*
        5. Branchlets pubescent; buds pubescent; young stem wood lined beneath bark
          4. *S. cinerea*
    3. Habit a small or slender shrub; only a few buds may be as wide or wider than twig
      6. Twigs slender, mostly 2–3mm wide; leaf width usually over 1cm
          5. *S. humilis* var. *humilis*
      6. Twigs very slender, under 2mm wide; leaf width under 1cm 6. *S. humilis* var. *tristis*
  2. Twigs not grayish green and not closely, densely pubescent
    7. Buds hairy or silky
      8. Buds to 1cm long; bark furrowed or scaly; trunk often single, or trunks not clumplike
          7. *S. eriocephala*
      8. Buds to 6mm; bark mostly smooth; stems usually multiple, clumplike 8. *S. sericea*
    7. Buds glabrous or nearly so to naked eye
      9. Bud scale margin overlapping slightly, not fused; most buds sharply pointed, a few bluntly so
        10. Twigs yellow-brown, not brittle at joints, often drooping 9. *S. amygdaloides*
        10. Twigs brown or red-brown, brittle at joints, erect or not drooping
          11. Petioles glandular; leaves 3–5cm wide; rare FL tree 10. *S. floridana*
          11. Petioles not glandular; leaves under 3cm wide; common in SE

        12. Buds slightly appressed to twig; twig may bear a few hairs; a shrub or small tree
            11. *S. caroliniana*
        12. Buds stick out from twig; all parts glabrous, glossy; a tree   12. *S. nigra*
    9. Bud scale margins fused; many buds bluntly pointed
        13. Bark of mature plants smooth on most of trunk, particularly upper portions
            14. Twigs yellow-brown to greenish; buds not as wide as twigs; leaves narrowly lanceolate
                15. Habit clumplike, well-branched; petiole over 5mm   13. *S. petiolaris*
                15. Habit colonial, narrow-branched; petiole under 5mm   14. *S. exigua*
            14. Twigs red-brown or orange-brown; some buds wider than twig; leaves mostly elliptic
                16. Buds purplish; petiole not glandular; leaves glaucous below   15. *S. discolor*
                16. Buds yellow to blackish; petiole glandular; leaves shiny green below
                    16. *S. pentandra*
        13. Bark of mature plants furrowed or scaly on most of trunk
            17. Twigs brittle at the joints and not pendulous; buds shiny or resinous; leaves lustrous above, green beneath
                18. Buds shiny; twigs diverge from branchlet at low angle; a native tree of n. parts of the SE   17. *S. lucida*
                18. Buds not shiny, but resinous; twigs diverge from branchlet at angle close to 90 degrees; naturalized exotic   18. *S. fragilis*
            17. Twigs pendulous, or erect and not brittle-jointed; buds not as above; leaves not lustrous, underside much paler green or glaucous
                19. Twigs long-pendulous, greenish; branchlets brown   19. *S. babylonica*
                19. Twigs erect, or yellowish if pendulous; branchlets olive- or yellow-brown
                    20. Leaves sharply glandular-serrate, glabrous; petioles mostly under 6mm
                        20. *S. matsudana*
                    20. Leaves serrulate, with some silky hairs below; petioles mostly over 6mm
                        21. Twigs erect or spreading   21. *S. alba*
                        21. Twigs long-drooping   22. *S. alba* var. *vitellina*

1. *Salix purpurea* L.   PURPLE WILLOW   *Figure 471*

A shrub or small, bushy tree, to 6m tall. Native to Europe; cultivated in earlier years for basket making materials. Naturalized sporadically in the SE, mostly in n. sections. Twigs purplish or reddish, glabrous, very slender, not brittle at joints. Petioles not glandular; leaves under 2cm wide. Bark thin, smooth or shallowly fissured.

2. *Salix bebbiana* Sarg.   BEBB WILLOW

A shrub or small tree, to 6m tall. Widespread across the n. parts of N. Amer., but reaching the SE perhaps only in IN, or n. WV, w. MD. Twigs thickly pubescent, grayish, not brittle-jointed. Petioles not glandular; leaves reticulate-veined below, usually hairy, 3–8cm long, 1–3cm wide. Bark grayish, smooth, fissured near trunk base.

3. *Salix caprea* L.   GOAT WILLOW

A wide-headed shrub or small tree, to 15m tall. Native to Europe; cultivated for the pussy willow–type showy spring staminate flowers. Widely naturalized in moist lowlands. Twigs densely pubescent, branchlets mostly glabrous, brownish, not brittle-jointed; buds nearly glabrous. Petioles not glandular; leaves pale beneath, shiny green above, 4–10cm long, usually hairy below. Bark grayish, smooth. Staminate flowers of catkin begin flowering from base upward.

4. *Salix cinerea* L. GRAY WILLOW   *Figure 472*

Similar to above species, to about 10m tall. Native to Europe; another "pussy willow" of cultivation. Widely naturalized in moist lowlands and waste areas, sometimes forming thickets. Branchlets hold on to more of the dense pubescence of the twigs and are dark or blackish. Twigs not brittle-jointed; buds hairy. Petioles not glandular; leaves grayish hairy on both sides, paler below, 2–10cm long. Bark grayish, smooth. Young stems bear ridges or lines, which are visible when bark is stripped away. Staminate flowers of catkin begin flowering from tip downward. The subspecies *oleifolia* (Smith) Macreight bears some reddish- or rusty-colored hairs on leaf undersides.

5. *Salix humilis* Marsh. PRAIRIE WILLOW

A shrub, to about 3m tall. Widespread across most of the SE. Sporadic in wet or moist lowlands where there is ample sunlight in the coastal plain. Occurs in drier sites in the interior and is fairly common in high elevation openings in the Appalachians. Twigs and buds pubescent; twigs not brittle-jointed. Petioles not glandular; leaves 5–8cm long, 1–2cm wide, underside pale, hairy, prominently veined, with conspicuous stipules. Bark grayish, smooth. An important variety that is sometimes considered a separate species follows.

6. *Salix humilis* var. *tristis* (Ait.) Griggs DWARF PRAIRIE WILLOW

A low shrub, to about 1m. Scattered from coastal VA to Piedmont and Appalachian areas of GA, MS, and in the interior west to MO, OK. Typically in dry, open sites such as plains, glades, roadsides, slopes. Differs from above species in very slender twigs with leaves 2–5cm long, 5–9mm wide; stipules usually lacking from leaves. [*S. humilis* var. *microphylla* (Anderss.) Fern]

7. *Salix eriocephala* Michx. MISSOURI WILLOW   *Figure 473*

A coarse shrub or small tree, potentially to 15m tall. Distributed mostly west of the Appalachians to MO, KS, and northward, though also occurs sporadically in w. VA, GA, AL, FL, and MS. Found in swamps and along streams. Twigs hairy when young, becoming glabrous or nearly so by winter, not brittle-jointed; buds large, hairy. Petioles not glandular; leaves 8–15cm long, 2–3cm wide; stipules large, half-round, persistent. Bark shallowly fissured or with thin, narrow, scaly ridges. (*S. rigida* Muhl.; *S. cordata* Muhl.)

8. *Salix sericea* Marsh. SILKY WILLOW

A large shrub, to 4m tall, usually clumplike and multiple-stemmed. Primarily a n. species, ranging into the n. half of the SE, south as far as n. SC to n. AL, n. AR. Occurs along streams, in boggy soils and moist openings. Twigs silky to nearly glabrous, slightly brittle at basal joint; buds silky-hairy, appressed to twig. Petioles not glandular; leaves 5–10mm long, 10–25mm wide, pale and with appressed, silky hairs below. Bark mostly grayish, smooth; fissured or scaly only on base of old trunks.

9. *Salix amygdaloides* Anderss. PEACHLEAF WILLOW

A small tree, to 15m tall. Mostly a Plains and midwestern species, reaching the SE in KY, IL, MO. Occurs near streams and other bodies of water. Twigs glabrous, yellow-brown, drooping, not brittle-jointed; buds pointed. Petioles not glandular, to 3cm; leaves 5–15mm long, mostly ovate with a prolonged tip, pale beneath. Bark grayish, fissured, with flat or slightly scaly elongated ridges with age.

Figure 469. *Sabal minor*

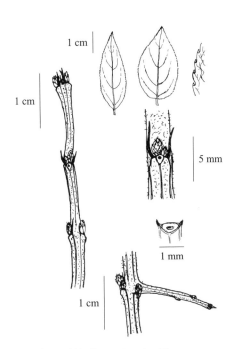

Figure 470. *Sageretia minutiflora*

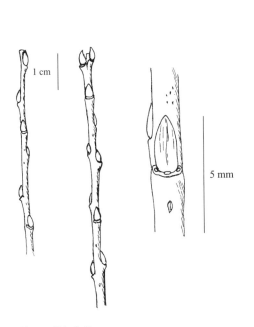

Figure 471. *Salix purpurea*

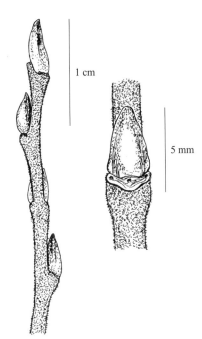

Figure 472. *Salix cinerea*

10. *Salix floridana* Chapm. FLORIDA WILLOW

A shrub or small tree, to 5m tall. Rare and local in the coastal plain of GA and n. FL. Occurs in wet woodlands under taller trees, having a tolerance for shade that is unusual among willows. Twigs brownish, slightly hairy to glabrous, brittle-jointed. Petioles glandular; leaves 8–16cm long, 4–5cm wide, pale and hairy below. Bark gray to brown, with scaly ridges.

11. *Salix caroliniana* Michx. COASTAL PLAIN WILLOW

A tall shrub or small tree, to 10m tall, sometimes with multiple stems. Widespread across the SE in general, though uncommon in large parts of the region. Most common in the coastal plain from NC to AL, and in MO, AR. Occurs along streams, sandbars of rivers, and wet ditches. Twigs brown, slightly hairy, brittle-jointed. Petioles not glandular; leaves 6–14cm long, 1–3cm wide, pale beneath. Bark grayish, with scaly ridges.

12. *Salix nigra* Marsh. BLACK WILLOW   Figure 474

A tree, to 20m tall, but usually shorter, ordinarily with a single trunk. Widespread across the entire SE, except ne. and central FL and high elevations of the Blue Ridge. Occurs in moist lowland soils, usually in habitats with a combination of nearness to water and ample sunlight. Twigs red-brown, glabrous, brittle-jointed. Petioles not glandular; leaves 6–16cm long, 5–20mm wide, glossy green on both sides. Bark brown, grayish, or darker, with scaly plates.

13. *Salix petiolaris* J. E. Smith MEADOW WILLOW

A shrub, to about 3m tall, usually clumplike and with multiple, widely branching stems. Of a n. distribution, rare in the SE and entering the region perhaps only in s. IN or OH. Typically in wet meadows or openings along streams. Twigs brown, glabrous, not brittle at joints. Petioles not glandular; leaves 5–10mm long, 5–20mm wide, pale and glabrous below. Bark mostly smooth, greenish to gray; fissured or scaly only on base of old trunks. (*S. gracilis* Anderss.)

14. *Salix exigua* Nutt. SANDBAR WILLOW

A slender shrub, to 6m tall, stoloniferous and forming colonies of closely spaced vertical stems. Distributed widely across n. N. Amer. and the Plains states, reaching the SE mostly west of the Blue Ridge from w. VA to LA, TX. Typically forming thickets along rivers and on sandbars. Twigs very slender, yellow-brown, not brittle at joints. Petioles not glandular, only 5mm or less in length; leaves 4–14cm long, 3–10mm wide, hairy below, margin finely to remotely toothed. Bark mostly smooth. (*Salix interior* Rowlee)

15. *Salix discolor* Muhl. PUSSY WILLOW

A shrub or small tree, to 8m tall. A n. species reaching the SE in its southernmost range in WV, n. VA, KY, IL. Occurs in moist soils of lowlands or mountain boggy sites. Twigs glabrous or nearly so, not brittle-jointed; buds lustrous, purplish or reddish. Petioles not glandular; leaves 5–12cm long, 15–30mm wide, glaucous beneath, glabrous. Bark grayish, mostly smooth or with wide scales on lower sections of old trunks. The staminate catkins emerging in spring account for the name, suggesting downy gray fur of a cat or cat's paw.

16. *Salix pentandra* L. BAY-LEAVED WILLOW

A shrub or tree, to 10m tall. Native to Eurasia; known to sporadically naturalize in WV, VA, NC; perhaps expected in other n. areas of the SE. Twigs glabrous, lustrous, not brittle-jointed;

buds yellow to dark, lustrous, resinous when crushed. Petioles glandular; leaves 4–10cm long, 2–3cm wide, shiny. Bark mostly smooth, fissured on lower trunk.

17. *Salix lucida* Muhl. SHINING WILLOW

A shrub or small tree, to 10m tall. A n. species barely reaching the SE, in MD, WV, VA. Occurs in moist to wet lowlands and openings along upland lakes, bogs, and streams. Twigs dark brown, lustrous, brittle-jointed; buds dark, lustrous. Petioles glandular; leaves 5–15cm long, 2–5cm wide, shiny above, green beneath, tip narrowly elongated. Bark grayish, with thick, scaly ridges.

18. *Salix fragilis* L. CRACK WILLOW

A tree, to 30m tall, with a well-developed trunk. Native to Eurasia; sporadically naturalized near sources of its cultivation. Twigs yellow-brown to dark brown, glabrous, brittle-jointed, divergent from the branchlet at near right angles; buds sticky or slightly resinous. Petioles glandular; leaves 6–15cm long, 2–4cm wide, pale green and glabrous below. Bark fissured, with thick, slightly scaly ridges.

19. *Salix babylonica* L. WEEPING WILLOW

A tree, to 25m tall, with conspicuously pendulous branchlets and twigs. Native to China; long cultivated as an ornamental. Naturalizes sporadically, usually by fallen twigs taking root after deposition on mud or soil by water. Twigs greenish, glabrous, brittle-jointed. Petioles glandular, mostly 4–6mm; leaves 5–15cm long, 5–15mm wide, serrulate on margin, paler green below and glabrous. Bark grayish or dark, with rough plates or scaly ridges. This species has been confused considerably in the horticultural trade with weeping forms of other willow species and hybrids, and there is taxonomic disagreement regarding its ambiguous status in N. Amer. One such weeping hybrid is *S. babylonica* × *fragilis* (*S.* × *pendulina* Wenderoth), tending to have glabrous young twigs and leaves; another is *S. babylonica* × *alba* (*S.* × *sepulcralis* Simonkai), with silky hairs on the young twigs and leaves, as influenced by the parentage of *S. alba*. A majority of the weeping types of arborescent willows cultivated in the SE involve the following species.

20. *Salix matsudana* Koidz. CORKSCREW WILLOW

A tree, to 15m tall, with erect or spreading branchlets and twigs, but the contorted twigs of the twisted or "curly" cultivar is more typically seen. There is also a form with pendulous twigs. A native of China; cultivated as an ornamental and occasionally naturalizing in moist soils near planted specimens through seed or shed twigs. Twigs usually yellow-green, glabrous, slightly brittle-jointed; branchlets yellow-green or olive-colored. Petioles glandular, mostly 3–6mm; leaves 5–9cm long, 1–2cm wide, sharply glandular-serrate on margin, glabrous when mature, pale beneath. Bark grayish, scaly or with scaly plates. The cultivar 'Tortuosa' Rehd. bears twisted branchlets and twigs; the cultivar 'Pendula' Schneid. has pendulous twigs; both of these are more favored in cultivation than the normal species.

21. *Salix alba* L. WHITE WILLOW

A tree, to 25m tall, usually with a short trunk and widely divergent branches. Native to Eurasia. Occasionally naturalizing near old home sites or moist areas near cultivated specimens. Twigs yellow-brown to dark brown, slightly hairy to glabrous to the naked eye, not brittle-jointed. Twigs erect to divergent in the typical species, though weeping forms are also commonly planted (see following var.). Petioles usually glandular, mostly 6–12mm; leaves 5–12cm long,

1–3cm wide, serrulate on margin, silky-hairy and pale beneath. Bark grayish to brown, with thick, scaly ridges.

22. *Salix alba* var. *vitellina* (L.) Stokes   GOLDEN WEEPING WILLOW

A tree generally similar to the above species in all respects, except twigs are yellower and, in some forms, droop conspicuously for lengths of 4dm or more. Other forms may droop only slightly or not at all; such variation is complicated by frequent hybridization with similar species and other cultivars, both intentional and sporadic. The cultivar 'Tristis' is also known as golden weeping willow.

Sambucus L.   ELDERBERRY   Family Caprifoliaceae

Two native species in the SE, both shrubs with lenticellate bark. Twigs have opposite leaf scars, each with several (usually 5) bundle scars; no terminal buds; laterals with several pairs of imbricate scales. A flat terminal scar between 2 lateral buds is commonly present where fruiting clusters have dropped or where twig tip has died back. Inner bark with a strong, rank odor reminiscent of boxwood and peeled peaches. The fruit is berrylike, about 5mm diam., borne in clusters. Leaves pinnately compound, with 5 to 11 leaflets; sometimes on vigorous shoots the lower leaflets are divided.

1. Buds widest near base, with 4 to 6 pairs of scales; leaf scar lower margin slightly thickened; branchlet pith white   1. *S. canadensis*
1. Buds constricted at base, widest near middle, with 3 to 4 pairs of scales; leaf scar margin prominently raised; pith brown   2. *S. racemosa* ssp. *pubens*

1. *Sambucus canadensis* L.   AMERICAN ELDERBERRY   *Figure 475*

A coarse shrub with clumplike or rhizomatous habit. Widespread, throughout the SE. Common in moist soils near forest openings, streams, roadsides, disturbed areas. The prominently raised lenticels adorn the twigs and stems; otherwise bark is mostly smooth. Fruits dark purplish-black or black, borne in flat-topped clusters, mature in autumn. The var. *laciniata* Gray occurs in the coastal plain of GA, FL, with leaflets commonly divided or cut.

2. *Sambucus racemosa* ssp. *pubens* (Michx.) House   EASTERN RED ELDERBERRY
Figure 476

A coarse shrub usually with multiple stems or clumplike habit. The species is circumboreal, distributed widely across the n. regions of N. Amer. and Europe. The subspecies *pubens* is the e. US form, extending from the North down the higher elevations of the Appalachians to n. GA. Fruits red, borne in elongated or racemose clusters, usually mature by midsummer. (*S. pubens* Michx.)

Sapindus L.   SOAPBERRY   Family Sapindaceae

One or 2 native species occur in the SE, depending on taxonomic viewpoint. Two species are considered here. Twigs of both lack terminal buds; lateral buds small, rounded, sometimes superposed; large leaf scars have 3 bundle scars. The fruit is drupelike, about 12–20mm wide, with a yellowish, semitransparent, leathery or semifleshy outer layer and a single rounded seed; poisonous if eaten. Macerated rinds of the fruit yield a lather in water. Outer bark grayish, scaly on oldest trunk portions and sometimes revealing the reddish-brown inner bark layers. Leaves pinnately compound, 6 to 18 leaflets.

1. Leaves subevergreen, usually without a terminal leaflet; native east of Mississippi River
1. *S. saponaria*
1. Leaves deciduous, usually with a terminal leaflet; native west of Mississippi River
2. *S. drummondii*

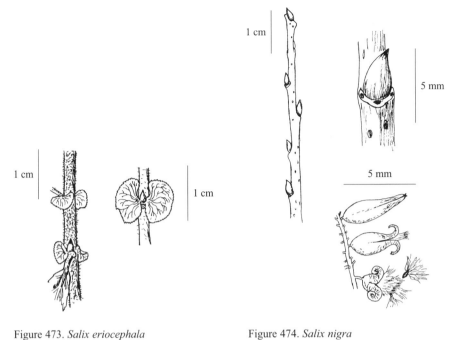

Figure 473. *Salix eriocephala*

Figure 474. *Salix nigra*

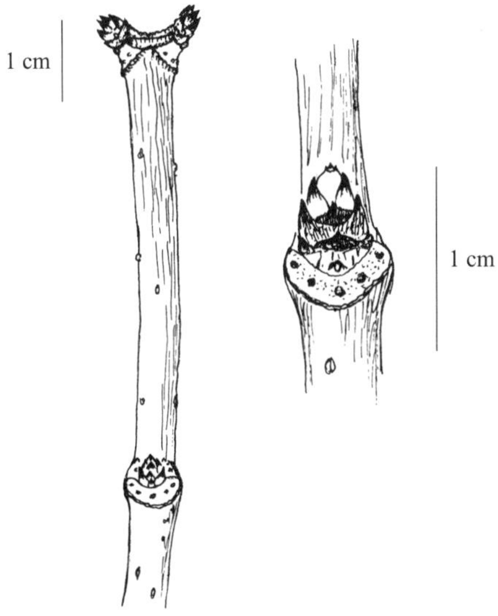

Figure 475. *Sambucus canadensis*

Figure 476. *Sambucus racemosa* ssp. *pubens*

1. *Sapindus saponaria* L.   FLORIDA SOAPBERRY   Figures 477 and 478

A small tree, rarely over 10m tall. Distributed in the coastal plain mainly in coastal and peninsular FL, but extends north along the Atlantic coast to s. SC. Typically in calcareous hammocks and shell middens. Two differing leaf variations have been 1 basis for separation into 2 species in the past: leaflets with obtuse or rounded tips and the rachis winged has been called *S. saponaria* L., most common in s. FL; those plants with leaflets acuminate at the tip and rachis not winged have been called *S. marginatus* Willd., of n. parts of the range. The former tends to have globose fruits; the latter, fruits longer than wide and keeled on the back, though these fruit characters are not always reliable. Fruits of this soapberry are up to 2cm wide, the yellowish translucent flesh wrinkling and darkening slightly after maturity.

2. *Sapindus drummondii* Hook. & Arn.   WESTERN SOAPBERRY

A small tree, to 15m tall. Distributed mostly in the SW and s. plains, extending eastward into the SE as far as MO, AR, e. LA. Mostly found in calcareous uplands and prairie streamsides in TX, OK, and bottomlands in e. parts of the range. Leaflets acuminate at the tip, rachis not winged. Fruits to 15mm diam., globose, often persistent over winter, flesh translucent yellow to red-brown. This species is considered by some to be a w. form of the earlier-described Florida soapberry, thus designated *S. saponoria* var. *drummondii* (Hook. & Arn.) L. Benson.

Sapium sebiferum (L.) Roxb.   TALLOW-TREE   Family Euphorbiaceae   Figure 479

An exotic tree, to 15m tall, native to China. Naturalized in the coastal plain from NC to FL, west to TX. Commonly seen in maritime forests, near marshes and shell middens, disturbed areas, and along streams inland. Twigs greenish, sometimes growing so late in the season that they do not lignify and die back from their tips in cold weather. Terminal buds lacking; laterals small, with 2 or 3 visible scales and stipular projections; 3 bundle scars; sap milky. The fruit is a capsule about 1cm long, splitting to reveal 3 white seed that remain connected together for a period of time in early winter (accounting for the alternate colloquial name of "popcorn-tree"). Bark light brown, mostly smooth but shallowly fissured on old trunks.

Sarcocornia perennis (P. Mill.) A. J. Scott   GLASSWORT   Family Chenopodiaceae

Figure 480

A semiwoody, succulent-stemmed tidewater shrub of brackish coastal soils, usually creeping or with erect stems to 4dm tall. Widely distributed in coastal areas of the entire SE, from MD to TX; also along the West Coast to AK, perhaps the same species that occurs in the West Indies, Europe, and Africa. Common in salt marshes, tidal flats, brackish marshes. The peculiar jointed green stems lack spreading leaves, and are salty to taste. (*Salicornia perennis* P. Mill.)

Sasa palmata (Milford) E. G. Camas   BAMBOO   Family Poaceae   Figure 481

An exotic, evergreen bamboo, rarely over 2m tall and 1cm diam. Usually forms extensive stands of closely spaced culms, spreading from points of cultivation. Native to Japan; cultivated in the SE for its dense foliage and low, dense habit. Leaves to 35cm long, 9cm wide, deep glossy green above, pale below, clustered near the shoot tips so as to resemble a large, palmate leaf; minutely serrulate on both margins; petiole thick, swollen at sheath junction. Culms green at first, brown-blotched with age; culm sheaths pale green to yellowish-brown, often persistent; branches mostly single per node. In n. sections of SE, may freeze back or have browned leaves in winter.

Sassafras albidum (Nutt.) Nees.   SASSAFRAS   Family Lauraceae   Figure 482

A native tree, rarely over 15m tall. Sometimes mature when only 3 or 4m height; sometimes attaining height of 25m. Widespread across the entire SE, and fairly common in a range of

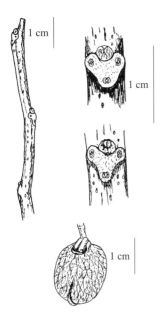

Figure 477. *Sapindus saponaria*

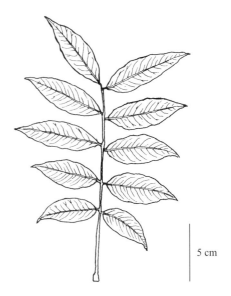

Figure 478. *Sapindus saponaria*

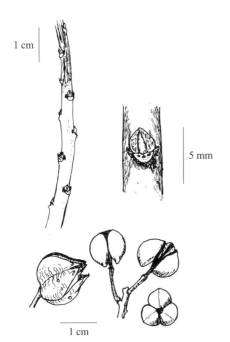

Figure 479. *Sapium sebiferum*

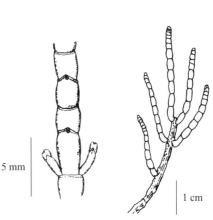

Figure 480. *Sarcocornia perennis*

habitats. Most often seen in moist soils in hardwood forests. Twigs, branchlets, and other young stems green, with inner bark spicy-aromatic. Terminal buds present, much larger than laterals; terminal flower buds plump. Leaf scars with 1 bundle scar. Twigs can be hairy or nearly glabrous. Side branches on the twigs are common; these derived from the same year's growth as the main twig, so no bud scale scars are evident at their base; called lammas shoots. Fruit a blue-black drupe on a red stalk, mature in August. Bark grayish to brown, thickening with age, fissured and with elongated ridges broken horizontally by additional cracks or in blocklike, scaly plates.

Schinus terebinthifolius Raddi   BRAZILIAN PEPPER TREE   Family Anacardiaceae

Figure 483

An evergreen, exotic shrub most common in coastal peninsular FL and throughout s. FL, but occasionally seen near the coast in n. FL. The pinnate leaves are about 7–15cm long, with 5 to 9 blunt-tipped leaflets, aromatic when crushed. Twigs with aromatic inner bark, 3 bundle scars in leaf scars; older stems brown, mostly smooth or shallowly fissured. The fruit is a red drupe, borne in clusters. Native to S. Amer.; 1 of the most invasive exotic shrubs naturalizing in s. FL. It has proliferated in many habitats, overgrowing native vegetation.

Schisandra glabra (Bickn.) Rehd.   MAGNOLIA-VINE   Family Schisandraceae   *Figure 484*

A twining native vine climbing high in the tree canopy or sprawling over the ground. Mostly found in the coastal plain from NC to FL, west to LA; also in the interior to ne. GA, w. TN, e. AR. Uncommon overall and sporadic in distribution, occurring in mesic woodlands, rich slopes, and ravines. Twigs red-brown, lenticellate; buds large, elongate, with red-brown imbricate scales; terminal buds may appear present. The fruit is a red drupe, borne in sparse, elongate clusters hanging by a long stem; matures by late summer. Leaves ovate, glabrous, remotely toothed, veins fairly indistinct.

Sebastiania fruticosa (Bartr.) Fern.   SEBASTIAN-BUSH   Family Euphorbiaceae

Figures 485 and 486

A slender shrub, to 3m tall, sparsely branched. Native to the coastal plain from NC to FL, west to TX. Occurs mostly in understories of mesic or swamp forests, though occasionally seen in drier upland forest habitats. Twigs slender, sap milky; no terminal bud; laterals small, obscurely scaled; stipular appendages persistent; leaf scars raised, sometimes with lines below, 3 bundle scars sometimes indistinct. A conspicuous branch scar may be at the tip of the twig. The fruit is a 3-lobed capsule, about 8mm long. [*S. ligustrina* (Michx.) Muel.]

Serenoa repens (Bart.) Small   SAW PALMETTO   Family Arecaceae   *Figure 487*

A native palm with repent stems or stems creeping just under the ground surface. Distributed in the outer coastal plain from se. SC to FL, west to LA. May form a dominant shrubby layer in pine flatwoods, hammocks, or other coastal forests from moist to dry. The evergreen leaves are deeply palmately divided, about 6–8dm wide, dark green or sometimes glaucous and blue-green; petiole 5–10dm long, sharply toothed along the margins (saw-edged). Fruit drupelike, oblong, to 25mm, dark purplish or black when mature.

Sesbania Scopoli   RATTLEBOX   Family Fabaceae

Several species are native and naturalized in the se. coastal plain, but only 3 are woody enough to warrant inclusion here. The greenish twigs usually have terminal buds, if not killed by freezing; these are much larger than lateral buds, with loose, elongated outer scales extending to tip of bud or beyond. Lateral buds are often hidden behind a raised ledge above the leaf scar; bundle scars 3, closely crowded. Bark greenish to brown, thin and smooth. Leaves pinnately

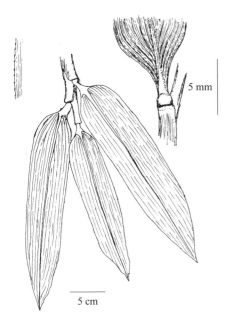

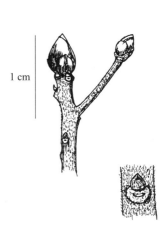

Figure 481. *Sasa palmata*

Figure 482. *Sassafras albidum*

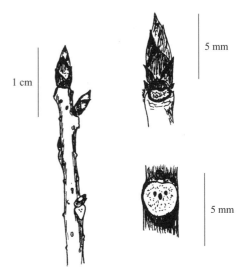

Figure 483. *Schinus terebinthifolius*

Figure 484. *Schisandra glabra*

324 · Sesbania

compound, with 20 to 40 small leaflets. Twigs alone cannot be used to identify the species in winter; fruits are more diagnostic, if present. Fruit a brown, dry legume 2–8cm long, with ridges or wings on margins; often persistent into winter, with the hardened seed loose and rattling inside when the fruit is shaken.

1. Legume margin with 2 thick ridges, a shallow groove between them   1. *S. virgata*
1. Legume with 4 thin wings, a wide trough between them
    2. Wings parallel and under 6mm; legume light brown; mature plant habit taller than wide
                                                                                                     2. *S. drummondii*
    2. Wings divergent and over 6mm; legume dark brown; habit usually wider than tall
                                                                                                     3. *S. punicea*

1. *Sesbania virgata* (Cav.) Poir.   MEXICAN RATTLEBOX   *Figure 488*

A slender, coarsely branched shrub, to 4m tall, native to Mexico. Naturalized along the Gulf Coast from w. FL to TX, mostly in disturbed and waste places, roadsides, and near water sources. Bark greenish. Legume about 1cm wide, 3–6cm long, dark brown, the edges 2-ridged; body of legume not constricted between the seed nor bulged over them. Flowers yellow.

2. *Sesbania drummondii* (Rydb.) Cory   YELLOW RATTLEBOX   *Figure 489*

A slender, coarsely branched shrub, to 5m tall, taller than wide; often forming colonies. Found in the outer coastal plain from w. FL to TX. Presumed native to Mexico; perhaps native along parts of the Gulf Coast. Occurs in similar places as above species. Bark greenish, becoming brown. Legume 10–15mm wide, 2–6cm long, light brown, with 2 wings along the edges; body of legume often constricted between and bulged around seed. Flowers yellow. (*Daubentonia drummondii* Rydb.)

3. *Sesbania punicea* (Cav.) Benth.   RATTLEBOX   *Figures 490 and 491*

A coarsely branched shrub, to 3m, usually wider than tall. Native to S. Amer.; naturalized in the coastal plain from NC to FL, west to TX. Seen along roads, ditches, shores, wet depressions, waste areas. Legume 10–15mm wide, 5–8cm long, dark brown, with 4 wings; body of legume slightly constricted and bulged between and around seed. Flowers orange-red. [*Daubentonia punicea* (Cav.) DC.]

Sibbaldiopsis tridentata (Ait.) Rydb.   THREE-TOOTHED CINQUEFOIL   Family Rosaceae
                                                                                                                      Figure 492

A diminutive rhizomatous shrub with erect stems, rarely over 5cm tall. Often forms matlike colonies or scattered stems protruding through moss. Widespread across n. N. Amer., extending into the SE along the high elevations of the Appalachians to n. GA. Occurs in balds, near rock outcrops, and similarly open areas of the Blue Ridge. The evergreen trifoliate leaves have about 3 large teeth at the tip, shiny green above. Fruit is a cluster of seed (achenes) on an erect stalk, remnants of which may persist in winter. (*Potentilla tridentata* Ait.)

Sideroxylon L.   BUMELIA   Family Sapotaceae

At least 6 species are native to the southeast region covered by this guide. Within these species, a few additional subspecies or aberrant forms are noted here since some of these have been assigned species status in the past and their recognition may be of interest. The nomenclatural change from *Bumelia* Swartz to *Sideroxylon* may cause confusion, since many earlier texts have used the former name and "Bumelia" has become familiar in colloquial use, as well. *Sideroxylon* is actually the older name and has priority in use, according to recent taxonomic studies. Twig features alone will not separate the species in the SE. All may have spinose twig

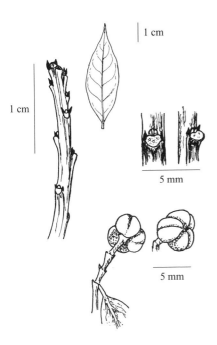

Figure 485. *Sebastiania fruticosa*

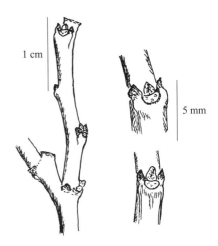

Figure 486. *Sebastiania fruticosa*

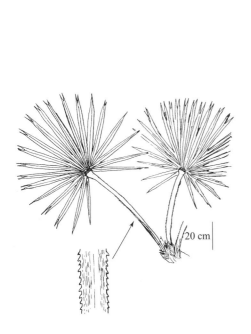

Figure 487. *Serenoa repens*

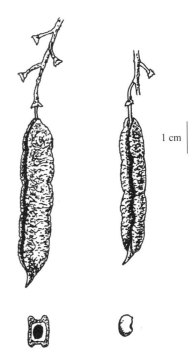

Figure 488. *Sesbania virgata*

tips or spur shoots, or spines at the nodes, short or rounded buds, and barely discernible bundle scars (normally 3 per leaf scar). Sap is milky during the growing season. Trunk bark is fissured, blocky, or scaly. Habit varies from prostrate and sprawling to treelike. The fruit is drupelike, about 1cm long, with a shiny black, fleshy skin and a single, egg-shaped seed. Leaves simple, alternate, clustered on spur shoots, untoothed, sometimes late-deciduous.

1. Twigs hairy, at least near apex; bud scales densely hairy
  2. Twigs overall densely white-hairy; leaf undersides with loose, white hairs
                                                                                             1. *S. lanuginosum*
  2. Twigs red-hairy, with appressed hairs or sparsely hairy; leaves appressed hairy or nearly glabrous beneath
    3. Twigs rusty-tomentose; colonial shrubs with many slender vertical stems under 1m tall
                                                                     7. *S. reclinatum* ssp. *rufotomentosum*
    3. Twigs brown-hairy near tips, more glabrous near base; habit not as above, well horizontally branched
                                                                                  2. *S. tenax*
1. Twigs glabrous; bud scales glabrous or nearly so
  4. Leaf undersides densely silvery-silky, the hairs appressed; rare shrub of FL only
                                                                               3. *S. alachuense*
  4. Leaf undersides green or hairs more sparse, not as above; in FL and GA or more widespread
    5. Leaves glabrous beneath, larger ones over 8cm long; fruit over 10mm; bark scaly; tall shrub or small tree; widespread   4. *S. lycioides*
    5. Leaves with some hairs beneath, largest leaves under 8cm; fruit 10mm or less; bark smooth or shallowly fissured; low or spindly shrubs of n. FL, s. GA
      6. Leaves densely pubescent beneath; fruit 8–10mm   5. *S. thornei*
      6. Leaves sparsely hairy beneath; fruit 5–8mm   6. *S. reclinatum*

1. *Sideroxylon lanuginosum* Michx.   GUM BUMELIA   *Figure 493*

A shrub or more commonly a small tree, to 20m tall, usually with a single trunk. Distributed in the se. states west of the Mississippi River and in the coastal plain east of the river to central FL, se. SC. Occurs in sandy soils near the coast and inland, widely scattered. More common in the w. parts of its range and forming several subspecies there. Twigs usually densely white-hairy, though sometimes grayish to brown in winter; hairs spreading and not appressed to twig or leaves. Buds densely hairy. Leaves variable, length from 1–12cm but mostly 4–8cm, densely pubescent beneath; tip blunt to rounded. Bark brown, furrowed and narrow-ridged, with reddish-brown inner bark. Fruit 6–12mm long. The ssp. *albicans* (Sarg.) Kartesz & Gandhi occurs in e. TX and LA, with silvery-white hairs under the leaves and fruit stems longer than fruit and is the tallest form of any of our bumelias, to 25m; ssp. *oblongifolium* (Nutt.) T. D. Pennington occurs in e. and n. TX to MO, AL, with grayish hairs and fruit stems short; ssp. *rigidum* (Gray) T. D. Pennington of OK, LA, and east-central TX west to AZ, with small leaves and short fruit stems, and short, thorny habit, usually over limestone. [*Bumelia lanuginosum* (Michx.) Pers.]

2. *Sideroxylon tenax* L.   TOUGH BUMELIA   *Figures 494 and 495*

A shrub or small tree, to 10m tall, usually upright, with a single trunk. Sometimes shrubby or even sprawling in certain habitats. Native to the coastal plain from se. NC to w. FL. Typically in maritime forests and sandy soils near the coast and in upland pinelands and scrub. Twigs hairy to nearly glabrous, but nearly always finely hairy near the tip and over the buds; 1 form has more glabrous, gray twigs. Bark brown, furrowed and narrow-ridged; inner bark reddish-brown. Leaves 2–7cm long, varying from spatulate to elliptic, tip obtuse or rounded; undersurface densely clothed with appressed, silky hairs rendering a silvery, coppery, or brown sheen. Fruit 10–14mm long. [*Bumelia tenax* (L.) Willd.] A significant variation of this species seems to be the prostrate or sprawling entity earlier named *Bumelia lacuum* Small, sometimes with a matlike habit in scrub habitats of FL. It has leaves obovate, sparsely brown-hairy or nearly glabrous beneath, the twigs very spiny.

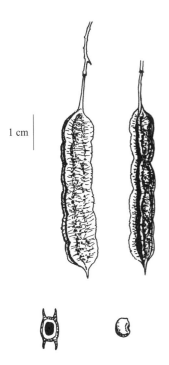

Figure 489. *Sesbania drummondii*

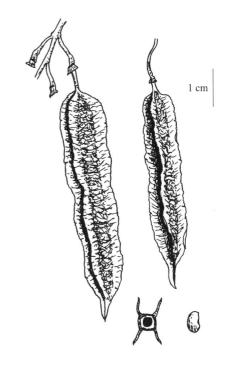

Figure 490. *Sesbania punicea*

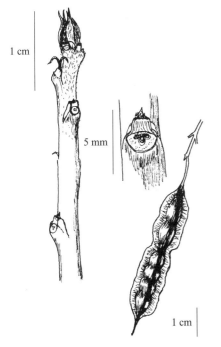

Figure 491. *Sesbania punicea*

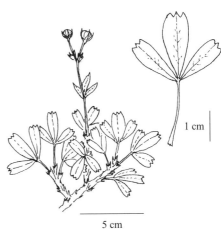

Figure 492. *Sibbaldiopsis tridentata*

3. *Sideroxylon alachuense* L. C. Anders.   ALACHUA BUMELIA

A shrub or small tree, to about 3m, usually with a single trunk. Rare and localized, known only from north-central and peninsular FL, the northernmost site barely reaching the area of coverage of this guide. Occurs in hardwood hammocks and mesic slopes, primarily in vicinity of sinkholes. Named for its site of original discovery in the Alachua Sink area near Gainesville. Twigs glabrous, gray or gray-green, with spinose spur shoots and axillary thorns. Leaves mostly ovate or elliptic, undersurface with a silvery sheen of appressed hairs. [*Sideroxylon anomala* Sarg.; *Bumelia anomala* (Sarg.) R. B. Clark]

4. *Sideroxylon lycioides* L.   BUCKTHORN-BUMELIA   *Figures 496, 497, and 498a*

A shrub or small tree, to 15m tall, usually with a single trunk. Widespread across the SE in the coastal plain and Piedmont from se. VA to n. FL, west to e. TX, scattered northward in the interior to MO, IL, IN, KY. Occurs in sandy to mesic forests near the coast, in bottomlands or swamps, on rock outcrops in basic to calcareous soils, in glades, and on alkaline prairies. Twigs glabrous, grayish. Bark scaling, the grayish strips or plates revealing red-brown underlying patches. Leaves glabrous, variable in size from 6–14cm long, usually 8–10cm and oblong-lanceolate, with prominent veins and blunt to acute tips. Fruit 10–15mm long. [*Bumelia lycioides* (L.) Pers.]

5. *Sideroxylon thornei* (Cronq.) T. D. Pennington   THORNY BUMELIA   *Figure 498b*

A crooked or spindly shrub, to 5m tall. Native to the coastal plain of GA, n. FL, in moist woodlands near streams, ponds, or depressions. Twigs slender, hairy at first, usually glabrous by winter; spines numerous. Bark dark, shallowly fissured. Leaves 1–7cm long, variously shaped from obovate to oblanceolate but normally the latter; underside woolly or densely pubescent. Fruit 8–10mm long. (*Bumelia thornei* Cronq.)

6. *Sideroxylon reclinatum* Michx.   SMOOTH BUMELIA   *Figure 499*

A shrub, variously low and spreading near the ground or more upright, to 5m, though usually crooked, thorny, or multiple-stemmed. Distributed from peninsular FL to sw. GA. Occurs in mesic woodlands near streams or bluffs and over calcareous rock. Twigs slender, mostly glabrous, with many spines or spiny spur shoots. Bark gray, mostly smooth to shallowly fissured. Leaves variable in shape, from nearly round to oblanceolate, 1–7cm long, with a notched, rounded, or obtuse tip; mostly glabrous when mature. Fruit 5–8mm long. [*Bumelia reclinata* (Michx.) Vent.] One subspecies with unique appearance follows:

7. *S. reclinatum* ssp. *rufotomentosum* (Small) Kartesz & Gandhi   RUFOUS BUMELIA

A colonial shrub, sprouting from a spreading root system to form many short stems up to 1m tall. Known only from FL in sandy pinelands and scrub. Twigs covered with spreading rusty hairs. Leaves similar to the species, though red-hairy when young. Fruit 8–12mm long. (*Bumelia rufotomentosa* Small)

Smilax L.   GREENBRIER   Family Smilacaceae   *Figure 500*

Nine woody species are native to the SE, all but 1 being vines that regularly climb with tendrils on the leaf petioles. The petioles and tendrils are persistent beyond abscission of the blade if attachment is made to adjacent host stems beforehand. The base of the petiole covers the buds, so no leaf scar is visible. Prickles are usually present on the main stems but may be sparse or lacking on mature twigs of some species, bristlelike in one species, and totally lacking in 1 species. Stems are angled or squared in most. Leader shoots arise from a tuberous underground rhizome

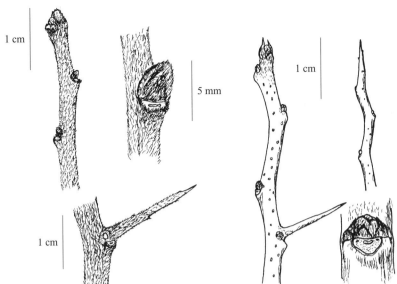

Figure 493. *Sideroxylon lanuginosum*

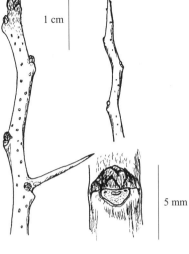

Figure 494. *Sideroxylon tenax*

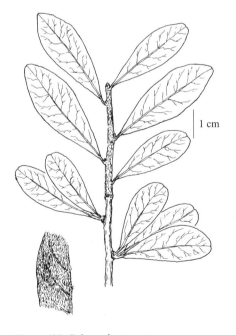

Figure 495. *Sideroxylon tenax*

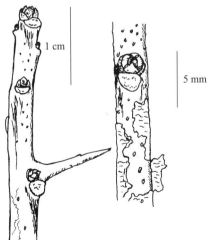

Figure 496. *Sideroxylon lycioides*

330 · Smilax

nearly every year in most species, these being robust and fast growing. Bark of twigs and older stems of most species is green. Many species that are deciduous in n. parts of the SE may be evergreen in the Deep South. The leaves are alternate, variable in their shape and their persistence into winter. The fruit is a berry, 5–10mm long, with a thin skin and smooth, reddish seed, 1 to 4 per fruit, depending on species; borne in umbellate clusters in the axils of leaves. Plants dioecious, so staminate plants bear no fruit.

1. Twigs and leaves densely hairy; plant trailing on ground, with short, erect stems; no prickles
       1. *S. pumila*
1. Twigs and leaves mostly glabrous; plant climbing or scrambling well aboveground; prickles present
   2. Leaves evergreen, thick (coriaceous); margin flat, with few undulations; leaf shape mostly narrowly ovate to oblong-elliptic; base mostly cuneate to rounded; immature fruit may be present over entire winter season
      3. Leaf tips gradually narrowed and acute; submarginal vein not evident   2. *S. smallii*
      3. Leaf tips rounded or abruptly narrowed; submarginal vein present, visible on underside
         4. Midvein of leaf distinct, others relatively indistinct; fruit 1-seeded   3. *S. laurifolia*
         4. Midvein and main laterals about equally distinct; fruit 2- or 3-seeded   4. *S. auriculata*
   2. Leaves subevergreen or deciduous, thin to firm (membranaceous to subcoriaceous); margins usually undulating; leaf shape mostly broadly ovate or pandurate; base mostly truncate to cordate; no immature fruit in winter
      5. Twigs or main stems with slender, bristlelike prickles   5. *S. tamnoides*
      5. Twigs or main stems with broad-based prickles
         6. Leaves glaucous beneath; fruit glaucous, black under the bloom   6. *S. glauca*
         6. Leaves pale to shiny green beneath; fruit not glaucous
            7. Prickles and adjacent stem surface scurfy on leader shoots; leaf margin thickened, often minutely spinose   7. *S. bona-nox*
            7. Prickles and stems glabrous; leaf margins not as above
               8. Leaves shiny green beneath, margins rarely revolute, sometimes minutely toothed; fruit black; common, widespread species   8. *S. rotundifolia*
               8. Leaves dull green beneath, margins usually revolute, entire; fruit red; species limited to wet lowland habitats   9. *S. walteri*

1. **Smilax pumila** Walt. DWARF GREENBRIER   *Figure 501*

An evergreen, trailing species with numerous erect stems 1–5dm tall. Native to the coastal plain from SC to FL, west to TX, s. AR. Occurs in sandy soils under maritime and mesic forests, and in upland oak scrub. Leaves densely hairy below, mostly ovate, veins distinct. Stems lack prickles. Fruit pointed on end, orange to red after maturity; 1-seeded.

2. **Smilax smallii** Morong. JACKSON-BRIER   *Figure 502a*

An evergreen, climbing vine with dark green, horizontal to drooping foliage. Native to the coastal plain from se. VA to FL, west to e. TX, s. AR. Occurs in moist forests and in rich, upland oak or oak-pine woodlands, roadside thickets. Leaves mostly lanceolate, tapered to the acute tip, light green below, not as thick as next 2 species. Twigs and mature fruiting stems often lack prickles. Fruit with a reddish or brownish tinge and glaucous when near maturity, black afterward; ripen in late spring nearly a year after flowering the previous summer; 1- to 3-seeded.

3. **Smilax laurifolia** L. LAUREL GREENBRIER   *Figure 502b*

An evergreen, climbing vine with mostly ascending foliage on the twigs. Distributed from MD to FL, west to e. TX, s. AR, inland to sw. NC, n. GA, s. TN. Most common in the coastal plain, but occasionally in the Piedmont and rarely in low bogs in the s. Appalachians. Typically in wet thickets, pocosins, bogs, bays, swamps. Leaves with obscure veins, except for midvein; margins slightly revolute, with a submarginal vein close to edge. Fruit glaucous near maturity,

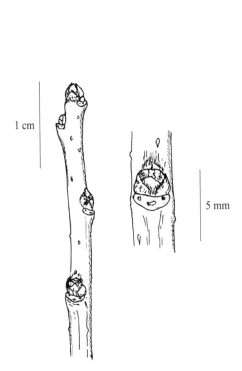

Figure 497. *Sideroxylon lycioides*

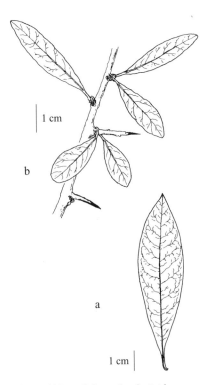

Figure 498. a: *Sideroxylon lycioides;* b: *S. thornei*

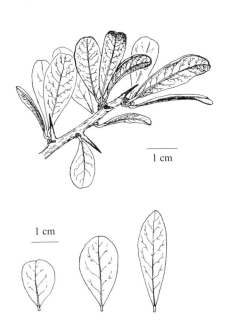

Figure 499. *Sideroxylon reclinatum*

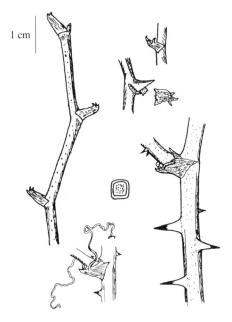

Figure 500. *Smilax*

shiny black afterward; ripe and persistent in autumn, more than a year after flowering in the summer of previous year; 1-seeded.

4. *Smilax auriculata* Walt.   DUNE GREENBRIER   *Figure 502c*

An evergreen, climbing vine with ascending foliage on the twigs. Native to the outer coastal plain from NC to FL, west to LA. Typically near the coast in thickets near dunes or marshes and openings of maritime forests or sandy scrublands. Lower portions of main stems often pinkish-tinted. Leaves mostly oblong-elliptic or narrowly ovate, some leaves pandurate; margin tightly revolute, with 1 or 2 submarginal veins close to the edge. Fruit glaucous, reddish to purplish near maturity, black and sometimes glaucous afterward; ripe in autumn, more than a year after flowering the summer of previous year; 2- to 3-seeded.

5. *Smilax tamnoides* L.   BRISTLY GREENBRIER   *Figure 503a*

A deciduous to subevergreen vine with numerous slender needlelike or bristlelike prickles, especially on lower portions of main stems. Widespread across the SE. Occurs in a wide variety of habitats, though most common in moist soils. Leaves variable in shape, from ovate to nearly orbicular or pandurate, the latter shape most common in coastal plain plants. Leaves green underneath; margins may have minute serrulate points. Fruit black, mature in autumn after previous spring flowering; 1- to 2-seeded. Stems are often angular, with 4 to 6 ridges or sides. (*S. hispida* Muhl.)

6. *Smilax glauca* Walt.   GLAUCOUS GREENBRIER   *Figure 503b*

A deciduous to subevergreen vine. Widespread across the SE in various habitats. Leaves glaucous beneath. Fruit glaucous near maturity, remaining so afterward or becoming shiny black on occasion; ripe in autumn, after spring flowering; 1- to 3-seeded.

7. *Smilax bona-nox* L.   CATBRIER   *Figure 503c*

A deciduous to subevergreen vine. Widespread across much of the SE, though absent from most of the Appalachians in VA, WV, MD. Occurs in a variety of habitats. Stems are scurfy on and about the prickle bases; this trait is most noticeable on the leader shoots of the last growing season. Leaf margins thickened, often with minute, serrulate points; undersides light green; leaf shape sometimes pandurate. Fruit dull to shiny black, ripe in autumn and persistent into winter; 1-seeded.

8. *Smilax rotundifolia* L.   COMMON GREENBRIER   *Figure 503d*

A deciduous to subevergreen vine. Widespread throughout the SE in various habitats. Leaf margins thin, sometimes with serrulate points; underside light green, glossy. Fruit glaucous to dull black, ripe in autumn and persistent into winter; 1- to 3-seeded.

9. *Smilax walteri* Pursh   CORAL GREENBRIER

A deciduous vine with sparsely prickled mature stems. Native to the coastal plain and Piedmont from MD to FL, west to e. TX, and sporadically in the interior to TN, AR. Occurs in wet soils of flood-prone forests and bogs, often in areas subject to inundation for parts of the year. Leaves similar in shape to *S. glauca,* mostly ovate-lanceolate, pale green below; margin usually closely or tightly revolute and entire. Fruit globose to egg-shaped, red after maturity in autumn, persist into winter; 2- to 4-seeded.

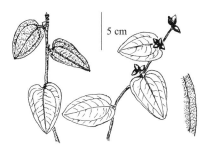

Figure 501. *Smilax pumila*

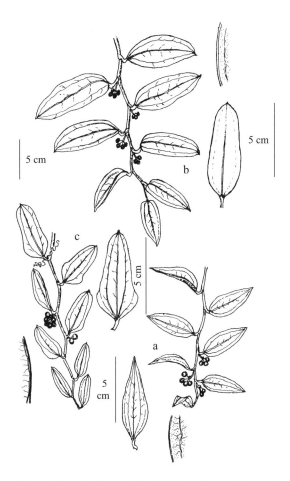

Figure 502. a: *Smilax smallii;* b: *S. laurifolia;* c: *S. auriculata*

Solanum dulcamara L. BITTER NIGHTSHADE   Family Solanaceae   *Figure 504*

An exotic vine, clambering through and over other plants for support and access to light. Native to Europe; widely naturalized across the SE. Most often seen near urban areas and waste places. Twigs brown, grayish, or greenish-gray, supple; terminal bud lacking; laterals blunt and few-scaled; bundle scar 1. Inner bark is rankly odorous. Fruit is a red, fleshy, oval berry about 1cm long, acrid and poisonous if eaten. Flowers violet or purplish, appear in summer. Also known as "bittersweet," thus sometimes confused colloquially with *Celastrus,* though unrelated botanically.

Sophora affinis Torr. & Gray   TEXAS SOPHORA   Family Fabaceae   *Figure 505*

An uncommon, small tree, to about 10m tall. Native to central and e. TX primarily, with scattered populations in se. OK, sw. AR, nw. LA. Occurs in calcareous uplands and along streams. The greenish twigs have no terminal bud; laterals sunken and additionally concealed by whitish hairs along the inner edge of a surrounding leaf scar. Legume 4–8cm long, black, constricted or pinched between the seed, not splitting after maturity. Bark brown or red-brown, scaly or with thin, scaly plates.

Sorbus L.   MOUNTAIN-ASH   Family Rosaceae

One native species and 1 sparingly naturalized exotic are treated here for the SE. In both, twigs show crescent-shaped alternate leaf scars, terminal buds that are much larger than laterals, and several bundle scars. Scraped twigs have a strong, cherrylike odor. Bark grayish and conspicuously lenticellate on young trunks. Leaves pinnately compound, with toothed leaflets. Fruit a small pome, reddish, in wide, flat-topped clusters. The species of mountain-ash are treated as members of *Pyrus* in some texts.

1. Twigs and buds glabrous; native species   1. *S. americana*
1. Twigs or buds hairy; exotic species   2. *S. aucuparia*

1. *Sorbus americana* Marsh.   AMERICAN MOUNTAIN-ASH   *Figure 506*

A small tree, to 10m tall. A n. species reaching the SE in its s. range along the high elevations of the Appalachians from WV and VA to n. GA. Usually in high mountain openings, balds, rocky outcrops, and disturbed slopes. Twigs stout, grayish brown to purplish-brown; terminal buds conical, 15–20mm long, with 3 or 4 reddish imbricate scales; resinous and aromatic within. Bark smooth, becoming scaly on old trunks. Fruit about 5mm wide, red.

2. *Sorbus aucuparia* L.   EUROPEAN MOUNTAIN-ASH   *Figure 507*

A small tree, to 15m tall. Native to Europe; planted extensively in n. sections of the SE, rarely naturalized near cultivated specimens. Twigs stout, hairy near nodes or apex; terminal bud with hairy scales, not resinous within. Fruit orange to orange-red, 8–12mm wide. This is the common "rowan tree" that most frequently is cultivated in urban and lawn environments in the SE.

Spiraea L.   SPIREA   Family Rosaceae

Five species are native to the SE. One additional exotic (*S. japonica*) is known to naturalize successfully and appears to be fairly tenacious in gaining a position in native habitats. These 6 species are not difficult to identify in winter, so the first key offered below includes only these species. Additional exotics that are known to occasionally naturalize or persist after cultivation to the point of appearing naturalized complicate matters. Many species, hybrids, and cultivars

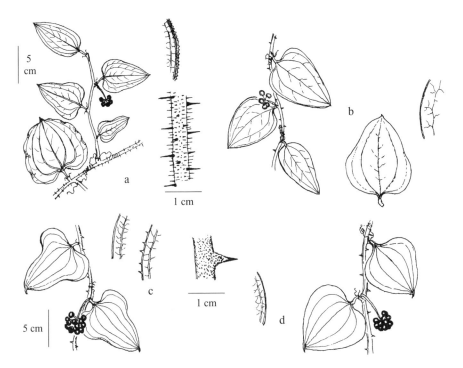

Figure 503. a: *Smilax tamnoides;* b: *S. glauca;* c: *S. bona-nox;* d: *S. rotundifolia*

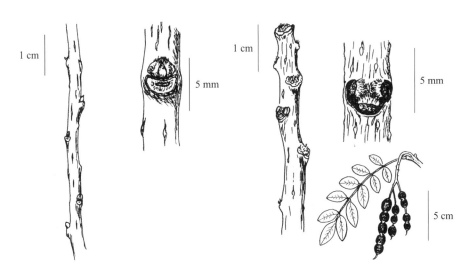

Figure 504. *Solanum dulcamara*

Figure 505. *Sophora affinis*

of exotic spireas are cultivated, and local naturalization may occur. These may not persist as long-term members of the flora of the SE, so they are included in a 2d key. These are best identified in flower and leaf, so the second key will not be highly useful in winter, especially if dried remnants of the fruit clusters are lacking. Twigs of all spireas are slender, with size of buds and leaf scars progressively smaller toward the tapering tips of the twigs. Terminal buds lacking; laterals with several brown imbricate scales; hairy and appearing naked in 1 species. Leaf scars alternate, often raised, with 1 tiny bundle scar, or bundle scars not discernible; lines or ridges may extend down twigs from sides of leaf scars. The dried fruits (capsules) often persist on twigs during the winter; they give valuable clues as to the type of branched inflorescence, either flat-topped (corymbose), elongated (paniculate), or radiating from 1 node (umbellate).

Key to native and most commonly naturalized exotic spireas

1. Twigs densely brown-hairy or tomentose near tips 1. *S. tomentosa*
1. Twigs glabrous or only slightly hairy
  2. Buds rounded or bluntly pointed
    3. Leaves coarsely toothed; low shrub of dry, rocky uplands 2. *S. corymbosa*
    3. Leaves remotely toothed or nearly entire; tall shrub of streambanks 3. *S. virginiana*
  2. Buds acute or sharply pointed
    4. Twigs lined; fruit in panicles; native
      5. Twigs yellow-brown or light brown; leaves finely toothed, rarely over 15mm wide
        4. *S. alba*
      5. Twigs red-brown; leaves coarsely toothed, usually over 15mm wide 5. *S. latifolia*
    4. Twigs barely lined or not lined; fruit in corymbs; exotic 6. *S. japonica*

Expanded key to native and occasionally naturalized exotic spireas using flowers and leaves

1. Inflorescence and fruit clusters in well-branched clusters, not umbels
  2. Inflorescence in flat-topped corymbs; section *Calospira* K. Koch
    3. Leaves blunt to acute at tip; buds blunt or rounded; native species
      4. Leaves coarsely toothed; low shrub of dry, rocky uplands 2. *S. corymbosa*
      4. Leaves remotely toothed or nearly entire; tall shrub of streambanks 3. *S. virginiana*
    3. Leaves acuminate at tip; buds acute or sharply pointed; exotic species
      5. Leaves hairy below; twigs barely lined or not lined; flowers pink 6. *S. japonica*
      5. Leaves glabrous below; twigs distinctly lined; flowers white or pink 7. *S.* × *bumalda*
  2. Inflorescence in panicles, or clusters taller than wide; section *Spiraria* Seringe.
    6. Flowers white, in loose panicles; buds glabrous, pointed
      7. Twigs yellow-brown or light brown; leaves finely toothed, rarely over 15mm wide
        4. *S. alba*
      7. Twigs red-brown; leaves coarsely toothed, usually over 15mm wide 5. *S. latifolia*
    6. Flowers pink, in dense pyramidal panicles; buds blunt or hairy
      8. Leaves glabrous beneath 8. *S. salicifolia*
      8. Leaves densely hairy beneath
        9. Leaves with impressed veins (rugose) above, usually bluntly serrate; rarely over 1m tall; fruits hairy; native species 1. *S. tomentosa*
        9. Leaves not rugose, sharply serrate; to 2m tall; fruits glabrous; exotic 9. *S.* × *billardii*
1. Inflorescence and fruit clusters in simple umbels; section *Chamaedryon* Seringe.
  10. Umbels sessile; leaves finely toothed
    11. Leaves ovate, 10mm or more wide; flowers usually double 10. *S. prunifolia*
    11. Leaves narrowly lanceolate, mostly under 10mm wide; flowers not double
      11. *S. thunbergii*

10. Umbels with a short stalk; leaves coarsely or deeply toothed, or slightly lobed
    12. Leaves mostly lanceolate; main side veins 2 or more per side; flowers usually double
                                                                                           12. *S. cantoniensis*
    12. Leaves ovate or obovate; usually with 1 pair of main side veins; flowers not double
                                                                                           13. *S.* × *vanhouttei*

Native species

1. *Spiraea tomentosa* L. STEEPLEBUSH SPIREA   *Figure 508*

A slender, rhizomatous shrub, rarely over 1m tall. Widespread from the NE to e. VA, n. GA, west to AR. Occurs in moist to wet soils of bogs, ditches, meadows. Often overlooked due to grazing or mowing that reduces the plant to many slender, sparsely branched stems of 2–4dm height. Twigs brown, lined, tomentose. Leaves 3–7cm long, 1–3cm wide, crenate-serrate, tomentose beneath, veins impressed above. Flowers normally pink, in a paniculate cluster. In var. *rosea* (Raf.) Fern., flowers deeper pink or rose-colored and not closely crowded as typical for the species.

2. *Spiraea corymbosa* Raf. DWARF SPIREA

A low, rhizomatous shrub, rarely to 1m tall, usually 5dm or less. Distributed in the Appalachians from MD and WV to n. AL. Not common; rare south of VA and KY. Occurs in rocky, open woodlands, over granitic outcrops, talus slopes, and sunny road banks. Twigs brown, slightly lined below nodes. Leaves 3–8cm long, 2–5cm wide, mostly oval, rounded to blunt at the tip, margin coarsely toothed above the middle. Fruit 2 or 3mm long. This plant is 1 of a trio of closely related low shrubs found in N. Amer. and e. Asia, all of which are considered varieties of *S. betulifolia* Pallas in some texts. The original *S. betulifolia* is native to ne. Asia and Japan, with more pointed buds and leaves, and heavier lined twigs. The w. N. Amer. *S. lucida* Greene [*S. betulifolia* var. *lucida* (Douglas ex Greene) C. L. Hitchcock] has 2–6cm leaves more often blunt to acute at apex and glossy above. The dwarf spirea of the Appalachians is then sometimes considered as *S. betulifolia* var. *corymbosa* (Raf.) Maxim. Flowers white.

3. *Spiraea virginiana* Britton VIRGINIA SPIREA   *Figure 509*

A weakly rhizomatous shrub, to 2m tall. Endemic to the Appalachian region, from WV and VA to n. GA. Rare and sporadic along streams and flood-scoured thickets. Twigs yellow-brown or grayish, faintly lined below nodes. Leaves 4–8cm long, 15–25mm wide, acute at the tip, margin remotely toothed beyond middle, glaucous below. Flowers white.

4. *Spiraea alba* DuRoi MEADOWSWEET   *Figure 510*

A clumped or rhizomatous shrub, to 2m tall. Distributed widely in the North, ranging southward into the SE mainly in the Appalachians of WV, VA, and NC. Occurs along streams, in bogs, and at similar wet sites. Twigs yellow-brown, angled or lined below leaf scars. Leaves 5–7cm long, 10–15mm wide, finely and sharply toothed. Flowers white.

5. *Spiraea latifolia* (Ait.) Borkh. BROADLEAF MEADOWSWEET

A rhizomatous shrub, to 2m tall. Distributed from the North southward into the coastal plain of VA and across the Piedmont to the Appalachians of NC. Occurs along streams, ditches, in bogs, and at similar wet sites. Twigs brown, angled or lined below leaf scars. Leaves 3–7cm long, 15–40mm wide, coarsely and sharply toothed. Flowers white.

Exotic species

6. Spiraea japonica L. f. JAPANESE SPIREA   Figure 511

A spreading shrub, to 1.5m tall, coarsely rhizomatous. Native to Japan; naturalized sporadically in the SE, sometimes locally abundant. Twigs brown, slightly lined to unlined. Leaves 6–12cm long, 3–4cm wide, acute to acuminate, doubly toothed, usually pubescent below. Flowers pink.

7. Spiraea × bumalda Burven. ANTHONY WATERER SPIREA

A clumping shrub, to 1m tall. Of hybrid origin [*S. japonica* L. × *albiflora* (Miq.) Zab.], the most commonly cultivated form in the SE being the selection 'Anthony Waterer'. Rare as an escape, or persistent after cultivation. Leaves lanceolate, sharply toothed, similar to *S. japonica*, but glabrous, also differing in twigs distinctly lined, flowers white to pink.

8. Spiraea salicifolia L. BRIDEWORT

A suckering and rhizomatous shrub, to 2m tall. Native to Eurasia; rarely naturalizing near plantings or appearing naturalized due to persistence at old home sites. Twigs yellow-brown, lined below nodes. Leaves 4–7cm long, 10–18mm wide, sharply toothed. Flowers pink.

9. Spiraea × billardii Herincq. BILLARD SPIREA

A suckering shrub, to 2m tall. Of hybrid origin (*S. douglasii* Hook × *salicifolia* L.); rarely naturalized but spreading from cultivation through sprouts. Twigs pubescent, lined, brown. Leaves oblong, 5–8cm, doubly toothed, grayish and hairy beneath. Flowers dark pink, in narrow, pyramidal panicles. Fruits hairless.

10. Spiraea prunifolia Sieb. & Zucc. BRIDAL WREATH SPIREA

A spreading shrub with arching branches, to 2m tall. Native to e. Asia; rarely naturalized near plantings or persistent at old sources of cultivation. Twigs brown, lined. Leaves ovate, finely toothed, 2–4cm long. Flowers white, commonly double-flowered (more than 1 set of petals), in sessile umbels, each flower stem pubescent.

11. Spiraea thunbergii Sieb. ex Blume CHINESE SPIREA

A spreading or clumplike shrub, to 2m tall. Native to e. Asia; rarely naturalizing near plantings, or appearing naturalized due to persistence at old home sites. Twigs brown, lined, very slender. Leaves narrow and willowlike, toothed, 2–4cm long, 5–8mm wide. Flowers white, in a sessile umbel; individual flower stems glabrous.

12. Spiraea cantoniensis Lour. REEVES SPIREA

A spreading shrub, to 2m tall, with arching branches. Native to China and Japan. Rarely naturalized, weakly persisting after cultivation. Twigs slightly lined; buds short-pointed. Leaves mostly lanceolate, 3–5cm long, deeply toothed beyond middle, blue-green below. Flowers white, about 1cm wide, in sessile, glabrous umbels; double-flowered in the var. *lanceata* Zab., which is the normally encountered form cultivated in the SE, more common than the typical species.

13. Spiraea × vanhouttei (C. Briot) Carriere HYBRID BRIDAL WREATH SPIREA

A spreading or mounded shrub, to 2m tall. Of hybrid origin (*S. cantoniensis* Lour. × *trilobata* L.), extensively cultivated across the SE. Occasionally spreading by suckers from cultivated

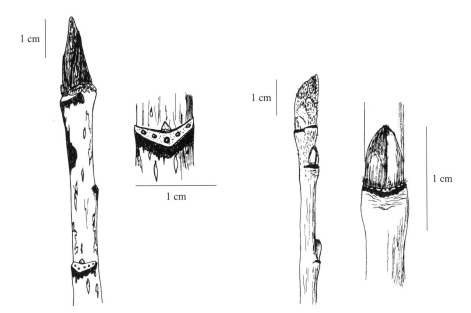

Figure 506. *Sorbus americana*

Figure 507. *Sorbus aucuparia*

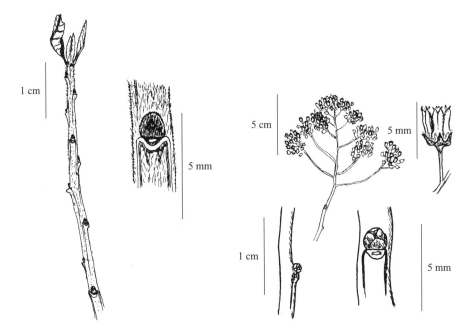

Figure 508. *Spiraea tomentosa*

Figure 509. *Spiraea virginiana*

specimens, and often long-persistent once planted. Twigs slightly lined, slender and arching. Leaves rhombic-ovate to obovate, blue-green beneath, 2–4cm long, large-toothed to slightly lobed beyond middle; usually with 3 main veins. Flowers white, in a short-stalked umbel.

Stachydeoma graveolens (Chapm. ex Gray) Small   MOCK-PENNYROYAL   Family Labiatae *Figure 512*

A semiwoody, minty-scented shrub, rarely over 3dm tall, with semierect branches and very pubescent twigs. Native to the FL panhandle, where it occurs in pine flatwoods on slightly elevated or better-drained soils. Leaves opposite, sessile, 1cm or less long, dotted, with rounded or nearly heart-shaped bases. Larger leaves may have a few large, blunt teeth; margins curled downward. The fruit is composed of the persistent calyx with nutlets inside, similar to fruit of *Conradina*. (*Hedeoma graveolens* Chapm. ex Gray)

Staphylea trifolia L.   BLADDERNUT   Family Staphyleaceae *Figure 513*

A shrub, rarely a small tree, to 6m tall. Widely scattered across the SE, except uncommon to absent in the outer coastal plain. Occurs in rich soils of bottomland or mesic forests, shaded slopes, ravines, and over calcareous rocks. Twigs greenish, glabrous; leaf scars opposite, with 3 or more obscure bundle scars; terminal bud lacking; laterals with 2 to 5 visible imbricate scales; stipule scars present. Bark gray, smooth, often white-striped on some young stems. The fruit is a 3-lobed, inflated or bladderlike pod about 2–5cm long, persistent into winter, with a few hard, 5–8mm seed within.

Stewartia L.   STEWARTIA   Family Theaceae

Two native species occur in the SE. Twigs have alternate leaf scars with a single, raised bundle scar and conspicuous buds with 2 to 3 visible silky-hairy, imbricate scales. Terminal buds are present. The fruit is a 4- to 5-parted capsule, bulged or angled toward the sutures.

1. Bundle trace peglike, nearly as wide as leaf scar; fruit with rounded lobes, seed plump, glossy 1. *S. malacodendron*
1. Bundle trace raised but not as above; fruit with sharply angled lobes, seed flat, dull brown 2. *S. ovata*

1. *Stewartia malacodendron* L.   SILKY STEWARTIA *Figure 514*

A shrub or small, wide-crowned tree, to 5m tall. Native in the coastal plain and Piedmont from se. VA to w. FL, west to se. TX, and in the interior to se. NC, n. AL, s. AR. Uncommon and sporadic; seen mostly in mesic forests, on rich slopes and in ravines. Twigs red-brown; terminal bud visible before leaves fall as petiole does not envelope bud. Bark thin, brown, smooth on young stems but scaling off in thin plates to reveal paler, grayish inner layers on older trunks. Fruit 1–2cm long and as wide, the 4 or 5 lobes rounded; seed about 6mm long, dark brown to black, smooth, not winged.

2. *Stewartia ovata* (Cav.) Weatherby   MOUNTAIN STEWARTIA *Figure 515*

A small tree, to 10m tall in optimum sites, but usually 4–6m tall over most of its range. Mostly concentrated in the s. Appalachians and interior plateaus of e. KY south to n. AL, n. GA and adjacent NC, SC, though also ranging into the Piedmont of NC and the coastal plain of VA. Uncommon to rare over most of its range; occurring at rich, moist ravines and slopes, mesic forests, and acidic forest understories in the Blue Ridge. Twigs red-brown; terminal bud nearly enclosed by the winged petiole of the endmost leaf, prior to leaf fall. Bundle scar single, raised, less than ½ as wide as leaf scar. Bark brown or gray-brown, fissured, with narrow ridges. Fruit

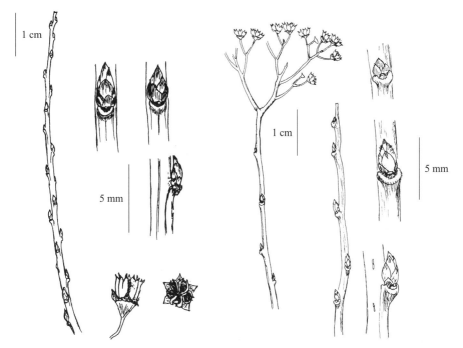

Figure 510. *Spiraea alba*

Figure 511. *Spiraea japonica*

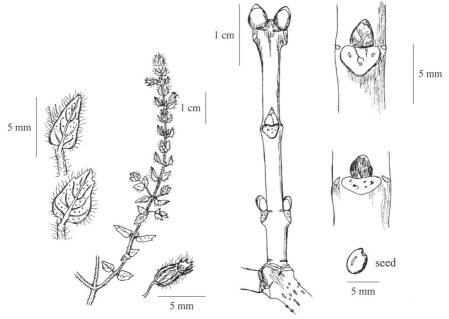

Figure 512. *Stachydeoma graveolens*

Figure 513. *Staphylea trifolia*

15–25mm long, less wide, pointed, with angular lobes; seed flat, brown, to 1cm long, slightly winged.

Stillingia aquatica Chapm.   SHRUB-CORKWOOD   Family Euphorbiaceae

Figures 516 and 517

A slender, sparingly branched semiaquatic shrub, to 1.5m tall. Main stem slender and straight, the few branches occurring at the top of the plant. Usually found in water, where it may form colonies or groves. Native to the coastal plain from SC to FL, west to AL. Occurs in wet ditches, depressions, pond margins, and similar wetlands where water is present for most of the year. Twigs bear many tiny buds with slightly fleshy, loose scales; no terminal buds. Stipules tiny, persistent; leaf scars darkened, with 1 bundle scar barely discernible; sap milky. The fruit is a 3-lobed capsule. The wood is very light, like cork. Leaves narrowly lanceolate, with small, thickened teeth; often crowded at twig tip, sometimes persistent into winter, usually turning red before falling.

Styrax L.   SNOWBELL   Family Styracaceae

Two species are native to the SE, both shrubby understory species and widespread in range. Twigs of both have superposed, scurfy buds appearing naked or 2-scaled, no terminal buds, leaf scar with 1 raised, roughened bundle scar, and often bear epidermal cracks or slits. Nodes are often swollen, or twigs zigzag. Fruits are drupelike, rounded, with a dry, grayish to light brown, scurfy skin that may split to release several plump red-brown seed 5–8mm long; may persist into early winter. Flowers white, pendent, accounting for the common name.

1. Twigs brownish, slender, internodes mostly 1–1.5mm diam.; leaves 4cm wide or less
1. *S. americanus*
1. Twigs grayish, moderately slender, internodes mostly 2–3mm diam.; leaves mostly over 4cm wide
2. *S. grandifolius*

1. *Styrax americanus* Lam.   AMERICAN SNOWBELL   *Figure 518*

A large suckering shrub or small tree, to 4.5m tall. Widely distributed across the SE, mostly in the coastal plain and Piedmont from se. VA to e. TX, and north in the Mississippi Embayment to s. OH. Typically in moist to wet, acidic soils. Twigs usually light brown and essentially glabrous, but in the var. *pulverulentus* (Michx.) Perkins ex Rehd., the twigs are grayish-scurfy; this variety occurs in the coastal plain of SC to TX. Leaves mostly 3–7cm long, 4cm wide or less, glabrous except in var. *pulverulentus*. Fruit 5–8mm diam. Bark grayish, thin and smooth.

2. *Styrax grandifolius* Ait.   BIGLEAF SNOWBELL

A shrub or small tree, to 6m tall. Widespread across most of the SE, though scarce or absent in WV and w. KY. Occurs in a variety of habitats across its range; in bottomlands, slightly elevated soils within floodplains, rich slopes, and mesic woodlands in the south and west, dry upland ridges and pine-oak forests in the north and east. Twigs stouter than in *S. americanus*, with nodes more conspicuously swollen, all probably in accommodation of the greater weight of the larger leaves. Leaves widely oval to suborbicular, mostly 8–12cm long, 4–10cm wide, pubescent beneath. Fruit 8–10mm diam. Bark gray, smooth on most stems, shallowly fissured on oldest trunks.

Symphoricarpos Duhamel   CORALBERRY   Family Caprifoliaceae

Two species are native to the SE. Both are shrubs that spread by rhizomes or root sprouts, often forming colonies. Twigs slender, with opposite leaf scars and buds; no terminal bud; and usually

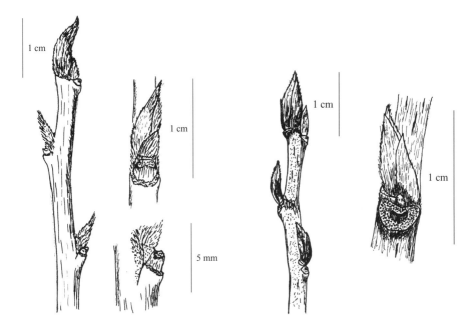

Figure 514. *Stewartia malacodendron*

Figure 515. *Stewartia ovata*

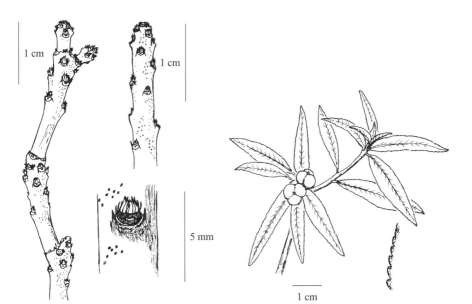

Figure 516. *Stillingia aquatica*

Figure 517. *Stillingia aquatica*

344 · Symphoricarpos

a prominent branch stub terminates the twig. Bundle scars single in each leaf scar, but may be obscure; a ridge or line joins the leaf scars at the nodes; pith small, spongy. The fruit is a berry with colorful but dryish flesh covering several tiny brown seed; borne in short clusters or spikes from axillary nodes or twig tips; these often persist into winter.

1. Buds fully exposed on twig, mostly 2–3mm long; bark tight on most stems; fruit white
   1. *S. albus*
1. Buds partially hidden, less than 2mm long; bark shreddy; fruit reddish
   2. *S. orbiculatus*

1. *Symphoricarpos albus* (L.) Blake   SNOWBERRY   *Figure 519*

A shrub with slender, arching stems to 1m tall. Widespread across n. N. Amer., reaching the SE in the Appalachians of WV and VA. Occurs in rocky often calcareous soils. Twigs smooth but lenticellate; bark flaking only on older stems; buds with 3 or 4 pairs of imbricate, keeled scales. Fruit white, 6–10mm diam. The var. *laevigatus* (Fernald) Blake is also known in the region, sporadically naturalized after its introduction from w. N. Amer.; it is taller (to 2.5m), coarser in habit, with fruit 12–20mm.

2. *Symphoricarpos orbiculatus* Moench   CORALBERRY   *Figure 520*

A suckering shrub, to 2m tall, with slender, arching branches. Widespread throughout the SE in dry woodlands, on road banks, rocky slopes, streamsides. Twigs brown, usually with shredding epidermis. Buds small, partially hidden behind petiole stubs or leaf scars of the swollen nodes; bud scales sharply tipped; stubs of fruiting stem remnants and collateral buds often present. Fruit reddish, coral pink, or purplish, 5–8mm long.

Symplocus tinctoria (L.) L'Her.   SWEETLEAF   Family Symplocaceae   *Figures 521 and 522*

A shrub or small tree, to 10m tall. Widespread in the SE, mainly in the coastal plain and Piedmont from se. VA to e. TX, and inland or north in the interior to w. NC, n. AL, central AR. Occurs mostly in moist to wet soils, though also seen in dry uplands in the Appalachians. Twigs glabrous or hairy, with alternate leaf scars and single bundle scars. Terminal bud present, larger than laterals, somewhat conical, with 3 to 6 elongate outer scales; lateral buds short, sometimes crowded near twig tip; lateral flower buds globular and larger than other laterals. The sweet taste that accounts for the name is not evident in twigs. Pith chambered. Leaves deciduous to subevergreen (more subevergreen in s. parts of range), entire to sparingly toothed, elliptic, variably sweet when chewed. Bark thin, smooth, grayish to brown or appearing green when wet, often with horizontal holes made by sap-feeding birds. Sometimes the bark shows pale longitudinal stripes. The fruit is an oblong drupe, pelletlike, purplish after maturity, 10–16mm long.

Syringa vulgaris L.   LILAC   Family Oleaceae   *Figure 523*

A tall, robust shrub or small tree, to 7m tall. Native to Europe; extensively and long cultivated for its fragrant flowers. Often seen persisting around old home sites or, rarely, spreading by root suckers. Twigs moderately stout, glabrous, with opposite leaf scars; bundle scar 1, sometimes obscure; lines or ridges below nodes. Terminal bud lacking; laterals with 1 to 4 pairs of imbricate greenish scales, these sharply tipped or keeled on larger buds. Bark gray, mostly smooth but lenticellate; may become fissured or slightly scaly on old trunks. Fruit a capsule about 2cm long, rarely seen.

Tamarix L.   TAMARISK   Family Tamaricaceae   *Figure 524*

All these large shrubs or small trees are exotic, and considerable difficulty can be expected in their identification unless they are examined when in flower. All have small, scalelike leaves

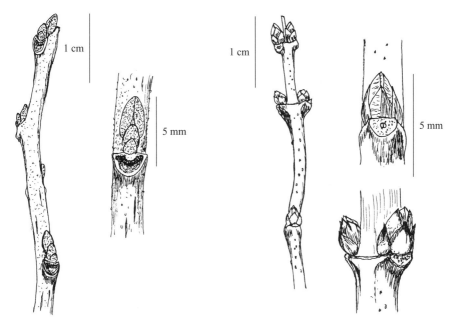

Figure 518. *Styrax americanus*

Figure 519. *Symphoricarpos albus*

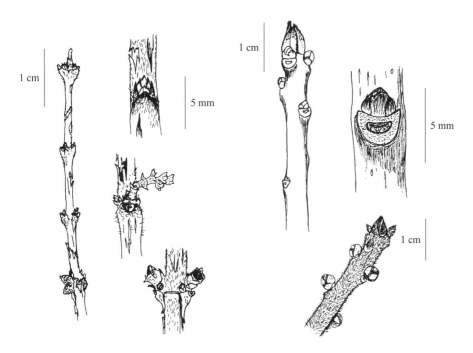

Figure 520. *Symphoricarpos orbiculatus*

Figure 521. *Symplocus tinctoria*

1–3mm long, appressed to slender green twigs. Leaves are more crowded on the side twigs that develop from the main twig; these are often deciduous in winter, similar to *Taxodium*. Buds are partially hidden beneath the wider-spaced, alternate, persistent leaves of the main twig. Flowers are small, white to pink-petaled, borne in narrow racemes at the tips of twigs or side twigs, sometimes at various times of the year. The fruit is a capsule 2–5mm long, splitting to release small seed tufted with hairs. Bark gray on small stems, becoming thick and furrowed with age and nearly black on old trunks, with rough or scaly ridges. These plants are most often seen naturalized near the coast or in sandy soils near lowland streams and waste places. The 3 species treated here are not the only species that may be encountered in the SE; several more are known in cultivation, and naturalization should be expected. Some of these less commonly naturalized species include: *T. canariensis* Willd., similar to *T. gallica* but calyx lobes deeply toothed; *T. africana* Poir., also similar to *T. gallica* but racemes broader, borne on lateral twigs of branchlets; *T. ramosissima* Ledeb., similar to *T. chinensis* but calyx lobes toothed and racemes borne on lateral twigs of branchlets; *T. pentandra* Pall., also similar to *T. chinensis* but racemes to 8cm long, in large, erect panicles.

1. Twigs and branchlets erect or spreading; flower petals fall quickly after anthesis; stamens on calyx lobe tips   1. *T. gallica*
1. Twigs and branchlets usually drooping; flower petals persist after anthesis; stamens arise between calyx lobes
    2. Calyx lobes 5; flower panicles terminal   2. *T. chinensis*
    2. Calyx lobes 4; flower clusters or racemes on lateral twigs from branchlet   3. *T. parviflora*

1. *Tamarix gallica* L.   FRENCH TAMARISK

A shrub or tree, to 10m tall. Native to the Mediterranean; naturalized in coastal areas of the SE. Seen near marshes, along roads, and near waste areas; sometimes locally abundant. Leaves blue-green or dull green, with scarious margins; branches erect or spreading. Racemes to 5cm long, 5mm wide, borne in loose, terminal panicles mostly erect or spreading; flowers with stamens arising on tip of entire-margined calyx lobes; petals fall after anthesis.

2. *Tamarix chinensis* Lour.   CHINESE TAMARISK

A shrub or small tree, to 8m tall. Native to China; naturalized in coastal areas of the SE, mainly SC to TX. Leaves blue-green, scarious-margined, keeled. Branchlets and twigs often drooping. Racemes to 6cm long, borne in loose, terminal panicles that often droop; each flower with stamens arising between entire-margined calyx lobes; petals usually persisting after anthesis.

3. *Tamarix parviflora* DC.   SMALLFLOWER TAMARISK

A shrub or tree, to 6m tall. Native to se. Europe; naturalized in the coastal plain of the SE; occasionally seen in brackish soils near the coast. Leaves blue-green or light green, with scarious margins; twigs and branchlets spreading or arching. Racemes to 4cm long, borne on twigs developing laterally from branchlets; flowers with stamens arising between calyx lobes; petals persist after anthesis; 4 petals and sepals (5 in all other species).

Taxodium L. C. Rich.   BALD-CYPRESS   Family Cupressaceae

Two species are native to the SE, though 1 of these is considered a variety of the other, in some texts. Both the native trees are well known and distinctive to many field workers not familiar with the taxonomic debate, so they are treated as separate species here. The needlelike leaves of these conifers are deciduous along with the side twigs that bear them. The main leader twigs that are available for winter inspection have short-pointed, imbricate-scaled terminal buds and may additionally have globular flower buds near the tip; other buds are sunken and visible only as lumps on the bark. Scales may persist over some of these lateral bud nodes. The fruit is a spherical cone, mostly 2–3cm diam., disintegrating after maturity in autumn and releasing

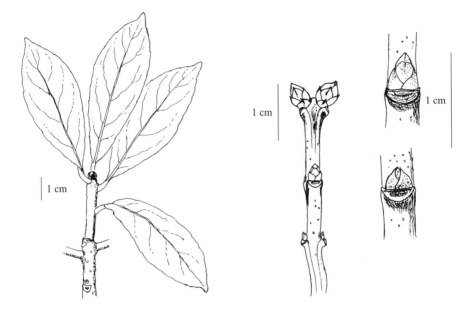

Figure 522. *Symplocus tinctoria*

Figure 523. *Syringa vulgaris*

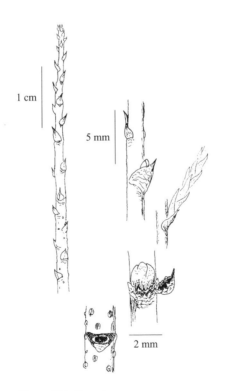

Figure 524. *Tamarix*

several sharply angled, red-brown seed. When growing in typical native habitat of wet or frequently inundated soils, the trunk base becomes fluted, and roots near the ground surface produce conical woody extensions called "knees" that may project upward to heights of 2m. Twigs cut from trees in winter cannot be used alone to separate the 2 entities; attention to bark and fallen twigs with attached leaves is more useful.

1. Leaves of mature plants acicular, appressed, overlapping on side twigs; bark grayish, with thick, scaly ridges (if a juvenile tree, leaves flat, spreading, linear; bark grayish and shallowly to moderately fissured and ridged)   1. *T. ascendens*
1. Leaves not appressed, more acicular and close to twigs in upper crown, linear and spreading on twigs of most of crown, giving featherlike appearance; bark brownish, shreddy or fibrous and shallowly ridged   2. *T. distichum*

1. *Taxodium ascendens* Brongn.   POND-CYPRESS

A tree, to 30m tall, usually with a buttressed base, even when young. Native to the coastal plain from NC to FL, west to se. LA. Occurs in frequently inundated depressions or ponds in pine flatwoods, edges of lakes, blackwater swamp ponds, and wet soils over calcareous clays. The deciduous side twigs are rather threadlike and upturned, with appressed leaves. Bark usually grayish, fissured into ridges early in youth, ridges or plates thickening with age. Knees usually short and stubby, rarely over 4dm tall, with bark uniformly thick over their top. Juvenile trees have foliage resembling the next species. Due to the presence of intermediates between this and the next species, pond-cypress is often considered only a variety of the bald-cypress. [*T. distichum* var. *imbricarium* (Nutt.) Croom; *T. distichum* var. *nutans* (Ait.) Sweet]

2. *Taxodium distichum* (L.) L. C. Rich.   BALD-CYPRESS   *Figure 525*

A tree, to 50m tall, developing a buttressed base with age. Native to the coastal plain from MD to FL, west to e. TX and north in the Mississippi Embayment to s. IL, w. KY. Occurs in river swamps, along sloughs and drainages, and in or along many stream-flooded lowlands. The deciduous side twigs are mostly featherlike, with flat, spreading needles; those from high in the crown may resemble the pond-cypress, though needles not as appressed. Bark usually light brown to dark brown, fibrous or shreddy on young trunks, thickening with age but remaining thinner than in pond-cypress, with elongated ridges on old trunks and fibrous upper portions. Knees to 2m tall, usually narrowly conical, with thinner bark over the reddish tips.

Taxus L.   YEW   Family Taxaceae   *Figure 526*

There are 2 native species of yews in the SE. These are shrubby, rarely cultivated conifers with flat, evergreen needles and red, fleshy fruits. The needles (leaves) are linear, flat, to 25mm long, abruptly pointed at the tip, green above, with 2 pale green stomatal bands beneath (where rows of stomates are present). Slender petioles connect each needle to the green, ridged twigs; terminal bud with green, imbricate scales; stalked flower buds may be visible in axils of some leaves. The fruit of yews is a brownish seed nearly surrounded by a scarlet aril, the flesh of the aril succulent, sweet, mucilaginous, edible; seed, however, are poisonous if chewed. Bark is reddish or purplish, with peeling outer scales or plates. A few species of exotic yews are commonly cultivated, but none are known to naturalize in the SE.

1. Habit low and sprawling; bark reddish; seed to 5mm long; n. species   1. *T. canadensis*
1. Habit upright; bark purplish-brown; seed to 6mm long; limited to FL   2. *T. floridana*

1. *Taxus canadensis* Marsh.   CANADA YEW

A sprawling or spreading shrub, to 2m high. A ne. conifer ranging into the SE in s. OH, KY, and in the Appalachians from WV and VA to NC, TN. Occurs in boggy soils of coniferous forests and over rocky slopes or bluffs. Uncommon and localized in the SE. Usually monoecious.

2. *Taxus floridana* Nutt. ex Chapm.   FLORIDA YEW

A shrub or small tree, to 8m tall, with vertical or leaning stems. Rare; confined to the FL panhandle in a few mesic ravines and slopes near the Apalachicola River and in a white-cedar swamp. Most of the remaining populations are not vigorous in health and are likely relictual of habitats influenced by past climatic periods. Dioecious.

Thuja occidentalis L.   NORTHERN WHITE-CEDAR   Family Cupressaceae   Figure 527

An evergreen coniferous tree, rarely over 20m tall. Distributed mainly in the NE, but scattered in the SE from WV and VA to NC, TN, KY, OH. Occurs over calcareous rocks and talus, and in swampy soils with near neutral pH. Leaves scalelike, opposite, appressed, acute, with a bump-like gland near tip; foliage in general flattened in a plane or fanlike spray, the upper side dark, glossy green, the lower light green. Crushed parts very aromatic, resembling turpentine. Cones about 1cm long, elliptic or ovoid, with 2 to 4 pairs of brown scales, usually held erect. Bark is gray or brown, fibrous, shallowly fissured into narrow, interlacing ridges. Often planted, and may produce seedlings nearby. Many horticultural cultivars have been selected and are used in the nursery trade.

Tilia L.   BASSWOOD   Family Tiliaceae

The native basswoods have a long history of taxonomic difficulty, with disagreement among botanists as to the number of valid species. Recent trends favor lumping all under the single species of *T. americana*. This approach is adopted here, with some reluctance. In my opinion, there are at least 3 types of easily recognizable basswoods in the SE, so at least varietal rank is deserved. Leaves of juvenile wood (sprouts, lower limbs, young trees, all incapable of flower production) are generally nearly glabrous beneath and do not have the conspicuous hair features of leaves from mature wood. For this reason, fallen leaves from the upper crown are more useful for identification. Twigs of all are greenish or reddish, with alternate leaf scars and several bundle scars; terminal buds lacking; laterals plump, red or greenish, 2- or 3-scaled; stipule scars present. All types cannot be separated by twigs alone. Bark is furrowed, often with sprouts at the base of the single or multiple trunks. The round, drupaceous fruits are 5–10mm diam., scurfy, leathery, and borne as a cluster of 3 to 10 suspended from a leaflike bract; these sometimes persist into winter or lie on the ground nearby.

1. Twigs and buds glabrous; buds mostly 6–7mm long; leaf teeth acute or acuminate
    2. Leaves from mature wood green and mostly glabrous beneath, except for scant and scattered stellate hairs   1. *T. americana*
    2. Leaves from mature wood whitened beneath due to dense stellate-tomentose hairs
                                                          2. *T. americana* var. *heterophylla*
1. Twigs or buds with some straight and stellate hairs; buds usually 4 or 5mm long; leaf teeth obtuse to acute
    3. Leaves from mature wood usually with brownish hue beneath, with scattered tufts of stellate hairs over most of undersurface   3. *T. americana* var. *caroliniana*
    3. Leaves from mature wood usually light, dull green below, glabrous or with a few stellate hair tufts in vein axils   *T. floridana* (= var. *caroliniana*) see 3

1. *Tilia americana* L.   AMERICAN BASSWOOD

A tree, to 25m tall. Scattered in the se. states of NC, TN, n. AR, and northward, becoming more common northward. Occurs in rich and moist soils in the SE, but on drier and thinner soils to the north. Leaves sharply toothed, with acuminate teeth; general shape mostly wide and cordate, base slightly unequal (cordate/truncate). Underside with distinct veins, few to no hairs, green. Bark often light brown to reddish-brown and ridges thicker than in next basswood, which partly shares the range in the SE.

2. Tilia americana var. heterophylla (Vent.) Loudon   WHITE BASSWOOD   *Figure 528*

A tree, to 25m tall. Most common in the Appalachian region of the SE from WV to AL, and west to central TN, w. KY, but also occurs in adjacent Piedmont of VA to AL and coastal plain of w. GA, w. FL, s. AL; scattered in s. IL to MO, AR. Typically in moist, fertile soils of mesic forests. Leaves sharply toothed, mostly with acute teeth; general shape widely ovate, with base conspicuously unequal (cordate/truncate). Underside whitened on mature wood (branches of flowering age). Bark gray to light brown, furrowed and narrowly ridged.

3. Tilia americana var. caroliniana (P. Mill.) Castig.   CAROLINA BASSWOOD   *Figure 529*

A tree, to 20m tall, but usually smaller in the Piedmont. Native to the coastal plain and Piedmont of NC to FL, west to OK, e. TX. Occurs in north-facing and mesic slopes and hammocks, maritime forests, river banks. Leaves rather coarsely and bluntly toothed; general shape like the above. Underside dull brownish-green, with tufts of stellate hairs along veins and between. Bark ridged, with reddish-brown inner layers often visible in furrows. A form of this basswood with leaves dull green beneath and mostly glabrous occurs in the coastal plain from VA to FL, west to e. TX, and inland to s. MO; called the Florida basswood (*Tilia floridana* Small) since it is most common throughout FL. Differs little except in the leaf undersides, and since intermediate trees exist that are difficult to separate, it is generally no longer considered distinct from Carolina basswood (though it may appear distinct in its typical form, in FL).

Tillandsia usneoides (L.) L.   SPANISH-MOSS   Family Bromeliaceae   *Figure 530*

An epiphyte, growing without roots on the branches and bark of other plants. Moisture and nutrients for life are obtained from humidity and dust in the air, so these plants only survive in humid climates. Similar in appearance to hanging moss or lichens, though actually an evergreen, flowering plant. Distributed in the coastal plain from se. VA to FL, TX, and southward into S. Amer. Typically seen in festooning masses on trees in humid habitats. Leaves elongate, mostly 3–4cm, gray-scurfy (more greenish when wet), alternate but appearing opposite due to inner leaf and shoot being held in sheath of larger, outer leaf. Twigs twisting, gray-scurfy; the membranous peltate scales are better seen with magnification. The fruit is an elongate, 3-parted capsule 15–30mm long, mature in summer but remnants persistent to autumn or longer. Flowers are small, yellowish. Other species in this large genus occur southward; the bunched forms that begin their appearance in n. FL are omitted in this treatment, since they do not obviously bear a leafy, woody stem with buds as does *T. usneoides*.

Torreya taxifolia Arn.   FLORIDA TORREYA   Family Taxaceae   *Figure 531*

An evergreen tree potentially to 18m tall, though rarely over 10m in its native range. Native only to a small area of the FL panhandle and adjacent GA, near the Apalachicola River. Occurs on mesic slopes and in ravines, but many plants have died from fungal blight and the plant is now an endangered species. Cultivated specimens outside the native range are often more vigorous and healthier, with some seedling regeneration observed in their vicinity. The needles are flat, to 35mm long, sharply tipped, with a short petiole, shiny green above, with 2 narrow pale stomatal bands below. Crushed foliage and twigs with a strong odor reminiscent of green peppers and citrus, but in accordance with varying noses, accounting for another name, "stinking cedar." Twigs green, ridged below leaf attachment; terminal bud conical, with green scales, resinous; some lateral buds similar. Bark brown, shallowly fissured and fibrous on young trunks, developing thick ridges and plates with age. Fruit 25–35mm long; a large, light brown seed enclosed within a fleshy greenish aril with purplish stripes, strongly aromatic and resinous, suggesting a drupe in general appearance. Plants presumed dioecious, though seedlings sometimes appear near singular old specimens.

Figure 525. *Taxodium distichum*

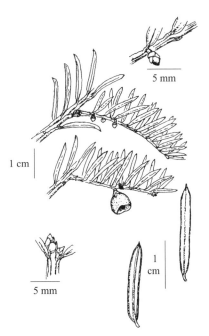

Figure 526. *Taxus*

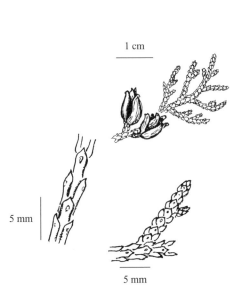

Figure 527. *Thuja occidentalis*

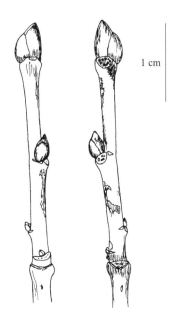

Figure 528. *Tilia americana* var. *heterophylla*

Toxicodendron P. Mill.   POISON-IVY, POISON-SUMAC   Family Anacardiaceae

Four native species are treated here for the SE. One of these, (*T. rydbergii*), is sometimes considered only a variety of *T. radicans* in other texts. These plants have an oil in the sap with compounds that cause skin inflammation and itching in many people from contact with foliage or other parts. The sap is clear when fresh, blackening with drying. The twigs of all have terminal buds, either scaled or appearing naked. The leaf scars have several bundle scars. The fruits are pale brown or ivory-colored drupes, different from the red drupes of genus *Rhus*. The members of *Toxicodendron* are included under *Rhus* in some texts, but differ by having the flower clusters axillary (terminal in *Rhus*), sap clear and poisonous (milky and usually nonirritating in *Rhus*), and fruit pale and in drooping clusters (red and erect in *Rhus*).

1. Twigs and buds glabrous; bud scales reddish, valvate; a tall shrub or tree of wet sites
   1. *T. vernix*
1. Twigs or buds hairy; bud scales brownish or buds naked; a vine or low shrub; in dry to moist soils
   2. Aerial rootlets lacking on stems; a shrub
      3. Twigs hairy; fruit pubescent; leaf undersides hairy; common in dry habitats
         2. *T. pubescens*
      3. Twigs and fruit sparsely hairy to glabrous; leaves mostly glabrous; rocky to moist sites
         4. Leaves and leaf scars clustered at twig tip; terminal leaflet nearly orbicular; n. species, rare in SE
            3. *T. rydbergii*
         4. Leaves and leaf scars more alternate; terminal leaflet ovate; widespread   4. *T. radicans*
   2. Aerial rootlets present; a climbing vine   4. *T. radicans*

1. *Toxicodendron vernix* (L.) Kuntz   POISON-SUMAC   *Figure 532*

A shrub or small tree, to 7m tall, usually with a single trunk. In the SE, distributed from MD and WV to TN, GA in the Appalachians, and in areas east and south, west to e. TX. Most common in the coastal plain, as wetland habitat is scarcer in the Appalachians. Typically in wet soils near streams or depressions, swampy thickets, bays, bogs, in peaty or acidic soils. Twigs fairly stout, grayish to brown, glabrous, with a large terminal bud; bud scales reddish, valvate or overlapping, outer ones long as the entire bud. Leaf scars shield-shaped, pale, with many dark bundle scars. Bark gray, smooth. Fruit ivory to pale brownish-colored, about 5mm; hang in elongate clusters into winter in pistillate plants (dioecious). (*Rhus vernix* L.)

2. *Toxicodendron pubescens* P. Mill.   POISON-OAK

A low, rhizomatous shrub, rarely to 1m tall, slender and few-branched or not branched. Widespread nearly throughout the SE, but most common in the coastal plain and Piedmont. Typically in dry woods or sandy uplands, or near rock outcrops. Twigs brownish, pubescent at least near tip; terminal bud hairy, appearing naked or unscaled; leaf scars with several bundle scars. A nonclimbing species normally with distinctly hairy leaves and petioles, the leaflets with large, coarse teeth. Fruit hairy, 6–8mm. [*Rhus toxicodendron* L.; *Toxicodendron toxicodendron* (L.) Britton; *T. toxicarium* Gillis]

3. *Toxicodendron rydbergii* (Small ex Rydb.) Greene   NORTHERN POISON-IVY

A low, rhizomatous shrub similar to above species in habit, but leaves tend to cluster at twig tips. A n. and w. species barely reaching the SE in the mountains of WV, VA. Occurs in rocky woodlands. Twigs similar to above, but more glabrous; buds hairy. Leaflets widely ovate to nearly orbicular, dentate, hairy below; petiole usually glabrous. Fruit 4–7mm, usually glabrous. [*T. radicans* var. *rydbergii* (Small) Erskine; *Rhus radicans* var. *rydbergii* (Small) Rehd.]

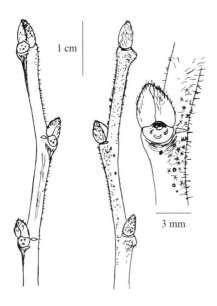

Figure 529. *Tilia americana* var. *caroliniana*

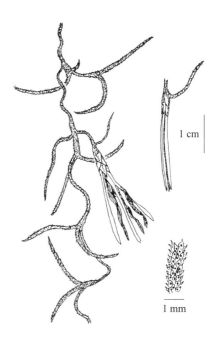

Figure 530. *Tillandsia usneoides*

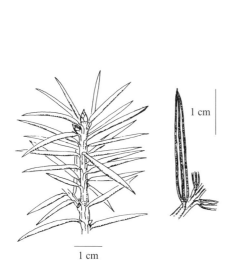

Figure 531. *Torreya taxifolia*

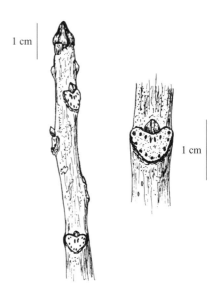

Figure 532. *Toxicodendron vernix*

4. Toxicodendron radicans (L.) Kuntz   POISON-IVY   *Figure 533*

A vine climbing by aerial rootlets or sometimes rhizomatous and more shrublike, with many stems produced to about 1m tall. Usually such clonal patches will produce some stems with aerial rootlets, if some object is reached on which they can climb. Widespread throughout the SE. Occurs in many habitats, including open and dense forest understories, roadsides, disturbed areas, rocky slopes, moist to fairly dry soils. Twigs and buds similar to above species, but aerial rootlets usually present or potentially present. Bark of mature stems gray, covered progressively by the brown, hairlike aerial rootlets with age. Leaflets normally ovate, entire or with a few blunt teeth on margin. Fruit usually slightly hairy in the coastal plain and Piedmont forms, glabrous in var. *negundo* (Greene) Reveal, which is prevalent in the Appalachians and regions west. (*Rhus radicans* L.)

Trachelospermum difforme (Walt.) Gray   CLIMBING DOGBANE   Family Apocynaceae
Figure 534

A native vine with slender twining stems and milky sap. Widely distributed across the entire coastal plain of the SE, occasional in the Piedmont and in scattered sites in the interior, north to MO, KY, IN. Typically in low, moist soils of bottomlands, swamps, streambanks, and thickets in lowlands. Twigs slender, reddish; no terminal buds; lateral buds dark, scurfy; leaf scars opposite, connected by a line or ridge, with 1 bundle scar. Twigs may die back extensively in winter. Bark smooth. The fruit is a long, slender follicle about 2mm wide and 1–2dm long, usually borne in pairs; splitting to release downy-tufted seed in autumn; thereafter the twisted remnants of the fruit may persist into winter. Leaves very narrow on juvenile growth, ovate on mature stems.

Tsuga Carr.   HEMLOCK   Family Pinaceae

There are 2 native species in the SE, both evergreen coniferous trees. They occur principally in the Appalachians and adjacent areas. The needles (leaves) are flat, rounded to notched at the apex, attached to the slender twigs by a short petiole; 2 whitish stomatal bands (where rows of stomates are present) are beneath. Twigs slender, roughened by peglike nodes where leaves are attached; buds small, rounded, with hairy brown scales. Foliage is not poisonous, despite colloquial name associated with a poisonous herb in the parsley family, also called hemlock.

1. Needles spread mostly in 1 horizontal plane on twig, minutely toothed at tip; cones to 25mm long   1. *T. canadensis*
1. Needles tend to spread in all directions on twig, entire; cones to 35mm, reddish-brown   2. *T. caroliniana*

1. *Tsuga canadensis* (L.) Carr.   EASTERN HEMLOCK   *Figure 535*

To 30m tall, widespread in the NE and extending into the SE in the Appalachian and other plateaus in the interior, from VA to AL, west to central TN, KY. Also occurs sporadically in the Piedmont of VA, NC. Typically in acidic soils of moist slopes, streamsides, coves, ravines, gorges; common in mountainous terrain, favoring sites with cooler, moist soils in lower elevations. Needles mostly under 15mm long, minutely toothed near tip (use lens); spreading out to form flattened, featherlike sprays; smaller needles held close to twig visible on top and bottom of spray. Twigs slender, hairy; buds hairy. Bark brown to red-brown, scaly, becoming thick with age and forming large plates or ridges separated by deep furrows; inner bark often with purplish tinge. Cones ovoid, gray to light brown, 12–25mm long but usually under 20mm, the scales broadly rounded and reflexing when dried only to an angle of about 60 degrees, or less.

2. *Tsuga caroliniana* Engelm.   CAROLINA HEMLOCK   *Figure 536*

Potentially to about 30m, but typically shorter in its native habitat. A s. Appalachian region endemic, ranging in the Blue Ridge of VA, NC, TN, SC, n. GA, and sporadic in a few rocky peaks

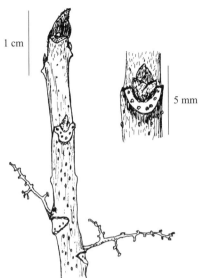

Figure 533. *Toxicodendron radicans*

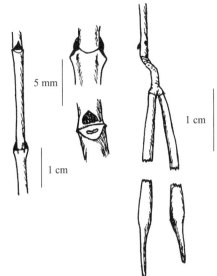

Figure 534. *Trachelospermum difforme*

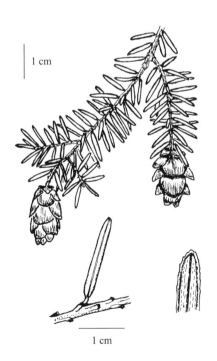

Figure 535. *Tsuga canadensis*

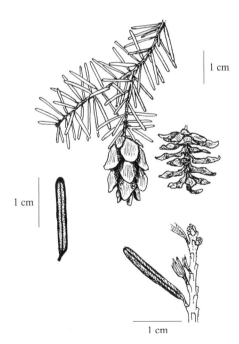

Figure 536. *Tsuga caroliniana*

in the w. Piedmont of NC, VA. Scattered and uncommon across its range, though may be locally abundant at some sites. Often in rocky soils of thinner and drier aspect than the former species tolerates, such as ridges, gorges, cliffs, rock outcrops. Needles commonly over 15mm, to 18mm long, entire on margin, sticking out in all directions from twig. Twigs slender, brown, slightly hairy to glabrous; buds hairy. Bark dark reddish-brown with thick, scaly plates or short ridges, thickening with age. Cones ovoid, red-brown, mostly 25–35mm long, the scales blunt and reflexing when dried to 90° angle or more.

Ulmus L.   ELM   Family Ulmaceae

Six species of elms are native to the SE, and 2 additional exotics have naturalized. Other exotic elms are cultivated, but these seem not to be naturalized to any appreciable extent in the SE. All elms have imbricate-scaled buds and lack terminal buds; in some species, the bud scales are in 2 rows. Twigs bear slitlike stipule scars; bundle scars usually 3, but sometimes more. Leaves (and leaf scars) are 2-ranked on twigs, which means occurring in 2 rows when viewed with twig apex pointed at eye. The fruit is a flat samara, the wing surrounding the seed. One exotic elm not included here is the English elm, *U. procera* Salisb., with scabrous leaves and corky branches; it is only rarely naturalized around planted parent trees. The 2 exotic elms more widely naturalized that are treated here appear first in the key:

1. Buds all blunt or rounded; bark furrowed, blackened   1. *U. pumila*
1. Buds pointed, or only the flower buds rounded
  2. Collateral buds usually present; bark strongly mottled, with thin scales   2. *U. parvifolia*
  2. Collateral buds lacking; bark grayish, furrowed, thickly scaly and not mottled (native elms)
    3. Bud scales dark, distinctly red-hairy; twigs gray   3. *U. rubra*
    3. Bud scales brown or dark on margin only, not red-hairy; twigs brownish
      4. Buds near twig tip often 6–8mm long; scales slightly hairy   4. *U. thomasii*
      4. Buds 3–6mm long; scales mostly glabrous
        5. Lenticels pale; lowermost bundle scar often divided (bundle scars 4 to 6)
          6. Twigs sometimes slightly hairy, no corky wings; bud scales distinctly dark-edged; flowers and fruits in spring   5. *U. americana*
          6. Twigs glabrous, sometimes winged; bud scales dark red-brown, only slightly dark-edged; flowers and fruits in autumn   6. *U. serotina*
        5. Lenticels orange; bundle scars usually 3 (twigs alone will not readily separate the next 2 species; both can have winged branches or not)
          7. Leaves sharply toothed, tip acute; flowers and fruits in spring; widespread and common   7. *U. alata*
          7. Leaves bluntly toothed, tip blunt to rounded; flowers and fruits in autumn; mostly w. parts of SE, but also in FL   8. *U. crassifolia*

1. *Ulmus pumila* L.   SIBERIAN ELM   *Figure 537*

A tree, to 35m tall, with conspicuously ashy-gray branchlets. Native to Asia; widely cultivated and naturalized in urban areas and near cultivated specimens. Most commonly seen in the Midwest and Plains states. Twigs slender, gray, often with tiny black dots and larger, pale lenticels. Buds small, blunt or rounded, with red-brown scales; flower buds larger, globular, with white hairs near tip. Bark furrowed, with gray ridges or rough, blocky plates, becoming nearly black on main trunk. Leaves 2–10cm long, teeth blunt, rarely doubly toothed. Flowers and fruits in spring. Fruit nearly round, glabrous.

2. *Ulmus parvifolia* Jacq.   CHINESE ELM   *Figure 538*

A tree, to 18m tall. Native to China; widely cultivated and naturalized near sources of its cultivation. Twigs red-brown; buds often paired (collateral), with 1 larger; bud scales red-brown, with darker margin. Bark thin, scaling in small irregular plates, the orange-brown inner layers revealed when the gray outer layers fall. Leaves 1.5–8cm long, teeth blunt. Flowers in late summer; fruits mature in autumn. Fruit oval, glabrous.

3. *Ulmus rubra* Muhl.   SLIPPERY ELM   *Figure 539*

A tree, to 35m tall, with coarse branches and an open, wide crown. Distributed across much of the SE, except parts of the coastal plain in SC and se. GA, and s. AL to se. TX. Occurs in rich, mesic forests of bottomlands and uplands, most commonly in circumneutral to calcareous soils. Twigs grayish, often roughened by minute grooves or lines; buds 5–6mm long, with 6 or more dark scales, rusty-hairy near tip; flower buds larger, globular, rusty-woolly on top half; leaf scars red-brown or pinkish; bundle scars white, variably 5 per leaf scar. Bark brown, furrowed, with scaly ridges. Inner bark slick, slimy when chewed, accounting for the common name. Leaves scabrous above, hairy below, 8–16cm long. Flowers and fruits in spring. Fruit circular, glabrous. (*U. fulva* Michx.)

4. *Ulmus thomasii* Sarg.   ROCK ELM   *Figure 540*

A tree, to 35m tall, with a narrow crown. Distributed in the SE from WV, s. OH to s. KY, and in n. AR. Typically over calcareous rocks and along streams in the SE, in deeper and sandier soils northward. Twigs brown; buds sharply pointed, larger ones 6–8mm long, with brown, slightly hairy scales. Corky growth usually forms on branches. Bark brownish, with thick, scaly ridges. Leaves 7–16cm long, coarsely double-toothed. Flowers and fruits in spring. Fruit oval, hairy.

5. *Ulmus americana* L.   AMERICAN ELM   *Figure 541*

A tree, to 40m tall. Distributed throughout the SE, though uncommon to absent in parts of the Blue Ridge. Occurs in a wide variety of habitats, including bottomlands and swamps, mesic forests, calcareous soils, glades. Twigs brown, finely hairy or glabrous; buds mostly 4–5mm long, scales brown, dark-edged. Branches not developing corky growth. Bark grayish to light brown, with scaly ridges. Leaves 7–15cm long, coarsely double-toothed, base usually distinctly unequal. Flowers and fruits in spring. Fruit oval, hairy-fringed on margin. (*U. floridana* Chapm.)

6. *Ulmus serotina* Sarg.   SEPTEMBER ELM   *Figure 542*

A tree, to 25m tall. Distributed sporadically west of the Blue Ridge, from KY to nw. GA, west to s. IL, AR, OK. Occurs on limestone or circumneutral to calcareous soils in mesic forests. Twigs brown, glabrous; buds sharply pointed, scales brown, darker-margined, glabrous. Branches sometimes with corky growth. Bark brown, with narrow, scaly ridges. Leaves 4–10cm long, coarsely double-toothed, base often distinctly unequal. Flowers and fruits in autumn. Fruit oval, with hairy-fringed margin.

7. *Ulmus alata* Michx.   WINGED ELM   *Figure 543*

A tree, to 35m tall. Distributed across most of the SE, uncommon to absent in middle and higher elevations of the Blue Ridge and in se. GA. Occurs in a wide variety of habitats from floodplains to upland clay soils and over limestone. Twigs slightly hairy to glabrous; buds sharply pointed, 3–5mm long, scales brown. Branches usually with corky growth ("winged"), but some trees lack such cork. Bark gray to light brown, with narrow, scaly ridges. Leaves 3–8cm long, sharply double-toothed, tip acute. Flowers and fruits in spring. Fruit narrowly oval, hairy-fringed on margin.

8. *Ulmus crassifolia* Nutt.   CEDAR ELM   *Figure 544*

A tree, to 30m tall. Distributed mostly west of the Mississippi River, AR and LA to s. OK, central TX; ranges east of the river in the adjacent floodplain; a disjunct population is found in

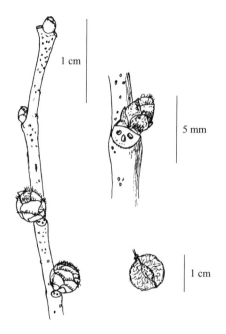

Figure 537. *Ulmus pumila*

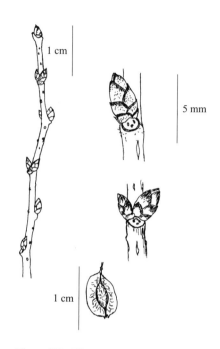

Figure 538. *Ulmus parvifolia*

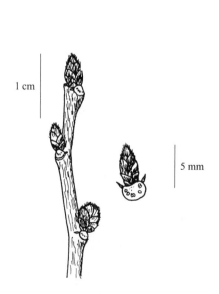

Figure 539. *Ulmus rubra*

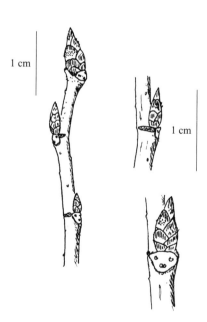

Figure 540. *Ulmus thomasii*

n. FL. Occurs in bottomlands and uplands over limestone. Twigs similar to above species, though buds overall smaller, mostly 2–4mm. Branches usually corky, but occasionally not corky. Bark gray or light brown, with scaly ridges or plates. Leaves 2–5cm long, double-toothed but teeth somewhat blunt; apex of leaf blunt to rounded. Flowers and fruits in autumn. Fruit oval, with short hairs on margin.

Vaccinium L. BLUEBERRY   Family Ericaceae

This genus includes a taxonomically complex collection of species, varieties, hybrids, and geographic variations. Within a plethora of opinions on exactly how many valid species occur in the SE, 21 species are here presented as native to the region, along with a few important variations. Within these species, 5 are true evergreens across their entire range; at least 2 more are partially evergreen in s. parts of their range. All are shrubs or, rarely, small trees, the entities with mature heights about 1m or less known generally as "lowbush blueberries" and most taller ones as "highbush blueberries." Leaves are alternate; terminal buds lacking; 1 bundle scar. Twigs and branchlets are predominately greenish, though reddish or grayish in a few. The twig surface of those with green coloration often reveals minute bumps (verrucose). The fruit is an edible berry, with typically more than 10 seed within; palatability varies from tasteless to sour, bitter, or sweet, depending on species. In winter, identification by twigs alone is rarely reliable in the deciduous forms, so the following key addresses the necessity of examining fruit or at least examining a few fallen leaves. Following the key, species are discussed in alphabetical order.

1. Leaves evergreen *and* under 2cm long 2
1. Leaves deciduous or if evergreen then over 2cm long 6
  2. Habit erect, with a stem and crown; of s. coastal plain only 3
  2. Habit prostrate, with creeping or trailing stems; wider range 4
3. Leaves glaucous, blue-green, glabrous below 8. *V. darrowi*
3. Leaves usually shiny green, glandular or with glandular hairs below 15. *V. myrsinites*
  4. Leaf tips mucronate; light green beneath; berry black 6. *V. crassifolium*
  4. Leaf tips blunt, or leaves pale beneath; berry red (cranberries) 5
5. Leaves white beneath; berry 5–8mm; of mountains in VA, WV 17. *V. oxycoccos*
5. Leaves pale beneath; berry 10–20mm; of mountains and coastal plain south to NC 14. *V. macrocarpon*
  6. Habit reclining or creeping; rare; native to SC only 7. *V. crassifolium* ssp. *sempervirens*
  6. Habit erect and not trailing; wider range 7
7. Bud scales 2, valvate 10. *V. erythrocarpum*
7. Bud scales more than 2, imbricate 8
  8. Buds short, rarely longer than wide; twigs reddish; branchlets gray, not dotted or speckled in appearance 9
  8. Buds pointed, usually longer than wide; twigs greenish or reddish on 1 side; branchlets greenish, with dotted or wart-speckled appearance (verrucose) 10
9. Trunk usually well defined; inner bark and smooth portions of stem reddish-brown; leaf tip blunt to rounded, upperside glossy 3. *V. arboreum*
9. Trunk weakly defined; bark gray throughout; leaf tip acute, upperside dull 20. *V. stamineum*
  10. Twigs pubescent or hirsute 11
  10. Twigs glabrous or nearly so 14
11. Shrub often over 1.5m high; pubescence dingy; bud scales dark; berry black 12. *V. fuscatum*
11. Shrub usually 1m or less; hairs whitish; bud scales brownish or green; berry black or glaucous 12
  12. Twigs glandular-pubescent, the hairs close; fruit smooth, black; leaves stipitate-glandular below; of coastal plain 21. *V. tenellum*
  12. Twigs velvety or hirsute, glandular or not; fruit glaucous or hirsute; n. and Appalachian species 13

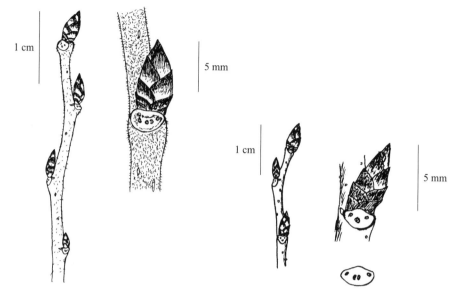

Figure 541. *Ulmus americana*

Figure 542. *Ulmus serotina*

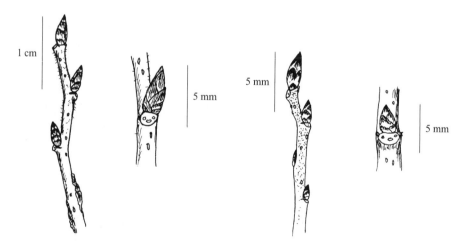

Figure 543. *Ulmus alata*

Figure 544. *Ulmus crassifolia*

13. Twigs velvety; bud scales with long, bristlelike points; fruit glaucous; n. species
   16. *V. myrtilloides*
13. Twigs hirsute; bud scales not as above; fruit hirsute; of Appalachians in NC, TN, GA
   13. *V. hirsutum*
   14. Leaves mostly under 3cm long, shiny green above and below 15
   14. Leaves often over 3cm long, dull green or glaucous beneath 16
   15. Leaves rounded at base; fruit shiny black; s. species of Piedmont and coastal plain
      9. *V. elliottii*
   15. Leaves mostly tapered at base; fruit glaucous; n. species, only in high mountains in SE
      2. *V. angustifolium*
      16. Habit a low shrub, usually under 1m high   18. *V. pallidum*
      16. Habit taller, usually 1–2m or more, or leaves strongly glaucous and blue-green 17
      17. Leaves glaucous above and below 18
      17. Leaves glaucous only below, or not glaucous 19
         18. Leaves often over 2cm wide; habit to 1.5m tall, in dense clonal colonies; of mountains
            1. *V. altomontanum*
         18. Leaves mostly under 2cm wide; habit 2–3m, not in dense colonies; of coastal plain
            4. *V. caesariense*
      19. Leaves stipitate-glandular below; of coastal plain   22. *V. virgatum*
      19. Leaves glabrous below, or not glandular; of coastal plain or mountains   20
      20. Twigs olive-green; fruit shiny black; of mountains   19. *V. simulatum*
      20. Twigs bright green; fruit glaucous; various provinces   21
         21. Leaves slightly revolute on margin, widest below middle, often subevergreen in Deep South   11. *V. formosum*
         21. Leaves not revolute, usually widest at or above middle, thin and deciduous
            5. *V. corymbosum*

1. *Vaccinium altomontanum* Ashe   MOUNTAIN DRYLAND BLUEBERRY

A rhizomatous shrub, to 1m tall, usually forming dense clonal patches of bushy-branched stems. Native to the Appalachian region of WV, e. KY, e. TN, w. VA, w. NC, n. GA. Most conspicuous in the higher elevations of the Blue Ridge, though scattered in other lower elevations. Typically in shrub balds, high meadows and grassy balds, disturbed forests. Twigs green, mostly glabrous, generally similar to those of *V. pallidum*, with which this species is sometimes combined; branchlets dark or dull green; older stems mostly reddish-brown. Leaves deciduous, glaucous and blue-green, mostly ovate or elliptic, 3–5cm long, 15–30mm wide, glabrous, margin entire, tip acute. Berry 5–8mm, glaucous, sweet.

2. *Vaccinium angustifolium* Ait.   NORTHERN LOWBUSH BLUEBERRY

A low, rhizomatous shrub, to 5dm tall, bushy-branched. Native mostly to the NE, but reaching the SE in the higher Appalachian elevations of WV, VA. Occurs in acidic soils among rocky, peaty, or disturbed sites in the mountains. Twigs slender, green, glabrous. Leaves deciduous, lanceolate, mostly 3cm long or less, under 15mm wide; margin finely serrulate, with gland-tipped teeth; bright green, glabrous, not glaucous below. Berry 5–8mm, glaucous, sweet. The uncommon var. *nigrum* (Wood) Dole has black fruit and glaucous leaves.

3. *Vaccinium arboreum* Marsh.   SPARKLEBERRY   Figure 545

A tall shrub or small tree, to 10m tall, usually with a distinct and stout trunk and bushy crown. Widespread over much of the SE, from VA to FL, west to e. TX, OK, north to MO, s. IN, KY; scarce or absent from most of the Blue Ridge of NC and other mountainous provinces of VA, WV. Typically in dry uplands, of sandy pinelands and scrub, rocky or dry woods, ridges and sandstone outcrops, and even over calcareous rocks in w. parts of its range. Twigs reddish-brown, slender, zigzag, hairy or glabrous; buds wide, blunt or rounded, usually hairy; branchlets grayish. Bark scaly, the grayish outer bark intermittently revealing red-brown inner bark; upper, smooth trunk sections mostly red-brown. Leaves deciduous to subevergreen, mostly 2–6cm long, 1–4cm wide, widely ovate to nearly round, margin entire or with minute serrulate points; usually glossy green but glaucous in var. *glaucescens* (Greene) Sargent. Berry 5–8mm, on stalks 1–2cm long, black, lustrous, with a dry or mealy flesh; usually persistent into winter.

4. *Vaccinium caesariense* Mackenz.   NEW JERSEY HIGHBUSH BLUEBERRY   Figure 546

A bushy-branched shrub, to 3m tall. Distributed in the SE mostly in the coastal plain from MD to n. FL in bogs, swampy thickets, peaty soils. Twigs similar to those of *V. corymbosum*. Leaves deciduous, glabrous, glaucous, mostly 3–5cm long, 15–20mm wide, margin entire. Berry 5–8mm, glaucous.

5. *Vaccinium corymbosum* L.   SMOOTH HIGHBUSH BLUEBERRY   Figure 547

A bushy-branched shrub, to 5m tall. Widely distributed in the interior of the SE, becoming less common eastward and southward into the Piedmont and coastal plain. Occurs in moist to boggy soils, high elevation balds, acidic forests. Twigs green, glabrous, verrucose; branchlets green; buds pinkish to red or red-brown, with slightly keeled and long-tipped imbricate scales. Leaves deciduous, green or slightly glaucous, mostly widely elliptic, 3–8cm long, 2–4cm wide; margin usually entire. Berry 5–12mm, glaucous, sweet. Considerable taxonomic disagreement exists regarding this species and closely related highbush blueberries. Difficulty in separating many of the highbush blueberries and uncertainty regarding whether they deserve separate species status due to widespread hybridization and complex genetic variation has resulted in some texts combining them under *V. corymbosum*. The mountain species *V. constablaei* Gray is now regarded in synonymy; its typical form having narrower leaves with acuminate tips and serrulate margins, otherwise similar to *V. corymbosum*. In other texts, the species *V. simulatum, V. caesariense, V. formosum, V. fuscatum*, and *V. virgatum* are also synonymized under *V. corymbosum*, though treated separately in this guide.

6. *Vaccinium crassifolium* Andrews   CREEPING BLUEBERRY   Figure 548a

A creeping shrub, rarely with erect stems to 1dm tall. Native to the coastal plain and adjacent Piedmont of se. VA to SC. Occurs in sandy or peaty soils of pinelands. Twigs slender, reddish. Leaves evergreen, glossy above, light green below, mostly elliptic or ovate, 1–2cm long, 3–10mm wide; margin thickened, with glandular-crenate teeth; base often rounded. Berry about 5mm, black, glossy. A subspecies with larger leaves follows.

7. *Vaccinium crassifolium* ssp. *sempervirens* (Rayn. & Hend.) Kirkm. & Ball.   RAYNOR'S BLUEBERRY   Figure 548b

Similar to the above species, except evergreen leaves to 4cm long, mostly obovate, base acute, marginal serrations more distinct. May have ascending stems to 3dm high, but mostly prostrate. Known only from a few boggy sites in the Sandhills of SC. Fruit to 8mm. (*V. sempervirens* Rayn. & Hend.)

8. *Vaccinium darrowi* Camp   GLAUCOUS BLUEBERRY   Figure 549

A low shrub, rarely over 5dm high, the small crown bushy-branched. Native to the coastal plain from s. GA, FL, west to se. TX. Inhabits sandy pinelands and scrub, edges of bays, bogs, and depressions in flatwoods. Twigs very slender, green, pubescent; buds with red-brown imbricate scales. Leaves evergreen, usually glaucous, blue-green, mostly 5–15mm long, 3–8mm wide, larger near twig base, smaller near tip; underside glabrous; margin entire to remotely toothed, slightly revolute. Berry glaucous, about 5mm diam.

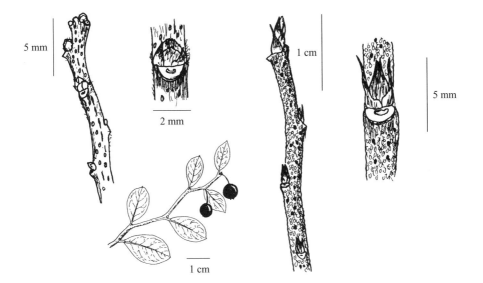

Figure 545. *Vaccinium arboreum*

Figure 546. *Vaccinium caesariense*

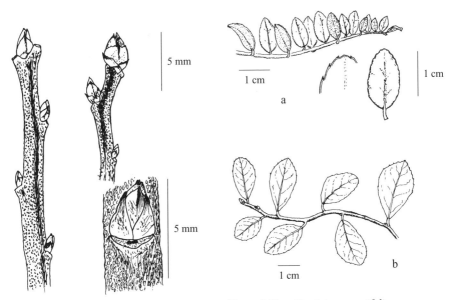

Figure 547. *Vaccinium corymbosum*

Figure 548. a: *Vaccinium crassifolium;* b: *V. crassifolium* ssp. *sempervirens*

364 · *Vaccinium*

9. *Vaccinium elliottii* Chapm. MAYBERRY   *Figure 550*

A shrub, to 4m tall, usually 2 or 3m, with a bushy-branched or twiggy crown. Native to the coastal plain from se. VA to FL, west to TX, AR, also in adjacent Piedmont of NC to AL. Occurs in moist sands near riverbanks and higher grounds in swamps and floodplains, in sandy pinelands, thin hardwood forests or forest edges. Twigs slender, green, zigzag, hairy or glabrous. Leaves deciduous to subevergreen in the Deep South, glossy green, 15–30mm long, 5–15mm wide, margin finely serrulate, thickened, with spinose or gland-tipped teeth. Berry 5–10mm, usually black and glossy.

10. *Vaccinium erythrocarpum* Michx. MOUNTAIN-CRANBERRY   *Figure 551*

A shrub, to about 2m tall, with 1 or several stems. Endemic to the Appalachian region of WV and VA to n. GA. Occurs in high elevation forests, balds, boggy soils, rock outcrops. Twigs red or greenish on 1 side, sparsely hairy on nodes or in a row on green side of twig; buds reddish, 2-scaled, appressed. Leaves deciduous, glabrous, 3–8cm long, 1–3cm wide, pale green below, dark, dull green and somewhat reticulate-veined above; margin finely serrate; tip acute to acuminate. Berry dark red to black, glossy, to 1cm diam., with succulent red juice but little flavor; borne singly in axils of leaves.

11. *Vaccinium formosum* H. C. Andrews SOUTHERN HIGHBUSH BLUEBERRY *Figure 552*

A shrub, to 5m tall, single or multiple-stemmed. Distributed mostly in the coastal plain from VA to FL, west to s. AL. Occurs in moist, sandy soils of pineland depressions, blackwater swamps, boggy sites. Twigs green, glabrous; buds with reddish or red-brown imbricate scales with long tips. Leaves deciduous, sometimes subevergreen in the Deep South, light green below, 4–10cm long, 3–5cm wide, mostly ovate; margin entire, slightly revolute. Berry 8–12mm, glaucous.

12. *Vaccinium fuscatum* Ait. BLACK HIGHBUSH BLUEBERRY

A shrub, to 5m tall, single or multiple-stemmed. Widespread across most of the SE, mostly in moist soils such as near streams, bogs, swampy depressions, but also in acidic upland woods. Twigs hairy, with an olive-green or dirty-green color; bud scales dark red-brown. Leaves deciduous, 4–8cm long, 2–3cm wide, pale green and hairy below; margin entire, tip acute. Berry 5–8mm, black, glossy. [*V. atrococcum* (Gray) Heller]

13. *Vaccinium hirsutum* Buckl. HAIRY BLUEBERRY

A rhizomatous shrub, to 1m tall. Endemic to the Blue Ridge of sw. NC, se. TN, n. GA, where it occurs in acidic understories of oak-forested slopes. Twigs with long, pale, straight hairs (hirsute). Leaves deciduous, mostly 3–6cm long, 2–3cm wide, ovate; margin entire; tip acute; underside pale green, hairy. Berry 5–8mm, black, hirsute.

14. *Vaccinium macrocarpon* Ait. CRANBERRY   *Figure 553a*

A creeping and prostrate shrub, with flowering stems erect to about 1dm. Native mostly to the NE, but ranges into the SE along the higher Appalachians from WV and VA to TN, NC, and in the coastal plain to se. NC. Occurs in boggy sites, seeps, acidic sands, and peaty soils. Leaves evergreen, elliptic, mostly 5–15mm long, 2–8mm wide; margin entire, slightly revolute; underside pale; tip rounded. Berry globose, 1–2cm diam., red, glossy, acidic and sour.

15. *Vaccinium myrsinites* Lam. SHINY BLUEBERRY   *Figure 554*

A low shrub, rarely over 5dm tall, the small crown bushy-branched. Native to the coastal plain from se. SC to FL, west to sw. AL. Inhabits sandy pinelands and scrub, edges of bays, bogs, and

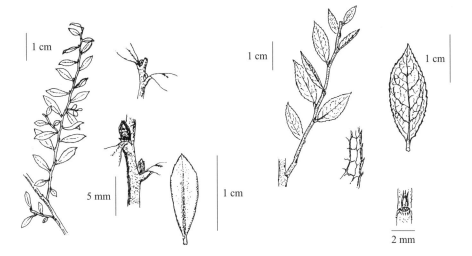

Figure 549. *Vaccinium darrowi*

Figure 550. *Vaccinium elliottii*

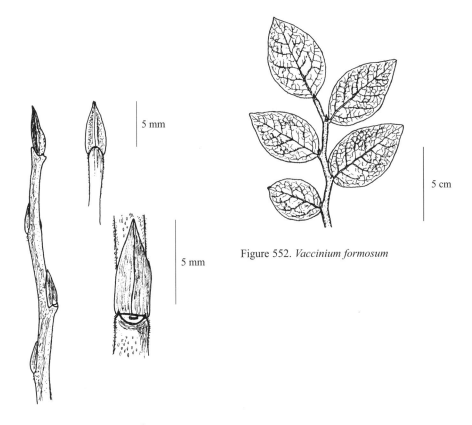

Figure 552. *Vaccinium formosum*

Figure 551. *Vaccinium erythrocarpum*

flatwoods depressions. Twigs very slender, green, minutely hairy to glabrous. Leaves evergreen, usually shiny but with a glaucous form known, mostly 5–15mm long, 3–8mm wide, larger near twig base, smaller near tip; underside with darkened glands or gland-tipped hairs; margin entire to remotely toothed, slightly revolute. Berry 6–8mm diam., usually black, though glaucous on those plants with glaucous leaves.

16. *Vaccinium myrtilloides* Michx.   VELVET-LEAF BLUEBERRY

A low shrub, rarely to 1m tall, bushy-branched. Widespread across n. N. Amer., reaching the SE in the higher Appalachians of WV and VA. Occurs in moist woods near rock outcrops, cliffs. Twigs densely hairy, otherwise greenish. Leaves deciduous, mostly lanceolate or elliptic, 2–4cm long, 5–15mm wide, pale green, downy-hairy below; margin entire; tip acute. Berry 6–10mm, glaucous.

17. *Vaccinium oxycoccos* L.   SMALL CRANBERRY   *Figure 553b*

A creeping and prostrate shrub, with erect flowering stems rarely 1dm. Widespread across n. N. Amer., entering the SE only in the high mountains of WV and VA. Occurs in boggy or peaty soils of high elevations. Leaves evergreen, mostly ovate, 3–12mm long, 1–5mm wide; margin strongly revolute; underside whitish; tip acute. Berry globose, red, 5–10mm diam., glossy, acidic and sour.

18. *Vaccinium pallidum* Ait.   SOUTHERN LOW BLUEBERRY

A well-branched, rhizomatous, low shrub, to about 5dm tall. Widespread across the interior of the SE, but scarce to absent in the coastal plain of SC to TX. Common in understories of acidic upland woods, mostly in dry soils but occasionally on moister slopes in higher elevations of the Appalachians. Twigs green, glabrous; branchlets green; buds with long-tipped scales. Leaves deciduous, slightly glaucous below, mostly ovate, 2–5cm long, 15–30mm wide; margin entire to serrulate; tip obtuse to acute. Berry 5–8mm, glaucous. A variation of this lowbush blueberry that has been called a separate species in the past is *V. vacillans* Kalm. ex Torr. (dryland blueberry); more common in lower elevations in dry oak or oak-pine woods, with mostly oval or obovate leaves 20–35mm long; tip obtuse or rounded.

19. *Vaccinium simulatum* Small   UPLAND HIGHBUSH BLUEBERRY

A bushy-branched shrub, to 4m tall. Endemic to the Appalachians from e. KY and adjacent VA to n. AL, n. GA. Occurs in middle and high elevations in acidic soils of moist forests or oak and pine woods, swampy or boggy sites, near streams, balds, rock outcrops. Twigs olive-green or dull green, mostly glabrous. Leaves deciduous, mostly narrowly ovate or elliptic, 4–7cm long, 2–3cm wide, light green or slightly glaucous below; margin entire to minutely serrulate; tip acuminate. Berry 5–10mm, black.

20. *Vaccinium stamineum* L.   DEERBERRY   *Figure 555*

A branching shrub, to 5m tall, usually multistemmed, sometimes forming colonies from a suckering root system. Widely distributed throughout the SE. Occurs in a variety of habitats, from shrub balds of high Appalachian elevations to dry or moist woodlands throughout other provinces and extending to coastal forests. Most common and typical in dry, upland woods. Twigs reddish-brown, usually hairy; buds wide and bluntly pointed; branchlets grayish. Leaves deciduous, mostly 3–8cm long, 15–40mm wide, acute at tip, usually entire on margin; usually glaucous and hairy beneath, but variable. Berry 5–15mm, varying from green to pink, brown, purplish-red, or nearly black, glaucous or not, succulent but sour or bitter to taste. Several forms and varieties of this species have been described based on variations in leaves, fruit, and bracts in the inflorescence. Leaves are whitened beneath in var. *candicans* (Small) C. Mohr, with small bracts in inflorescence; this variety most common in the Appalachian region. Also with glaucous leaves is var. *caesium* (Greene) Ward, with large bracts; most common in the coastal plain. Leaves are weakly glaucous to green beneath in the typical var. *stamineum*, with glabrous fruit, and similarly in var. *sericeum* (Mohr) Ward, which bears pubescent, dark fruit and grows sporadically in the coastal plain (var. *melanocarpum* Mohr).

21. *Vaccinium tenellum* Ait.   SLENDER BLUEBERRY

A rhizomatous shrub, to 3 or 4dm tall, forming colonies of slender, erect, few-branched stems. Native to the coastal plain and Piedmont from se. VA to ne. FL, sw. GA; scattered and uncommon westward to se. MS. Occurs in sandy pinelands and dry upland woods. Twigs hairy, green. Leaves deciduous, glossy, 15–35mm long, 5–15mm wide, green and glandular beneath; margin minutely toothed, the teeth often gland-tipped. Berry 5–10mm, black, lustrous.

22. *Vaccinium virgatum* Ait.   SMALLFLOWER BLUEBERRY

A shrub, to 5m tall, most commonly 2–3m, with a bushy-branched crown. Native to the coastal plain from NC to FL, west to MS. Scattered in sandy woods and pinelands, pocosins, shrub bays. Twigs hairy, green. Leaves deciduous, glossy, mostly 2–4cm long, 6–18mm wide, green and stipitate-glandular beneath; margin minutely toothed, the teeth often glandular. Berry 8–15mm, black, lustrous. (*V. amoenum* Ait.)

Vernicia fordii (Hemsl.) Airy-Shaw   TUNG-OIL-TREE   Family Euphorbiaceae   *Figure 556*

A small tree, to about 10m tall, with a broad, open crown. Native to China; sparingly naturalized in the southernmost parts of the SE, usually near urban and waste areas, roadsides, or close to cultivated specimens. Twigs stout, glabrous, green or brownish-green; terminal buds large, 10–12mm long, with several elongate, greenish scales that partially cover underlying folded leaves; lateral buds much reduced. Leaf scars large, with indention on top margin; bundle scars several; stipule scars large, slitlike. Pith rather small for twig size, about 2–4mm diam., white, homogenous. Scraped twigs with strong "cornstalk" odor; ranker to some noses. Bark thin, gray, smooth but with many short vertical rows of conspicuous lenticels. The capsular fruit is 4–8cm wide, not as long, reddish or green, splitting along 3 to 5 sutures to release 3 to 7 large seed each about 3–5cm long. Seed poisonous if consumed, but yield a commercially important oil (tung oil) that was once the principal reason for widespread cultivation in the South. (*Aleurites fordii* Hemsl.)

Viburnum L.   VIBURNUM   Family Caprifoliaceae

Of the 14 species treated here for the SE, only 1 is an exotic. Many other exotic species are cultivated, but they do not appear to naturalize in the SE. All viburnums are characterized by opposite leaf scars and 3 bundle scars; most have terminal buds. Three of our native species may form a small tree (*V. lentago, V. prunifolium, V. rufidulum*), but most are shrubs. The fruit is a drupe, changing from pink or red to bluish or black when ripe, often persisting into winter; these are important food sources for birds. Cut or scraped bark may have a disagreeable odor in many species, as does the wood when burned.

1. Terminal bud with 2 or fewer scales, or terminal bud lacking
  2. Largest buds with 1 or 2 smooth, lustrous scales
    3. Terminal bud usually absent, occasionally present; glands on petiole concave 1. *V. opulis*
    3. Terminal bud often present, occasionally absent; glands on petiole convex 2. *V. trilobum*
  2. Largest buds with 2 scurfy scales, or naked
    4. Terminal buds naked, or at least terminal flower buds naked

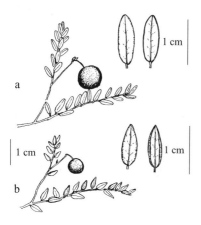

Figure 553. a: *Vaccinium macrocarpon;* b: *V. oxycoccos*

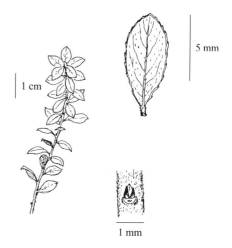

Figure 554. *Vaccinium myrsinites*

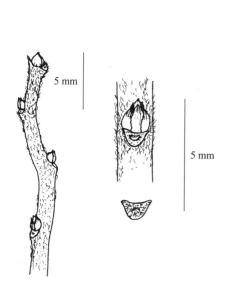

Figure 555. *Vaccinium stamineum*

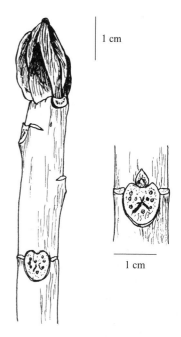

Figure 556. *Vernicia fordii*

    5. Terminal buds under 1cm long; coastal plain species   3. *V. obovatum*
    5. Terminal buds 1cm or more; of high mountain elevations in SE   4. *V. lantanoides*
  4. Terminal buds with 2 valvate scales
    6. Terminal bud length 2 to 3 times its width; trunk bark blocky
      7. Buds gray to light reddish-brown   5. *V. prunifolium*
      7. Buds dark red-brown or rusty   6. *V. rufidulum*
    6. Terminal bud length over 3 times its width; trunk bark smooth, or blocky only on lower trunk
      8. Buds gray or pink; swollen base of flower buds usually completely covered by the 2 outer scales; n. species, rare in SE   7. *V. lentago*
      8. Buds brown or red-brown; flower buds not completely covered by 2 scales; common species, widespread in SE
        9. Lateral buds of at least 1 node elongated, length over 3 times the width; habit usually clumplike   8. *V. cassinoides*
        9. Lateral buds rarely elongated; habit often rhizomatous   9. *V. nudum*
1. Terminal buds with 4 or more imbricate scales
  10. Lowermost bud scales rarely exceed midpoint of bud; shrub, rarely over 1m tall
         10. *V. acerifolium*
  10. Lowermost bud scales often exceed midpoint of bud; shrub, usually over 1.5m tall; the "arrowwood complex"
    11. Bark of main stems shreddy/papery; mostly of Cumberland Plateau and Ozark regions
         11. *V. molle*
    11. Bark smooth, gray, or shallowly fissured on old stems, not papery
      12. Terminal bud base often tapered or necked, or twigs distinctly ridged or lined between nodes
        13. Leaf petioles 2–10mm, stipules often persisting; fruit ellipsoid
         12. *V. rafinesquianum*
        13. Leaf petioles over 10mm, stipules deciduous; fruit globose
         14. *V. dentatum* var. *lucidum*
      12. Terminal bud base not as above; twigs rounded or very faintly ridged
        14. Twigs pubescent; fruits globose, 6–8mm; common species
         13. *V. dentatum* var. *dentatum*
        14. Twigs nearly glabrous; fruits ellipsoid, 10–12mm; rare species   15. *V. bracteatum*

1. *Viburnum opulis* L.   GUELDER-ROSE   Figure 557

A large shrub, to 4m tall, with fairly stiff branches and twigs. Native to Eurasia; cultivated widely and naturalized sporadically. Twigs moderately stout, grayish, often ridged or lined below nodes; leaf scars raised; buds reddish, the scales tight, lustrous, edges obscure. Fruit 8–10mm, globose, red, bitter, persist into winter. Leaves toothed, mostly 3-lobed, reminiscent of a maple leaf; petiole with stalked reddish glands with concave tops. A more popularly and widely cultivated variety is "snowball viburnum," *V. opulis* var. *roseum* L., with wholly sterile, globular clusters of white flowers and no fruit.

2. *Viburnum trilobum* Marsh.   CRANBERRY VIBURNUM   Figure 558

A coarse shrub, to 4m tall, rarely over 2m in the SE. Widely distributed across n. N. Amer., but barely reaching the SE in the high elevations of WV. Occurs in cool, moist forest understories, on rocky slopes and shores. Twigs grayish, usually round and unlined; terminal buds present except where flowers or fruit were borne on twig tips; bud scales red or greenish, 2-scaled, but scale margins obscure. Fruit 8–10mm, subglobose, red, juicy and sour when ripe, resembling cranberry flavor; usually persistent into winter. Leaves similar to above species but petiole glands not concave. This species is sometimes considered only a N. Amer. variety of a circumpolar species, *V. opulis* var. *americanum* Ait.

3. *Viburnum obovatum* Walt. WALTER VIBURNUM   *Figures 559 and 560*

A shrub or small tree, to 5m tall. Native to the coastal plain from SC to FL, west to se. AL. Typically in moist to wet soils in bottomlands, swamps, near streams, drainages, depressions. Twigs reddish-scurfy, at least near apex, ridged below leaf scars, usually short or stubby; terminal vegetative buds valvate or naked, about 2mm long; flower buds naked, stalked, about 3mm long, with 2 to 4 folded, scurfy leaves around the perimeter. Leaves subevergreen in the Deep South, 2–5cm long, sessile or short-petioled, blunt at tip, remotely toothed or entire; margins slightly revolute, dotted beneath, glossy above. Bark grayish to black, shallowly fissured. Fruit ellipsoid, 5–10mm, glossy black when mature, borne in sessile clusters at twig tips.

4. *Viburnum lantanoides* Michx. HOBBLEBUSH   *Figure 561*

A coarse shrub, to 4m tall. Mainly a ne. species, ranging into the SE along the higher Appalachians from WV to n. GA. Occurs in moist, cool habitats of rocky slopes, streamsides, balds, and boreal zone forest understories. Twigs stout, brown, round and unlined, scurfy near tip, lenticellate; pith white, homogenous. Terminal buds 15–25mm long, scurfy, stalked, covered by 2 embryonic folded leaves instead of scales; terminal flower buds lumpy and flanked by scurfy bracts and naked, folded leaves. Bark thin, smooth to lightly flaky or shallowly fissured. Fruit ellipsoid, about 1cm long, glossy black when ripe. (*V. alnifolium* Marsh.)

5. *Viburnum prunifolium* L. BLACKHAW   *Figure 562*

A large shrub or small tree, to 8m tall, usually with a well-developed trunk but prone to root sprouting. Widely distributed across the n. half of the SE; sparse or lacking in the coastal plain from GA, FL to s. AR, TX. Occurs in a variety of woodlands and forest edges, roadsides, fencerows. May form colonies, mostly from root sprouts. Twigs grayish brown, often with short spur shoots; terminal buds grayish or brown, scurfy, 5–10mm long; terminal flower buds larger, plump, not covered completely by 2 valvate scales. Bark blocky. Leaves finely toothed, dull green, rather thin. Fruit ellipsoid, 8–14mm, black or bluish when ripe, often persistent into winter.

6. *Viburnum rufidulum* Raf. RUSTY BLACKHAW   *Figure 563*

A large shrub or small tree, to 10m tall, usually with a single, central trunk. Distributed throughout most of the SE, excluding the Blue Ridge of NC and much of VA, WV. Occurs in various upland wooded habitats, but most common in calcareous or dry habitats. Twigs grayish, lenticellate; buds dark red-brown and scurfy or densely rusty-pubescent. Bark dark gray to blackish, blocky. Leaves glossy, rather thick, finely toothed. Fruit ellipsoid, to 15mm long, glaucous blue-black when ripe.

7. *Viburnum lentago* L. NANNYBERRY   *Figure 564*

A large shrub or small tree, to 10m tall. A n. species uncommon in the SE, ranging into this region only in s. OH, MD, WV, w. VA. Occurs along mountain streams and moist woodlands. Twigs grayish to brown; terminal buds elongated, pinkish to gray, covered by 2 valvate scales. Bark grayish, shallowly furrowed to slightly flaky, blocky on lower part of main trunk. Fruit ellipsoid, blue-black when ripe, 10–15mm.

8. *Viburnum cassinoides* L. WITHEROD   *Figure 565*

A clumping shrub, to 5m tall. Distributed in the SE primarily in the Appalachians and interior, MD to n. AL, west to TN, KY, OH. Occurs in moist to wet woodlands, streamside thickets, swamps, high elevation openings. Twigs and fruit similar to above species, but lateral buds often much elongated, as with terminal buds. Leaves dull green, often toothed beyond middle; tip

Figure 557. *Viburnum opulis*

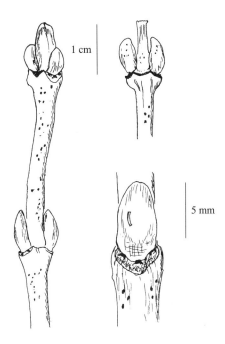

Figure 558. *Viburnum trilobum*

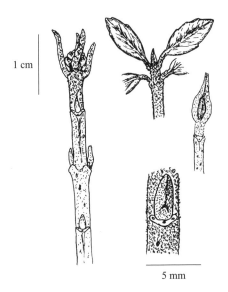

Figure 559. *Viburnum obovatum*

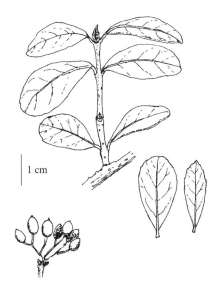

Figure 560. *Viburnum obovatum*

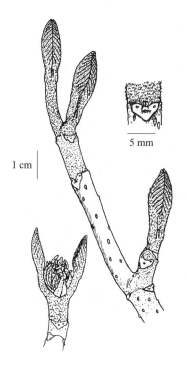

Figure 561. *Viburnum lantanoides*

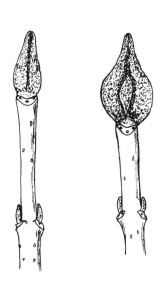

Figure 562. *Viburnum prunifolium*

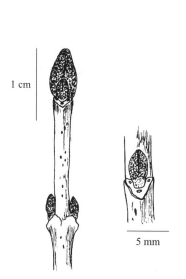

Figure 563. *Viburnum rufidulum*

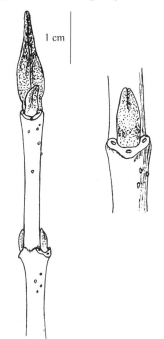

Figure 564. *Viburnum lentago*

acuminate; when wilting in autumn emitting a foul odor. Considered by some botanists to be only an upland variant form of *V. nudum* [*V. nudum* var. *cassinoides* (L.) Torr. & Gray]

9. *Viburnum nudum* L.   POSSUMHAW VIBURNUM   *Figure 566*

A rhizomatous shrub, to 5m, loosely clumplike or with many slender erect stems originating from the spreading root system. Widely distributed across much of the SE; most common in the coastal plain and Piedmont from VA to FL, west to TX, and scattered in the interior to AR, w. KY, w. VA. Typically in wet soils near water, bottomland forests, bogs, pineland depressions, swampy thickets. Twigs brownish; terminal buds rusty-brown, elongated; bud scales 2, valvate, scurfy, only partially covering base of flower buds. Bark thin, smooth. Fruit subglobose, 6–10mm, glaucous blue-black when mature. Leaves mostly elliptic, glossy above, 3–7cm wide; tip acute; margin entire to remotely toothed, slightly revolute. The var. *angustifolium* T. & G. bears narrow leaves to 3cm wide.

10. *Viburnum acerifolium* L.   MAPLE-LEAF VIBURNUM   *Figure 567*

A loosely branched, rhizomatous shrub, to 2m tall, but usually 1m or less. Widely distributed across most of the SE, but not common in the coastal plain. Typically in moist understories of mesic and rocky woods, though scattered in a variety of habitats where forest cover is available. Twigs brownish, lenticellate, rounded at the internodes, pubescent in the typical form but glabrous in the var. *glabrescens* Rehder, which is more common in the Appalachians and adjacent uplands in the interior of the SE. Buds with brown or brown-edged imbricate scales; leaf scars connected by a ridge. Pith brown in twigs, white in branchlets. Fruit ellipsoid, 6–8mm, lustrous black, usually on reddish stems. Several varieties have been described, based mainly on differences in foliage shape, degree of pubescence, and flower or fruit variances; none of these distinguishable in winter except perhaps var. *glabrescens*. The toothed and predominately 3-lobed leaves suggest those of a maple.

11. *Viburnum molle* Michx.   KENTUCKY ARROWWOOD   *Figure 568*

A spreading shrub, to 4m tall, usually clumplike and many-stemmed. Distributed from IN and KY to MO, AR. Occurs in rocky woods, along streams, near bluffs, mostly in calcareous soils. Twigs ashy gray, rounded and unlined; terminal buds usually with 4 scales brown to nearly black, keeled on backs; pith brownish-white. Twig epidermis begins to split in 2d or 3d year; papery and shreddy on older, main stems. Fruit ellipsoid, about 1cm long, blue-black. Leaf petiole 15–30mm long.

12. *Viburnum rafinesquianum* J. A. Schultes   SHORTSTALK ARROWWOOD

A spreading or suckering shrub, to 3m tall. A n. species ranging southward to the Piedmont of NC and west of the Blue Ridge to n. GA, MO. Typically in calcareous or circumneutral soils in dry or rocky woodlands. Twigs grayish, often ridged or lined, pubescent to glabrous; terminal buds often with a tapered "waist"; bud scales 4 or 6, with brown edges. Leaves with petioles only 2–10mm long, often with narrow stipules persisting. Fruit ellipsoid, 6–10mm, blue-black.

13. *Viburnum dentatum* L.   SOUTHERN ARROWWOOD   *Figure 569*

A spreading shrub, to 4m tall. Widespread throughout the SE. Occurs in a wide variety of sites, from dry and sandy forests to mesic or swampy habitats. Twigs brownish, pubescent; the stellate hairs either persist in winter or leave the twig rough with their remnants after falling; twigs rarely very distinctly lined or ridged at the internodes; buds with dark brown scales. Fruit globose, bluish or blue-black when mature, 5–8mm. Leaves ovate to subcordate, dentate, mostly pubescent beneath and on petiole; petioles 10–25mm. Numerous forms and varieties of this predominately pubescent species have been described in the past, considered minor and

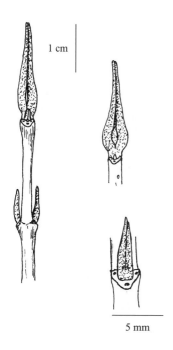

Figure 565. *Viburnum cassinoides*

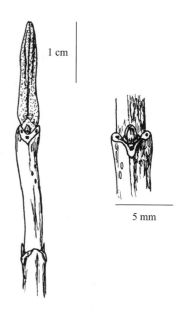

Figure 566. *Viburnum nudum*

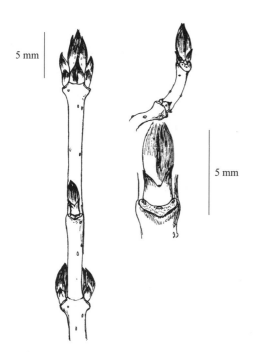

Figure 567. *Viburnum acerifolium*

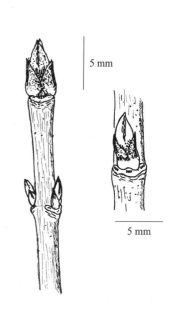

Figure 568. *Viburnum molle*

synonymous here. [*V. pubescens* (Ait.) Pursh; *V. semitomentosum* (Michx.) Rehd.; *V. scabrellum* (T. & G.) Chapm.] Two other glabrous entities exist in the SE: *V. pubescens* var. *deamii* (Rehd.) Fern., with leaves and twigs nearly glabrous and stipules often persisting on petioles (*V. carolinianum* Ashe), and *V. dentatum* var. *lucidum* Ait. described below.

14. *Viburnum dentatum* var. *lucidum* Ait.   NORTHERN ARROWWOOD   *Figure 570*

A shrub, to 4m tall, distributed mostly in the Piedmont and coastal plain from MD to s. SC, and scattered in the interior west to TN, OH, and northward. Typically in damp soils, floodplains, swampy sites, though sometimes in more upland woods. Twigs grayish, often ridged or lined in internodes, glabrous. Leaves mostly glabrous, usually glossier and with teeth more acute than in typical *V. dentatum*. Fruit subglobose, bluish or blue-black, 4–8mm. (*V. recognitum* Fern.)

15. *Viburnum bracteatum* Rehd.   GEORGIA ARROWWOOD

A spreading shrub, to 3m tall, capable of forming clonal colonies from root sprouts. Known only from a few upland locations in n. GA and n. AL. Occurs over limestone or in circumneutral soils of rich bluffs, streamsides, and mesic forest understories. Twigs grayish, rounded at internodes, hairy when young but usually nearly glabrous by winter. Most easily identified in flower, when it bears conspicuous bracts at the base of the inflorescence. The rather large, cordate leaves are light green, glossy, dentate, with petioles 15–20mm long, and stipules persistent until leaf fall.

Vinca L.   PERIWINKLE   Family Apocynaceae

Two exotic woody species are naturalized in the SE. Both are trailing shrubs that have been cultivated widely as groundcovers. Leaves are evergreen or subevergreen, simple, opposite, dark green above and light green below, with entire margins. Stems are green, mostly training over the ground, but with short vertical stems bearing the blue to violet flowers. Fruit is a follicle, borne in pairs.

1. Leaves widely ovate or triangular, truncate at base, mostly over 2cm wide, margin finely ciliate   1. *V. major*
1. Leaves elliptic, tapered to base, mostly under 2cm wide, margin not ciliate   2. *V. minor*

1. *Vinca major* L.   LARGE PERIWINKLE   *Figure 571*

A trailing, vinelike shrub with erect stems to 3dm high. Native to s. Europe, w. Asia. Spreading from cultivation into woodland understories and roadsides; sporadic in the SE. Leaves mostly evergreen in the South, but may be only subevergreen in n. portions of the region.

2. *Vinca minor* L.   PERIWINKLE   *Figure 572*

A trailing, vinelike shrub with erect stems rarely over 1dm tall. Native to Europe, w. Asia. Widely cultivated, and commonly spreading into woodland understories, waste areas, along roadsides, and persistent at old home sites, throughout the SE. Leaves evergreen, 2–4cm long.

Vitex agnus-castus L.   CHASTE-TREE   Family Verbenaceae   *Figure 573*

An exotic shrub, to 4m tall, often with dieback of twigs or larger stems in cold winter climates. Native to Eurasia; cultivated throughout the SE and persistent or spreading sparingly into nearby disturbed land. Twigs velvety-hairy, strongly aromatic when scraped, slightly 4-angled; opposite leaf scars U-shaped, connected by a line; a single bundle scar; pith white, homogenous; terminal buds lacking, usually a withered branch terminating the twig; lateral buds naked, superposed. Leaves palmately compound. Fruit bladderlike, about 3mm, with a single seed, pungent when crushed.

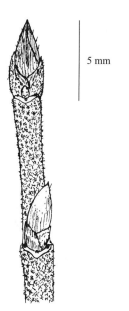

Figure 569. *Viburnum dentatum*

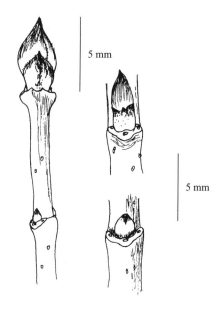

Figure 570. *Viburnum dentatum* var. *lucidum*

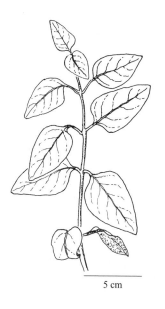

Figure 571. *Vinca major*

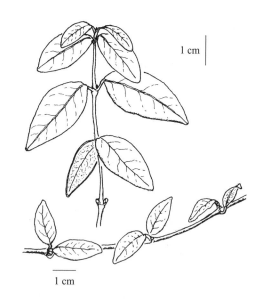

Figure 572. *Vinca minor*

Vitis L.   GRAPE   Family Vitaceae

Ten species are native to the SE, along with several significant varieties. Grapes are vines climbing by tendrils, with alternate, simple leaves. Winter determination of grape species using only twigs is impossible in many cases, and identification is best assured using leaf and fruit characters. For this reason, fallen leaves or remnant fruits are valuable aids. The twigs of all grapes lack terminal buds, and laterals are partially sunken in the twig. Leaf scars show several bundle scars, sometimes partially in 3 groups. Tendrils arise from some of the nodes, usually opposite a bud. Pith is brown or paler, homogenous in the internodes, but in most species it is interrupted in the nodes by a diaphragm. This is seen by dividing the twig through the node with a sharp blade; the thickness of the diaphragm is often a valuable aid in identification of a few species. The fruit is a berry, 1–4 seeded, usually ripening in autumn, edible in all but varying in palatability.

1. Bark lenticellate, not shreddy; pith continuous through nodes; tendrils not forked
   1. *V. rotundifolia*
1. Bark shreddy, not lenticellate; pith interrupted by a diaphragm at nodes; tendrils usually forked
   2. Leaves felty below (leaf surface between veins obscured); fruit mostly over 10mm diam.
      3. Tendrils at 3 or more consecutive nodes; widespread in Appalachians and easternmost states of SE, north of s. GA   2. *V. labrusca*
      3. Tendrils at 2 or fewer consecutive nodes; range restricted to w. states of SE or to FL
         4. Twigs slightly tomentose, with brownish hairs; pith diaphragms at nodes about 10mm thick; some leaves with reddish undersides; native to peninsular FL   3. *V. shuttleworthii*
         4. Twigs white-hairy; pith diaphragms at nodes 3–5mm thick; all leaves gray or white-felty below; native to TX, OK, rare in LA, AL   4. *V. mustangensis*
   2. Leaves not felty below, may have cobwebby hairs but these not obscuring leaf surface; fruit mostly 10mm or less
      5. Twigs prominently angled and pubescent in winter, with whitish or gray hairs
         5. *V. cinerea*
      5. Twigs weakly angled or not angled, mostly glabrous or with a few reddish hairs
         6. Leaf underside glaucous, or largest leaves to 20cm wide, most 10–15cm wide
            6. *V. aestivalis*
         6. Leaf underside greenish, largest leaves 15cm or less wide, most 5–10cm wide
            7. Leaves wider than long; habit usually trailing, with few tendrils   7. *V. rupestris*
            7. Leaves longer than wide; habit usually climbing, tendrils usually present
               8. Twigs reddish; leaves very long-tipped; mostly near MS River   8. *V. palmata*
               8. Twigs brown, gray, or greenish; leaves not as above; widespread, common
                  9. Fruit glaucous; pith nodal diaphragms 1–2mm thick; sides of leaf teeth straight or concave   9. *V. riparia*
                  9. Fruit shiny black; pith nodal diaphragms 2–6mm thick; sides of leaf teeth convex
                     10. *V. vulpina*

1. *Vitis rotundifolia* Michx.   MUSCADINE   Figure 574

A high-climbing vine with gray bark not becoming shreddy, but often with hanging adventitious roots developing in humid habitats. Distributed throughout most of the SE, except middle and higher elevations of the Appalachians of NC, VA, KY, WV. Occurs in a variety of habitats from moist lowlands and riverbanks to dry or sandy uplands. Twigs grayish, ridged or angled; tendrils opposite 2 consecutive nodes, not forked; pith continuous through the nodes, lacking a diaphragm. Leaves lustrous green on both sides; margin with large teeth; blades about 8cm long and wide. Fruit 10–25mm diam., purplish to black when ripe, dotted, with a thick skin, ripening and falling Aug–Sept. The var. *munsoniana* (Simpson ex Munson) Moore occurs in FL and s. GA, differing in smaller leaves and fruit (fruit 5–10mm). A form with bronze fruit is popularly known as "Scuppernong."

2. *Vitis labrusca* L. FOX GRAPE

A high-climbing vine with dark brown, shreddy bark. Distributed in the SE mainly from WV to coastal VA, south to central GA, AL, n. MS, also to central TN, KY. Occurs in bottomlands, along rivers, and in mesic woods or thickets; most abundant in Piedmont and mountains of NC, VA. Twigs with tendrils opposite nearly every node. Leaves densely close-velvety below, pale, tawny, or brownish, mostly 10–15cm wide, largest to 20cm. Fruit 15–25mm, deep purplish-black to black when ripe, dull and not conspicuously dotted, with a thick skin, ripening Aug–October. This is a parent of the Concord grape, *Vitis* × *labruscana* Bailey, and thus of many other horticultural derivatives of the Concord. Some of these may spread from sources of cultivation, but leaf undersides and fruit appearance are generally similar to *V. labrusca*.

3. *Vitis shuttleworthii* House CALUSA GRAPE

A climbing vine with brown, shreddy bark. Native to peninsular FL; barely ranging far enough north to be included in the se. region covered by this book. Occurs in various woodland understories and thickets. Twigs slightly hairy; tendrils at 2 consecutive nodes; pith interrupted by distinct diaphragms in nodes, some 10mm or more thick. Leaves pale to reddish-velvety beneath, blades mostly 8–15cm wide. Fruit 10–16mm, dark reddish to nearly black, dull and not dotted.

4. *Vitis mustangensis* Buckl. MUSTANG GRAPE

A high-climbing vine with shreddy bark. Distributed primarily in e. TX and OK; sporadic eastward into AR, LA, and south-central AL. Occurs in calcareous soils, on rocky slopes, near streams, and amid thickets near bottomlands. Twigs white-tomentose, hairs oriented toward twig tip; pith nodal diaphragms 3–5mm thick. Leaf undersides and petioles white-woolly; blades mostly 10–15cm wide. Fruit 15–20mm, purplish-black, with a tough and pungent-flavored skin, ripening in Sept and persisting; palatability of fruit sweeter in var. *diversa* (Bailey) Shinners.

5. *Vitis cinerea* (Engelm.) Engelm. ex Millard. PIGEON GRAPE

A high-climbing vine with shreddy bark. Widely distributed across most of the SE, with 3 varieties described. Occurs in moist lowland soils, especially along streams and in bottomlands. Twigs angled, with pale grayish or white hairs; pith nodal diaphragms 3–5mm thick; nodes often red-banded during the growing season in plants of the more w. varieties. Leaves green below but with short, straight hairs (hirtellous) and/or with cobwebby hairs; blades mostly 8–15cm wide. Fruit 3–9mm, mostly black, not glaucous. The var. *cinerea* mainly occurs in the Mississippi Embayment and eastward to AL, w. FL; it has mostly hirtellous hairs on twigs and leaves, and some cobwebby hairs. The var. *baileyana* (Munson) Comeaux is distributed mostly in the interior regions and east to the Piedmont; twigs and leaves mostly with pale, cobwebby hairs. The var. *floridana* Munson occurs in the coastal plain from VA to FL, MS; twigs and leaves with tawny to reddish cobwebby hairs.

6. *Vitis aestivalis* Michx. SUMMER GRAPE   Figure 575

A high-climbing vine with shreddy bark. Widespread throughout the SE, common in a wide variety of habitats from moist lowlands to fairly dry uplands. Twigs glabrous, or hairy when young; pith nodal diaphragms 1–4mm thick. Leaves usually glaucous beneath but sometimes green, varying in degree and color of cobwebby or clustered hairs; blades mostly 10–15cm wide. Fruit 5–12mm, black or glaucous blue-black. The var. *bicolor* Deam is more strongly glaucous beneath and less hairy, with fruit and pith diaphragms on the lesser size range of variation than the typical species; it is most common in the Appalachian region.

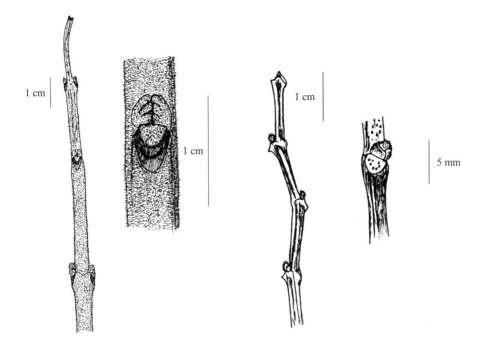

Figure 573. *Vitex agnus-castus* Figure 574. *Vitis rotundifolia*

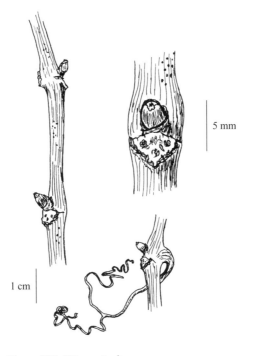

Figure 575. *Vitis aestivalis*

7. *Vitis rupestris* Scheele SAND GRAPE

A sprawling vine, bushy and rarely climbing. Sporadic in its range from MD and the mountains of VA, WV west and south to TN, AR, MO, TX. Typically in rocky, calcareous soils of slopes and streamsides. Twigs with few tendrils, those present are red and mostly opposite endmost nodes of fruiting twigs; pith nodal diaphragms 2–3mm thick. Bark eventually shreddy. Leaves wider than long, usually folded or trough-shaped; blades 4–12cm wide. Fruit 6–10mm, black or slightly glaucous, ripe July–early Sept and soon falling.

8. *Vitis palmata* Vahl CAT GRAPE

A high-climbing vine with shreddy bark. Distributed mostly in the Mississippi Embayment area and sporadic eastward to s. GA, n. FL. Occurs in bottomland forests and river swamps. Twigs slender, glabrous; pith nodal diaphragms 2–5mm thick. Leaves lustrous green on both sides, with conspicuously elongate tips; blades 4–9cm wide. Fruit 5–8mm, black, with thin flesh.

9. *Vitis riparia* Michx. RIVERBANK GRAPE

A high-climbing vine with shreddy bark. Widely distributed over much of the SE, except the coastal plain from SC to s. MS. Occurs along streams, in lowland forests and other moist soils. Twigs glabrous; pith nodal diaphragms mostly less than 1mm thick. Leaves green and mostly glabrous beneath; margin with minute cilia, basal portion of each tooth straight or concave on sides; blades mostly 8–12cm wide, usually distinctly 3-lobed. Fruit 8–12mm, very glaucous.

10. *Vitis vulpina* L. FROST GRAPE

A high-climbing vine with shreddy bark. Distributed throughout the SE in a variety of habitats from moist lowlands to upland forests, often along streams and roadsides. Twigs glabrous; pith nodal diaphragms 1–5mm thick. Leaves green and mostly glabrous beneath; margin not ciliate, base of teeth mostly convex on sides; blades mostly 5–12cm wide, obscurely lobed or unlobed. Fruit 3–10mm, black, shining, not glaucous, ripe in Sept and persistent.

Weigela florida (Bunge) A. DC. WEIGELA Family Caprifoliaceae *Figure 576*

An exotic shrub, to 3m tall, sparingly naturalized or persistent after cultivation. Native to n. China and Korea; commonly cultivated for its 3cm tubular flowers that may be white, pink, or red. Twigs gray-brown; leaf scars opposite, appear sunken due to raised, irregular lower margin; a row of hairs extends down the twig between the leaf scars; bundle scars 3. Terminal buds present, unless flowering has occurred on the twig tip; bud scales paired, imbricate, lower pair usually fused at base. The fruit is an elongate capsule about 2cm long.

Wisteria Nutt. WISTERIA Family Fabaceae *Figure 577*

Four species are treated here for the SE, 2 native and 2 exotics. All are twining vines with grayish, mostly smooth bark with conspicuous lenticels. Twigs grayish, with prominent branch scars and no terminal buds; lateral buds with 2 or 3 loosely imbricate and long-tipped scales. Dried, straplike stipules may persist. Leaf scars raised, with lumplike projections commonly on the sides; bundle scars indistinct, appearing as a fibrous mass. Pith whitish, homogenous. The fruit is a legume, maturing in autumn and persisting into early winter. Wisterias are popularly cultivated vines due to the showy nature of the flowers, which hang in racemes in spring; color usually violet, but white forms of most species occur. Winter identification by twigs alone is not possible, so the key by necessity uses other characters.

1. Legume glabrous, seed pelletlike; main stems slender, not often high-climbing; native species

2. Inflorescences 5–12cm long; common and widespread species   1. *W. frutescens*
2. Inflorescences 12–30cm long; native mostly west of Blue Ridge   2. *W. macrostachya*
1. Legume velvety; seed orbicular and flattened; main stems robust, habit often high-climbing; exotic species
   3. Leaflets 13 to 19 per leaf; inflorescence to 50cm, flowers opening gradually from base to tip   3. *W. floribunda*
   3. Leaflets 7 to 13 per leaf; inflorescence to 30cm, flowers opening nearly at the same time
                                                                                              4. *W. sinensis*

1. *Wisteria frutescens* (L.) Poir.   AMERICAN WISTERIA

A slender vine, twining into shrubs or low trees, less commonly into high canopy. Distributed mostly in the coastal plain from VA to GA, FL, west to e. TX, and north in the interior to KY. Occurs near streams, swampy thickets in bottomlands, and similar moist sites. Legume mostly 8–10cm long. Leaves with 9 to 15 leaflets.

2. *Wisteria macrostachya* Nutt.   KENTUCKY WISTERIA

A slender vine very similar to the above species and often considered only a variety of it. Distributed mainly in the Mississippi Embayment region of s. IL, w. KY, south to TX, LA, and sporadic eastward in the coastal plain to FL, NC. Occurs in moist or rich woodlands and thickets, and is more often chosen for cultivation due to its larger flower clusters.

3. *Wisteria floribunda* (Willd.) DC.   JAPANESE WISTERIA

A robust vine capable of forming stems or trunks over 1dm diam. Native to Asia; cultivated for its flowers and often maintained shrublike as a lawn specimen in the South. Climbs into trees by twining. Occasionally naturalizing, mostly near urban areas and vicinity of cultivation. Legume 10–15cm long. Flowers sweetly fragrant.

4. *Wisteria sinensis* (Sims) DC.   CHINESE WISTERIA

A robust vine similar to above, but more commonly and extensively naturalized throughout the SE. Native to Asia; cultivated for its flowers and often naturalizing in urban and waste areas. Frequently climbs high into tree canopies. Legume 10–15cm long. Flowers slightly fragrant.

Xanthorhiza simplicissima Marsh.   SHRUB-YELLOWROOT   Family Saxifragaceae
Figure 578

A slender shrub, rarely to 1m tall, rhizomatous and forming colonies of many vertical, few-branched stems. Distributed throughout most of the SE, though most common in the interior. Occurs along shaded streamsides and in moist woodlands. Twigs moderately stout, grayish, lenticellate, with bright yellow inner bark. Terminal bud large, conical, with thin, greenish, imbricate scales; lateral buds much reduced. Leaf scars nearly encircle the twig and have several bundle scars. The fruit is a cluster of small follicles, each with a lustrous seed about 1mm long.

Yucca L.   YUCCA   Family Agavaceae

As a group, all have evergreen straplike or swordlike leaves averaging 40–70cm long, ending in a pointed tip or spine and with parallel venation. The fruit is normally a dry, dehiscent capsule, though in 3 cases is indehiscent or fleshy and berrylike. Nine entities are treated here, though taxonomic validity of several is debatable; as few as 4 species for the SE are recognized by some taxonomists. The taxonomy is complicated by intermediate forms, unexplainable variation, and supposed hybridization. Origins and natural ranges of many are also unclear as

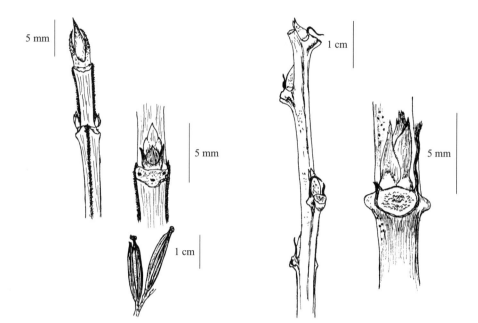

Figure 576. *Weigela florida*

Figure 577. *Wisteria*

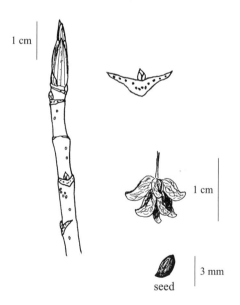

Figure 578. *Xanthorhiza simplicissima*

they may have been cultivated and spread by native peoples prior to recorded history. These 9 yuccas can be generalized as 3 groups:

the "glauca" complex (*glauca, arkansana, louisianensis*), which occurs in the westernmost sections of the SE, in LA, AR, TX, OK; the members often with narrow, white-edged leaves with some filaments on margins, all acaulescent.

the "filamentosa" complex (*filamentosa, flaccida, smalliana*), widespread in the eastern/southeastern sections of the region, from VA to FL, MS, with the more inland form of *flaccida* reaching the Appalachians; all acaulescent yuccas with quite noticeable filamentous-edged leaves and a flower panicle held well above the limits of the reach of the leaves.

the "caulescent" complex (*aloifolia, gloriosa, recurvifolia*), of sandy soils very near the coast, NC to MS; these have the capability to form trunks over 1m high; leaf margins bear few filaments; fruits indehiscent; flower panicles reached by leaves.

1. Leaves with many threadlike fibers curling from margins; fruit erect, dry, split when mature; plants mostly acaulescent
  2. Leaves mostly 25mm or less wide
    3. Leaves rigid, 5–10mm wide, glaucous   1. *Y. glauca*
    3. Leaves flexible, limp, or recurved, 10–25mm wide, light to dark green
      4. Flower stalks glabrous; leaf tips reach level of lower flowers   2. *Y. arkansana*
      4. Flower stalks pubescent, at least central axis; all flowers held well above leaf limits
        5. Leaf margin distinctly white-striped; adaxial surface not rough   3. *Y. louisianensis*
        5. Leaf margin not white-striped; adaxial surface rough   5. *Y. flaccida*
  2. Leaves mostly over 25mm wide
    6. Leaves stiff, abruptly narrowed at tip, adaxially scabrous; flower panicle glabrous
           4. *Y. filamentosa*
    6. Leaves pliable, often recurved, long-tapered at tip, rough but not scabrous; panicles not all glabrous
      7. Flower panicle branches pubescent; flower petals broad, mostly acute, style elongate
           5. *Y. flaccida*
      7. Flower panicle branches nearly glabrous; flower petals acuminate, style short
           6. *Y. smalliana*
1. Leaves with few or no marginal fibers; fruit drooping, not splitting, dry to fleshy; plants often caulescent (with a trunk)
  8. Leaves rigid, toothed on edge; older ones bent downward from near base   7. *Y. aloifolia*
  8. Leaves stiff to floppy, older ones not toothed; bent or recurved from near middle
    9. Leaves stiff, often blue-green; fruit leathery to semifleshy   8. *Y. gloriosa*
    9. Leaves thin, pliable, green; fruit dry   9. *Y. recurvifolia*

1. *Yucca glauca* Nutt.   SOAPWEED   *Figure 579a*

Mostly a w. species of plains and prairies, ranging eastward to the westernmost sections of the SE in OK, AR, TX. Leaves rigid, mostly 6–12cm wide, with a gradually tapering tip, grayish to glaucous green, a few threads along the white-striped margin. Flower raceme unbranched; lower flowers reached by leaves. Fruit dry, dehiscent; seed glossy.

2. *Yucca arkansana* Trel.   ARKANSAS YUCCA   *Figure 579b*

Distributed in grassy prairies, rocky and often calcareous soils and openings in AR, OK, TX. Leaves stiff to pliable, mostly 10–25mm wide, long-tapered, dark green, slightly scabrous on the back, finely filamentous on white-striped margin. Flower raceme glabrous, reached by leaves. Fruit dry, dehiscent; seed dull.

3. Yucca louisianensis Trel. LOUISIANA YUCCA *Figure 579c*

Similar to above species and sometimes considered as a variety of it or of hybrid origin. Distributed in AR, LA, TX in sandy soils and open ground. Leaves similar to *Y. arkansana*, but to 30mm wide. Flower raceme pubescent to glabrous, held well above reach of leaves. Fruit dry, dehiscent, very angular when young; seed dull. (*Y. arkansana* var. *paniculata* McKelvey)

4. Yucca filamentosa L. ADAM'S-NEEDLE *Figure 580a*

Native presumably to the coastal plain from VA to FL, west to MS, and sporadically inland to the Appalachians, but widely planted and naturalizing from cultivation. Leaves stiff, 25–70mm wide, abruptly narrowed and often cusped or with concavity on inner surface near tip, green to glaucous, scabrous on back; margin with curly filaments. Flower panicle glabrous, held beyond reach of leaves, open Apr–July. Fruit dry, dehiscent; seed glossy.

5. Yucca flaccida Haw. WEAKLEAF YUCCA *Figure 580b*

Similar to the above species and often considered only a form of it. Additionally, the following species, *Y. smalliana*, is sometimes treated as synonymous with *Y. flaccida*. Occurs in the lower elevations of the Appalachians from VA to AL, and eastward and southward to the coastal plain. Leaves pliable to limp, older ones drooping (recurved), 15–40mm wide, long-tapered, green, rough but not scabrous; marginal filaments often straight. Flower panicle pubescent, held well above reach of leaves; open Apr–June. Fruit dry, dehiscent; seed dull.

6. Yucca smalliana Fern. BEAR-GRASS *Figure 580c*

This species is sometimes considered a form of *Y. flaccida* or of *Y. filamentosa*. Believed native to the coastal plain from SC to FL. Similar to *Y. flaccida*, but leaf filaments often curlier; flower raceme pubescent only on central axis; flowers smaller and with narrower petals. Blooms mostly in June–Sept. [*Y. filamentosa* var. *smalliana* (Fern.) Ahles]

7. Yucca aloifolia L. SPANISH-DAGGER *Figure 581a*

Native to coastal fringes of the coastal plain from NC to FL, but often planted elsewhere. Typically seen near brackish or salt marshes, shell middens, dunes. Habit caulescent, to 3m. Leaves thick, rigid, 3–5cm wide, stiffly spinose-tipped; margins not filamentous but minutely notched and sharp-edged, green to yellow-green, slightly concave, bending downward from base with age. Flower panicle reached by leaves; open June–July. Fruit fleshy, pendulous, indehiscent.

8. Yucca gloriosa L. MOUND-LILY YUCCA *Figure 581b*

Native to the vicinity of coastal NC to FL, but often planted. Often seen near or on coastal dunes and in sandy maritime forests. Habit caulescent, to 2.5m. Leaves stiff, 25–60mm wide, spinose-tipped, margin yellow to brown-lined and only slightly filamentous when old, green or glaucous, bending over (recurved) from near middle with age. Flower panicles reached by leaves; open Aug–Oct. Fruit leathery to slightly fleshy, pendulous, indehiscent.

9. Yucca recurvifolia Salisb. CURVE-LEAF YUCCA *Figure 581c*

This species is suspected to be of hybrid origin (*aloifolia* × *flaccida*?). It was first described from the GA coastal plain. Habit caulescent, to 2m. Leaves thin, pliable, 25–45mm wide, margin thin, light brown, sometimes filamentous when old, green, recurved from near middle with age. Flower panicle barely reached by leaves; open Apr–May or Aug–Oct. Fruit dry, indehiscent, erect to pendulous.

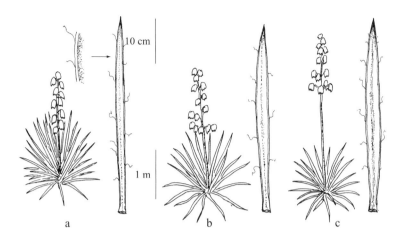

Figure 579. a: *Yucca glauca;* b: *Y. arkansana;* c: *Y. louisianensis*

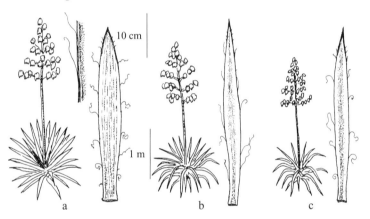

Figure 580. a: *Yucca filamentosa;* b: *Y. flaccida;* c: *Y. smalliana*

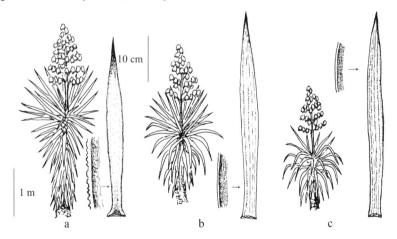

Figure 581. a: *Yucca aloifolia;* b: *Y. gloriosa;* c: *Y. recurvifolia*

Zamia integrifolia Ait.   COONTIE   Family Cycadaceae   *Figure 582*

A native gymnosperm with an underground stem and pinnate, leathery evergreen leaves. Distributed primarily in peninsular FL, but ranging northward to ne. FL and se. GA. Leaves crowded at top of stem, close to the ground, to 1m long in some robust forms from s. parts of the range, but mostly 3–10dm long in n. FL; leaflets 10 to 30 per side. Aboveground portion of stem short, but much enlarged below the soil surface; this used as starchy food source by native people. The cones are mostly 5–15cm long, borne amid the leaf bases on the stem; containing many large seed covered by a reddish, fleshy aril. Plants are dioecious, the fruit borne only on female plants. Variations of this species with wide leaves are most common near the coast of ne. FL, GA. (*Z. pumila* L; *Z. floridana* A. DC.; *Z. silvicola* Small)

Zanthoxylum L.   PRICKLY-ASH   Family Rutaceae

Two native species occur in the SE. Twigs usually bear prickles, small reddish buds, and citruslike, aromatic inner bark. Inner bark and crushed leaves, if chewed, cause harmless tingling and numbing sensations. Leaf scars pale, alternate, with 3 bundle scars that may be obscure or appear as 4 when lower group is divided. Pith white, homogenous. The fruits are reddish, bumpy pods (follicles) about 5–8mm long, which split to release a shiny black seed. Leaves pinnately compound, deciduous. This genus has been alternately spelled *Xanthoxylum* in the past.

1. Twigs reddish-brown or gray; prickles mostly paired at nodes; trunk bark brownish; habit shrubby   1. *Z. americanum*
1. Twigs greenish-gray; prickles more scattered; trunk bark gray, with conical growths; habit treelike   2. *Z. clava-herculis*

1. *Zanthoxylum americanum* P. Mill.   PRICKLY-ASH   *Figure 583*

A shrub, rarely over 4m tall in the SE, though capable of reaching 10m. Often forming colonies from root sprouts. Widespread in the SE but of sporadic occurrence; more common in the n. states; ranging from WV, VA, to MO, south to AL, n. FL, se. SC. Occurs near rock outcrops, calcareous soils, or mesic woodlands. Twigs usually with many fine longitudinal lines, glabrous, with flat, broad-based prickles paired near the nodes. Terminal bud rusty-brown, hairy or granular-surfaced, no distinct scales, often with small, peglike bases where small leaves were attached. Main trunk bark brownish, sometimes gray, often finely fissured, sometimes with a few remnant prickles. Fruits borne at nodes of branchlets.

2. *Zanthoxylum clava-herculis* L.   HERCULES-CLUB   *Figures 584 and 585*

A large shrub or small tree, often 4–10m tall, usually with a single trunk. Sometimes trunk is short and forks close to the ground, but plants tend to be single and not forming colonies of root sprouts. Distributed in the coastal plain from se. VA to FL, west to TX, AR, OK. Occurs in maritime forests, shell middens near marshes, dunes, and in calcareous soils of uplands or bottomlands in the w. portions of its range. Twigs usually greenish-gray, often with red lenticels or pitlike depressions near apex; prickles not as broad-based as with *Z. americanum,* more scattered over twig or in irregular groups near nodes; sparse to absent in some plants. Terminal bud warty, reddish, often with minute spinose projections. Trunk bark gray, thin, mostly smooth but often with conical or similarly moundlike outgrowths of cork, these with or without the remnant prickle at their apex. Fruits borne in terminal clusters on twig tips.

Zenobia pulverulenta (Bartr. ex Willd.) Pollard   HONEY-CUPS   Family Ericaceae
Figure 586

A rhizomatous or clumplike shrub, rarely over 1.5m tall. Endemic to the coastal plain from se. VA to e. GA. Typically in acidic, moist sands and peaty soils in pinelands and pocosins. Twigs

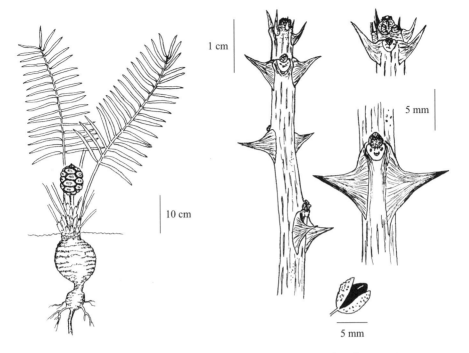

Figure 582. *Zamia integrifolia*

Figure 583. *Zanthoxylum americanum*

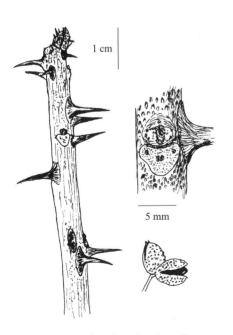

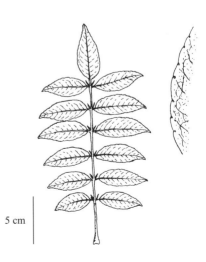

Figure 584. *Zanthoxylum clava-herculis*

Figure 585. *Zanthoxylum clava-herculis*

gray or reddish-tinted on 1 side, zigzag, slightly flattened or angled, with a ridge below each leaf scar; no terminal bud; laterals blunt, angled sharply away from twig, with several imbricate scales more reddish near bud tip. Leaf scar with thickened or wrinkled lower margin and a single bundle scar. Pith rather large, green, spongy. Fruit a persistent, flattish capsule with 5 cells. Leaves deciduous, margin crenately toothed, slightly revolute, glabrous and lustrous green on most plants near the outer coastal plain, but glaucous forms are common farther inland. Sometimes twigs as well as leaves are nearly white with a heavy waxy bloom on these inland plants. Flowers fragrant, white, bell-shaped.

Zizyphus jujuba P. Mill.   CHINESE-DATE   Family Rhamnaceae   *Figure 587*

A small tree, to 10m tall, with a slender, forked trunk and arching to drooping branches. Native to Eurasia; occasionally cultivated for its fruit and sparingly naturalizing nearby. Twigs red-brown, slightly hairy near apex; branchlets may bear numerous stubby spur shoots on older plants, whereas young trees usually have slender stipular spines at the nodes. Terminal bud lacking, but may appear present due to larger lateral bud situated near twig apex; bud scales reddish, their bases somewhat fleshy, tips long and spinose. Lammas shoots (2d growth flush on the twig, at nodes) commonly present, these sometimes dying back in winter or falling to leave a large, flat, raised scar or stub alongside a lateral bud. Leaf scar smaller than bud; bundle scars tiny, 3 or obscure. Bark gray, scaly. The fruit is a red-brown drupe 2–4cm long, with an elongated pit and sweet pulp (datelike flavor). Several cultivar selections are known with various fruit shapes.

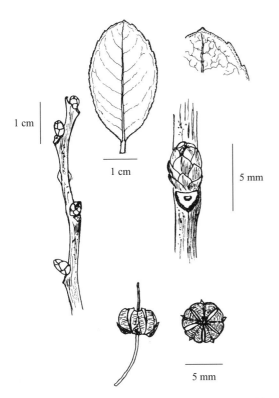

Figure 586. *Zenobia pulverulenta*

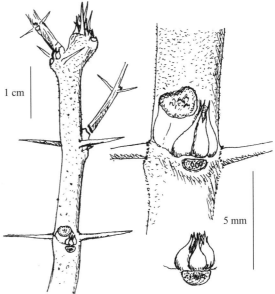

Figure 587. *Zizyphus jujuba*

Glossary of Terms Used in This Text

Abaxial. Pertaining to the side away from an axis; as applied to leaves, the lower surface.

Abscission (leaf). Separation of the leaf from the twig through tissue changes, such as in the lower petiole of deciduous plants' leaves in autumn.

Acaulescent. Lacking a true leaf-bearing stem aboveground.

Accessory (buds). Additional lateral buds occurring around a primary lateral bud.

Achene. A small, hard fruit with a thin, dry covering, one-seeded and one-locular, derived from a single carpel or simple pistil.

Acicular. Needle-shaped.

Acidic (soils). Having a pH of less than 7.0; most commonly ranging between 4.0 and 5.5.

Acuminate. Tapering gradually to a narrow point.

Acute. Sharply tapered to a point.

Adaxial. Pertaining to the side toward an axis; as applied to leaves, the upper side.

Adnate. Defines a fusion of unlike structures or organs, such as stamens to petals, ovaries to calyx, or stipules to leaf petioles.

Adventitious. Root or bud development in an abnormal or irregular pattern, such as roots arising from aboveground stems.

Aerial rootlet. Roots arising from the stems of climbing vines, mostly as adaptations for clinging to objects.

Aggregate (fruit). A cluster of separate fruits derived from separate ovaries or carpels within a single flower.

Alkaline (soils). Having a pH over 7.0; most commonly ranging between 7.5 and 9.0.

Alkaloid. One of many bitter-tasting alkaline ring compounds found in plants.

Alluvium. Sediments and soil particles deposited by flowing waters.

Alternate. Arrangement of one member (such as a leaf) on an axis at a distinct distance above and below another.

Angiosperm. A plant bearing ovules within an ovary and subsequently seed borne within a pericarp, or matured ovary.

Angled. In outline or cross section, having distinct sides or points, not rounded or oval.

Anthesis. Time of flower expansion.

Apiculate. Having an abrupt, distinct, stiff point.

Appressed. Lying closely against.

Arboreal. Growing or dwelling within trees.

Arborescent. Treelike in habit or size.

Arcuate. Arching or curving.

Areole. In application to cacti, the circular patches that correspond to nodes of the cactus stem or joint where buds, leaves, spines, and glochids are borne.

Aril. An appendage or enlarged covering on a seed derived from the ovule stalk.

Armed. Bearing sharp spines or thorns.

Auriculate. Bearing auricles, which are earlike lobes.

Awn. A bristle or stiff hairlike appendage, usually terminal on fruit or leaf points.

Axillary. Borne in an axil, or the angle between two members or structures.

Barren. As applied to plant communities, an area of low soil fertility and sparse tree cover.

Basic (soils). Having a neutral pH of 7.0 or nearly so.

Bay. As applied to plant communities, a depression in coastal plain lowlands filled with water and ringed by dense shrubby growth or entirely filled with shrubby thickets; as applied geographically, a large body of water in an indented shoreline; also a member of a generalized group of plants having leathery, evergreen leaves suggestive of the size and shape of the bay leaves of commerce.

Berry. A fleshy fruit with several immersed seed, collectively derived from a single pistil.

Bipinnate (leaf). Twice pinnate, with pinnae attached to a rachis.

Bitter-almond (odor). A moderately rank odor associated with a majority of members in the genus *Prunus,* due partially to presence of prussic acid in the sap.

Blocky (bark). Furrowed bark with particularly thick, short, or squarish plates.

Bloom. A glaucous, waxy or powdery covering.

Bog. A wetland area with soft, spongy ground, the soils usually high in organic matter.

Boreal. Northern; or of a northern character in some cases of vegetation description.

Brackish. Having a mixture of fresh- and saltwater.

Bract. A modified leaf or leaflike appendage subtending flowers or inflorescences.

Bracteole. A small bractlet.

Bractlet. A secondary or reduced bract, such as those borne on individual flower stalks or pedicels within an inflorescence.

Bramble. A generalized term for members of the genus *Rubus,* though sometimes applied to any coarse prickly shrub or vinelike shrub.

Branch scar. A scar occurring at the point where the growing twig tip withered and fell away, usually near a distal lateral bud.

Branchlet. Divisions of a branch one season older than the twigs; a second-year stem.

Bristle. A stiff hair or similarly shaped trichome.

Bud scale. A modified and reduced leaf or an immature bract that covers part or all of a dormant bud.

Bulbous. Having a shape like a bulb; having bulbs.

Bundle scar (or bundle trace). A scar within a leaf scar caused by abscission of the leaf and separation of the vascular bundles.

Buttressed. Having a pronounced widening or swelling of the trunk base.

Caducous. Falling early; not persisting for very long.

Calcareous (soils). Of a limey character or associated with limestone or calcite; containing high proportions of calcium.

Calcophiles. Plants growing in and adapted to grow in calcareous soils.

Calyx. A basal or outer part of a flower; considered with the corolla as the perianth.

Calyx lobe. Distal extensions of the calyx.

Calyx tube. The prolonged basal portion of a calyx, when it is tubular in shape.

Capsule. A dry, dehiscent fruit with several locules, derived from a compound pistil.

Carpel. A simple pistil or one member of a compound pistil.

Carpellate. Having one or more carpels; derived from a carpel.

Catkin. A type of spike or spikelike inflorescence; an ament.

Catkin bud. A dormant bud, usually elongate, that forms a catkin upon opening.

Caulescent. Having an aboveground leaf-bearing stem.

Cespitose. Growing in a tufted or small clumplike habit.

Chambered (pith). An excavated pith with numerous membranaceous, horizontal walls.

Ciliate. Having cilia, or spaced hairs, along the margin.

Circumboreal. Around the northern regions of the earth.

Circumneutral (soil). Around a basic or neutral pH, or between 6.5 and 7.5 pH.

Circumpolar. Around the polar region of the earth.

Clambering. Climbing into and over other plants or objects in a vinelike habit but rather weakly attached to supports; climbing with difficulty or reduced efficiency.

Clasping (leaf). Having the base partly or fully surrounding the stem.

Cleft. Cut deeply toward the middle; forming distinct lobes with acute sinuses.

Clinal. Relating to a gradual change in characteristics between individuals of a species among a series of adjacent populations.

Clonal. Derived from vegetative, asexual reproduction and existing as a clone of a previous plant's root system or from an adjoining plant.

Clone. A descendent from an individual without involvement of sexual reproduction; in plants, an asexually produced plant that is essentially a genetic duplication of the parent.

Collateral (bud). An accessory bud positioned beside the primary lateral bud at a node.

Colonial. Forming colonies, usually from spreading root systems or underground stems.

Colony. A group or population of plants derived from vegetative or seed regeneration.

Compound. In application to a leaf or pistil, being made up of two or more similar parts united together; as applied to a fruit, derived from several carpels.

Cone. A fruit composed of a spiral arrangement of bracts or scales with seed between.

Conifer. A cone-bearing plant, mostly defining a taxonomic order of the gymnosperms.

Coniferous. Cone bearing; belonging to the conifer group of plants; in descriptions of forests, consisting primarily of conifers.

Connate. United or joined, such as the bases of the last pair of opposite leaves on the twigs of some *Lonicera*.

Cordate. Heart-shaped.

Coriaceous. Having a leathery texture.

Corolla. The inner part of the perianth, often colored and lying interior to the calyx, called the "petals" of a flower when lobed or divided into units.

Corolla tube. The prolonged basal portion of a corolla, when it is tubular in shape.

Corymb. A broad, domelike or flat-topped racemose flower cluster, anthesis progressing from outermost flowers toward the inner.

Crenate. With shallow, rounded teeth.

Crescent-shaped. Resembling the shape of the moon in its first or last quarter.

Culm. The upright or flowering stem of grasses and sedges, including bamboo.

Cultivar. A subdivision of a species or a selection of a hybrid between species, originating in cultivation or maintained through cultivation, replicated by asexual reproduction.

Cuneate. Wedge-shaped.

Cupped. Recurved so as to cause opposing convex and concave surfaces.

Deciduous. Falling away; not persistent.

Dehiscent (fruit). Opening along one or more slits or seams to release seed.

Deltoid. Shape suggestive of an equilateral triangle.

Dentate. With marginal teeth that point outward.

Denticulate. Minutely dentate.

Diaphragmed (pith). A homogenous pith with horizontal partitions or thick membranes scattered throughout.

Dichotomous. Branching or forking regularly in pairs.

Dicot. A plant bearing two cotyledons in its seed.

Dioecious. Having functional pistillate and staminate flowers occurring on separate plants of the same species, each plant being unisexual.

Disjunct. A population or occurrence of a species separated widely from the bulk of the range.

Doubly serrate. Having smaller marginal serrations borne on margins of larger teeth.

Drupe. A fleshy fruit, usually one-seeded, with a stony seed coat or endocarp.

Dry (fruit). A fruit that is nonsucculent when mature.

Eglandular. Without glands.

Ellipsoid. A solid with an elliptical outline.

Elliptic. A shape widest at the middle, tapering to rounded ends.

Emarginate. Having a small notch at the apex or end.

Endemic. Native range restricted to a particular geographical area.

Endocarp. The inner layer of a matured ovary wall; the inner pericarp.

Entire. A margin having no serrations.

Entity. A plant existing as a distinct taxonomic unit or as a separate, identifiable condition.

Epidermal. Of or occurring on the epidermis.

Epidermis. The outer or superficial layer of cells.

Epiphyte. A plant growing on another plant, but not parasitic.

Ericad. A plant of the Ericaceae family, a majority of which favor acidic soils.

Eurasia. Europe and Asia considered together as one continent.

Evergreen. Leaves persistent into a second growing season since their formation.

Excavated (pith). A pith seemingly absent, the central channel of the twig hollow.

Exfoliating. Peeling off in thin layers.

Exotic. Not native to the southeastern region before European settlement.

Exserted. Projecting beyond an enclosing part.

Fastigiate. A habit of growth in trees or shrubs that is columnar or narrow, with branches throughout the crown held compactly close to the trunk.

Filament. The stalk of the stamen that attaches to the anther.

Fleshy (fruit). A fruit having succulence or juicy tissues upon maturity.

Floral tube. A tubular structure supporting parts of the flower, becoming part of the fruit.

Floricane. The flowering cane of the genus *Rubus* that is in its second season of growth after its initial appearance as a primocane.

Flowering plant. A plant that is a member of the Angiosperm group.

Foliaceous. Leaflike in appearance or texture.

Follicle. A dry, one-chambered fruit derived from a simple pistil; dehiscent on one side.

Forma. A taxonomic category within a species that describes an aberrant entity other than a variety.

Frutescent. Shrublike.

Fruticose. Shrubby or related to shrubs.

Geniculate. Bent, similar to a knee; having a bent or zigzag pattern.

Genus. A taxonomic category defining a group of species that share sufficient morphological characteristics so they appear closely related.

Glabrous. Surface smooth, not pubescent.

Glade. An open or sparsely forested area within a forest; when of natural occurrence often originating from periodic fire, soil variations, or water fluctuations.

Gland dot. A dotlike structure caused by a minute gland secreting a substance.

Glandular. Having glands or with the secretions of glands visible.

Glaucous. Having a pale or lightened color due to a waxy or powdery bloom.

Globose. A solid with a globular or spherical shape.

Globular. Suggesting a globose shape.

Glochid. A barbed trichome or bristle.

Groundcover. A plant having a prostrate or reclining habit upon the ground, with ascending shoots lacking or not exceeding 2dm in height.

Gummy. Having a sticky or resinous sap or secretion; glutinous.

Gymnosperm. A plant bearing ovules not enclosed within an ovary.

Habit. The growth form or general appearance of a plant.

Hammock. An elevated, fertile area of hardwood forest surrounded by wetland, pinelands, or scrubland in the lowlands of the coastal plain.

Hip. A peculiar fruit of the genus *Rosa* that is an aggregate of achenes with a partially fleshy receptacle forming the outer wall.

Hirsute. Bearing coarse or stiff hairs.

Hirtellous. Finely or minutely hirsute.

Hispid. Bearing bristles or rigid hairs.

Homogenous (pith). Having a uniform, rather solid appearance and texture throughout.

Hybrid. An individual derived from sexual reproduction between two separate species.

Hydric. Of or pertaining to a wet condition.

Hypanthium. A floral tube or extension of the receptacle bearing the perianth and subsequently forming part of the mature fruit (especially when it is adherent to the ovary).

Imbricate. Overlapping, similar to shingles of a roof.

Incised. Having deep marginal cuts; deeply cleft nearly to the midrib.

Indehiscent (fruit). Remaining closed; not opening along any seams or splits.

Inflorescence. The arrangement or style of flower production in a plant; flower cluster.

Infrutescence. The arrangement or style of fruit production; the fruit cluster.

Inner bark. The layer of living tissue beneath the outer corky portion of bark; the phloem and cork cambium layers outside the vascular cambium.

Inodorous. Lacking an odor.

Internodal. Between the nodes or on the zone between nodes.

Internode. A zone of the twig between the nodes.

Invasive. Known to exhibit rapid or extensive growth and capable of domination over the majority of other plants within certain ecosystems or growing conditions.

Involucre. A cluster or layer of bracts subtending or enclosing a flower or inflorescence.

Joints (cacti). The peculiar widened stem section of some cacti that appear as segments or pads.

Juvenile. As applied toward age of plants, the leaf or twig-bearing wood is immature, since flowering age of the plant has not been attained.

Keeled. Having a fine ridge or raised longitudinal line.

Lammas shoot. A secondary twig extension from active meristem in leaf axils during the same season of growth as the main twig. This lateness in appearance during the growing season is artificially associated with Lammas Day, celebrated on August 1 in England.

Lanate. Having long, curly, intertwining hairs; woolly.

Lanceolate. Shaped like a lance; broadest near the base and long tapered.

Late-deciduous. Leaves fall during the course of a winter season, leaving twigs bare before the advent of spring growth.

Lateral bud. Any bud situated in a position on the twig other than true terminal.

Leaflet. A single blade or unit of a compound leaf.

Legume. A dry, one-locular fruit derived from a simple pistil; usually dehiscent on two sides, peculiar to the family Fabaceae.

Lenticel. A corky spot on the thin or young bark of plants that originates as an epidermal stoma.

Lenticellate. Bearing conspicuous lenticels.

Lignify. To become woody; relates to the formation of *lignin* as the chief hardening substance of wood, typifying *ligneous* or woody plant stems.

Linear. Long and very narrow, with parallel sides.

Lined. Having fine longitudinal ridges or raised lines.

Lobed. Having lobes or extensions of a margin resulting in protruding segments.

Locular. Having one or more locules.

Locule. A compartment within an ovary and subsequently within the fruit.

Loess. A loamy or silty soil deposited by wind; usually yellowish and calcareous.

Marl. A sedimentary clay soil deposited along with calcium carbonate, very calcareous.

Marsh. An area with hydric or extensively wet soil, usually devoid of trees.

Membranaceous. Having a thin, pliable texture.

Mesic. Remaining relatively moderate in moisture availability throughout the year.

Metamorphic. As applied to rocks, a structural change and composition brought about by intense pressure and heat.

Midrib. The central or main rib of a leaf or similar broadened structure.

Moderately slender (twig). Generally having a diameter between 2 and 4mm across the distal few internodes; between slender and moderately stout.

Moderately stout (twig). Generally having a diameter between 4 and 5mm across the distal few internodes; between moderately slender and stout.

Monocot. A plant having one cotyledon within each seed.

Monoecious. Having functional pistillate and staminate flowers occurring on the same individual plant.

Morphological. Pertaining to the form and structure.

Mottled (bark). Having a differentiated color pattern caused by pieces of older bark falling away to reveal different colors or shades beneath.

Mucronate. Having a short, abrupt tip, including stiff and spinose tips.

Multiple (fruit). A compound fruit developing from a close cluster of separate flowers in an inflorescence.

Naked (bud). A bud having no specialized bud scales; the first embryonic leaves to be expanded usually visible and folded or entirely concealed by hairs or granular pubescence.

Native. Naturally occurring in the southeastern region before European colonization.

Naturalized. Not native to the area, but established and reproducing without aid of cultivation.

Needles. A generalized term describing a leaf of a linear or acicular shape.

Netted venation. A leaf venation pattern whereby the veins successively and visibly branch into a smaller network.

Neutral (soil). Having a pH of 7.0 or very nearly so; basic soil.

Node. The area of a twig or other stem where vascular tissue exits into a leaf.

Nut. A dry, hardened, one-seeded fruit; usually one or more borne in a husk or involucre.

Nutlet. A small nut or nutlike fruit, held by an involucre or associated with a bract.

Oblanceolate. Shaped in reverse of lanceolate, being widest beyond the middle, tapering abruptly on the tip, and gradually tapering to the base.

Oblong. Having nearly parallel sides and two to three times longer than broad.

Obovate. Shaped in reverse of ovate, being broader beyond the middle.

Obovoid. A solid with an outline that is obovate.

Obtuse. Having a blunt tip, nearly rounded.

Ocreae. A sheathlike stipule, as seen in *Polygonella*.

Opposite (leaf). Borne in pairs, each leaf on the opposing side of the twig from another.

Orbicular. Circular-shaped or nearly so.

Ovary. The basal part of a pistil containing one or more ovules.

Ovate. Shaped like the outline of an egg, with the wider end being the base.

Ovoid. An egg-shaped solid; also with an ovate outline.

Ovule. The egg-producing structure within an ovary, which after fertilization matures into a seed.

Palmate. Having three or more parts originating from a common point; digitate.

Pandurate. Suggesting a fiddle shape, the margin on each side twice widened with a narrow section between.

Panicle. A loose, irregular inflorescence having compound branching and stalked flowers, resembling a branching raceme.

Pappus bristles. Modified calyx lobes in some members of the family Asteraceae that appear as silky or downy material attached to the base of the fruits or achenes.

Parallel venation. Main veins appear to run parallel and unbranched from base to tip of a leaf, mostly in monocots such as grass, lily, and palm families.

Parasitic. A condition of one organism deriving nutrition from the host organism to which it is attached.

Pectinate. Divided into narrow, equal segments suggestive of a comb.

Pedicel. A stalk of a single flower in an inflorescence.

Peduncle. A stalk supporting a solitary flower not in an inflorescence, or the main stalk of an inflorescence.

Peltate. Having a stalk attached to the lower surface inside the margin.

Peltate scales. Trichomes modified as shield-shaped structures with peltate attachment.

Pericarp. The wall of a matured ovary, as in the fruit.

Persistent. As applied in this text, remaining attached through part or all of the winter.

Petiolate. Having a petiole.

Petiole. The stalk of a leaf that joins it to the twig.

pH. A logarithmic scale of hydrogen ion concentration used to describe acidity or alkalinity of a solution on a scale of 0 to 14, with 7 being neutral or basic; acidity increasing exponentially with increments below 7.0 and alkalinity increasing exponentially with increments over 7.0.

Pinna. One of the side or primary divisions of a bipinnate leaf, or a leaflet of a pinnate leaf.

Pinnate. Having an elongate main axis with divergent lateral members; as applied to leaves, the main axis is a rachis, laterally bearing separated blades called leaflets.

Pistil. The portion of a flower bearing an ovary and ovules, forming, after pollination and fertilization, the fruit and seed.

Pistillate. Having pistils; flowers having pistils and no stamens; plants bearing no staminate flowers.

Pith. The central tissue of a stem, containing soft, thin-walled cells.

Plane. Having a flat surface.

Plate (bark). A raised, relatively flat-topped section of outer bark, its margins defined by longitudinal and horizontal furrows.

Pneumatophores. Modified elongate root structures projecting out of water-saturated soils; thought to aid in aeration of the plant's root system.

Pocosin. A wetland community in the coastal plain dominated by dense shrubby growth, with few trees and with soils high in peat and organic matter; a shrub bog.

Pod. A generalized name applied to any dry fruit with one or more chambers; traditionally used as an alternate name for a legume.

Pome. A fleshy fruit derived from an ovary with fused carpels and a floral tube, the fleshy portion being the enlarged floral tube.

Prickle. A spinose structure borne on the epidermal tissue or outer bark; easily detached if forced to one side.

Primocane. The first-year cane in the genus *Rubus,* which usually does not flower or branch until its second growing season, then called a floricane.

Prostrate. Lying flat on the ground.

Pruinose. Having a bloom, a pale waxy or powdery covering.

Pseudoterminal bud. A lateral bud, positioned at the apex of a twig after the twig has self-pruned (branch scar present), that may resemble a terminal bud; a false-terminal bud.

Pubescent. Having hairs or trichomes, especially short and soft hairs.

Punctate. Having dots on the surface caused by minute depressions or glands.

Pyriform. Pear-shaped.

Raceme. An inflorescence with a central axis and stalked flowers.

Rachis. The central axis that bears leaflets in a compound leaf, or the central axis in an inflorescence.

Raised (leaf scar). A condition in which the leaf scar is elevated or protruding on the twig.

Rank (leaf). As applied to leaf arrangement, the way in which leaves (and nodes) are borne on a spiral along the twig, whether two-ranked (distichous) and in two rows, or three-ranked or more where in additional rows; all rows discernible if the view is directed down at the twig apex; leaf phyllotaxy.

Rank (odor). A pungent odor, or any unpleasant odor.

Reflexed. Abruptly bent or curved backward or downward.

Remotely toothed. Having serrations that are widely spaced.

Repent. Creeping or prostrate over the ground and rooting at the nodes, as in stems.

Resin dots. Minute dots on a surface caused by glands or glandular secretions.

Resin globules. Small exudations of resin or glandular secretions persistent on a surface.

Reticulate. Having a network of veins; a visibly conspicuous netted appearance.

Revolute. Having the margin rolled backward; in the case of leaves, the margin is rolled toward the undersurface.

Rhizomatous. Having rhizomes or prone to produce them.

Rhizome. A prostrate or underground stem, rooting at the nodes.

Rhombic. Suggestive of a diamond shape; essentially having four equilateral sides and two oblique angles.

Riparian. Associated with or growing by rivers or streams.

Robust (twigs). Having a stout appearance or large circumference.

Rugose. Wrinkled; with veins impressed into a surface.

Samara. A winged, indehiscent, achenelike fruit.

Sandbar. A deposit of sand and other sediments within a stream channel or its floodplain.

Sapling. A juvenile tree at least 1m in height; any small, treelike plant with a stem diameter of 1–10cm.

Savanna. An area of flat terrain with lush herbaceous growth and scattered trees.

Scabrous. Very rough to the touch, with a harsh or gritty texture caused by stiff trichomes pointing collectively in one direction.

Scarious. Thin, dry, and membranaceous, usually not green.

Scalelike (leaf). A small, subulate or similarly shaped sessile leaf mostly appressed to the twig.

Scaly (bark). A condition of the outer bark in which aged pieces lift or curl away from underlying layers.

Scrub. A generalized term applied to plant communities containing short, contorted, or dwarfed trees and shrubs in poor or dry soils.

Scurfy. Having a granular, scaly, or flaky surface.

Sedimentary. Pertaining to sediments or the deposition of soil and other organic materials by water or wind.

Seepage. Water flowing from subterranean sources through soil.

Sepal. A unit or division of the calyx.

Serotinous. Produced late in the season; as applied to cones of pines, remaining closed for an indefinite period after maturity (usually opened by heat from fires).

Serrated. Having a toothed margin, with teeth generally pointing forward.

Serrulate. Minutely and finely serrate.

Sessile. Lacking a stalk.

Sheath. An enveloping structure or tubular covering, as over the lower part of the leaf of palms or grasses (including bamboos).

Shell midden. An accumulation of discarded shells, resulting from human consumption of shellfish, that increases amounts of calcium carbonate in nearby soil as weathering occurs.

Shrub. A woody plant with a habit smaller than a tree, the ultimate size usually less than 4m tall and 10cm diameter, commonly with multiple main stems.

Shrub bay. A wetland depression of the coastal plain dominated by shrubby thickets and a few scattered trees.

Simple. Of one unit, unbranched; as applied to leaves, having one blade, not compound; as applied to pistils, having a single carpel; as applied to fruit, derived from one pistil in one flower.

Sinus. The marginally recessed space between two lobes or divisions.

Slender (twig). Generally having a diameter of less than 2mm across the distal few internodes.

Slough. A canal or drainage channel in a lowland with slow-moving water or deposits of mud.

Solid (pith). An alternate term for homogenous pith, which has no large cavities or horizontal partitions.

Spatulate. Generally oblong in shape, but tapering toward the base and widening toward the rounded tip, similar to the shape of a druggist's spatula.

Species. The basic category of taxonomic classification defining a group of organisms similar in appearance and reproducing consistently.

Spine. As applied in this text, any woody or rigid outgrowth that ends in a sharp point.

Spinose. Resembling a spine or having spines.

Spongy (pith). A generally homogenous pith perforated with holes, resembling a sponge.

Spray (leaf). A portion of the twigs, branchlets, or small branches of a plant with the leaves attached.

Spur shoot. A short lateral shoot of a branch or branchlet, rarely on a twig, sometimes spinose, with crowded leaf scars, and typically giving rise to flowers and fruit.

Stamen. The portion of a flower bearing an anther that produces pollen and is attached to another part of the flower or inflorescence by a stalk or filament.

Staminate. Having stamens; flowers having stamens and no pistil; plants bearing no pistillate flowers.

Stellate. Starlike; as applied to trichomes, branches of the hair radiate out from a central point.

Stipitate. Having a stipe or short stalk.

Stipule. An appendage at the base of a leaf or petiole, or at the node near the petiole; usually in pairs but sometimes fused.

Stolon. A stem or branch elongating and taking root at its tip, borne either above- or below-ground.

Stoloniferous. Having stolons or prone to produce them.

Stoma. A minute opening or pore in the epidermis, controlled by guard cells and functioning in the respiration of plants; plural = "stomata."

Stout (twig). A size generalization describing twigs having a diameter over 5mm across the distal few internodes.

Striated. Having stripes or fine longitudinal lines.

Strigose. Having stiff, sharp, straight hairs appressed to a surface or collectively pointing in one direction.

Strobile. A type of inflorescence or fruit with imbricated bracts or scales, most often applied in describing the fruit of *Betula*.

Style. An extended portion of the pistil that connects the stigma (which receives pollen) to the ovary.

Subcoriaceous. Having a somewhat leathery texture; nearly coriaceous.

Suberin. A waxlike substance within the cell walls of bark cork cells, giving a resilient and water-resistant nature; most highly developed and familiar in the true cork bark of commerce.

Subevergreen. At least some leaves persist through the winter, but most fall before or with the new spring growth.

Subglobose. A solid that is nearly globose in shape.

Subopposite. Nearly opposite, with an opposing member slightly off center of true opposite, but not spaced widely enough to typify an alternate arrangement.

Subsessile. Nearly sessile, with a very short stalk.

Subshrub. A morphological generalization describing very low or diminutive semiwoody plants; suffrutescent.

Subspecies. A taxonomic subdivision of a species, often defining a naturally occurring geographical or ecological variance recognizable by morphological characteristics; most often used in taxonomic classification of animal species, but essentially equivalent to "variety" in plants.

Subtending. Situated in a position below and close to.

Subulate. Awl-shaped; tapering gradually to a point from the wide, often sessile base.

Sucker. A sprout or vertically elongate shoot from a subterranean origin.

Suckering. Having suckers or prone to produce them.

Suffruticose. A small woody plant resembling a diminutive shrub.

Sunken (buds). Being partially or wholly concealed by being imbedded within the twig.

Superposed. Situated above and close to; one above the other.

Suture. A line or seam where dehiscence occurs.

Synonym. An unused or superceded name.

Synonymy. The series of discarded names applied to the same entity.

Talus. A slope composed mostly of stones and rock fragments.

Taxon. A named member of a taxonomic category; plural = "taxa."

Tendril. A slender twining or clasping appendage of a stem or leaf enabling climbing.

Terete. Circular or very nearly so in cross section.

Terminal bud. A bud formed on the apex of a twig.

Thorn. A spinose outgrowth of wood, or a stem modified as such.

Tomentose. Having dense, woolly pubescence, the hairs matted and soft.

Tree. A woody plant with a well-developed main trunk at least 10cm diameter at 1.35m (about 4.5ft) above the ground at maturity.

Trichome. A hairlike outgrowth of the epidermis in plants.

Trifoliate. Bearing three leaflets.

Truncate. Having an abrupt end; not tapered, as if cut off squarely.

Turbinate. Shaped like a top or an inverted cone.

Twining. Twisting around other objects to enable climbing.

Umbel. An inflorescence with flower stalks arising from a common point.

Utricle. A small, bladderlike, one-seeded fruit; a bladderlike body.

Valvate. Meeting at the margins and not overlapping.

Variety. A taxonomic subdivision of a species based on some naturally occurring hereditary and morphological difference in characteristics; commonly used in plant taxonomy and equivalent to "subspecies" in taxonomic classification of animals.

Vascular bundle. A close association of conductive tissue (xylem and phloem) that passes between twig and leaf leaving a bundle scar in the leaf scar after leaf abscission.

Verrucose. Covered with small bumps or wartlike elevations.

Villous. Having fine, long, soft hairs that are not interwoven.

Vine. An elongating plant, having no rigid stem capable of supporting its own weight, that climbs or trails onto and over other plants and objects.

Viscid. Sticky; glutinous, with a sticky exudation.

Warty. Having wartlike bumps; as applied to bark, having outgrowths of cork that resemble warts.

Wavy. Having undulations along the margin or surface.

Whorled (leaves). Three or more leaves borne at one node.

Winged. Having a membranaceous, coriaceous, or woody extension; having corky outgrowths on the twigs or branches.

Wintergreen (odor). A pleasant odor peculiar to members of the genus *Gaultheria* and some species of *Betula*.

Woody. In the strict sense, a plant having woody tissue or secondary xylem from a vascular cambium; in the working application for this text, a plant having visible dormant parts remaining alive in winter and with buds for shoot growth present above the surface of the ground.

Xeric. Of or pertaining to an arid or dry condition.

References

Adams, W. P. 1962. Taxonomic and distributional observations on North American taxa of *Hypericum*. *Rhodora* 64:231–242.
Ahrendt, L. W. 1961. *Berberis* and *Mahonia*: A taxonomic revision. *J. Linn. Soc., Bot.* 57:1–410.
Anderson, L. C. 1985. *Forestiera godfreyi* (Oleaceae), a new species from Florida and South Carolina. *Sida* 11:1–5.
Argus, G. W. 1986. The genus *Salix* (Salicaceae) in the southeastern United States. *Syst. Bot. Mono.* 9:1–170.
Baas, P. 1984. Vegetative anatomy and the taxonomic status of *Ilex collina* and *Nemopanthus* (Aquifoliaceae). *J. Arnold Arb.* 65:243–250.
Baum, B. R. 1967. Introduced and naturalized tamarisks in the United States and Canada (Tamaricaceae). *Baileya* 15:19–25.
Benson, L. 1982. *The Cacti of the United States and Canada.* Stanford CA: Stanford Univ. Press.
Blakeslee, A. F., and C. D. Jarvis. 1972. *Northeastern Trees in Winter.* New York: Dover Publications.
Bozeman, J. R., and J. F. Logue. 1968. A range extension for *Hudsonia ericoides* in the southeastern United States. *Rhodora* 70:289–292.
Braun, E. L. 1969. *The Woody Plants of Ohio.* New York: Hafner Publishing.
Brizicky, G. K. 1963. The genera of Sapindales in the southeastern United States. *J. Arnold Arb.* 44:462–501.
———. 1964. The genera of Rhamnaceae in the southeastern United States. *J. Arnold Arb.* 45:439–463.
———. 1965. The genera of Vitaceae in the southeastern United States. *J. Arnold Arb.* 46:48–67.
Burckhalter, R. E. 1992. The genus *Nyssa* (Cornaceae) in North America: A revision. *Sida* 15 (2):323–342.
Camp, W. H. 1945. The North American blueberries with notes on other groups of Vacciniaceae. *Brittonia* 5:203–275.
Chapman, A. W. 1883. *Flora of the Southern United States.* New York: American Book Co.
Chester, E. W., B. E. Wofford, and R. Kral. 1997. *Atlas of Tennessee Vascular Plants.* Vol. 2, *Angiosperms: Dicots.* Misc. Pub. No. 13. Clarksville TN: Center for Field Biology, Austin Peay State University.
Chester, E. W., B. E. Wofford, R. Kral, H. R. DeSelm, and A. M. Evans. 1993. *Atlas of Tennessee Vascular Plants.* Vol. 1, *Pteridophytes, Gymnosperms, Angiosperms: Monocots.* Misc. Pub. No. 9. Clarksville TN: Center for Field Biology, Austin Peay State University.
Clark, R. B. 1942. A revision of the genus *Bumelia* in the United States. *Ann. Mo. Bot. Gard.* 29:155–182.
Clark, R. C. 1971. The woody plants of Alabama. *Ann. Mo. Bot. Gard.* 58:99–242.
———. 1974. *Ilex collina*, a second species of *Nemopanthus* in the southern Appalachians. *J. Arnold Arbor.* 55:425–440.
Clewell, A. F. 1966. Identification of the lespedezas in North America. *Bull. Tall Timbers Research Sta.* No. 7.
———. 1985. *Guide to the Vascular Plants of the Florida Panhandle.* Tallahassee: Florida State Univ. Press.

Cliburn, J., and G. Klomps. 1980. *A Key to Missouri Trees in Winter.* Jefferson City: Missouri Dept. of Conservation.

Coker, W. C., and H. R. Totten. 1945. *Trees of the Southeastern States.* 3d ed. Chapel Hill: Univ. of North Carolina Press.

Core, E. L., and N. P. Ammons. 1958. *Woody Plants in Winter.* Pittsburgh: Boxwood Press.

Correll, D. S., and M. C. Johnston. 1970. *Manual of the Vascular Plants of Texas.* Renner: Texas Research Foundation.

Culwell, D. E. 1970. A taxonomic study of the section *Hypericum* in the eastern United States. Ph.D. diss., University of North Carolina, Chapel Hill.

Dryer, G. D., L. M. Baird, and C. Fickler. 1987. *Celastrus scandens* and *Celastrus orbiculatus*: Comparisons of reproductive potential between a native and an introduced woody vine. *Bull. of Torrey Bot. Club* 114:260–264.

Duncan, W. H. 1975. *Woody Vines of the Southeastern United States.* Athens: Univ. of Georgia Press.

Duncan, W. H., and M. B. Duncan. 1988. *Trees of the Southeastern United States.* Athens: Univ. of Georgia Press.

Duncan, W. H., and J. T. Kartesz. 1981. *Vascular Flora of Georgia: An Annotated Checklist.* Athens: Univ. of Georgia Press.

Ebinger, J. E. 1974. A systematic study of the genus *Kalmia* (Ericaceae). *Rhodora* 76:315–398.

Elias, T. S. 1971. The genera of Myricaceae in the Southeastern United States. *J. Arnold Arb.* 52:305–318.

———. 1980. *The Complete Trees of North America.* New York: Van Nostrand Reinhold.

Eyde, R. H. 1966. The Nyssaceae in the Southeastern United States. *J. Arnold Arb.* 47:117–125.

Faircloth, W. R. 1970. An occurrence of *Elliotia* in central south Georgia. *Castanea* 35:58–61.

Farrelly, D. 1995. *The Book of Bamboo.* San Francisco: Sierra Club Books.

Ferguson, I. K. 1966a. The Cornaceae in the Southeastern United States. *J. Arnold Arb.* 47:106–116.

———. 1966b. The genera of Caprifoliaceae in the Southeastern United States. *J. Arnold Arb.* 47:33–59.

Fernald, M. L. 1950. *Gray's Manual of Botany.* 8th ed. New York: American Book Co.

Flora of North America Editorial Committee, eds. 1993–1997. *Flora of North America North of Mexico.* Vols. 2–3. New York: Oxford Univ. Press.

Furlow, J. J. 1990. The genera of Betulaceae in the southeastern United States. *J. Arnold Arb.* 71:1–67.

Gillis, W. T. 1971. The systematics and ecology of poison-ivy and the poison-oaks (Toxicodendron, Anacardiaceae). *Rhodora* 73.

Gleason, H. A. 1952. *The New Britton and Brown Illustrated Flora of the Northeastern United States and Adjacent Canada.* 3 vols. New York: New York Botanical Garden.

Gleason, H. A., and A. Cronquist. 1991. *Manual of Vascular Plants of Northeastern United States and Adjacent Canada.* 2d ed. Bronx: New York Botanical Garden.

Godfrey, R. K. 1988. *Trees, Shrubs, and Woody Vines of Northern Florida and Adjacent Georgia and Alabama.* Athens: Univ. of Georgia Press.

Godfrey, R. K., and J. W. Wooten. 1981. *Aquatic and Wetland Plants of Southeastern United States: Dicotyledons.* Athens: Univ. of Georgia Press.

Graves, A. H. 1952. *Illustrated Guide to Trees and Shrubs.* Wallingford CT: Published by the author.

Hardin, J. W. 1971a. Studies of the southeastern United States flora: 1. Betulaceae. *J. Elisha Mitch. Sci. Soc.* 87:39–41.

———. 1971b. Studies of the southeastern United States flora: 2. The gymnosperms. *J. Elisha Mitch. Sci. Soc.* 87:43–50.

———. 1973. The enigmatic chokeberries (*Aronia*, Rosaceae). *Bull. Torrey Bot. Club* 100:178–184.

———. 1979. *Quercus prinus* L. *-nomen ambiguum. Taxon* 28:355–357.
———. 1990. Variation patterns and recognition of varieties in *Tilia americana* s.l. *Systematic Bot.* 15:33–48.
Harlow, W. M. 1966. *Fruit Key and Twig Key to Trees and Shrubs*. New York: Dover Publications.
Harvill, A. M., Jr., et al. 1992. *Atlas of the Virginia Flora*. 3d ed. Burkeville: Virginia Botanical Associates.
Hickock, L. G., and J. C. Anway. 1972. A morphological and chemical analysis of geographical variation in *Tilia* of eastern North America. *Brittonia* 24: 2–8.
Hicks, R. H., Jr., and G. K. Stephenson, comps. and eds. 1978. *Woody Plants of the Western Gulf Region*. Dubuque IA: Kendall/Hunt Pub.
Hu, Shiu-ying. 1954–56. A monograph of the genus *Philadelphus. J. Arnold Arb.* 35:276–333; 36:52–109; 37:15–90.
Isely, D. 1990. *Vascular Flora of the Southeastern United States*. Vol. 3, pt. 2, *Leguminosae (Fabaceae)*. Chapel Hill: Univ. of North Carolina Press.
Isely, D., and F. J. Peabody. 1984. *Robinia* (Leguminosae: Papilionoideae). *Castanea* 49:187–202.
Jones, G. N. 1963. Flora of Illinois. *Am. Midl. Nat. Monogr.* 7.
Judd, W. S. 1981. A monograph of *Lyonia* (Ericaceae). *J. Arnold Arb.* 62:1–209.
———. 1982. A taxonomic revision of *Pieris* (Ericaceae). *J. Arnold Arb.* 63:103–144.
Judd, W. S., and K. A. Kron. 1995. A revision of *Rhododendron* VI. Subgenus Pentanthera. *Edinb. J. Bot.* 52(1):1–54.
Kartesz, J. T. 1994. *A Synonymized Checklist of the Vascular Plants of the United States, Canada, and Greenland*. 2d ed. Portland OR: Timber Press.
Kirkman, W. B., and J. R. Ballington. 1990. Creeping blueberries (Ericaceae: *Vaccinium* sect. *Herpothamnus*): A new look at *V. crassifolium* including *V. sempervirens*. Syst. Bot. 15(4): 679–699.
Kral, R. 1960. A revision of *Asimina* and *Deeringothamnus* (Annonaceae). *Brittonia* 12:233–278.
Kron, K. A., and M. Creel. 1999. A new species of deciduous azalea (*Rhododendron* Section Pentanthera; Ericaceae) from South Carolina. *Novon* 9:377–380.
Kurtz, H., and R. K. Godfrey. 1962. *Trees of Northern Florida*. Gainesville: Univ. of Florida Press.
Lakela, O. 1963. The identity of *Bumelia lacuum* Small. *Rhodora* 65:280–282.
Lance, R. 1994a. *The Hawthorns of the Southeastern United States*. Asheville NC: Published by the author.
———. 1994b. *Woody Plants of the Blue Ridge*. Clyde NC: Haywood Community College Press.
Lance, R., and J. B. Phipps. 2000. *Crataegus harbisonii* Beadle rediscovered and amplified. *Castanea* 65:291–296.
Little, E. L., Jr. 1971. *Atlas of United States Trees*. Vol. 1, *Conifers and Important Hardwoods*. USDA Forest Service. Misc. Pub. 1146. Washington DC.
———. 1977. *Atlas of United States Trees*. Vol. 4, *Minor Eastern Hardwoods*. USDA Forest Service. Misc. Pub. 1342. Washington DC.
———. 1979. *Checklist of United States Trees (Native and Naturalized)*. USDA Handbook 541. Washington DC.
———. 1981. *Atlas of United States Trees*. Vol 6, supp. USDA. Pub. 1410. Washington DC.
Manning, W. E. 1950. A key to the hickories north of Virginia with notes on the two pignuts, *Carya glabra* and *C. ovalis. Rhodora* 52:188–199.
McClure, F. A. 1973. Genera of bamboos native to the New World. *Smithsonian Contr. Bot.* 9:1–148.
Miller, H. M., and S. Lamb. 1985. *Oaks of North America*. Happy Camp CA: Naturegraph Publishers.

Moore, M. O. 1991. Classification and systematics of eastern North American *Vitis* L.(Vitaceae) north of Mexico. *Sida* 14:339–367.
Muenscher, W. C. 1969. *Keys to Woody Plants*. Ithaca: Cornell University Press.
Nicely, K. A. 1965. A monographic study of the Calycanthaceae. *Castanea* 30:38–81.
Ogle, D. W. 1991a. *Spiraea virginiana* Britton: 1. Delineation and distribution. *Castanea* 56(4): 287–296.
———. 1991b. *Spiraea virginiana* Britton: 2. Ecology and species biology. *Castanea* 56(4): 297–303.
Ogle, D. W., and P. M. Mazzeo. 1976. *Betula uber,* the Virginia round-leaf birch, rediscovered in southwest Virginia. *Castanea* 41:248–256.
Palmer, E. J. 1925. Synopsis of North American Crataegi. *J. Arnold Arb.* 6:5–128.
Petrides, G. A. 1958. *A Field Guide to Trees and Shrubs*. Boston: Houghton Mifflin.
Phipps, J. B. 1988a. Crataegus (Maloideae, Rosaceae) of the southeastern United States: 1. Introduction and series Aestivales. *J. Arnold Arb.* 69:401–431.
———. 1988b. The re-assessment of *Crataegus flava* Aiton and its implications for the *Crataegus* serial name Flavae (Loud.) Rehd. and its sectional equivalent. *Taxon* 37:108–113.
Phipps, J. B., P. G. Robertson, P. G. Smith, and J. Rohrer. 1990. A checklist of the subfamily Maloideae (Rosaceae). *Can. J. Bot.* 68:2209–2269.
Preston, R. J., Jr., and V. G. Wright. 1988. *Identification of Southeastern Trees in Winter*. Raleigh: NC Agric. Ext. Service.
Price, R. A. 1990. The genera of Taxaceae in the southeastern United States. *J. Arnold Arb.* 71:69–91.
Radford, A. E., H. E. Ahles, and C. R. Bell. 1968. *Manual of the Vascular Flora of the Carolinas*. Chapel Hill: Univ. of North Carolina Press.
Reed, C. F. 1975. *Betula uber* (Ashe) Fernald rediscovered in Virginia. *Phytologia* 32:302–311.
Rehder, A. 1927. *Manual of Cultivated Trees and Shrubs Hardy in North America*. New York: Macmillan.
Robertson, K. R. 1974. The genera of Rosaceae in the southeastern United States. *J. Arnold Arb.* 55.
Sargent, C. S. 1922. *Manual of the Trees of North America (Exclusive of Mexico)*. 2d ed. New York and Boston: Houghton Mifflin.
Shinners, L. H. 1962. *Calamintha* (Labiatae) in the southern United States. *Sida* 1:69–75.
Small, J. K. 1933. *Manual of the Southeastern Flora*. Chapel Hill: Univ. of North Carolina Press.
Smith, E. B. 1985. *An Atlas and Annotated List of the Vascular Plants of Arkansas*. 2d ed. Fayetteville: Univ. of Arkansas Press.
Southall, R. M., and J. W. Hardin. 1974. A taxonomic revision of *Kalmia* (Ericaceae). *J. Elisha Mitchell Sci. Soc.* 90:1–23.
Spongberg, S. A. 1972. The genera of Saxifragaceae in the southeastern United States. *J. Arnold Arb.* 53:409–498.
———. 1974. A review of deciduous-leaved species of *Stewartia* (Theaceae). *J. Arnold Arb.* 55:182–214.
Steyermark, J. A. 1949. *Lindera melissaefolia*. *Rhodora* 51:153–162.
———. 1963. *Flora of Missouri*. Ames: Iowa State Univ. Press.
Strausbaugh, P. D., and E. L. Core. 1952–1964. Flora of West Virginia. *West Virginia Univ. Bull.*
Symonds, G. 1958. *The Tree Identification Book*. New York: William Morrow.
———. 1963. *The Shrub Identification Book*. New York: M. Barrows.
Thomas, J. L. 1960. A monographic study of the Cyrillaceae. *Contr. Gray Herb.* 186:1–114.
Trelease, W. 1925. *Winter Botany*. Urbana IL: Published by the author.
Tucker, G. C. 1988. The genera of Bambusoideae (Gramineae) in the southeastern United States. *J. Arnold Arb.* 69:239–273.

———. 1990. The genera of Arundinoideae (Gramineae) in the southeastern United States. *J. Arnold Arb.* 71:145–177.
Tucker, G. E. 1976. A guide to the woody flora of Arkansas. Ph.D. diss., Univ. of Arkansas, Fayetteville.
Uttal, L. J. 1986. Updating the genus *Vaccinium* L. (Ericaceae) in West Virginia. *Castanea* 51:197–201.
———. 1987. The genus *Vaccinium* L. (Ericaceae) in Virginia. *Castanea* 52:231–255.
Vander Kloet, S. P. 1988. *The Genus* Vaccinium *in North America*. Publication 1828. Ottawa: Research Branch, Agriculture Canada.
Vines, R. A. 1960. *Trees, Shrubs, and Woody Vines of the Southwest*. Austin: Univ. of Texas Press.
Watts, M., and T. Watts. 1982. *Winter Tree Finder*. Berkeley CA: Nature Study Guild.
Weakley, A. S. 1996. Flora of the Carolinas and Virginia. Working draft. The Nature Conservancy, Southeast Regional Office, Chapel Hill NC.
Wharton, M. E., and R. W. Barbour. 1973. *Trees and Shrubs of Kentucky*. Lexington: Univ. Press of Kentucky.
Wilbur, R. L. 1970. Taxonomic and nomenclatural observations on the eastern North American genus *Asimina* (Annonaceae). *J. Elisha Mitchell Sci. Soc.* 86:88–96.
———. 1975. A revision of the North American genus *Amorpha* (Leguminosae-Psoraleae). *Rhodora* 77:337–409.
———. 1994. The Myricaceae of the United States and Canada. *Sida* 16(1):93–107.
Wofford, B. E. 1983. A new *Lindera* (Lauraceae) from North America. *J. Arnold Arb.* 64:325–331.
Wofford, B. E., and R. Kral. 1993. Checklist of the vascular plants of Tennessee. *Sida, Bot. Misc.* 10:1–66.
Wood, C. E., Jr. 1961. The genera of Ericaceae in the southeastern United States. *J. Arnold Arb.* 42:10–80.
———. 1976. The genera of Guttiferae (Clusiaceae) in the southeastern United States. *J. Arnold Arb.* 57:74–90.

Index

ABELIA, GLOSSY, 43
Abelia × grandiflora, 43
Abies, 43
  balsamea, 43
    var. phanerolepsis, 43
  fraseri, 43
ACACIA, 44
  PRAIRIE, 44
  SMALL'S, 44
  SWEET, 44
Acacia, 44
  angustissima var. hirta, 44
  farnesiana, 44
  minuta, 44
  smallii, 44
Acer, 44
  barbatum, 46
  floridanum, 46
  leucoderme, 46
  negundo, 48
    var. texanum, 48
  nigrum, 48
  pensylvanicum, 46
  platanoides, 48
  rubrum, 48
    var. drummondii, 48
    var. trilobum, 48
  saccharinum, 48
  saccharum, 48
    ssp. *floridanum,* 46
    ssp. *leucoderme,* 46
    ssp. *nigrum,* 48
  spicatum, 46
Aceraceae family, 44
ADAM'S–NEEDLE, 384
Aesculus, 50
  flava, 52
  glabra, 50
    var. arguta, 50
  hippocastanum, 50
  octandra, 52
  parviflora, 50
  pavia, 52
    var. flavescens, 52
  sylvatica, 52
Agarista populifolia, 52
Agavaceae family, 381
AILANTHUS, 52

Ailanthus altissima, 52
AKEBIA, 52
Akebia quinata, 52
ALABAMA
  AZALEA, 291
  BLACK CHERRY, 257
  CROTON, 126
ALACHUA BUMELIA, 328
Albizia julibrissin, 52
ALDER, 52
  EUROPEAN, 54
  GREEN, 54
  SEASIDE, 54
  SMOOTH, 54
  SPECKLED, 54
  WITCH-, 142, 144
ALDERLEAF BUCKTHORN, 284
Aleurites fordii, 367
ALLEGHENY
  BLACKBERRY, 309
  CHINQUAPIN, 92
  PLUM, 252
  SERVICEBERRY, 56
  -SPURGE, 226
Alnus, 52
  glutinosa, 54
  incana ssp. rugosa, 54
  maritima, 54
  serrulata, 54
  viridis ssp. crispa, 54
ALTERNATE-LEAF DOGWOOD, 112
ALTHEA, SHRUB-, 160
Amelanchier, 54
  arborea, 56
    var. alabamensis, 56
    var. austromontana, 56
    var. *laevis,* 56
  bartramiana, 56
  canadensis, 56
  laevis, 56
  obovalis, 56
  sanguinea, 56
  spicata, 58
  stolonifera, 58
AMERICAN
  AMPELOPSIS, 61
  BARBERRY, 74
  BASSWOOD, 349

AMERICAN (*continued*)
  BEECH, 138
  BITTERSWEET, 96
  CHESTNUT, 90
  ELDERBERRY, 318
  ELM, 357
  FLY-HONEYSUCKLE, 199
  HAZELNUT, 114
  HOLLY, 172
  HORNBEAM, 84
  MOUNTAIN-ASH, 334
  PLUM, 252
  SMOKETREE, 114
  SNOWBELL, 342
  SYCAMORE, 243
  WISTERIA, 381
Amorpha, 58
  canescens, 59
    var. glabrata, 59
  fruticosa, 59
  georgiana, 59
    var. confusa, 59
  glabra, 59
  herbacea, 58
  laevigata, 61
  nitens, 59
  ouachitensis, 59
  paniculata, 59
  roemeriana, 61
  schwerinii, 59
  texana, 61
Ampelopsis, 61
  arborea, 61, 62
  brevipedunculata, 61
  cordata, 61
AMUR
  HONEYSUCKLE, 199
  PRIVET, 194
Anacardiaceae family, 114, 243, 292, 322
Andrachne phyllanthoides, 61
Andromeda polifolia var. glaucophylla, 61
ANISE-TREE, 176
  FLORIDA, 176
  YELLOW, 176
Annonaceae family, 67
ANTHONY WATERER SPIREA, 338
APALACHICOLA ROSEMARY, 110
Apocynaceae family, 354, 375
Appalachian Mountains, 5
APPLE, 210
  COMMON, 210
  GOPHER-, 190
  HYBRID, 212
Aquifoliaceae family, 170
Aralia spinosa, 63

Araliaceae family, 63
ARBUTUS, TRAILING, 134
Arctostaphylos uva-ursi, 63
Arecaceae family, 284, 310, 322
Aristolochia, 63
  durior, 63
  macrophylla, 63
  tomentosa, 63
Aristolochiaceae family, 63
ARKANSAS
  OAK, 266
  YUCCA, 383
ARNOT BRISTLY LOCUST, 300
Aronia, 63
  arbutifolia, 65
  melanocarpa, 65
  prunifolia, 65
ARROW BAMBOO, 257
ARROWWOOD, 373, 375
  GEORGIA, 375
  KENTUCKY, 373
  NORTHERN, 375
  SHORTSTALK, 373
  SOUTHERN, 373
Arundinaria, 65
  gigantea, 65
    ssp. tecta, 67
Asclepiadaceae family, 126
ASH, 144
  BILTMORE, 148
  BLACK, 146
  BLUE, 146
  GREEN, 148
  MOUNTAIN-, 334
  PRICKLY-, 386
  PUMPKIN, 146
  red, 148
  WATER, 146
  WHITE, 146
ASHE
  HAWTHORN, 125
  MAGNOLIA, 208
ASHE'S JUNIPER, 182
ASIATIC PACHYSANDRA, 226
Asimina, 67
  angustifolia, 68
  grandiflora, 68
  incana, 68
  longifolia, 68
    var. spathulata, 68
  obovata, 68
  parviflora, 67
  pygmaea, 68
  reticulata, 68
  speciosa, 68

triloba, 67
ASPEN, 246
  BIGTOOTH, 246
  European, 246
  QUAKING, 246
Aster carolinianus, 68
Asteraceae family, 68, 70, 78, 102, 148, 176
ATLANTIC WHITE-CEDAR, 100
ATLAS CEDAR, 96
AUTUMN-OLIVE, 134
Avicennia germinans, 68
Avicenniaceae family, 68
AZALEA, 284, 286
  ALABAMA, 291
  CUMBERLAND, 291
  DWARF, 290
  EASTMAN, 291
  ELECTION-PINK, 290
  FLAME, 290
  FLORIDA, 290
  OCONEE, 291
  PIEDMONT, 290
  PINK-SHELL, 288
  PINXTER-FLOWER, 291
  PLUMLEAF, 291
  SWAMP, 290
  SWEET, 290

Baccharis, 70
  angustifolia, 70
  dioica, 70
  glomeruliflora, 70
  halimifolia, 70
BALD-CYPRESS, 346, 348
BALLOONBERRY, 307
BALM-OF-GILEAD, 248
BALSAM POPLAR, 248
BAMBOO, 234, 320
  ARROW, 257
  FISH-POLE, 234
  HEAVENLY-, 218
  JAPANESE, 234
  YELLOW-GROVE, 234
BARBERRY, 74
  AMERICAN, 74
  COMMON, 74
  haw, 121
  JAPANESE, 74
bark, terminology for, 13
BASIL, 80
  GEORGIA, 80
  SCARLET, 80
  TOOTHED, 80
BASSWOOD, 349
  AMERICAN, 349

  CAROLINA, 350
  Florida, 350
  WHITE, 350
Bataceae family, 70
Batis maritima, 70
BAY, 230
BAYBERRY, 218, 216
  NORTHERN, 218
  ODORLESS, 2216
  POCOSIN, 218
BAY-LEAVED WILLOW, 316
BEACH
  -ELDER, 180
  -HEATHER, WOOLLY, 160
  PLUM, 252
  -TEA, 125
BEAKED HAZELNUT, 114
BEAR OAK, 270
BEARBERRY, 63
BEAR-GRASS, 384
BEAUTYBERRY, 80
  PURPLE, 82
BEBB WILLOW, 313
BEECH, AMERICAN, 138
Befaria racemosa, 70
Berberidaceae family, 74, 210, 218
Berberis, 74
  canadensis, 74
  thunbergii, 74
  vulgaris, 74
Berchemia scandens, 74
Betula, 74
  alleghaniensis, 76
  cordifolia, 78
  lenta, 76
  nigra, 76
  papyrifera, 78
    var. cordifolia, 78
  pendula, 76
  populifolia, 76
  uber, 76
Betulaceae family, 52, 74, 84, 114, 226
BIGELOW OAK, 280
BIG-FLOWER PAWPAW, 68
BIGLEAF
  MAGNOLIA, 208
  SNOWBELL, 342
Bignonia capreolata, 78
Bignoniaceae family, 78, 82, 92, 204
BIGTOOTH ASPEN, 246
BILLARD SPIREA, 338
BILTMORE ASH, 148
BIRCH, 74
  EUROPEAN WHITE, 76
  GRAY, 76

414 · Index

BIRCH (*continued*)
  MOUNTAIN PAPER, 78
  PAPER, 78
  SWEET, 76
  VIRGINIA ROUND-LEAF, 76
  YELLOW, 76
BITTER NIGHTSHADE, 334
BITTERNUT HICKORY, 86
BITTERSWEET, 96, 334
  AMERICAN, 96
  ORIENTAL, 96
BLACK
  ASH, 146
  CHERRY, 257
  CHOKEBERRY, 65
  HICKORY, 88
  HIGHBUSH BLUEBERRY, 364
  HUCKLEBERRY, 150
  LOCUST, 298
  MANGROVE, 68
  MAPLE, 48
  MULBERRY, 216
  OAK, 282
  RASPBERRY, 307
  WALNUT, 180
  WILLOW, 316
BLACKBERRY, 305, 309
  ALLEGHENY, 309
  BLANCHARD'S, 309
  BRISTLY, 309
  COMMON, 309
  CUT-LEAVED, 310
  HIGHBUSH, 310
  HIMALAYAN, 310
  NORTHERN, 310
  SAND, 309
  SMOOTH, 309
BLACKGUM, 220
  SWAMP, 220
BLACKHAW, 370
  RUSTY, 370
BLACKJACK OAK, 274
BLACKWATER ST. JOHN'S-WORT, 167
BLADDERNUT, 340
BLANCHARD'S BLACKBERRY, 309
BLUE ASH, 146
BLUE RIDGE ST. JOHN'S-WORT, 167
BLUEBERRY, 359
  BLACK HIGHBUSH, 364
  CREEPING, 362
  GLAUCOUS, 362
  HAIRY, 364
  haw, 120
  MOUNTAIN DRYLAND, 361

  NEW JERSEY HIGHBUSH, 362
  NORTHERN LOWBUSH, 361
  RAYNOR'S, 362
  SHINY, 364
  SLENDER, 367
  SMALLFLOWER, 367
  SMOOTH HIGHBUSH, 362
  SOUTHERN HIGHBUSH, 364
  SOUTHERN LOW, 366
  UPLAND HIGHBUSH, 366
  VELVET-LEAF, 366
BLUEJACK OAK, 270
BLUFF OAK, 266
BOG
  HUCKLEBERRY, 152
  -ROSEMARY, 61
  SPICEBUSH, 194
  ST. JOHN'S-WORT, 168
BORDER PRIVET, 194
Borrichia frutescens, 78
BOX HUCKLEBERRY, 150
BOXELDER, 48
BOYNTON
  LOCUST, 300
  OAK, 266
BRAMBLE, 305
BRAZILIAN PEPPER TREE, 322
BRIDAL WREATH SPIREA, 338
BRIDEWORT, 338
BRISTLY
  BLACKBERRY, 309
  GREENBRIER, 332
  LOCUST, 300
  ROSE, 304
  SWAMP CURRANT, 297
BROADLEAF MEADOW-SWEET, 337
Bromeliaceae family, 350
BROOM, SCOTCH-, 128
Broussonetia papyrifera, 78
Brunnichia ovata, 78
BUCKBERRY, 150
BUCKEYE, 50
  BOTTLEBRUSH, 50
  OHIO, 50
  PAINTED, 52
  RED, 52
  YELLOW, 52
Buckleya distichophylla, 78
BUCKTHORN, 144, 282
  ALDERLEAF, 284
  -BUMELIA, 328
  CAROLINA, 144
  DAHURIAN, 284
  EUROPEAN, 284

GLOSSY, 144
NARROW-LEAF, 284
SHELLMOUND-, 312
BUCKWHEAT-TREE, 106
Buddleja, 80
  davidii, 80
  lindleyana, 80
BUFFALO
  CURRANT, 296
  -NUT, 258
BUMELIA, 324
  ALACHUA, 328
  BUCKTHORN, 328
  GUM, 326
  RUFOUS, 328
  SMOOTH, 328
  THORNY, 328
  TOUGH, 326
Bumelia, 324
  anomala, 328
  lacuum, 326
  lanuginosum, 326
  lycioides, 328
  reclinata, 328
  rufotomentosa, 328
  tenax, 326
  thornei, 328
BUNCHBERRY, 112
BUNCHLEAF ST. JOHN'S-WORT, 167
BUR OAK, 272
BUSH-HONEYSUCKLE, 130, 199
  HAIRY, 130
  NORTHERN, 130
  SOUTHERN, 130
BUTTERFLY-BUSH, 80
  LINDLEY, 80
BUTTERNUT, 180
BUTTONBUSH, 98
Buxaceae family, 226

CABBAGE PALMETTO, 310
Cactaceae family, 134, 220
CACTUS
  PRAIRIE CORY, 134
  PRICKLY-PEAR, 220
Calamintha, 80
  coccinea, 80
  dentata, 80
  georgiana, 80
CALIFORNIA PRIVET, 192
CALLERY PEAR, 260
Callicarpa, 80
  americana, 82
  dichotoma, 82

Calluna vulgaris, 82
CALUSA GRAPE, 378
Calycanthaceae family, 82
Calycanthus, 82
  floridus, 82
    var. glaucous, 82
    var. *laevigatus,* 82
Calycocarpum lyonii, 82
CAMPHOR-TREE, 102
Campsis radicans, 82
CANADA
  PLUM, 255
  YEW, 348
CANE, 65
  LARGE, 65
  SMALL, 67
Caprifoliaceae family, 130, 195, 318, 342, 367, 380
CAROLINA
  BASSWOOD, 350
  BUCKTHORN, 144
  HEMLOCK, 354
  HOLLY, 174
  JESSAMINE, 152, 154
  LAURELCHERRY, 255
  RHODODENDRON, 288
  ROSE, 304
  SHEEP-LAUREL, 184
  SILVERBELL, 158
Carpinus, 84
  caroliniana, 84
    var. virginiana, 84
Carya, 84
  alba, 86
  aquatica, 86
  carolinae-septentrionalis, 88
  cordiformis, 86
  floridana, 88
  glabra, 88
    var. leiodermis, 90
    var. megacarpa, 90
  illinoinensis, 86
  laciniosa, 88
  myristiciformis, 86
  ovalis, 90
  ovata, 88
    var. *australis,* 88
  pallida, 88
  texana, 88
  tomentosa, 86
Caryophyllaceae family, 226
Cassandra calyculata, 100
Castanea, 90
  alnifolia, 92

Castanea (*continued*)
    var. floridana, 92
    dentata, 90
    mollissima, 92
    ozarkensis, 92
    pumila, 92
        var. ashei, 92
CAT GRAPE, 380
CATALPA, 92
    NORTHERN, 94
    SOUTHERN, 92
Catalpa, 92
    bignonioides, 92
    speciosa, 94
CATAWBA RHODODENDRON, 288
CATBRIER, 332
CAT'S-CLAW-VINE, 204
Ceanothus, 94
    americanus, 94
        var. intermedius, 94
    herbaceus, 94
    microphyllus, 94
CEANOTHUS, LITTLELEAF, 94
CEDAR, 94
    ATLANTIC WHITE-, 100
    ATLAS, 96
    DEODAR, 96
    EASTER RED-, 182
    ELM, 357
    NORTHERN WHITE-, 349
    -OF-LEBANON, 96
    stinking, 350
Cedrus, 94
    atlantica, 96
    deodara, 96
    libani, 96
        ssp. *atlantica,* 96
Celastraceae family, 96, 136, 228
Celastrus, 96
    orbiculatus, 96
    scandens, 96
Celtis, 98
    laevigata, 98
    occidentalis, 98
    tenuifolia, 98
Cephalanthus occidentalis, 98
Ceratiola ericoides, 100
Cercis canadensis, 100
Chamaecyparis thyoides, 100
Chamaedaphne calyculata var. angustifolia, 100
CHAPMAN
    OAK, 268
    RHODODENDRON, 288

ST. JOHN'S-WORT, 167
CHASTE-TREE, 375
Chenopodiaceae family, 320
CHEROKEE ROSE, 302
CHERRY, 250, 255, 257
    ALABAMA BLACK, 257
    BLACK, 257
    CHOKE, 257
    FIRE, 255
    MAZZARD, 257
    NANKING, 255
    PERFUMED, 255
    SAND, 255
    SOUR, 257
CHERRYBARK OAK, 276
CHESTNUT, 90
    AMERICAN, 90
    CHINESE, 92
    HORSE-, 50
    OAK, 274
CHICKASAW PLUM, 254
Chimaphila, 100
    maculata, 100
    umbellata ssp. cisatlantica, 102
CHINABERRY, 212
CHINA-FIR, 126
CHINESE
    CHE, 126
    CHESTNUT, 92
    -DATE, 388
    ELM, 356
    EUONYMUS, 138
    HOLLY, 172
    MATRIMONY-VINE, 202
    PISTACHIO, 243
    PRIVET, 192
    SPIREA, 338
    TAMARISK, 346
    WISTERIA, 381
CHINQUAPIN
    ALLEGHENY, 92
    -OAK, 274
    OZARK, 92
Chionanthus, 102
    pygmaeus, 102
    virginicus, 102
Choenomeles speciosa, 102
CHOKEBERRY, 63
    BLACK, 63
    PURPLE, 63
    RED, 63
CHOKECHERRY, 257
CHRISTMAS-BERRY, 202
Chrysobalanaceae family, 190

Chrysoma pauciflosculosa, 102
Cinnamomum camphora, 102
CINQUEFOIL
  SHRUB-, 230
  THREE-TOOTHED, 324
Cissus
  incisa, 104
  trifoliata, 102
Cistaceae family, 160
Citrus aurantium, 104
Cladrastis
  kentukea, 104
  lutea, 104
CLAMMY LOCUST, 298
CLAW-VINE, CAT'S-, 204
CLEMATIS, 104
  catesbyana, 104
  terniflora, 104
  virginiana, 104
  YAM-LEAF, 104
Clerodendrum, 106
  japonicum, 106
  trichotomum, 106
Clethra, 106
  acuminata, 106
  alnifolia, 106
Clethraceae family, 106
Cliftonia monophylla, 106
CLIMBING
  ASTER, 68
  DOGBANE, 354
  FIG, 140
  HYDRANGEA, 128
Clusiaceae family, 162
COASTAL
  DOG-HOBBLE, 190
  PLAIN WILLOW, 316
  PRICKLY-PEAR, 224
coastal plain, 3
COASTAL ROUGHLEAF DOGWOOD, 114
Cocculus carolinus, 108
cockspur hawthorn, 120
COFFEETREE, KENTUCKY, 156
COMMON
  APPLE, 210
  BARBERRY, 74
  BEARBERRY, 63
  BLACKBERRY, 309
  FIG, 138
  GREENBRIER, 332
  INDIGO-BUSH, 59
  JUNIPER, 182
  MOCK-ORANGE, 232
  PEAR, 260

WINTERBERRY, 174
Comptonia peregrina, 108
Concord grape, 378
Conradina, 108
  brevifolia, 108
  canescens, 108
  etonia, 108
  glabra, 110
  grandiflora, 110
  verticillata, 110
COONTIE, 386
CORAL
  -BEAN, EASTERN, 134
  -BERRY, 342, 344
  GREENBRIER, 332
  HONEYSUCKLE, 197
CORKSCREW WILLOW, 317
CORKWOOD, 188
  SHRUB-, 342
Cornaceae family, 110, 218
Cornus, 110
  alternifolia, 112
  amomum, 112
    var. obliqua, 112
  asperifolia, 114
  canadensis, 112
  drummondii, 112
    form priceae, 112
  florida, 112
  foemina, 114
  racemosa, 112
  rugosa, 112
  sericea, 114
  stolonifera, 114
  stricta, 114
CORY CACTUS, PRAIRIE, 134
Corylus, 114
  americana, 114
  cornuta, 114
    var. californica, 114
Coryphantha missouriensis, 136
Cotinus obovatus, 114
COTTONWOOD
  EASTERN, 248
  plains, 250
  SWAMP, 248
CRABAPPLE, 212
  PRAIRIE, 212
  SOUTHERN, 212
  SWEET, 212
CRACK WILLOW, 317
CRANBERRY, 364
  MOUNTAIN-, 364
  SMALL, 366

CRANBERRY (*continued*)
  -VIBURNUM, 369
CRAPE-MYRTLE, 184
Crataegus, 116
  aestivalis, 120
  allegheniensis, 121
  alma, 123
  ancisa, 123
  aprica, 121
  ashei, 125
  austrina, 123
  austromontana, 125
  basilica, 124
  berberifolia, 121
  biltmoreana, 122
  boyntonii, 122
  brachyacantha, 120
  buckleyi, 122
  calpodendron, 122
  chapmanii, 122
  coccinea, 120
  coccinioides, 121
  collina, 124
  communis, 122
  contrita, 123
  cordata, 120
  crus-galli, 120
    var. inermis, 121
    var. pyracanthifolia, 120
  dilatata, 121
  dispar, 122
  engelmanii, 121
  extraria, 121
  flabellata, 124
  flava, 121
  floridana, 122
  gattingeri, 123
  harbisonii, 125
  ignava, 121
  illustris, 123
  impar, 122
  intricata, 122
  iracunda, 124
  lacrimata, 121
  lanata, 122
  lepida, 122
  macrosperma, 124
  margaretta, 124
  marshallii, 120
  mendosa, 123
  meridiana, 122
  meridionalis, 123
  michauxii, 122
  mollis, 123
  munda, 122
  neofluvialis, 122
  nitida, 125
  opaca, 120
  opima, 123
  phaenopyrum, 120
  pruinosa, 123
    var. virella, 123
  pulcherrima, 123
  punctata, 124
  ravenellii, 122
  robur, 123
  rubella, 122
  rufula, 120
  sargentii, 122
  schuettii, 124
  senta, 122
  spathulata, 122
  straminea, 122
  succulenta, 122
  texana, 123
  tomentosa, 122
  triflora, 125
  uniflora, 123
  viburnifolia, 123
  viridis, 125
    'Winter King', 125
  youngii, 120
CREEK PLUM, 254
CREEPER
  NORTHERN-, 228
  TRUMPET-, 82
  VIRGINIA-, 228
  WINTER-, 136
CREEPING
  BLUEBERRY, 362
  SNOWBERRY, 150
  ST. ANDREW'S-CROSS, 165
  ST. JOHN'S-WORT, 167
CROOKEA, 165
Crookea microsepala, 165
CROSS-LEAVED HEATH, 134
CROSSVINE, 78
CROTON, 125
  ALABAMA, 126
  SILVER, 126
Croton, 125
  alabamensis, 126
  argyranthemus, 126
  punctatus, 125
CUCUMBER-TREE, 208
  YELLOW, 208
Cudrania tricuspidata, 126
CUMBERLAND

AZALEA, 291
ROSEMARY, 110
Cumberland Plateau, 5
Cunninghamia lanceolata, 126
Cupressaceae family, 100, 126, 182, 346, 349
CUPSEED, 82
CURRANT
  BRISTLY SWAMP, 297
  BUFFALO, 296
  EASTERN BLACK, 296
  EUROPEAN BLACK, 296
  northern red, 297
  RED GARDEN, 296
  SKUNK, 296
  SWAMP RED, 296
CURVE-LEAF YUCCA, 384
CUT-LEAVED BLACKBERRY, 310
Cycadaceae family, 386
Cynanchum, 126
  angustifolium, 126
  scoparium, 128
CYPRESS
  BALD-, 346, 348
  POND-, 348
Cyrilla, 128
  parvifolia, 128
  racemiflora, 128
Cyrillaceae family, 106, 128
Cytissus scoparius, 128

DAHOON, 172
  MYRTLE, 172
DAHURIAN BUCKTHORN, 284
DAMASK ROSE, 304
DAMSON PLUM, 254
DANGLEBERRY, 152
  DWARF, 152
  HAIRY, 152
DARLINGTON OAK, 270
DATE, CHINESE-, 388
Daubentonia
  drummondii, 324
  punicea, 324
Decodon verticillatus, 128
Decumaria barbara, 128
DEERBERRY, 366
DELTA POST OAK, 280
DENSE ST. JOHN'S-WORT, 170
DEODAR CEDAR, 96
DEUTZIA, 128
  ROUGHLEAF, 130
  SLENDER, 130
  SMALLFLOWER, 130
Deutzia, 128

gracilis, 130
parviflora, 130
scabra, 130
DEVIL'S WALKINGSTICK, 63
DEVILWOOD, 224
DEWBERRY, 305, 307
  NORTHERN, 309
  SAND, 307
  SOUTHERN, 307
  SWAMP, 307
Diapensiaceae family, 260
Diervilla, 130
  lonicera, 130
  rivularis, 130
  sessilifolia, 130
Diospyros virginiana, 132
Dirca palustris, 132
DOG ROSE, 302
DOGBANE, CLIMBING, 354
DOG-HOBBLE, 190
  COASTAL, 190
  MOUNTAIN, 190
DOGWOOD, 110
  ALTERNATE-LEAF, 112
  COASTAL ROUGHLEAF, 114
  FLOWERING, 112
  GRAY, 112
  RED-OSIER, 114
  ROUGH-LEAF, 112
  ROUND-LEAF, 112
  SILKY, 112
  SWAMP, 114
dotted hawthorn, 124
DOWNY
  hawthorn, 123
  INDIGO-BUSH, 59
  MOCK-ORANGE, 232
  SERVICEBERRY, 56
DRYLAND ST. JOHN'S-WORT, 168
DUNE GREENBRIER, 332
  PRICKLY-PEAR, 222
DURAND OAK, 280
DUTCHMAN'S-PIPE, 63
DWARF
  AZALEA, 290
  BRISTLY LOCUST, 300
  CHINQUAPIN OAK, 278
  DANGLEBERRY, 152
  GREENBRIER, 330
  HUCKLEBERRY, 152
  INDIGO-BUSH, 58
  LIVE OAK, 274
  PALMETTO, 310
  PRAIRIE ROSE, 304

DWARF (*continued*)
  PRAIRIE WILLOW, 314
  RASPBERRY, 307
  SPIREA, 337
  WITCH-ALDER, 142

EASTERN
  BLACK CURRANT, 296
  CORAL-BEAN, 134
  COTTONWOOD, 248
  HEMLOCK, 354
  PRICKLY-PEAR, 222
  RED ELDERBERRY, 318
  RED-CEDAR, 182
  WHITE PINE, 239
EASTER-ROSE, 184
EASTMAN AZALEA, 291
Ebenaceae family, 132
Elaeagnaceae family, 132
Elaeagnus, 132
  angustifolia, 132
  multiflora, 132
  pungens, 132
  umbellata, 134
ELDER
  BEACH-, 180
  MARSH-, 176, 180
ELDERBERRY, 318
  AMERICAN, 318
  EASTERN RED, 318
ELECTION-PINK, 290
Elliottia racemosa, 134
ELM, 356
  AMERICAN, 357
  CEDAR, 357
  CHINESE, 356
  English, 356
  ROCK, 357
  SEPTEMBER, 357
  SIBERIAN, 356
  SLIPPERY, 357
  WINGED, 357
ENGLISH
  elm, 356
  IVY, 158
Epigaea repens, 134
Erica, 134
  carnea, 134
  tetralix, 134
Ericaceae family, 61, 63, 70, 82, 100, 134, 148, 150, 182, 188, 202, 214, 226, 236, 284, 359, 386
Erythrina herbacea, 134
Escobaria missouriensis var. similis, 134
ETONIA ROSEMARY, 108

EUONYMUS, 136
  CHINESE, 138
  JAPANESE, 136
  WINGED, 136
Euonymus, 136
  alatus, 136
  americanus, 136
  atropurpureus, 138
  bungeanus, 138
  europaeus, 138
  fortunei, 136
  japonicus, 136
  obovatus, 136
Euphorbiaceae family, 61, 125, 212, 320, 322, 342, 367
EUROPEAN
  ALDER, 54
  aspen, 246
  BLACK CURRANT, 296
  black poplar, 248
  BUCKTHORN, 284
  FLY-HONEYSUCKLE, 200
  GOOSEBERRY, 297
  larch, 186
  MOUNTAIN-ASH, 334
  PRIVET, 192
  SPINDLE-TREE, 138
  WHITE BIRCH, 76
Exochorda racemosa, 138

Fabaceae family, 58, 100, 102, 128, 134, 156, 176, 200, 226, 250, 258, 298, 322, 334, 380
Fagaceae family, 90, 138, 260
Fagus grandifolia, 138
FETTERBUSH, 204, 236
  MOUNTAIN, 236
Ficus, 138
  carica, 138
  pumila, 140
FIG, 138
  CLIMBING, 140
  COMMON, 138
  INDIAN-, 224
FIR, 43
  BALSAM, 43
  CHINA-, 126
  FRASER, 43
FIRE CHERRY, 255
FIRETHORN, 258
Firmiana simplex, 140
FISH-POLE BAMBOO, 234
FLAG PAWPAW, 68
FLAME AZALEA, 290
FLATWOODS

PAWPAW, 68
PLUM, 252
fleshy haw, 122
FLORIDA
  ANISE-TREE, 176
  AZALEA, 290
  basswood, 350
  -LEUCOTHOE, 52
  MAPLE, 46
  NEEDLE PALM, 284
  PRICKLY-PEAR, 222
  -PRIVET, 140
  SOAPBERRY, 320
  TORREYA, 350
  WILLOW, 316
  YEW, 349
FLOWERING
  DOGWOOD, 112
  JASMINE, 180
  QUINCE, 102
flowers, terminology for, 14
FLY-HONEYSUCKLE, 199
  AMERICAN, 199
  EUROPEAN, 200
Forestiera, 140
  acuminata, 140
  godfreyi, 140
  ligustrina, 140
  segregata, 140
Forsythia, 142
  × intermedia, 142
  suspensa, 142
  viridissima, 142
Fothergilla, 142
  gardenii, 142
  major, 144
FOX GRAPE, 378
FRAGRANT SUMAC, 292
Frangula, 144
  alnus, 144
  caroliniana, 144
Franklinia alatamaha, 144
FRASER
  FIR, 43
  MAGNOLIA, 208
Fraxinus, 144
  americana, 146
    var. biltmoreana, 148
  biltmoreana, 148
  caroliniana, 146
  nigra, 146
  pennsylvanica, 148
    var. subintegerrima, 148
  profunda, 146
  quadrangulata, 146

FRENCH
  ROSE, 304
  TAMARISK, 346
FRINGE-TREE, 102
FROST GRAPE, 380
frosted hawthorn, 123
fruit, terminology for, 14

GALLBERRY, 172
  LOW, 172
  TALL, 172
Garberia heterophylla, 148
GARDEN PLUM, 254
Gaultheria, 148
  hispidula, 150
  procumbens, 148
Gaylussacia, 150
  baccata, 150
  brachycera, 150
  dumosa, 152
    var. bigeloviana, 152
  frondosa, 152
    var. nana, 152
    var. tomentosa, 152
  mosieri, 152
  ursina, 150
Gelsemium, 152
  rankinii, 154
  sempervirens, 154
GEORGIA
  ARROWWOOD, 375
  BASIL, 80
  HACKBERRY, 98
  HOLLY, 174
  INDIGO-BUSH, 59
  OAK, 270
Ginkgo biloba, 154
Ginkgoaceae family, 154
GLADE-PRIVET, 140
GLASSWORT, 320
GLAUCOUS
  BLUEBERRY, 362
  GREENBRIER, 332
Gleditsia, 154
  aquatica, 154
  triacanthos, 154
    var. inermis, 156
GLORYBOWER, 106
  HARLEQUIN, 106
  JAPANESE, 106
GLOSSY
  BUCKTHORN, 144
  hawthorn, 125
  PRIVET, 192
GOAT WILLOW, 313

GODFREY'S FORESTIERA, 140
GOLDEN
   -HEATHER, 160
   ST. JOHN'S-WORT, 168
   WEEPING WILLOW, 318
GOOSEBERRY, 294
   EUROPEAN, 297
   GRANITE, 297
   MICCOSUKEE, 297
   MISSOURI, 297
   PRICKLY, 297
   ROUNDLEAF, 298
   SMOOTH, 297
GOPHER-APPLE, 190
Gordonia lasianthus, 156
GOUMI-BERRY, 132
GRANITE GOOSEBERRY, 297
GRAPE, 377
   CALUSA, 378
   CAT, 380
   Concord, 378
   FOX, 378
   FROST, 380
   -HOLLY, OREGON, 210
   -HONEYSUCKLE, 197
   MUSCADINE, 377
   MUSTANG, 378
   PIGEON, 378
   RIVERBANK, 380
   SAND, 380
   Scuppernong, 377
   SUMMER, 378
GRAY
   BIRCH, 76
   DOGWOOD, 112
   HYDRANGEA, 162
   POPLAR, 246
   WILLOW, 314
GREEN
   ALDER, 54
   ASH, 148
   hawthorn, 125
GREENBRIER, 328
   BRISTLY, 332
   COMMON, 332
   CORAL, 332
   DUNE, 332
   DWARF, 330
   GLAUCOUS, 332
   LAUREL, 330
Grossulariaceae family, 294
GROUNDSEL-BUSH, 70
   GULF, 70
   NARROW-LEAF, 70
   SOUTHERN, 70

GROUNDSEL-TREE, 70
GUELDER-ROSE, 369
GULF
   GROUNDSEL-BUSH, 70
   JOINTWEED, 243
   ST. JOHN'S-WORT, 168
GUM BUMELIA, 326
Gymnocladus dioicus, 156

habit, 7
HACKBERRY, 98
   GEORGIA, 98
HAIRY
   BLUEBERRY, 364
   BUSH-HONEYSUCKLE, 130
   DANGLEBERRY, 152
   MOCK-ORANGE, 232
   -WICKY, 184
Halesia, 156
   carolina, 158
   diptera, 156
      var. magniflora, 156
   parviflora, 158
   tetraptera, 158
      var. monticola, 158
Hamamelidaceae family, 142, 158, 195
Hamamelis, 158
   vernalis, 158
   virginiana, 158
Harbison hawthorn, 125
HARLEQUIN GLORYBOWER, 106
HARTWEG LOCUST, 300
haw
   barberry, 121
   blueberry, 120
   broadleaf, 121
   fleshy, 122
   parsley, 120
   riverflat, 120
   scarlet, 120
   three-flower, 125
   yellow, 121
HAWTHORN, 116
   Ashe, 125
   cockspur, 120
   dotted, 124
   downy, 123
   frosted, 123
   glossy, 125
   green, 125
   Harbison, 125
   Kansas, 121
   littlehip 181
   one-flowered, 123
   Washington, 120

weeping, 121
HAZEL, WITCH-, 158
HAZELNUT, 114
   AMERICAN, 114
   BEAKED, 114
HEATH, 134
   CROSS-LEAVED, 134
   SPRING, 134
HEATHER, 82, 160
   GOLDEN-, 160
   MOUNTAIN GOLDEN-, 160
   SCOTS-, 82
   WOOLLY BEACH-, 160
HEAVENLY-BAMBOO, 218
Hedeoma graveolens, 340
Hedera helix, 158
HEMLOCK, 354
   CAROLINA, 354
   EASTERN, 354
HERCULES-CLUB, 386
Hibiscus syriacus, 160
HICKORY, 84
   BITTERNUT, 86
   BLACK, 88
   MOCKERNUT, 86
   NUTMEG, 86
   PIGNUT, 88
   RED, 90
   SAND, 88
   SCRUB, 88
   SHAGBARK, 88
   SHELLBARK, 88
   SOUTHERN SHAGBARK, 88
   WATER, 86
HIGHBUSH BLACKBERRY, 310
hill thorn, 124
HIMALAYA-BERRY, 310
HIMALAYAN BLACKBERRY, 310
Hippocastanaceae family, 50
HOBBLEBUSH, 370
HOLLY, 170
   AMERICAN, 172
   CAROLINA, 174
   CHINESE, 172
   GEORGIA, 174
   JAPANESE, 172
   MOUNTAIN-, 174
   SERVICEBERRY, 176
   YAUPON, 172
HONEY MESQUITE, 250
HONEY-CUPS, 386
HONEY-LOCUST, 154
HONEYSUCKLE, 195
   AMERICAN FLY-, 199
   AMUR, 199

   BUSH-, 130, 199
   CORAL, 197
   EUROPEAN FLY-, 200
   GRAPE, 197
   JAPANESE, 197
   MORROW, 199
   MOUNTAIN, 197
   TATARIAN, 199
   YELLOW, 199
HOPHORNBEAM, 226
HOPTREE, 258
HORNBEAM, AMERICAN, 84
HORSE-CHESTNUT, 50
HORTULAN PLUM, 254
HUCKLEBERRY, 150
   BLACK, 150
   BOG, 152
   BOX, 150
   DWARF, 152
Hudsonia, 160
   ericoides, 160
   montana, 160
   tomentosa, 160
HYDRANGEA, 160
   CLIMBING, 128
   GRAY, 162
   OAK-LEAF, 162
   PEEGEE, 162
   SNOWY, 162
   WILD, 162
Hydrangea, 160
   arborescens, 162
   cinerea, 162
   paniculata, 162
   quercifolia, 162
   radiata, 162
Hydrangeaceae family, 128, 160, 230
Hypericaceae family, 162
Hypericum, 162
   apocynifolium, 168
   brachyphyllum, 165
   buckleyi, 167
   chapmanii, 167
   cistifolium, 168
   crux-andreae, 165
   densiflorum, 170
   dolabriforme, 168
   exile, 167
   fasciculatum, 167
   frondosum, 168
   galioides, 168
   hypericoides, 165
      ssp. multicaule, 165
   lissophloeus, 167
   lloydii, 167

Hypericum (*continued*)
  lobocarpum, 168
  microsepalum, 165
  myrtifolium, 167
  nitidum, 167
  nudiflorum, 168
  prolificum, 170
  reductum, 165
  sphaerocarpum, 168
  stragalum, 165
  suffruticosum, 165
  tetrapetalum, 165

Ilex, 170
  ambigua, 174
    var. *montana,* 174
  amelanchier, 176
  beadlei, 174
  buswellii, 174
  cassine, 172
  collina, 174
  coriacea, 172
  cornuta, 172
  crenata, 172
  cuthbertii, 176
  decidua, 174
  glabra, 172
  laevigata, 174
  longipes, 174
  montana, 174
  mucronata, 174
  myrtifolia, 172
  opaca, 172
    var. arenicola, 172
  verticillata, 174
  vomitoria, 172
Illiciaceae family, 176
Illicium, 176
  floridanum, 176
  parviflorum, 176
INDIAN
  -FIG, 224
  -OLIVE, 218
INDIGO, 176
INDIGO-BUSH, 58
  COMMON, 59
  DOWNY, 59
  DWARF, 58
  GEORGIA, 59
  MOUNTAIN, 59
  OUACHITA, 59
  PIEDMONT, 59
  ROEMER, 61
  SAVANNA, 59
  SHINING, 59
  SMOOTH, 61
  SWAMP, 59
Indigofera suffruticosa, 176
interior plateau, 5
INTERIOR ST. JOHN'S-WORT, 168
Itea virginica, 176
Iva, 176
  frutescens, 180
    var. oraria, 180
  imbricata, 180
IVY
  ENGLISH-, 158
  MARINE-, 102
  POISON-, 352, 354

JACKSON-BRIER, 330
JAPANESE
  BAMBOO, 234
  BARBERRY, 74
  EUONYMUS, 136
  GLORYBOWER, 106
  HOLLY, 172
  HONEYSUCKLE, 197
  Larch, 188
  LESPEDEZA, 188
  PRIVET, 192
  SPIREA, 338
  WISTERIA, 381
JASMINE, FLOWERING, 180
Jasminum nudiflorum, 180
JERUSALEM-THORN, 226
JESSAMINE, 152
  CAROLINA, 154
  SWAMP, 154
JETBEAD, 291
JOINTWEED, 243, 244
  GULF, 243
Juglandaceae family, 84, 180
Juglans, 180
  cinerea, 180
  nigra, 180
JUNIPER, 182
  ASHE'S, 182
  COMMON, 182
Juniperus, 182
  ashei, 182
  communis var. depressa, 182
  virginiana, 182
    var. silicicola, 182

Kalmia, 182
  angustifolia, 184
  carolina, 184
  cuneata, 184
  hirsuta, 184

latifolia, 184
Kansas hawthorn, 121
KELSEY LOCUST, 300
KENTUCKY
   ARROWWOOD, 373
   COFFEETREE, 156
   WISTERIA, 381
Kerria japonica, 184
keys
   to genera, 16–41
   master key to diagnostic, 15
   use and types of, 2
KUDZU, 258

Labiatae family, 80, 108, 340
LADIES' EARDROPS, 78
LADY LUPINE, 200
Lagerstroemia indica, 184
LANTANA, 186
   TEXAS, 186
   WEEPING, 186
Lantana, 186
   camara, 186
   horrida, 186
   montevidensis, 186
   ovatifolia, 186
LARCH, 186
   European, 186
   Japanese, 188
Lardizabalanaceae family, 52
LARGE PERIWINKLE, 375
LARGE-FLOWERED ROSEMARY, 110
Larix, 186
   decidua, 186
   laricina, 186
   leptolepis, 188
Lauraceae family, 102, 194, 195, 230, 320
LAUREL, 182
   -GREENBRIER, 330
   MOUNTAIN-, 184
   -OAK, 272
   SHEEP-, 184
LAURELCHERRY, CAROLINA, 255
Laurus nobilis, 230
leaf, terminology for, 7–10
LEATHERLEAF, 100
LEATHERWOOD, 132
Leiophyllum buxifolium, 188
Leitneria floridana, 188
Leitneriaceae family, 188
LESPEDEZA, 188
   JAPANESE, 188
   SHRUB-, 188
Lespedeza, 188
   bicolor, 188

   thunbergii, 188
Leucothoe, 188
   axillaris, 190
   fontanesiana, 190
   racemosa, 190
   recurva, 190
LEUCOTHOE, FLORIDA-, 52
Licania michauxii, 190
Ligustrum, 190
   amurense, 194
   japonicum, 192
   lucidum, 192
   obtusifolium, 194
   ovalifolium, 192
   quihoui, 192
   sinense, 192
   vulgare, 192
LILAC, 344
Lindera, 194
   benzoin, 194
   melissifolia, 194
   subcoriacea, 194
Linnaea borealis var. americana, 195
Liquidambar styraciflua, 195
Liriodendron tulipifera, 195
Litsea aestivalis, 195
LITTLE SILVERBELL, 158
littlehip hawthorn, 122
LITTLELEAF CEANOTHUS, 94
LIVE OAK, 282
LOBLOLLY
   -BAY, 156
   PINE, 241
LOCUST, 298
   ARNOT BRISTLY, 300
   BLACK, 298
   BOYNTON, 300
   BRISTLY, 300
   CLAMMY, 298
   DWARF BRISTLY, 300
   HARTWEG, 300
   HONEY-, 154
   KELSEY, 300
   WATER-, 154
Loganiaceae family, 152
LOMBARDY POPLAR, 248
LONGLEAF PINE, 241
LONGSTALK MOUNTAIN HOLLY, 174
Lonicera, 195
   × bella, 199
   canadensis, 199
   dioica, 197
      var. glaucescens, 199
   flava, 199
   flavida, 199

426 · Index

Lonicera (*continued*)
  fragrantissima, 199
  japonica, 197
  maackii, 199
  morrowii, 199
  prolifera, 197
  reticulata, 197
  sempervirens, 197
  tatarica, 199
  xylosteum, 200
Loranthaceae family, 232
LOUISIANA YUCCA, 384
LOW
  GALLBERRY, 172
  ST. JOHN'S-WORT, 165
  ST. PETER'S-WORT, 165
  STAGGERBUSH, 204
LUPINE, 200
  LADY, 200
  SCRUB, 200
  SMALL'S, 200
  SPREADING, 200
Lupinus, 200
  cumulicola, 200
  diffusus, 200
  villosus, 200
  westianus, 200
Lycium, 200
  barbarum, 202
  carolinianum, 202
  chinense, 202
Lyonia, 202
  ferruginea, 204
  fruticosa, 204
  ligustrina, 202
    var. foliosiflora, 202
  lucida, 204
  mariana, 204
LYONIA, TREE, 204
Lythraceae family, 128, 184

Macfadyena unguis-cati, 204
Maclura pomifera, 204
MAGNOLIA, 206
  ASHE, 208
  BIGLEAF, 208
  FRASER, 208
  PYRAMID, 210
  saucer, 206
  SOUTHERN, 206
  UMBRELLA, 208
  -VINE, 322
Magnolia, 206
  acuminata, 208
    var. cordata, 208
    var. subcordata, 208
  ashei, 208
  fraseri, 208
  grandiflora, 206
  macrophylla, 208
  pyramidata, 210
  × soulangeana, 206
  tripetala, 208
  virginiana, 206
Magnoliaceae family, 195, 206
Mahonia, 210
  aquifolium, 210
  bealei, 210
  repens, 210
MAIDEN-BUSH, 61
MALEBERRY, 202
Malus, 210
  angustifolia, 212
  coronaria, 212
  glabrata, 212
  glaucescens, 212
  ioensis, 212
  × platycarpa, 210, 212
  sylvestris, 210
Malvaceae family, 160
MANGROVE, BLACK, 68
Manihot, 212
  esculenta, 212
  grahamii, 212
MAPLE, 44
  BLACK, 48
  CHALK, 46
  FLORIDA, 46
  MOUNTAIN, 46
  NORWAY, 48
  RED, 48
  SILVER, 48
  STRIPED, 46
  SUGAR, 48
MAPLE-LEAF VIBURNUM, 373
MAPLE-LEAVED OAK, 265
MARINE-IVY, 104
MARSH
  -ELDER, 176, 180
  MILKVINE, 126, 128
MATRIMONY-VINE, 200
MAYBERRY, 364
mayhaw, 119
MAZZARD CHERRY, 257
MCCARTNEY ROSE, 300
MEADOW
  -SWEET, 337
  WILLOW, 316

MEDLAR, STEARN'S, 214
Melia azedarach, 212
Meliaceae family, 212
MEMORIAL ROSE, 300
Menispermaceae family, 82, 108, 212
Menispermum canadense, 212
Menziesia pilosa, 214
Mespilus canescens, 214
MESQUITE, HONEY, 250
MEXICAN
    PLUM, 252
    RATTLEBOX, 324
MICCOSUKEE GOOSEBERRY, 297
MICHAUX SUMAC, 294
MILKVINE, 126
MIMOSA, 52
MINNIE-BUSH, 214
Mississippi Embayment, 5
MISSOURI
    GOOSEBERRY, 297
    WILLOW, 314
MISTLETOE, 232
Mitchella repens, 214
MOCKERNUT HICKORY, 86
MOCK-ORANGE, 230
    COMMON, 232
    DOWNY, 232
    HAIRY, 232
    ODORLESS, 232
MOCK-PENNYROYAL, 340
MOONSEED, 212
Moraceae family, 78, 126, 138, 204, 214
MORROW HONEYSUCKLE, 199
Morus, 214
    alba, 214
    nigra, 216
    rubra, 214
MOSS, SPANISH-, 350
MOUND-LILY YUCCA, 384
MOUNTAIN
    -ASH, 334
    -CRANBERRY, 364
    DOG-HOBBLE, 190
    DRYLAND BLUEBERRY, 361
    FETTERBUSH, 236
    GOLDEN-HEATHER, 160
    -HOLLY, 174
    HONEYSUCKLE, 197
    INDIGO-BUSH, 59
    -LAUREL, 184
    -LOVER, 228
    MAPLE, 46
    PAPER BIRCH, 78
    PEPPERBUSH, 106

STEWARTIA, 340
SWEETBELLS, 190
WINTERBERRY, 174
MULBERRY, 214
    BLACK, 216
    PAPER-, 216
    RED, 214
    WHITE, 214
MULTIFLORA ROSE, 302
MUSCADINE, 377
MUSTANG GRAPE, 378
Myrica, 216
    cerifera, 216
        var. pumila, 216
    gale, 26
    heterophylla, 218
    inodora, 216
    pennsylvanica, 218
Myricaceae family, 108, 216
MYRTLE
    CRAPE-, 184
    DAHOON, 172
    OAK, 276
    SAND-, 188
    WAX-, 216
MYRTLELEAF ST. JOHN'S-WORT, 167

NAILWORT, 226
Nandina domestica, 218
NANKING CHERRY, 255
NANNYBERRY, 370
NARROW-LEAF
    BUCKTHORN, 284
    GROUNDSEL-BUSH, 70
    PAWPAW, 68
native plants, discussion of, 3
naturalized plants, discussion of, 3
NEEDLE PALM, FLORIDA, 284
Nemopanthus mucronatus, 174
Nestronia umbellula, 218
Neviusia alabamensis, 218
NEW JERSEY
    HIGHBUSH BLUEBERRY, 362
    TEA, 94
NIGHTSHADE, BITTER, 334
NINEBARK, 234
NORTHERN
    ARROWWOOD, 375
    BAYBERRY, 218
    BLACKBERRY, 309
    BUSH-HONEYSUCKLE, 130
    CATALPA, 94
    -CREEPER, 228
    DEWBERRY, 309

NORTHERN (*continued*)
  LOWBUSH BLUEBERRY, 361
  POISON-IVY, 352
  red currant, 297
  RED OAK, 278
  WHITE-CEDAR, 349
NORWAY
  MAPLE, 48
  POPLAR, 250
  SPRUCE, 236
NUTMEG HICKORY, 86
NUTTALL OAK, 282
Nyssa, 218
  aquatica, 220
  biflora, 220
  ogeche, 220
  sylvatica, 220
  ursina, 220

OAK, 260
  ARKANSAS, 266
  BEAR, 270
  BIGELOW, 280
  BLACK, 282
  BLACKJACK, 274
  BLUEJACK, 270
  BLUFF, 266
  BOYNTON, 266
  BUR, 272
  CHAPMAN, 268
  CHERRYBARK, 276
  CHESTNUT, 274
  CHINQUAPIN, 274
  DARLINGTON, 270
  DELTA POST, 280
  DURAND, 280
  DWARF CHINQUAPIN, 278
  DWARF LIVE, 274
  GEORGIA, 270
  LAUREL, 272
  -LEAF HYDRANGEA, 162
  LIVE, 282
  MAPLE-LEAVED, 265
  MYRTLE, 276
  NORTHERN RED, 278
  NUTTALL, 282
  OGLETHORPE, 276
  OVERCUP, 272
  PIN, 278
  POISON-, 352
  POST, 282
  RUNNER, 278
  SAND LIVE, 270
  SAND POST, 272
  SAWTOOTH, 266
  SCARLET, 268
  SHINGLE, 270
  SHUMARD, 280
  SOUTHERN RED, 268
  SWAMP CHESTNUT, 274
  SWAMP WHITE, 266
  TEXAS, 268
  TEXAS LIVE, 268
  TURKEY, 272
  WATER, 276
  WHITE, 266
  WILLOW, 278
OCONEE AZALEA, 291
OCTOBER-FLOWER, 244
ODORLESS
  BAYBERRY, 216
  MOCK-ORANGE, 232
OGEECHEE TUPELO, 220
OGLETHORPE OAK, 276
OHIO BUCKEYE, 50
OKLAHOMA PLUM, 254
Oleaceae family, 102, 140, 142, 144, 180, 190, 224, 344
OLIVE
  AUTUMN-, 134
  INDIAN-, 218
  RUSSIAN-, 132
one-flowered hawthorn, 123
Opuntia, 220
  engelmannii var. lindheimeri, 224
  ficus-indica, 224
  humifusa, 222
    var. ammophila, 224
    var. austrina, 222
  macrorhiza, 222
  monocantha, 224
  pusilla, 222
  stricta, 224
    var. dillenii, 224
  vulgaris, 224
ORANGE
  MOCK-, 230
  OSAGE-, 204
  SOUR, 104
  TRIFOLIATE, 244
OREGON GRAPE-HOLLY, 210
ORIENTAL
  BITTERSWEET, 96
  PEAR, 260
OSAGE-ORANGE, 204
Osmanthus americanus, 224
Ostrya virginiana, 226
OUACHITA INDIGO-BUSH, 59

Ouachita Mountains, 6
OVERCUP OAK, 272
Oxydendrum arboreum, 226
OZARK CHINQUAPIN, 92
Ozark Plateau, 6

PACHYSANDRA, 226
Pachysandra, 226
  procumbens, 226
  terminalis, 226
PALM, FLORIDA NEEDLE, 284
PALMETTO, 310
  CABBAGE, 310
  DWARF, 310
  SAW-, 322
  SCRUB, 310
PAPER
  BIRCH, 78
  -MULBERRY, 78
PARASOL-TREE, 140
Parkinsonia aculeata, 226
Paronychia virginica, 226
parsley haw, 120
Parthenocissus, 228
  inserta, 228
  quinquefolia, 228
  vitacea, 228
PARTRIDGE-BERRY, 214
Paulownia tomentosa, 228
PAWPAW, 67
  BIG-FLOWER, 68
  FLAG, 68
  FLATWOODS, 68
  NARROW-LEAF, 68
  PYGMY, 68
  SMALL-FLOWER, 67
  WOOLLY, 68
Paxistima canbyi, 228
PEACH, 250, 252
PEACHLEAF WILLOW, 314
PEAR, 258
  CALLERY, 260
  COMMON, 260
  ORIENTAL, 260
PEARLBUSH, 138
PECAN, 86
PEEGEE HYDRANGEA, 162
PENNYROYAL, MOCK-, 340
Pentaphylloides floribunda, 230
PEPPERBUSH, 106
  MOUNTAIN, 106
  SWEET, 106
PEPPERVINE, 61
PERFUMED CHERRY, 255

PERIWINKLE, 375
Persea, 230
  borbonia, 230
  humilis, 230
  palustris, 230
PERSIMMON, 132
Philadelphus, 230
  coronarius, 232
  floridus, 232
  gattingeri, 232
  gloriosus, 232
  grandiflorus, 232
  hirsutus, 232
  inodorus, 232
  laxus, 232
  pubescens, 232
    var. intectus, 232
  sharpianus, 232
Phoradendron leucarpum, 232
Phyllostachys, 234
  aurea, 234
  aureosulcata, 234
  bambusoides, 234
Physocarpus opulifolius, 234
Picea, 234
  abies, 236
  rubens, 236
PIEDMONT
  AZALEA, 290
  INDIGO-BUSH, 59
  RHODODENDRON, 288
Piedmont region, 5
Pieris, 236
  floribunda, 236
  phillyreifolia, 236
PIGEON GRAPE, 378
PIGNUT HICKORY, 88
PIN OAK, 278
Pinaceae family, 43, 94, 186, 234, 238, 354
Pinckneya bracteata, 238
PINE, 238
  EASTERN WHITE, 239
  LOBLOLLY, 241
  LONGLEAF, 241
  PITCH, 241
  POND, 241
  RED, 239
  SAND, 239
  SCOTS, 239
  SHORTLEAF, 241
  SLASH, 241
  SPRUCE, 239
  TABLE MOUNTAIN, 239
  VIRGINIA, 239

PINELAND ST. ANDREW'S-CROSS, 165
PINK-SHELL AZALEA, 288
Pinus, 238
   clausa, 239
   echinata, 241
   elliottii, 241
   glabra, 239
   palustris, 241
   pungens, 239
   resinosa, 239
   rigida, 241
   serotina, 241
   strobus, 239
   sylvestris, 239
   taeda, 241
   virginiana, 239
PINXTER-FLOWER, 291
pipevine, 63
PIPSISSEWA, 100
PIRATEBUSH, 78
PISTACHIO, CHINESE, 243
Pistacia chinensis, 243
PITCH PINE, 241
pith, 12, 13
PLAINS
   cottonwood, 250
   PRICKLY-PEAR, 222
Plains region, 6
Planera aquatica, 243
PLANER-TREE, 243
Platanaceae family, 243
Platanus occidentalis, 243
PLUM, 250, 252, 254, 255
   ALLEGHENY, 252
   AMERICAN, 252
   BEACH, 252
   CANADA, 255
   CHICKASAW, 254
   CREEK, 254
   DAMSON, 254
   FLATWOODS, 252
   GARDEN, 254
   HORTULAN, 254
   MEXICAN, 252
   OKLAHOMA, 254
   PURPLELEAF, 255
   WILDGOOSE, 254
PLUMLEAF AZALEA, 291
Poaceae family, 65, 234, 257, 320
POCOSIN BAYBERRY, 218
POISON
   -IVY, 352, 354
      NORTHERN, 352
   -OAK, 352

   -SUMAC, 352
Polygonaceae family, 78, 243
Polygonella, 243
   americana, 244
   macrophylla, 243
   polygama, 244
      var. croomii, 244
Poncirus trifoliata, 244
POND
   -BERRY, 194
   -CYPRESS, 348
   -PINE, 241
   -SPICE, 195
POPLAR, 244
   BALSAM, 248
   European black, 248
   GRAY, 246
   LOMBARDY, 248
   NORWAY, 250
   WHITE, 246
Populus, 244
   alba, 246
   balsamifera, 248
   × canadensis, 250
   × *candicans*, 248
   × canescens, 246
   deltoides, 248
      ssp. monolifera, 250
   × *euramericana*, 250
   × *gileadensis*, 248
   grandidentata, 246
   heterophylla, 248
   × jackii, 248
   nigra, 248
      var. italica, 248
   sargentii, 250
   tremula, 246
   tremuloides, 246
PORCELAIN-BERRY, 61
POSSUM-HAW, 174
POSSUMHAW VIBURNUM, 373
POST OAK, 282
Potentilla
   fruticosa, 230
   tridentata, 324
PRAIRIE
   CORY CACTUS, 134
   CRABAPPLE, 212
   ROSE, 302
   WILLOW, 314
PRICKLY
   -ASH, 386
   GOOSEBERRY, 297
   -PEAR, 220

COASTAL, 224
DUNE, 222
EASTERN, 222
FLORIDA, 222
PLAINS, 222
SCRUB, 224
SOUTH AMERICAN, 224
SPINELESS, 224
TEXAS, 224
PRINCE'S-PINE, 102
PRIVET, 190
  AMUR, 194
  BORDER, 194
  CALIFORNIA, 192
  CHINESE, 192
  EUROPEAN, 192
  FLORIDA-, 140
  GLADE-, 140
  GLOSSY, 192
  JAPANESE, 192
  SWAMP-, 140
  WAXLEAF, 192
Prosopis glandulosa, 250
Prunus, 250
  alabamensis, 257
  alleghaniensis, 252
  americana, 252
    var. lanata, 252
  angustifolia, 254
  avium, 257
  caroliniana, 255
  cerasifera, 255
    var. atropurpurea, 255
  cerasus, 257
  cuthbertii, 257
  domestica, 254
  gracilis, 254
  hortulana, 254
  insititia, 254
  mahaleb, 255
  maritima, 252
  mexicana, 252
  munsoniana, 254
  nigra, 255
  pensylvanica, 255
  persica, 252
  pumila var. susquehanae, 255
  rivularis, 254
  serotina, 257
  tomentosa, 255
  umbellata, 252
    var. injucunda, 252
  virginiana, 257
Pseudosasa japonica, 257

Ptelia trifoliata, 258
Pueraria montana var. lobata, 258
PUMPKIN ASH, 146
PURPLE
  CHOKEBERRY, 65
  -FLOWERING RASPBERRY, 306
  -LEAF PLUM, 255
  WILLOW, 313
PUSSY WILLOW, 316
PYGMY PAWPAW, 68
Pyracantha coccinea, 258
PYRAMID MAGNOLIA, 210
Pyrularia pubera, 258
Pyrus, 258
  calleryana, 260
  communis, 260
  pyrifolia, 260
Pyxidanthera barbulata, 260
  var. brevifolia, 260
PYXIE-MOSS, 260

QUAKING ASPEN, 246
Quercus, 260
  acerifolia, 265
  acutissima, 266
  alba, 266
  arkansana, 266
  austrina, 266
  bicolor, 266
  boyntonii, 266
  buckleyi, 268
  chapmanii, 268
  coccinea, 268
  drummondii, 282
  durandii, 280
  falcata, 268
  fusiformis, 268
  geminata, 270
  georgiana, 270
  hemisphaerica, 270
  ilicifolia, 270
  imbricaria, 270
  incana, 270
  laevis, 272
  laurifolia, 272
  lyrata, 272
  macrocarpa, 272
  margarettiae, 272
  marilandica, 274
    var. ashei, 274
  michauxii, 274
  minima, 274
  mississippiensis, 280
  montana, 274

Quercus (*continued*)
  muhlenbergii, 274
  myrifolia, 276
  nigra, 276
  nuttallii, 282
  oglethorpensis, 276
  pagoda, 276
  palustris, 278
  phellos, 278
  prinoides, 278
  prinus, 274
  pumila, 278
  rubra, 278
    var. *ambigua,* 278
    var. borealis, 278
  shumardii, 280
    var. schneckii, 280
  similis, 280
  sinuata, 280
    var. breviloba, 280
  stellata, 282
    var. *paludosa,* 280
  texana, 282
  velutina, 282
  virginiana, 282
QUINCE, FLOWERING, 102

Ranunculaceae family, 104
RASPBERRY, 305–307
  BLACK, 307
  DWARF, 307
  PURPLE-FLOWERING, 306
  RED, 306, 307
rat-stripper, 230
RATTLEBOX, 322, 324
  MEXICAN, 324
  YELLOW, 324
RAYNOR'S BLUEBERRY, 362
RED
  ash, 148
  -BAY, 230
  BUCKEYE, 52
  -BUD, 100
  -CEDAR
    EASTERN, 182
    SOUTHERN, 182
  CHOKEBERRY, 65
  ELDERBERRY, 318
  GARDEN CURRANT, 296
  HICKORY, 90
  MAPLE, 48
  MULBERRY, 214
  -OSIER DOGWOOD, 114
  PINE, 239
  RASPBERRY, 306, 307

  -ROOT, 94
  SPRUCE, 236
REEVES SPIREA, 338
Rhamnaceae family, 74, 94, 144, 282, 312, 388
Rhamnus, 282
  alnifolia, 284
  caroliniana, 144
  cathartica, 284
  davurica, 284
  frangula, 144
  lanceolata, 284
Rhapidophyllum histrix, 284
RHODODENDRON, 284
  CAROLINA, 288
  CATAWBA, 288
  CHAPMAN, 288
  PIEDMONT, 288
  ROSEBAY, 288
  subgenera, 284
Rhododendron, 284
  alabamense, 291
  arborescens, 290
  atlanticum, 290
  austrinum, 290
  bakeri, 291
  calendulaceum, 290
  canescens, 290
  carolinianum, 288
  catawbiense, 288
  chapmanii, 288
  coryi, 290
  cumberlandense, 291
  eastmanii, 291
  flammeum, 291
  maximum, 288
  minus, 288
  nudiflorum, 291
  oblongifolium, 290
  periclymenoides, 291
  prinophyllum, 290
  prunifolium, 291
  punctatum, 299
  roseum, 290
  serrulatum, 290
  speciosum, 291
  vaseyi, 288
  viscosum, 290
Rhodotypos scandens, 291
Rhus, 292, 352
  aromatica, 292
    var. illinoensis, 292
    var. serotina, 292
  copallinum, 292
    var. lanceolata, 292

var. latifolia, 292
glabra, 292
hirta, 294
michauxii, 294
radicans, 354
  var. *rydbergii,* 352
toxicodendron, 352
typhina, 294
vernix, 352
Ribes, 294
  americanum, 296
  aureum, 296
  curvatum, 297
  cynosbati, 297
  echinellum, 297
  glandulosum, 296
  grossularia, 297
  hirtellum, 297
  lacustre, 297
  missouriense, 297
  nigrum, 296
  odoratum, 296
  rotundifolium, 298
  rubrum, 297
  sativum, 296
  triste, 296
  uva-crispa, 297
Ridge and Valley Province, 5
RIVER BIRCH, 76
RIVERBANK GRAPE, 380
riverflat haw, 120
Robinia, 298
  boyntonii, 300
  elliottii, 300
  hispida, 300
    var. fertilis, 300
    var. kelseyi, 300
    var. nana, 300
    var. rosea, 300
  pseudoacacia, 298
  viscosa, 298
    var. hartwegii, 300
ROCK ELM, 357
Rosa, 300
  acicularis, 304
  arkansana, 304
  blanda, 305
  bracteata, 301
  canina, 302
  carolina, 304
  damascena, 304
  eglanteria, 302
  foliolosa, 305
  gallica, 304
  laevigata, 302

micrantha, 302
multiflora, 302
palustris, 304
rugosa, 302
setigera, 302
virginiana, 304
wichuraiana, 301
Rosaceae family, 54, 63, 102, 116, 138, 184, 210, 214, 218, 230, 234, 250, 258, 291, 300, 305, 324, 334
ROSE, 300
  BRISTLY, 304
  CAROLINA, 304
  CHEROKEE, 302
  DAMASK, 304
  DOG, 302
  DWARF PRAIRIE, 304
  EASTER-, 184
  FRENCH, 304
  GUELDER-, 369
  MCCARTNEY, 300
  MEMORIAL, 300
  MULTIFLORA, 302
  PRAIRIE, 302
  RUGOSA, 302
  SMALL-FLOWERED, 302
  SMOOTH, 305
  SWAMP, 304
  SWEETBRIAR, 302
  VIRGINIA, 304
  WHITE PRAIRIE, 305
ROSEBAY RHODODENDRON, 288
ROSEMARY, 100, 108
  APALACHICOLA, 110
  BOG-, 61
  CUMBERLAND, 110
  ETONIA, 108
  LARGE-FLOWERED, 110
  SHORTLEAF, 108
Rosmarinus, 108
ROUGH-LEAF
  DEUTZIA, 130
  DOGWOOD, 112
ROUND-FRUITED ST. JOHN'S-WORT, 168
ROUND-LEAF
  BIRCH, VIRGINIA, 76
  DOGWOOD, 112
  GOOSEBERRY, 298
  SERVICEBERRY, 56
rowan tree, 334
ROYAL PAULOWNIA, 228
Rubiaceae family, 98, 214, 238
Rubus, 305
  allegheniensis, 309
  argutus, 309

Rubus (*continued*)
  betulifolius, 309
  bifrons, 310
  canadensis, 309
  cuneifolius, 309
  discolor, 310
  enslenii, 307
  flagellaris, 309
  hispidus, 307
  idaeus, 306
    var. strigosus, 307
  illecebrosus, 307
  laciniatus, 310
  occidentalis, 307
  odoratus, 306
  orarius, 309
  ostryifolius, 310
  pensilvanicus, 309
  phoenicolasius, 307
  pubescens, 307
  setosus, 309
  trivialis, 307
RUFOUS
  BUMELIA, 328
  mayhaw, 120
RUGOSA ROSE, 302
RUNNER OAK, 278
RUNNING STRAWBERRY-BUSH, 136
RUSSIAN-OLIVE, 132
RUSTY BLACKHAW, 370
Rutaceae family, 104, 244, 258, 386

Sabal, 310
  etonia, 310
  louisiana, 312
  minor, 310
  palmetto, 310
Sageretia minutiflora, 312
Salicaceae family, 244, 312
Salicornia perennis, 320
Salix, 312
  alba, 317
    var. vitellina, 318
      'Tristis', 318
  amygdaloides, 314
  babylonica, 317
  bebbiana, 313
  caprea, 313
  caroliniana, 316
  cinerea, 314
    var. oleifolia, 314
  cordata, 314
  discolor, 316
  eriocephala, 314
  exigua, 316
  floridana, 316
  fragilis, 317
  gracilis, 316
  humilis, 314
    var. *microphylla,* 314
    var. tristis, 314
  interior, 316
  lucida, 317
  matsudana, 317
    'Pendula', 317
    'Tortuosa', 317
  nigra, 316
  × pendulina, 317
  pentandra, 316
  petiolaris, 316
  purpurea, 313
  rigida, 314
  × sepulcralis, 317
  sericea, 314
SALTBUSH, 70
SALTWORT, 70
Sambucus, 318
  canadensis, 318
    var. laciniata, 318
  pubens, 318
  racemosa ssp. pubens, 318
SAND
  -BAR WILLOW, 316
  BLACKBERRY, 309
  CHERRY, 255
  DEWBERRY, 307
  GRAPE, 380
  HICKORY, 88
  LIVE OAK, 270
  -MYRTLE, 188
  PINE, 239
  POST OAK, 272
  -VINE, 126
Santalaceae family, 78, 218, 258
Sapindaceae family, 318
Sapindus, 318
  drummondii, 320
  marginatus, 320
  saponaria, 320
    var. *drummondii,* 320
Sapium sebiferum, 320
Sapotaceae family, 324
Sarcocornia perennis, 320
Sasa palmata, 320
SASSAFRAS, 320
Sassafras albidum, 320
saucer magnolia, 253
SAVANNA INDIGO-BUSH, 59

SAW-PALMETTO, 322
SAWTOOTH OAK, 266
Saxifragaceae family, 176, 381
SCARLET
    BASIL, 80
    haw, 120
    OAK, 268
Schinus terebinthifolius, 322
Schisandra glabra, 322
Schisandraceae family, 322
SCOTCH-BROOM, 128
SCOTS
    HEATHER, 82
    PINE, 239
Scrophulariaceae family, 228
SCRUB
    HICKORY, 88
    LUPINE, 200
    PALMETTO, 310
    PRICKLY-PEAR, 224
Scuppernong grape, 377
SEA-OXEYE, 78
SEASIDE ALDER, 54
SEBASTIAN-BUSH, 322
Sebastiania, 322
    fruticosa, 322
    ligustrina, 322
SEPTEMBER ELM, 357
Serenoa repens, 322
SERVICEBERRY, 54
    ALLEGHENY, 56
    BARTRAM, 56
    COASTAL, 56
    DOWNY, 56
    -HOLLY, 176
    ROUNDLEAF, 56
    RUNNING, 58
    THICKET, 56
Sesbania, 322
    drummondii, 324
    punicea, 324
    virgata, 324
SHAGBARK HICKORY, 88
SHEEP-LAUREL, 184
    CAROLINA, 184
SHELLBARK HICKORY, 88
SHELLMOUND-BUCKTHORN, 312
SHINGLE OAK, 270
SHINING
    BLUEBERRY, 364
    INDIGO-BUSH, 59
    WILLOW, 317
SHORTLEAF
    PINE, 241

ROSEMARY, 108
ST. JOHN'S-WORT, 165
SHORTSTALK ARROWWOOD, 373
SHRUB
    -ALTHEA, 160
    -CINQUEFOIL, 230
    -CORKWOOD, 342
    -GOLDENROD, 102
    -LESPEDEZA, 188
    -YELLOWROOT, 381
SHRUBBY ST. JOHN'S-WORT, 170
SHUMARD OAK, 280
Sibbaldiopsis tridentata, 324
SIBERIAN ELM, 356
Sideroxylon, 324
    alachuense, 328
    anomala, 328
    lanuginosum, 326
        ssp. albicans, 326
        ssp. oblongifolium, 326
        ssp. rigidum, 326
    lycioides, 328
    reclinatum, 328
        ssp. rufotomentosum, 328
    tenax, 326
    thornei, 328
SILKBAY, 230
SILKY
    DOGWOOD, 112
    STEWARTIA, 340
    WILLOW, 314
SILVER
    CROTON, 126
    MAPLE, 48
SILVERBELL, 156
    CAROLINA, 158
    LITTLE, 158
    TWO-WING, 156
SILVERBERRY, 132
Simaroubaceae family, 52
SINKHOLE ST. JOHN'S-WORT, 167
SKUNK CURRANT, 296
SLASH PINE, 241
SLENDER
    BLUEBERRY, 367
    DEUTZIA, 130
    ST. JOHN'S-WORT, 167
SLIPPERY ELM, 357
SMALL CRANBERRY, 366
SMALLFLOWER
    BLUEBERRY, 367
    DEUTZIA, 130
    TAMARISK, 346
SMALL-FLOWER PAWPAW, 67

SMALL-FLOWERED ROSE, 302
SMALL'S LUPINE, 200
Smilacaceae family, 328
Smilax, 328
   auriculata, 332
   bona-nox, 332
   glauca, 332
   hispida, 332
   laurifolia, 330
   pumila, 330
   rotundifolia, 332
   smallii, 330
   tamnoides, 332
   walteri, 332
SMOKETREE, AMERICAN, 114
SMOOTH
   ALDER, 54
   BLACKBERRY, 309
   BUMELIA, 328
   GOOSEBERRY, 297
   HIGHBUSH BLUEBERRY, 362
   INDIGO-BUSH, 61
   ROSE, 305
   ST. JOHN'S-WORT, 168
   SUMAC, 292
   WINTERBERRY, 174
SNAILSEED, 108
snowball viburnum, 369
SNOWBELL, 342
   AMERICAN, 342
   BIGLEAF, 342
SNOWBERRY, 344
   CREEPING, 150
SNOW-WREATH, 218
SNOWY HYDRANGEA, 162
SOAPBERRY, 318
   FLORIDA, 320
   WESTERN, 320
SOAPWEED, 383
soils, discussion of, 6
Solanaceae family, 200, 334
Solanum dulcamara, 334
Sophora affinis, 334
Sorbus, 334
   americana, 334
   aucuparia, 334
SOUR
   CHERRY, 257
   ORANGE, 104
SOURWOOD, 226
SOUTH AMERICAN PRICKLY-PEAR, 224
Southeastern United States
   physiographic regions of, 3–6
   woody plant summaries of, 3
SOUTHERN

   ARROWWOOD, 373
   BUSH-HONEYSUCKLE, 130
   CATALPA, 92
   CRABAPPLE, 212
   DEWBERRY, 307
   GROUNDSEL-BUSH, 70
   HIGHBUSH BLUEBERRY, 364
   LOW BLUEBERRY, 366
   MAGNOLIA, 206
   RED OAK, 268
   RED-CEDAR, 182
   SHAGBARK HICKORY, 88
SPANISH-DAGGER, 384
SPANISH-MOSS, 350
SPARKLEBERRY, 361
SPECKLED ALDER, 54
SPICEBUSH, 194
   BOG, 194
SPINDLE-TREE, EUROPEAN, 138
SPINELESS PRICKLY-PEAR, 224
Spiraea, 334
   alba, 337
   betulifolia, 337
      var. *corymbosa,* 337
      var. *lucida,* 337
   × billardii, 338
   × bumalda, 338
      'Anthony Waterer', 338
   cantoniensis, 338
      var. lanceolata, 338
   corymbosa, 337
   japonica, 338
   latifolia, 337
   lucida, 337
   prunifolia, 338
   salicifolia, 338
   thunbergii, 338
   tomentosa, 337
      var. rosea, 337
   × vanhouttei, 338
   virginiana, 337
SPIREA, 334
   ANTHONY WATERER, 338
   BILLARD, 338
   BRIDAL-WREATH, 338
   CHINESE, 338
   DWARF, 337
   HYBRID BRIDAL-WREATH, 338
   JAPANESE, 338
   REEVES, 338
   STEEPLEBUSH, 337
   VIRGINIA, 337
SPOTTED WINTERGREEN, 100
SPREADING LUPINE, 200
SPRING HEATH, 134

SPRINGTIME WITCH-HAZEL, 158
SPRUCE, 234
   NORWAY, 236
   RED, 236
SPRUCE PINE, 239
ST. ANDREW'S-CROSS, 165
   CREEPING, 165
   PINELAND, 165
ST. JOHN'S-WORT, 165, 167
   BLACKWATER, 167
   BLUE RIDGE, 167
   BOG, 168
   BUNCHLEAF, 167
   CHAPMAN, 167
   CREEPING, 167
   DENSE, 170
   DRYLAND, 168
   GOLDEN, 168
   GULF, 168
   INTERIOR, 168
   LOW, 165
   MYRTLELEAF, 167
   ROUND-FRUITED, 168
   SHORTLEAF, 165
   SHRUBBY, 170
   SINKHOLE, 167
   SLENDER, 167
   SMOOTH, 168
   SWAMP, 168
ST. PETER'S-WORT, LOW, 165
Stachydeoma graveolens, 340
STAGGERBUSH, 202, 204
   LOW, 204
STAGHORN SUMAC, 294
Staphylea trifolia, 340
Staphyleaceae family, 340
STEARN'S MEDLAR, 214
STEEPLEBUSH SPIREA, 337
Sterculiaceae family, 140
STEWARTIA, 340
   MOUNTAIN, 340
   SILKY, 340
Stewartia, 340
   malacodendron, 340
   ovata, 340
Stillingia aquatica, 342
stinking-cedar, 350
STRAWBERRY-BUSH, 136
   RUNNING, 136
Styracaceae family, 156, 342
Styrax, 342
   americanus, 342
     var. pulverulentus, 342
   grandifolius, 342
SUGAR MAPLE, 48

SUGARBERRY, 98
SUMAC, 292
   FRAGRANT, 292
   MICHAUX, 294
   POISON-, 352
   SMOOTH, 292
   STAGHORN, 294
   WINGED, 292
SUMMER GRAPE, 378
SUPPLEJACK, 74
SWAMP
   AZALEA, 290
   -BAY, 230
   BLACKGUM, 220
   CHESTNUT OAK, 274
   COTTONWOOD, 248
   DEWBERRY, 307
   DOGWOOD, 114
   INDIGO-BUSH, 59
   JESSAMINE, 154
   -PRIVET, 140
   RED CURRANT, 296
   ROSE, 304
   ST. JOHN'S-WORT, 168
   SWEETBELLS, 190
   WHITE OAK, 266
SWEET
   ACACIA, 44
   AZALEA, 290
   BIRCH, 76
   -BREATH-OF-SPRING, 199
   CRABAPPLE, 212
   -FERN, 108
   PEPPERBUSH, 106
SWEETBAY, 206
SWEETBELLS, 190
   MOUNTAIN, 190
   SWAMP, 190
SWEETBRIAR ROSE, 302
SWEETGALE, 216
SWEETGUM, 195
SWEETLEAF, 344
SWEETSHRUB, 82
SWEETSPIRE, VIRGINIA, 176
SYCAMORE, AMERICAN, 243
Symphoricarpos, 342
   albus, 344
     var. laevigatus, 344
   orbiculatus, 344
Symplocaceae family, 344
Symplocus tinctoria, 344
Syringa vulgaris, 344

TABLE MOUNTAIN PINE, 239
TALL GALLBERRY, 172

TALLOW-TREE, 320
Tamaricaceae family, 344
TAMARISK, 344
  CHINESE, 346
  FRENCH, 346
  SMALLFLOWER, 346
Tamarix, 344
  africana, 346
  canariensis, 346
  chinensis, 346
  gallica, 346
  parviflora, 346
  pentandra, 346
  ramosissima, 346
TARFLOWER, 70
TATARIAN HONEYSUCKLE, 199
Taxaceae family, 348, 350
Taxodium, 346
  ascendens, 348
  distichum, 348
    var. *imbricarium*, 348
    var. *nutans*, 348
taxonomy, 1, 2
Taxus, 348
  canadensis, 348
  floridana, 349
TEA
  BEACH-, 125
  NEW JERSEY, 94
TEABERRY, 148
TEXAS
  LANTANA, 186
  LIVE OAK, 268
  OAK, 268
  PRICKLY-PEAR, 224
  SOPHORA, 334
Theaceae family, 144, 156, 340
THORNY
  BUMELIA, 328
  ELEAGNUS, 132
three-flower haw, 125
THREE-TOOTHED CINQUEFOIL, 324
Thuja occidentalis, 349
Thymeleaceae family, 132
Tilia, 349
  americana, 349
    var. caroliniana, 350
    var. heterophylla, 350
  floridana, 350
Tiliaceae family, 349
Tillandsia usneoides, 350
TITI, 128
TORREYA, FLORIDA, 350
Torreya taxifolia, 350

TOUGH BUMELIA, 326
Toxicodendron, 292, 352
  pubescens, 352
  radicans, 354
    var. negundo, 354
    var. *rydbergii*, 352
  rydbergii, 352
  toxicarium, 352
  toxicodendron, 352
  vernix, 352
Trachelospermum difforme, 354
TRAILING ARBUTUS, 134
TREE LYONIA, 204
TRIFOLIATE ORANGE, 244
TRUMPET-CREEPER, 82
Tsuga, 354
  canadensis, 354
  caroliniana, 354
TULIP-TREE, 195
TUNG-OIL-TREE, 367
TUPELO, 218
  OGEECHEE, 220
  WATER, 220
TURKEY OAK, 272
twigs, terminology for, 10–13
TWINFLOWER, 195
TWO-WING SILVERBELL, 156

Ulmaceae family, 98, 243, 356
Ulmus, 356
  alata, 357
  americana, 357
  crassifolia, 357
  floridana, 357
  fulva, 357
  parvifolia, 356
  procera, 356
  pumila, 356
  rubra, 357
  serotina, 357
  thomasii, 357
UMBRELLA MAGNOLIA, 208
UPLAND HIGHBUSH BLUEBERRY, 366

Vaccinium, 359
  altomontanum, 361
  amoenum, 367
  angustifolium, 361
    var. nigrum, 361
  arboreum, 361
    var. glaucescens, 362
  atrococcum, 364
  caesariense, 362
  constablaei, 362

corymbosum, 362
crassifolium, 362
    ssp. sempervirens, 362
darrowi, 362
elliottii, 364
erythrocarpum, 364
formosum, 364
fuscatum, 364
hirsutum, 364
macrocarpon, 364
myrsinites, 364
myrtilloides, 366
oxycoccos, 366
pallidum, 366
sempervirens, 362
simulatum, 366
stamineum, 366
    var. caesium, 367
    var. candicans, 367
    var. *melanocarpum,* 367
    var. sericeum, 367
tenellum, 367
vacillans, 366
virgatum, 367
VELVET-LEAF BLUEBERRY, 366
Verbenaceae family, 80, 106, 186, 375
Vernicia fordii, 367
VIBURNUM, 367
    CRANBERRY, 369
    MAPLE-LEAF, 373
    POSSUMHAW, 373
    snowball, 369
    WALTER, 370
Viburnum, 367
    acerifolium, 373
        var. glabrescens, 373
    alnifolium, 370
    bracteatum, 375
    carolinianum, 375
    cassinoides, 370
    dentatum, 373
        var. lucidum, 375
    lantanoides, 370
    lentago, 370
    molle, 373
    nudum, 373
        var. angustifolium, 373
        var. *cassinoides,* 373
    obovatum, 370
    opulis, 369
        var. *americanum,* 369
        var. roseum, 369
    prunifolium, 370
    pubescens, 375

    var. *deamii,* 375
    rafinesquianum, 373
    recognitum, 375
    rufidulum, 370
    scabrellum, 375
    semitomentosum, 375
    trilobum, 369
Vinca, 375
    major, 375
    minor, 375
VINE-WICKY, 236
VIRGINIA
    -CREEPER, 228
    PINE, 239
    ROSE, 304
    ROUND-LEAF BIRCH, 76
    SPIREA, 337
    SWEETSPIRE, 176
VIRGIN'S-BOWER, 104
Vitaceae family, 61, 102, 158, 228, 377
Vitex agnus-castus, 375
Vitis, 377
    aestivalis, 378
        var. bicolor, 378
    cinerea, 378
        var. baileyana, 378
        var. floridana, 378
    labrusca, 378
    × labruscana, 378
    mustangensis, 378
        var. diversa, 378
    palmata, 380
    riparia, 380
    rotundifolia, 377
        var. munsoniana, 377
    rupestris, 380
    shuttleworthii, 378
    vulpina, 380

WAHOO, 138
WALNUT, BLACK, 180
WALTER VIBURNUM, 370
Washington hawthorn, 120
WATER
    ASH, 146
    HICKORY, 86
    -LOCUST, 154
    OAK, 276
    TUPELO, 220
    -WILLOW, 128
WAXLEAF PRIVET, 192
WAX-MYRTLE, 216
WEAKLEAF YUCCA, 384
WEEPING

WEEPING (*continued*)
  hawthorn, 121
  LANTANA, 186
  WILLOW, 317, 318
WEIGELA, 380
Weigela floridana, 380
WESTERN
  mayhaw, 120
  SOAPBERRY, 320
WHITE
  ASH, 146
  BASSWOOD, 350
  BIRCH, EUROPEAN, 76
  -CEDAR, ATLANTIC, 100
  -CEDAR, NORTHERN, 349
  MULBERRY, 214
  OAK, 266
  PINE, EASTERN, 239
  POPLAR, 246
  PRAIRIE ROSE, 305
  -WICKY, 184
  WILLOW, 317
WICKY
  HAIRY-, 184
  VINE-, 236
  WHITE-, 184
WILD HYDRANGEA, 162
WILDGOOSE PLUM, 254
WILLOW, 312
  BAY-LEAVED, 316
  BEBB, 313
  BLACK, 316
  COASTAL PLAIN, 316
  CORKSCREW, 317
  CRACK, 317
  DWARF PRAIRIE, 314
  FLORIDA, 316
  GOAT, 313
  GOLDEN WEEPING, 318
  GRAY, 314
  MEADOW, 316
  MISSOURI, 314
  -OAK, 278
  PEACHLEAF, 314
  PRAIRIE, 314
  PURPLE, 313
  PUSSY, 316
  SANDBAR, 316
  SHINING, 317
  SILKY, 314
  WATER-, 128
  WEEPING, 317
  WHITE, 317
WINEBERRY, 307

WINGED
  ELM, 357
  EUONYMUS, 136
  SUMAC, 292
WINTER
  -BERRY, 174
  -CREEPER, 136
  -GREEN, SPOTTED, 100
WISTERIA, 380
Wisteria, 380
  floribunda, 381
  frutescens, 381
  macrostachya, 381
  sinensis, 381
WITCH-ALDER, 142, 144
WITCH-HAZEL, 158
  SPRINGTIME, 158
WITHEROD, 370
WOODBINE, 104
WOOLLY
  BEACH-HEATHER, 160
  PIPEVINE, 63
WOOLY PAWPAW, 68

Xanthorhiza simplicissima, 381
Xanthoxylum, 386

YAM-LEAF CLEMATIS, 104
YAUPON HOLLY, 172
YELLOW
  ANISE-TREE, 176
  BIRCH, 76
  BUCKEYE, 52
  -GROVE BAMBOO, 234
  haw, 121
  HONEYSUCKLE, 199
  RATTLEBOX, 324
YELLOWROOT, SHRUB-, 381
YELLOWWOOD, 104
YEW, 348
  CANADA, 348
  FLORIDA, 349
YUCCA, 381
  ARKANSAS, 383
  CURVE-LEAF, 384
  LOUISIANA, 384
  MOUND-LILY, 384
  WEAKLEAF, 384
Yucca, 381
  aloifolia, 384
  arkansana, 383
    var. *paniculata,* 384
  filamentosa, 384
    var. *smalliana,* 384

flaccida, 384
glauca, 383
gloriosa, 384
louisianensis, 384
recurvifolia, 384
smalliana, 384

Zamia, 386
floridana, 386

integrifolia, 386
pumila, 386
silvicola, 386
Zanthoxylum, 386
americanum, 386
clava-herculis, 386
Zenobia pulverulenta, 386
Zizyphus jujuba, 388